Advances in Mathematical Physics

Advances in Mathematical Physics

Edited by
Lucas Morgan

Willford Press
www.willfordpress.com

Published by Willford Press,
118-35 Queens Blvd., Suite 400,
Forest Hills, NY 11375, USA

ISBN: 978-1-64728-354-4

Cataloging-in-Publication Data

Advances in mathematical physics / edited by Lucas Morgan.
p. cm.
Includes bibliographical references and index.
ISBN 978-1-64728-354-4
1. Mathematical physics. 2. Physics. 3. Mathematics. I. Morgan, Lucas.
QC20 .A28 2022
530.15--dc23

For information on all Willford Press publications
visit our website at www.willfordpress.com

Willford Press

Contents

Preface

This book has been an outcome of determined endeavour from a group of educationists in the field. The primary objective was to involve a broad spectrum of professionals from diverse cultural background involved in the field for developing new researches. The book not only targets students but also scholars pursuing higher research for further enhancement of the theoretical and practical applications of the subject.

Mathematical physics is concerned with developing mathematical methods to apply to problems in physics. Mathematical physics is a broad field that is distinguished by the blending of physics and pure mathematics. Its primary focus is on the expansion and explanation of physical theories. There are various branches of mathematical physics such as classical mechanics, partial differential equations, quantum theory, relativity, and quantum relativistic theories and statistical mechanics. Fields like dynamical systems and Hamiltonian mechanics also belong to mathematical physics. This book provides comprehensive insights into the field of mathematical physics. While understanding the long-term perspectives of the topics, the book makes an effort in highlighting their impact as a modern tool for the growth of the discipline. Coherent flow of topics, student-friendly language and extensive use of examples make this book an invaluable source of knowledge.

It was an honour to edit such a profound book and also a challenging task to compile and examine all the relevant data for accuracy and originality. I wish to acknowledge the efforts of the contributors for submitting such brilliant and diverse chapters in the field and for endlessly working for the completion of the book. Last, but not the least; I thank my family for being a constant source of support in all my research endeavours.

Editor

1

The Johnson Equation, Fredholm and Wronskian Representations of Solutions, and the Case of Order Three

Pierre Gaillard

Université de Bourgogne, Institut de Mathématiques de Bourgogne, 9 avenue Alain Savary, BP 47870, 21078 Dijon Cedex, France

Correspondence should be addressed to Pierre Gaillard; pierre.gaillard@u-bourgogne.fr

Academic Editor: Giampaolo Cristadoro

We construct solutions to the Johnson equation (J) first by means of Fredholm determinants and then by means of Wronskians of order $2N$ giving solutions of order N depending on $2N-1$ parameters. We obtain N order rational solutions that can be written as a quotient of two polynomials of degree $2N(N+1)$ in x, t and $4N(N+1)$ in y depending on $2N-2$ parameters. This method gives an infinite hierarchy of solutions to the Johnson equation. In particular, rational solutions are obtained. The solutions of order 3 with 4 parameters are constructed and studied in detail by means of their modulus in the (x, y) plane in function of time t and parameters a_1, a_2, b_1, and b_2.

1. Introduction

The Johnson equation was introduced in 1980 by Johnson [1] to describe waves surfaces in shallow incompressible fluids [2, 3]. This equation was derived for internal waves in a stratified medium [4]. The Johnson equation is dissipative; it is well known that there is no solution with a linear front localized along straight lines in the (x, y) plane. This Johnson equation is, for example, able to explain the existence of the horseshoe-like solitons and multisoliton solutions quite naturally.

We consider the Johnson equation (J) in the following normalization:

$$\left(u_t + 6uu_x + u_{xxx} + \frac{u}{2t}\right)_x - 3\frac{u_{yy}}{t^2} = 0, \qquad (1)$$

where as usual subscripts x, y, and t mean partial derivatives.

The first solutions were constructed in 1980 by Johnson [1]. Other types of solutions were found in [5]. A new approach to solve this equation was given in 1986 [6] by giving a link between solutions of the Kadomtsev-Petviashvili (KP) [7] and solutions of the Johnson equation. In 2007, other types of solutions were obtained by using the Darboux transformation [8]. More recently, in 2013, other extensions have been considered as the elliptic case [9].

Here, we consider the famous Kadomtsev-Petviashvili (KPI), which can be written in the following form:

$$\left(4u_t - 6uu_x + u_{xxx}\right)_x - 3u_{yy} = 0. \qquad (2)$$

The KPI equation first appeared in 1970 [7] in a paper written by Kadomtsev and Petviashvili. This equation is considered as a model for surface and internal water waves [10] and in nonlinear optics [11].

In the following, we will use the KPI equation to construct solutions to the Johnson equation but in another way different from this used in [6]. Indeed, these last authors consider another representation of KPI equation given by

$$\left(u_t + 6uu_x + u_{xxx}\right)_x - 3u_{yy} = 0, \qquad (3)$$

and so the transformations between solutions of (3) and (1) are different from those we use to transform solutions to (2) in solutions to (1).

In fact, to obtain solutions to (1) from solutions to (2), we use the following transformation:

$$(x; y; t) \longmapsto \left(x_1 = -ix - i\frac{y^2 t}{12};\ y_1 = yt;\ t_1 = 4it\right) \qquad (4)$$

In this paper, we give solutions by means of Fredholm determinants of order $2N$ depending on $2N-1$ parameters and then by means of Wronskians of order $2N$ with $2N-1$ parameters. So we construct an infinite hierarchy of solutions to the Johnson equation, depending on $2N-1$ real parameters.

New rational solutions depending a priori on $2N-2$ parameters at order N are constructed, when one parameter tends to 0.

We obtain families depending on $2N-2$ parameters for the Nth order as a ratio of two polynomials of degree $2N(N+1)$ in x, t and of degree $4N(N+1)$ in y.

In this paper, we construct only rational solutions of order 3, depending on 4 real parameters; we construct the representations of their modulus in the plane of the coordinates (x, y) according to the four real parameters a_i and b_i for $1 \le i \le 2$ and time t.

2. Solutions to Johnson Equation Expressed by Means of Fredholm Determinants

Some notations are given. We define first real numbers λ_j such that $-1 < \lambda_\nu < 1$, $\nu = 1, \dots, 2N$; they depend on a parameter ϵ and can be written as

$$\lambda_j = 1 - 2\epsilon^2 j^2, \qquad \lambda_{N+j} = -\lambda_j, \qquad 1 \le j \le N. \tag{5}$$

Then, we define $\kappa_\nu, \delta_\nu, \gamma_\nu$, and $x_{r,\nu}$; they are functions of λ_ν, $1 \le \nu \le 2N$, and are defined by the following formulas:

$$\begin{aligned}
\kappa_j &= 2\sqrt{1-\lambda_j^2}, \\
\delta_j &= \kappa_j \lambda_j, \\
\gamma_j &= \sqrt{\frac{1-\lambda_j}{1+\lambda_j}};, \\
x_{r,j} &= (r-1)\ln\frac{\gamma_j - i}{\gamma_j + i}, \quad r = 1, 3, \\
\tau_j &= -12i\lambda_j^2\sqrt{1-\lambda_j^2} - 4i\left(1-\lambda_j^2\right)\sqrt{1-\lambda_j^2}, \\
\kappa_{N+j} &= \kappa_j, \\
\delta_{N+j} &= -\delta_j, \\
\gamma_{N+j} &= \gamma_j^{-1}, \\
x_{r,N+j} &= -x_{r,j}, \\
\tau_{N+j} &= \tau_j \qquad j = 1, \dots, N.
\end{aligned} \tag{6}$$

e_ν $1 \le \nu \le 2N$ are defined by

$$\begin{aligned}
e_j &= 2i\left(\sum_{k=1}^{1/2M-1} a_k (je)^{2k+1} - i\sum_{k=1}^{1/2M-1} b_K (je)^{2k+1}\right), \\
e_{N+j} &= 2i\left(\sum_{k=1}^{1/2M-1} a_k (je)^{2k+1} + i\sum_{k=1}^{1/2M-1} b_k (je)^{2k+1}\right), \qquad 1 \le j \le N,
\end{aligned} \tag{7}$$

$$a_k, b_k \in \mathbf{R}, \quad 1 \le k \le N.$$

ϵ_ν, $1 \le \nu \le 2N$ are defined by

$$\epsilon_j = 1, \qquad \epsilon_{N+j} = 0 \qquad 1 \le j \le N. \tag{8}$$

As usual I is the unit matrix and $D_r = (d_{jk})_{1 \le j,k \le 2N}$ is the matrix defined by the following:

$$\begin{aligned}
d_{\nu\mu} &= (-1)^{\epsilon_\nu}\prod_{\eta\ne\mu}\left(\frac{\gamma_\eta + \gamma_\nu}{\gamma_\eta - \gamma_\mu}\right) \\
&\cdot \exp\left(\kappa_\nu x + \left(\frac{\kappa_\nu y}{12} - 2\delta_\nu\right) yt + 4i\tau_\nu t + x_{r,\nu} + e_\nu\right).
\end{aligned} \tag{9}$$

Then we get the following theorem.

Theorem 1. *The function v defined by*

$$v(x,y,t) = -2\frac{|n(x,y,t)|^2}{d(x,y,t)^2} \tag{10}$$

with

$$n(x,y,t) = \det\left(I + D_3(x,y,t)\right), \tag{11}$$

$$d(x,y,t) = \det\left(I + D_1(x,y,t)\right), \tag{12}$$

and $D_r = (d_{jk})_{1 \le j,k \le 2N}$ is the matrix

$$\begin{aligned}
d_{\nu\mu} &= (-1)^{\epsilon_\nu}\prod_{\eta\ne\mu}\left(\frac{\gamma_\eta + \gamma_\nu}{\gamma_\eta - \gamma_\mu}\right) \\
&\cdot \exp\left(\kappa_\nu x + \left(\frac{\kappa_\nu y}{12} - 2\delta_\nu\right) yt + 4i\tau_\nu t + x_{r,\nu} + e_\nu\right)
\end{aligned} \tag{13}$$

is a solution to (1), depending on $2N-1$ parameters a_k, b_k, $1 \le k \le N-1$ and ϵ.

Proof. The solution v to the KPI equation can be written as follows by using [12]:

$$v(x,y,t) = -2\frac{|n(x,y,t)|^2}{d(x,y,t)^2} \tag{14}$$

where

$$n(x,y,t) = \det(I + D_3(x,y,t)), \tag{15}$$

$$d(x,y,t) = \det(I + D_1(x,y,t)), \tag{16}$$

and $D_r = (d_{jk})_{1\le j,k\le 2N}$ is the matrix

$$d_{\nu\mu} = (-1)^{\epsilon_\nu}\prod_{\eta\neq\mu}\left(\frac{\gamma_\eta+\gamma_\nu}{\gamma_\eta-\gamma_\mu}\right) \cdot \exp(i\kappa_\nu x - 2\delta_\nu y + \tau_\nu t + x_{r,\nu} + e_\nu). \tag{17}$$

where κ_ν, δ_ν, $x_{r,\nu}$, γ_ν, τ_ν, e_ν, and ϵ_ν are defined in (6), (5), (7), and (8).

The connection between the solutions to the Johnson equation and these to the KPI equation was already explained in [6] but with another expression of the KPI equation (3).

Here, the knowledge of a solution u to the KPI equation (2) gives a solution to the Johnson equation (1). Let us consider $u(x,y,t)$ a solution of the KPI equation (2), then the function

$$\tilde{u}(x_1,y_1,t_1) \tag{18}$$

for

$$x_1 = -ix - i\frac{y^2t}{12}, \quad y_1 = yt, \quad t_1 = 4it \tag{19}$$

is a solution to the KPI equation (2). Using this crucial transformation, the solution to the Johnson equation takes the form

$$v(x,y,t) = -2\frac{|\det(I + D_3(x,y,t))|^2}{\det(I + D_1(x,y,t))^2} \tag{20}$$

with the matrix D_r defined in (17).

So we get the solutions to (14) by means of Fredholm determinants. □

3. Solutions to the Johnson Equation by Means of Wronskians

We use the following notations:

$$\phi_{r,\nu} = \sin\Theta_{r,\nu}, \quad 1\le\nu\le N,$$
$$\phi_{r,\nu} = \cos\Theta_{r,\nu}, \quad N+1\le\nu\le 2N, \; r = 1,3, \tag{21}$$

with

$$\Theta_{r,\nu} = \frac{-i\kappa_\nu x}{2} + i\left(\frac{-\kappa_\nu y}{24} + \delta_\nu\right)yt - i\frac{x_{r,\nu}}{2} + 2\tau_\nu t + \gamma_\nu w - i\frac{e_\nu}{2}, \quad 1\le\nu\le 2N. \tag{22}$$

$W_r(w)$ is the Wronskian of the functions $\phi_{r,1},\dots,\phi_{r,2N}$ defined by

$$W_r(w) = \det\left[\left(\partial_w^{\mu-1}\phi_{r,\nu}\right)_{\nu,\mu\in[1,\dots,2N]}\right]. \tag{23}$$

We consider the matrix $D_r = (d_{\nu\mu})_{\nu,\mu\in[1,\dots,2N]}$ defined in (17).

Then we have the following result.

Theorem 2.

$$\det(I + D_r) = k_r(0)\times W_r(\phi_{r,1},\dots,\phi_{r,2N})(0), \tag{24}$$

where

$$k_r(y) = \frac{2^{2N}\exp\left(i\sum_{\nu=1}^{2N}\Theta_{r,\nu}\right)}{\prod_{\nu=2}^{2N}\prod_{\mu=1}^{\nu-1}(\gamma_\nu-\gamma_\mu)}. \tag{25}$$

Proof. First, we remove the factor $(2i)^{-1}e^{i\Theta_{r,\nu}}$ in each row ν in the Wronskian $W_r(w)$ for $1\le\nu\le 2N$.

Then

$$W_r = \prod_{\nu=1}^{2N} e^{i\Theta_{r,\nu}}(2i)^{-N}(2)^{-N}\times\widetilde{W}_r, \tag{26}$$

with

$$\widetilde{W}_r = \begin{vmatrix} \left(1-e^{-2i\Theta_{r,1}}\right) & i\gamma_1\left(1+e^{-2i\Theta_{r,1}}\right) & \dots & (i\gamma_1)^{2N-1}\left(1+(-1)^{2N}e^{-2i\Theta_{r,1}}\right) \\ \left(1-e^{-2i\Theta_{r,2}}\right) & i\gamma_2\left(1+e^{-2i\Theta_{r,2}}\right) & \dots & (i\gamma_2)^{2N-1}\left(1+(-1)^{2N}e^{-2i\Theta_{r,2}}\right) \\ \vdots & \vdots & \vdots & \vdots \\ \left(1-e^{-2i\theta_{r,2N}}\right) & i\gamma_{2N}\left(1+e^{-2i\Theta_{r,2N}}\right) & \dots & (i\gamma_{2N})^{2N-1}\left(1+(-1)^{2n}e^{-2i\Theta_{r,2N}}\right) \end{vmatrix} \tag{27}$$

The determinant $\widetilde{W}_r$ can be written as

$$\widetilde{W}_r = \det\left(\alpha_{jk}e_j + \beta_{jk}\right), \tag{28}$$

where $\alpha_{jk} = (-1)^k (i\gamma_j)^{k-1}$, $e_j = e^{-2i\Theta_{r,j}}$, and $\beta_{jk} = (i\gamma_j)^{k-1}$, $1 \le j \le N$, $1 \le k \le 2N$;

$\alpha_{jk} = (-1)^{k-1}(i\gamma_j)^{k-1}$, $e_j = e^{-2i\Theta_{r,j}}$, and $\beta_{jk} = (i\gamma_j)^{k-1}$, $N+1 \le j \le 2N$, $1 \le k \le 2N$.

We have to calculate $\widetilde{W}_r$. So, we use the following lemma.

Lemma 3. *Let $A = (a_{ij})_{i,j\in[1,\dots,N]}$, let $B = (b_{ij})_{i,j\in[1,\dots,N]}$, and let $(H_{ij})_{i,j\in[1,\dots,N]}$ be the matrix formed by replacing in A the jth row of A by the ith row of B. Then*

$$\det\left(a_{ij}x_i + b_{ij}\right) = \det\left(a_{ij}\right) \times \det\left(\delta_{ij}x_i + \frac{\det\left(H_{ij}\right)}{\det\left(a_{ij}\right)}\right) \tag{29}$$

Proof. Let $\widetilde{A} = (\widetilde{a}_{ji})_{i,j\in[1,\dots,N]}$ be the transposed matrix in cofactors of A. Then $A \times \widetilde{A} = \det A \times I$.

So $\det(\widetilde{A}) = (\det(A))^{N-1}$.

Then the general term of the product $(c_{ij})_{i,j\in[1,\dots,N]} = (a_{ij}x_i + b_{ij})_{i,j\in[1,\dots,N]} \times (\widetilde{a}_{ji})_{i,j\in[1,\dots,N]}$ can be expressed by

$$\begin{aligned} c_{ij} &= \sum_{s=1}^{N} \left(a_{is}x_i + b_{is}\right) \times \widetilde{a}_{js} = x_i \sum_{s=1}^{n} a_{is}\widetilde{a}_{js} + \sum_{s=1}^{n} b_{is}\widetilde{a}_{js} \\ &= \delta_{ij} \det(A)\, x_i + \det\left(H_{ij}\right). \end{aligned} \tag{30}$$

We obtain

$$\begin{aligned} \det\left(c_{ij}\right) &= \det\left(a_{ij}x_i + b_{ij}\right) \times (\det(A))^{N-1} \\ &= (\det(A))^{N} \times \det\left(\delta_{ij}x_i + \frac{\det\left(H_{ij}\right)}{\det(A)}\right). \end{aligned} \tag{31}$$

So $\det(a_{ij}x_i + b_{ij}) = \det(A) \times \det(\delta_{ij}x_i + \det(H_{ij})/\det(A))$. □

We use the notations $U = (\alpha_{ij})_{i,j\in[1,\dots,2N]}$ and $V = (\beta_{ij})_{i,j\in[1,\dots,2N]}$.

Using the preceding lemma, we get

$$\begin{aligned} \widetilde{W}_r &= \det\left(\alpha_{ij}e_i + \beta_{ij}\right) \\ &= \det\left(\alpha_{ij}\right) \times \det\left(\delta_{ij}e_i + \frac{\det\left(H_{ij}\right)}{\det\left(\alpha_{ij}\right)}\right) \\ &= \det(U) \times \det\left(\delta_{ij}e_i + \frac{\det\left(H_{ij}\right)}{\det(U)}\right), \end{aligned} \tag{32}$$

where $(H_{ij})_{i,j\in[1,\dots,N]}$ is the matrix obtained by replacing in U the jth row of U by the ith row of V defined previously.

U is the classical Vandermonde determinant that is equal to

$$\det(U) = i^{N(2N-1)} \prod_{2N\ge l>m\ge 1} \left(\gamma_l - \gamma_m\right). \tag{33}$$

We have to compute $\det(H_{ij})$ to evaluate the determinant $\widetilde{W}_r$. To do that, we study two cases.

(1) For $1 \le j \le N$, the matrix H_{ij} is a Vandermonde matrix, where the jth row of U in U is replaced by the ith row of V. Then we have

$$\det\left(H_{ij}\right) = (-1)^{N(2N+1)+N-1}\, (i)^{N(2N-1)} \times M, \tag{34}$$

with $M = M(m_1, \dots, m_{2N})$ being the determinant defined by $m_k = \gamma_k$ for $k \ne j$ and $m_j = -\gamma_i$. Thus we get

$$\begin{aligned} \det\left(H_{ij}\right) &= -(i)^{N(2N-1)} \times \prod_{2N\ge l>k\ge 1,} \left(m_l - m_k\right) \\ &= -(i)^{N(2N-1)} \times \prod_{2N\ge l>m\ge 1, l\ne j, m\ne j} \left(\gamma_l - \gamma_m\right) \\ &\quad \times \prod_{l<j} \left(-\gamma_i - \gamma_l\right) \times \prod_{l>j} \left(\gamma_l + \gamma_i\right), \\ &= (-1)^j (i)^{N(2N-1)} \\ &\quad \times \prod_{2N\ge l>m\ge 1, l\ne j, m\ne j} \left(\gamma_l - \gamma_m\right) \\ &\quad \times \prod_{l\ne j} \left(\gamma_l + \gamma_i\right). \end{aligned} \tag{35}$$

To compute $\widetilde{W}_r$, we have to simplify the quotient $q_{ij} := \det(H_{ij})/\det(U)$:

$$\begin{aligned} &q_{ij} \\ &= \frac{(-1)^j (i)^{N(2N-1)} \times \prod_{2N\ge l>m\ge 1, l\ne j, m\ne j} \left(\gamma_l - \gamma_m\right) \times \prod_{l\ne j} \left(\gamma_l + \gamma_i\right)}{i^{N(2N-1)} \prod_{2N\ge l>m\ge 1} \left(\gamma_l - \gamma_m\right)} \\ &= \frac{(-1)^j \prod_{l\ne j} \left(\gamma_l + \gamma_i\right)}{\prod_{l<j} \left(\gamma_j - \gamma_l\right) \prod_{l>j} \left(\gamma_l - \gamma_j\right)} = \frac{(-1)^j \prod_{l\ne j} \left(\gamma_l + \gamma_i\right)}{(-1)^{j-1} \prod_{l\ne j} \left(\gamma_l - \gamma_j\right)} \\ &= -\frac{\prod_{l\ne j} \left(\gamma_l + \gamma_i\right)}{\prod_{l\ne j} \left(\gamma_l - \gamma_j\right)}. \end{aligned} \tag{36}$$

q_{ij} is equal to r_{ij} defined by $-\prod_{l\ne j}(\gamma_l+\gamma_i)/\prod_{l\ne i}(\gamma_l-\gamma_i)$, because $\det(\delta_{ij}x_i + \det(q_{ij})/\det(A)) = \det(\delta_{ij}x_i + \det(r_{ij})/\det(A))$.

Thus r_{ij} can be written as

$$r_{ij} = (-1)^{\epsilon_i} \left(\prod_{l\ne j} \frac{\gamma_l + \gamma_i}{\gamma_l - \gamma_j}\right) = c_{ij} e^{-2i\Theta_{r,i}(0)}, \tag{37}$$

with the notations given in (17).

(2) We can do the same estimations for $N+1 \le j \le 2N$ are made; $\det H_{ij}$ is first as follows:

$$\det\left(H_{ij}\right) = (-1)^{N(2N+1)+N-1}\, (i)^{N(2N-1)} \times M, \tag{38}$$

with $M = M(m_1, \ldots, m_{2N})$ being the determinant defined by $m_k = \gamma_k$ for $k \neq j$ and $m_j = -\gamma_i$. Then we get

$$\det(H_{ij}) = (i)^{N(2N-1)} \times \prod_{2N\geq l>k\geq 1,} (m_l - m_k)$$
$$= (i)^{N(2N-1)} \times \prod_{2N\geq l>m\geq 1, l\neq j, m\neq j} (\gamma_l - \gamma_m)$$
$$\times \prod_{l<j} (-\gamma_i - \gamma_l) \times \prod_{l>j} (\gamma_l + \gamma_i),$$
$$= (-1)^{j-1} (i)^{N(2N-1)} \times \prod_{2N\geq l>m\geq 1, l\neq j, m\neq j} (\gamma_l - \gamma_m) \times \prod_{l\neq j} (\gamma_l + \gamma_i). \quad (39)$$

Then $q_{ij} := \det(H_{ij})/\det(U)$ can be expressed as

$$q_{ij} = \frac{(-1)^{j-1} (i)^{N(2N-1)} \times \prod_{2N\geq l>m\geq 1, l\neq j, m\neq j} (\gamma_l - \gamma_m) \times \prod_{l\neq j} (\gamma_l + \gamma_i)}{i^{N(2N-1)} \prod_{2N\geq l>m\geq 1} (\gamma_l - \gamma_m)} = \frac{(-1)^{j-1} \prod_{l\neq j} (\gamma_l + \gamma_i)}{\prod_{l<j} (\gamma_j - \gamma_l) \prod_{l>j} (\gamma_l - \gamma_j)}$$
$$= \frac{(-1)^{j-1} \prod_{l\neq j} (\gamma_l + \gamma_i)}{(-1)^{j-1} \prod_{l\neq j} (\gamma_l - \gamma_j)} = \frac{\prod_{l\neq j} (\gamma_l + \gamma_i)}{\prod_{l\neq j} (\gamma_l - \gamma_j)}. \quad (40)$$

q_{ij} is replaced by r_{ij} defined by $\prod_{l\neq j}(\gamma_l + \gamma_i)/\prod_{l\neq i}(\gamma_l - \gamma_i)$, for the same reason as previously exposed.

Then r_{ij} can be written as

$$r_{ij} = (-1)^{\epsilon_i} \prod_{l\neq j} \left(\frac{\gamma_l + \gamma_i}{\gamma_l - \gamma_i}\right) = (-1)^{\epsilon(i)} \frac{\prod_{l\neq j} (\gamma_l + \gamma_i)}{\prod_{l\neq i} (\gamma_l - \gamma_i)} \quad (41)$$
$$= c_{ij} e^{-2i\Theta_{r,i}(0)},$$

with notations given in (17).

e_i is replaced by $e^{-2i\Theta_{r,i}}$. Then $\det \widetilde{W}_r$ can be rewritten as

$$\det \widetilde{W}_r = \det(U) \times \det\left(\delta_{ij} e_i + \frac{\det(H_{ij})}{\det(U)}\right)$$
$$= \det(U) \times \det(\delta_{ij} e_i + r_{ij}) = \det(U) \quad (42)$$
$$\cdot \prod_{i=1}^{2N} e^{-2i\Theta_i} \det\left(\delta_{ij} + (-1)^{\epsilon(i)} \prod_{l\neq i} \left|\frac{\gamma_l + \gamma_i}{\gamma_l - \gamma_i}\right| e^{2i\Theta_{r,i}}\right).$$

We compute the two members of the last relation (42) in $y = 0$. Using (33), we get

$$\det \widetilde{W}_r(0)$$
$$= i^{N(2N-1)} \prod_{2N\geq l>m\geq 1} (\gamma_l - \gamma_m) \prod_{i=1}^{2N} e^{-2i\Theta_{r,i}(0)}$$
$$\times \det\left(\delta_{ij} + (-1)^{\epsilon(i)} \prod_{l\neq i} \left|\frac{\gamma_l + \gamma_i}{\gamma_l - \gamma_i}\right| e^{2i\Theta_{r,i}(0)}\right)$$
$$= i^{N(2N-1)} \prod_{j=2}^{2N} \prod_{i=1}^{j-1} (\gamma_j - \gamma_i) e^{-2i\sum_{i=1}^{2N} \Theta_{r,i}(0)} \det(\delta_{ij} + c_{ij})$$
$$= i^{N(2N-1)} \prod_{j=2}^{2N} \prod_{i=1}^{j-1} (\gamma_j - \gamma_i) e^{-2i\sum_{i=1}^{2N} \Theta_{r,i}(0)} \det(I + C_r)$$
$$= i^{N(2N-1)} \prod_{j=2}^{2N} \prod_{i=1}^{j-1} (\gamma_j - \gamma_i) e^{-2i\sum_{i=1}^{2N} \Theta_{r,i}(0)} \det(I + D_r). \quad (43)$$

Thus, the Wronskian W_r given by (26) can be rewritten as

$$W_r(\phi_{r,1}, \ldots, \phi_{r,2N})(0) = \prod_{j=1}^{2N} e^{i\Theta_{r,j}(0)} (2)^{-2N} (i)^{-N}$$
$$\times \widetilde{W}_r = \prod_{j=1}^{2N} e^{i\Theta_{r,j}(0)} (2)^{-2N} (i)^{-N}$$
$$\cdot i^{N(2N-1)} \prod_{j=2}^{2N} \prod_{i=1}^{j-1} (\gamma_j - \gamma_i) e^{-2i\sum_{i=1}^{2N} \Theta_{r,i}(0)} \det(I + D_r)$$
$$= (2)^{-2N} \prod_{j=2}^{2N} \prod_{i=1}^{j-1} (\gamma_j - \gamma_i) e^{-i\sum_{i=1}^{2N} \Theta_{r,i}(0)} \det(I + D_r). \quad (44)$$

Then

$$\det(I + D_r) = k_r(0) W_r(\phi_1, \ldots, \phi_{2N})(0). \quad (45)$$

This finishes the proof of Theorem 2.

Then the solution v to the Johnson equation can be rewritten as

$$v(x,y,t) = -2\frac{|\det(I + D_3(x,y,t))|^2}{(\det(I + D_1(x,y,t)))^2}. \quad (46)$$

With (24), the following link between Fredholm determinants and Wronskians is obtained:

$$\det(I + D_3) = k_3(0) \times W_3(\phi_{r,1}, \ldots, \phi_{r,2N})(0) \quad (47)$$

and

$$\det(I + D_1) = k_1(0) \times W_1(\phi_{r,1}, \ldots, \phi_{r,2N})(0). \quad (48)$$

As $\Theta_{3,j}(0)$ contains N terms $x_{3,j}$ $1 \leq j \leq N$ and N terms $-x_{3,j}$ $1 \leq j \leq N$, we have the relation $k_3(0) = k_1(0)$, and we get the following theorem.

Theorem 4. *The function v defined by*

$$v(x, y, t) = -2\frac{|W_3(\phi_{3,1}, \ldots, \phi_{3,2N})(0)|^2}{(W_1(\phi_{1,1}, \ldots, \phi_{1,2N})(0))^2} \quad (49)$$

is a solution of the Johnson equation which depends on $2N - 1$ *real parameters* a_k, b_k, *and* ϵ, *with* ϕ_ν^r *defined in (21).*

$$\begin{aligned}\phi_{r,\nu} &= \sin\left(\frac{-i\kappa_\nu x}{2} + i\left(\frac{-\kappa_\nu y}{24} + \delta_\nu\right)yt - i\frac{x_{r,\nu}}{2} + 2\tau_\nu t\right.\\ &\quad \left. + \gamma_\nu w - i\frac{e_\nu}{2}\right), \quad 1 \leq \nu \leq N,\\ \phi_{r,\nu} &= \cos\left(\frac{-i\kappa_\nu x}{2} + i\left(\frac{-\kappa_\nu y}{24} + \delta_\nu\right)yt - i\frac{x_{r,\nu}}{2} + 2\tau_\nu t\right.\\ &\quad \left. + \gamma_\nu w - i\frac{e_\nu}{2}\right), \quad N+1 \leq \nu \leq 2N, \; r = 1, 3,\end{aligned} \quad (50)$$

where κ_ν, δ_ν, $x_{r,\nu}$, γ_ν, *and* e_ν *are defined in (6), (5), and (7).*

4. Study of the Limit Case When ϵ Tends to 0

4.1. Rational Solutions of Order N Depending on $2N - 2$ *Parameters.* An infinite hierarchy of rational solutions to the Johnson equation depending on $2N - 2$ parameters is obtained. For this, we take the limit when the parameter ϵ tends to 0.

We get the following statement.

Theorem 5. *The function v*

$$v(x, y, t) = \lim_{\epsilon \to 0} -2\frac{|W_3(x, y, t)|^2}{(W_1(x, y, t))^2} \quad (51)$$

is a rational solution to the Johnson equation. It is a quotient of two polynomials $n(x, y, t)$ *and* $d(x, y, t)$ *depending on* $2N - 2$ *real parameters* $\tilde{a}_j$ *and* $\tilde{b}_j$, $1 \leq j \leq N-1$, *of degrees* $2N(N+1)$ *in x, t and* $4N(N+1)$ *in y.*

4.2. Families of Rational Solutions of Order 3 Depending on 4 Parameters. Here we construct families of rational solutions to the Johnson equation of order 3 explicitly; they depend on 4 parameters.

We only give the expression without parameters and we give it in the appendix because of the length of the solutions.

We construct the patterns of the modulus of the solutions in the plane (x, y) of coordinates in functions of parameters a_i, b, $1 \leq i \leq 2$, and time t.

The role of the parameters ai and b_i for the same integer i is the same one; one will be interested primarily only in parameters a_i.

The study of these configurations makes it possible to give the following conclusions. The variation of the configuration of the module of the solutions is very fast according to time t. When time t grows from 0 to 0, 01, one passes from a rectilinear structure with a height of 98 to a horseshoe structure with a maximum height equal to 4. The role played by the parameters a_i and b_i is the same for same index i. When variables x, y, and time tend towards infinity, the modulus of the solutions tends towards 2 in accordance with the structure of the polynomials which will be studied in a forthcoming article.

5. Conclusion

We have constructed solutions to the Johnson equation, starting from the solutions of the KPI equation, what makes it possible to obtain rational solutions. These solutions are expressed by means of quotients of two polynomials of degree $2N(N + 1)$ in x, t and $4N(N + 1)$ in y depending on $2N - 2$ parameters.

Here we have given a new method to construct solutions to the Johnson equation related to previous results [12–14].

We have given two types of representations of the solutions to the Johnson equation. An expression by means of Fredholm determinants of order $2N$ depending on $2N - 1$ real parameters is given. Another expression by means of Wronskians of order $2N$ depending on $2N-1$ real parameters is also constructed. Also rational solutions to the Johnson equation depending on $2N - 2$ real parameters are obtained when one of parameters (ϵ) tends to zero.

The patterns of the modulus of the solutions in the plane (x, y) and their evolution according to time and parameters have been studied in Figures 1, 2, 3, 4, and 5.

In another study, we will give a more general representation of rational solutions to the Johnson equation. It can be written without limit at order N depending on $2N - 2$ real parameters. We will prove that these solutions can be written as a quotient of polynomials of degree $2N(N + 1)$ in x, t and $4N(N + 1)$ in y.

Appendix

The solutions to the Johnson equation can be written as

$$\begin{aligned} v_3(x, y, t) &= -2\frac{|n_3(x, y, t)|^2}{(d_3(x, y, t))^2},\\ n_3(x, y, t) &= A_3(x, y, t) + iB_3(x, y, t),\\ d_3(x, y, t) &= C_3(x, y, t)\end{aligned} \quad (A.1)$$

with

$$\begin{aligned} A_3(x, y, t) &= \sum_{k=0}^{12} a_k(y, t)\,x^k,\\ B_3(x, y, t) &= \sum_{k=0}^{12} b_k(y, t)\,x^k,\\ C_3(x, y, t) &= \sum_{k=0}^{12} c_k(y, t)\,x^k,\end{aligned} \quad (A.2)$$

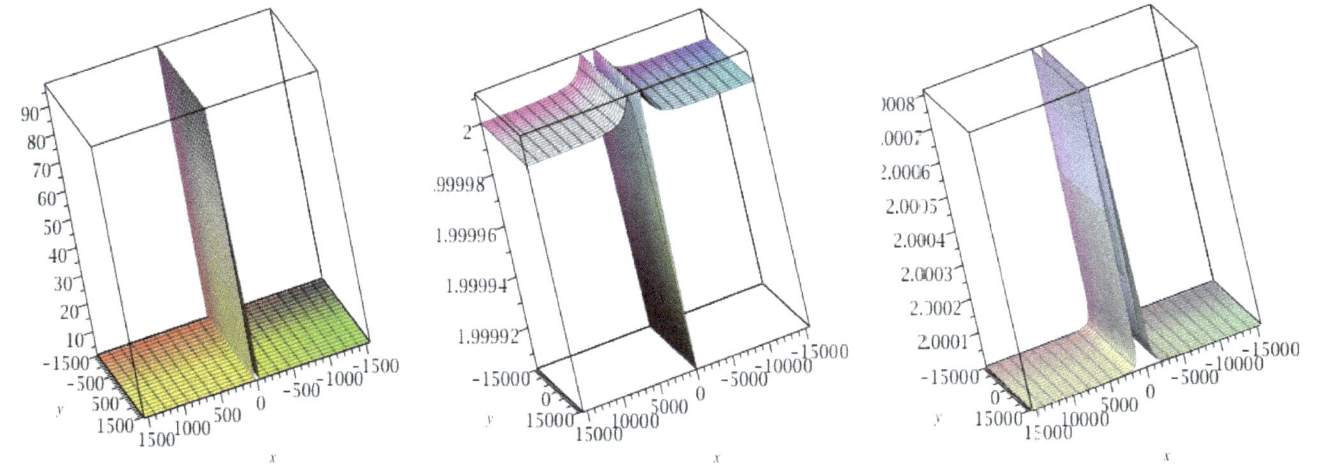

FIGURE 1: Solution of order 3 to (1), on the left for $t = 0$, $a_1 = 0$, $b_1 = 0$, $a_2 = 0$, and $b_2 = 0$; in the center for $t = 0$, $a_1 = 10^3$, $b_1 = 0$, $a_2 = 0$, and $b_2 = 0$; and on the right for $t = 0$, $a_1 = 0$, $b_1 = 10^3$, $a_2 = 0$, and $b_2 = 0$.

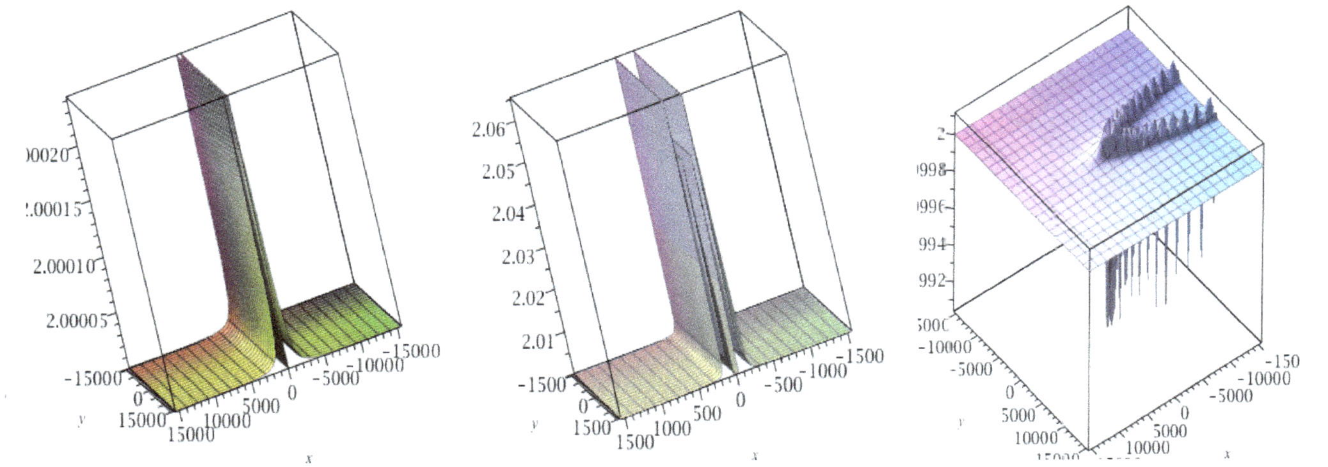

FIGURE 2: Solution of order 3 to (1), on the left for $t = 0$, $a_1 = 10^3$, $b_1 = 0$, $a_2 = 0$, and $b_2 = 0$; in the center for $t = 0$, $a_1 = 0$, $b_1 = 0$, $a_2 = 0$, and $b_2 = 10^6$; and on the right for $t = 0,01$, $a_1 = 0$, $b_1 = 10^3$, $a_2 = 0$, and $b_2 = 0$.

$$\mathbf{a}_{12} = 8916100448256,$$

$$\mathbf{a}_{11} = 8916100448256ty^2 + 2567836929097728t,$$

$$\begin{aligned}\mathbf{a}_{10} &= 4086546038784t^2y^4 \\ &\quad + 2139864107581440t^2y^2 \\ &\quad + 338954474640900096t^2 \\ &\quad + 40122452017152,\end{aligned}$$

$$\begin{aligned}\mathbf{a}_9 &= 1135151677440t^3y^6 + 802449040343040t^3y^4 \\ &\quad + 231105323618795520t^3y^2 \\ &\quad + 27116357971272007680t^3 \\ &\quad + 33435376680960ty^2 \\ &\quad + 13909116699279360t,\end{aligned}$$

$$\begin{aligned}\mathbf{a}_8 &= 212840939520t^4y^8 + 178322008965120t^4y^6 \\ &\quad + 69545583496396800t^4y^4 \\ &\quad + 14790740711602913280t^4y^2 \\ &\quad + 12538266255360t^2y^4 \\ &\quad + 1464283330448688414720t^4 \\ &\quad + 6419592322744320t^2y^2\end{aligned}$$

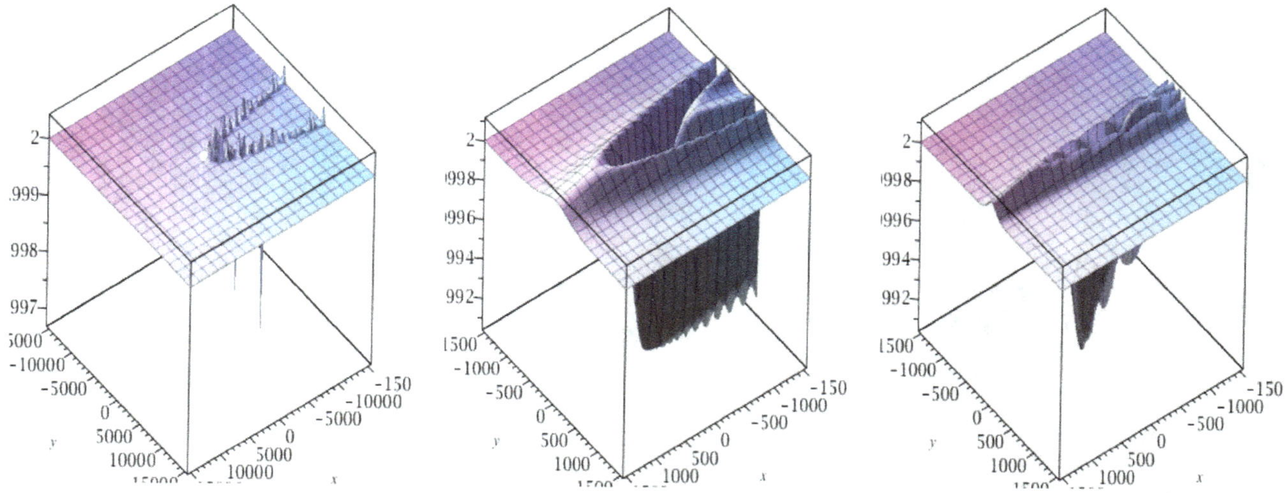

FIGURE 3: Solution of order 3 to (1), on the left for $t = 0, 01$, $a_1 = 0$, $b_1 = 0$, $a_2 = 10^6$, and $b_2 = 0$; in the center for $t = 0, 1$, $a_1 = 10^3$, $b_1 = 0$, $a_2 = 0$, and $b_2 = 0$; and on the right for $t = 1$, $a_1 = 10^3$, $b_1 = 0$, $a_2 = 0$, and $b_2 = 0$.

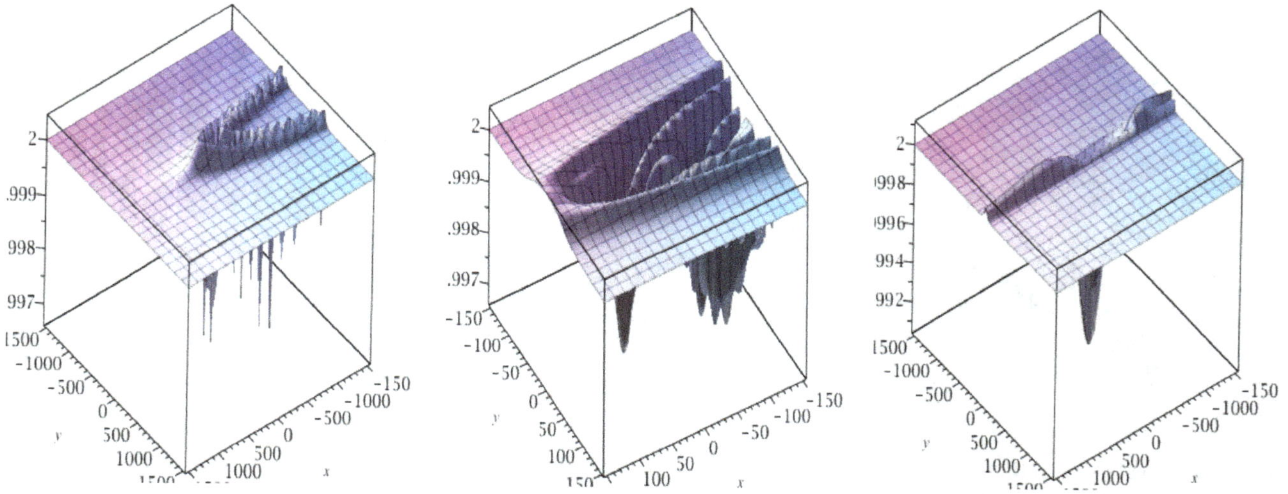

FIGURE 4: Solution of order 3 to (1), on the left for $t = 0, 1$, $a_1 = 0$, $b_1 = 0$, $a_2 = 10^6$, and $b_2 = 0$; in the center for $t = 1$, $a_1 = 0$, $b_1 = 0$, $a_2 = 10^6$, and $b_2 = 0$; and on the right for $t = 10$, $a_1 = 10^3$, $b_1 = 0$, $a_2 = 0$, and $b_2 = 0$.

$$+ 1964395250759761920t^2$$

$$- 125382662553600,$$

$\mathbf{a}_7$

$$= 28378791936t^5y^{10} + 26005292974080t^5y^8$$

$$+ 12125896609628160t^5y^6$$

$$+ 3492258223572910080t^5y^4$$

$$+ 2786281390080t^3y^6$$

$$+ 621211109887322357760t^5y^2$$

$$+ 802449040343040t^3y^4$$

$$+ 56228479889229635125248t^5$$

$$+ 539245755110522880t^3y^2$$

$$+ 155302777471830589440t^3$$

$$- 83588441702400ty^2$$

$$- 24073471210291200t,$$

$\mathbf{a}_6$

$$= 2759049216t^6y^{12} + 2600529297408t^6y^{10}$$

$$+ 1352275234652160t^6y^8$$

$$+ 467916951524474880t^6y^6$$

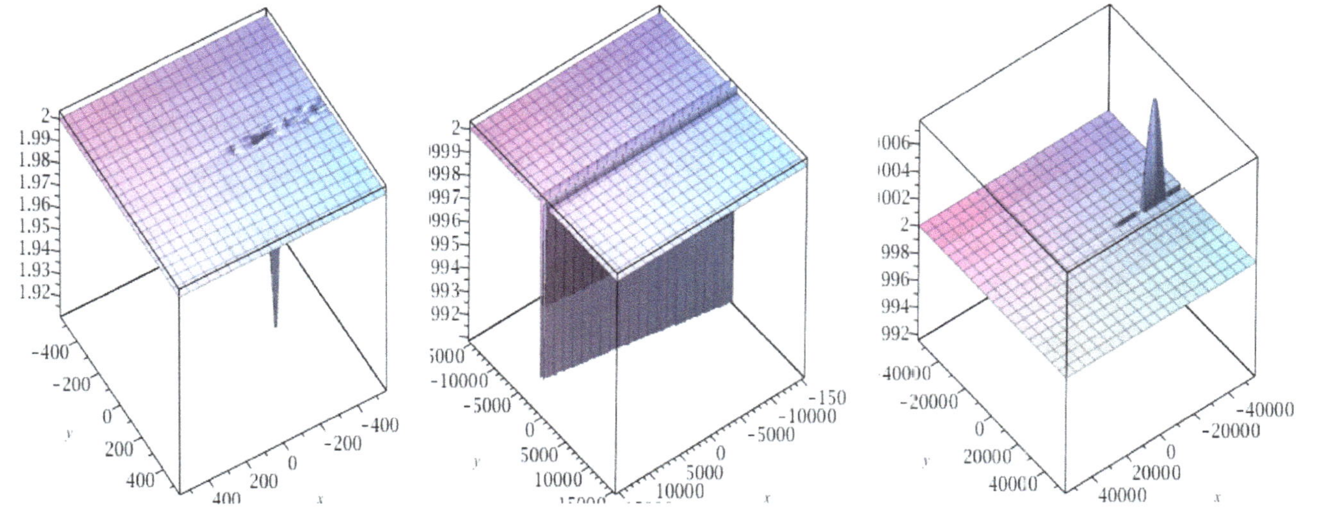

FIGURE 5: Solution of order 3 to (1), on the left for $t = 10$, $a_1 = 0$, $b_1 = 0$, $a_2 = 10^6$, and $b_2 = 0$; in the center for $t = 100$, $a_1 = 10^6$, $b_1 = 0$, $a_2 = 0$, and $b_2 = 0$; and on the right for $t = 10^3$, $a_1 = 10^5$, $b_1 = 10^3$, $a_2 = 0$, and $b_2 = 0$.

$$+ 406332702720t^4 y^8$$
$$+ 1121631170629887590 40t^6 y^4$$
$$- 104021171896320t^4 y^6$$
$$+ 1789087996475488390 3488t^6 y^2$$
$$- 18188844914442240t^4 y^4$$
$$+ 1574397436898429783506944t^6$$
$$+ 2588379624530509 8240t^4 y^2$$
$$- 24379962163200t^2 y^4$$
$$+ 7765138873591529472000t^4$$
$$- 18456327927889920t^2 y^2$$
$$- 1714031150172733440t^2$$
$$+ 376147987660800,$$

$\mathbf{a}_5$

$$= 197074944t^7 y^{14} + 180592312320t^7 y^{12}$$
$$+ 100553799499776t^7 y^{10}$$
$$+ 39230841972326400t^7 y^8$$
$$+ 40633270272t^5 y^{10}$$
$$+ 11298482488030003200t^7 y^6$$
$$- 45509262704640t^5 y^8$$
$$+ 2402016291564313116672t^7 y^4$$

$$- 24073471210291200t^5 y^6$$
$$+ 357817599295097678069760t^7 y^2$$
$$- 30300475763353190 40t^5 y^4$$
$$- 4063327027200t^3 y^6$$
$$+ 32387604416196269832142848t^7$$
$$+ 776513887359152947200t^5 y^2$$
$$- 5717449412444160t^3 y^4$$
$$+ 259417759488945816600576t^5$$
$$- 722204135308736000t^3 y^2$$
$$- 526920137850853785 60t^3$$
$$+ 188073993830400ty^2$$
$$+ 78238781433446400t,$$

$\mathbf{a}_4$

$$= 10264320t^8 y^{16} + 8599633920t^8 y^{14}$$
$$+ 5056584744960t^8 y^{12}$$
$$+ 2139864107581440t^8 y^{10}$$
$$+ 2821754830t^6 y^{12}$$
$$+ 697595699071549440t^8 y^8$$
$$- 6501323243520t^6 y^{10}$$
$$+ 177488888539234959360t^8 y^6$$

$$
\begin{aligned}
&- 3845068318310400t^6 y^8 \\
&+ 34787822153690052034560t^8 y^4 \\
&- 1044253684499742720t^6 y^6 \\
&- 423263232000t^4 y^8 \\
&+ 490721279033276815638528 0t^8 y^2 \\
&- 109389853179563212800t^6 y^4 \\
&- 947335672627200t^4 y^6 \\
&+ 4858140662429440474821427 20t^8 \\
&+ 149090666372957365862 40t^6 y^2 \\
&- 10231225264373 7600t^4 y^4 \\
&+ 5903990388369111688151040t^6 \\
&+ 1386631941712773 1200t^4 y^2 \\
&+ 39182082048000t^2 y^4 \\
&- 249593749508299161600t^4 \\
&+ 34104084214579200t^2 y^2 \\
&+ 96053150129061888 00t^2 \\
&+ 493694233804800,
\end{aligned}
$$

$\mathbf{a}_3$

$$
\begin{aligned}
&= 380160t^9 y^{18} + 268738560t^9 y^{16} \\
&\quad + 171992678400t^9 y^{14} \\
&\quad + 75951966781440t^9 y^{12} + 134369280t^7 y^{14} \\
&\quad + 27104945362698240t^9 y^{10} \\
&\quad - 511678218240t^7 y^{12} \\
&\quad + 7806224264457093120t^9 y^8 \\
&\quad - 288534917283840t^7 y^{10} \\
&\quad + 1814330860623290695680t^9 y^6 \\
&\quad - 100216969038397440t^7 y^8 \\
&\quad - 28217548800t^5 y^{10} \\
&\quad + 340778665995331121971200t^9 y^4 \\
&\quad - 20508457607060520960t^7 y^6 \\
&\quad - 91715095756800t^5 y^8 \\
&\quad + 44164915112994913407467520t^9 y^2 \\
&\quad - 1804470366815555420160t^7 y^4 \\
&\quad - 3789342690508800t^5 y^6 \\
&\quad + 5182016706591403173142855680t^9 \\
&\quad + 178908799647548839034880t^7 y^2 \\
&\quad + 8152882249885286400t^5 y^4 \\
&\quad + 4353564672000t^3 y^6 \\
&\quad + 9078343662115621089312768 0t^7 \\
&\quad + 135889930287851765760 0t^5 y^2 \\
&\quad + 5934779360870400t^3 y^4 \\
&\quad + 292856666089737682944 00t^5 \\
&\quad + 1420334801407180800t^3 y^2 \\
&\quad + 52614978677212446720 0t^3 \\
&\quad + 164564744601600t y^2 \\
&\quad + 119615060076134400t,
\end{aligned}
$$

$\mathbf{a}_2$

$$
\begin{aligned}
&= 9504t^{10} y^{20} + 4976640t^{10} y^{18} \\
&\quad + 3941498880t^{10} y^{16} + 1651129712640t^{10} y^{14} \\
&\quad + 4199040t^8 y^{16} + 653847366205440t^{10} y^{12} \\
&\quad - 23648993280t^8 y^{14} \\
&\quad + 205426954327818240t^{10} y^{10} \\
&\quad - 10990332149760t^8 y^{12} \\
&\quad + 54232715942544015360t^{10} y^8 \\
&\quad - 4695812902748160t^8 y^{10} \\
&\quad - 1175731200t^6 y^{12} \\
&\quad + 11359288866511037399040t^{10} y^6 \\
&\quad - 1230421861859328000t^8 y^8 \\
&\quad - 5224277606400t^6 y^{10} \\
&\quad + 2249139195569185405009920t^{10} y^4 \\
&\quad - 212000616866308423680t^8 y^6 \\
&\quad + 376147987660800t^6 y^8 \\
&\quad + 235546213935972871506493440t^{10} y^2 \\
&\quad - 13134177751903386992640t^8 y^4
\end{aligned}
$$

$$+ 1033554363961835520t^6 y^6$$
$$+ 272097792000t^4 y^8$$
$$+ 3731052028745810284662856089 6t^{10}$$
$$+ 122680319758319203909632 0t^8 y^2$$
$$+ 2468975207327465472 00t^6 y^4$$
$$+ 515462057164800t^4 y^6$$
$$+ 9053807598163957248530841 60t^8$$
$$+ 27954499944929506099200t^6 y^2$$
$$+ 141431643360460800t^4 y^4$$
$$+ 862595998300681902489600t^6$$
$$+ 4429518702693580800 0t^4 y^2$$
$$+ 20570593075200t^2 y^4$$
$$+ 1363059198703655976960 0t^4$$
$$- 10155995666841600t^2 y^2$$
$$+ 8350485326069760000t^2$$
$$- 246847116902400,$$

$\mathbf{a}_1$

$$= 144t^{11} y^{22} + 41472t^{11} y^{20} + 59719680t^{11} y^{18}$$
$$+ 17199267840t^{11} y^{16} + 77760t^9 y^{18}$$
$$+ 9906778275840t^{11} y^{14} - 604661760t^9 y^{16}$$
$$+ 2853152143441920t^{11} y^{12}$$
$$- 180592312320t^9 y^{14}$$
$$+ 821707817311272960t^{11} y^{10}$$
$$- 121358033879040t^9 y^{12} - 27993600t^7 y^{14}$$
$$+ 236651851385646612480t^{11} y^8$$
$$- 31028029559930880t^9 y^{10}$$
$$- 162855567360t^7 y^{12}$$
$$+ 34077866599533112197120t^{11} y^6$$
$$- 6881802969981911040t^9 y^8$$
$$+ 37614798766080t^7 y^{10}$$
$$+ 9814425580665536312770560t^{11} y^4$$
$$- 1242422219774644715520t^9 y^6$$
$$+ 472107518735155520t^7 y^8$$
$$+ 9069926400t^5 y^{10}$$
$$+ 5653109134463348916155842 56t^{11} y^2$$
$$- 170389332997665560985 60t^9 y^4$$
$$+ 285928642055032012 80t^7 y^6$$
$$+ 22348298649600t^5 y^8$$
$$+ 162809543072544443785288265728t^{11}$$
$$+ 36804095927495761172889 60t^9 y^2$$
$$+ 4252337954585837568 00t^7 y^4$$
$$+ 16216157690265600t^5 y^6$$
$$+ 5299789813559389608896102400t^9$$
$$+ 18849319952866752684032 0t^7 y^2$$
$$+ 3450530873475072000t^5 y^4$$
$$+ 1142810726400t^3 y^6$$
$$+ 1012112638006133432254464 0t^7$$
$$+ 654028065841191321600t^5 y^2$$
$$- 4184646362726400t^3 y^4$$
$$+ 164399083009466381107200t^5$$
$$- 675561785838796800t^3 y^2$$
$$+ 332358343529280307200t^3$$
$$- 41141186150400t y^2$$
$$- 59243308056576000t,$$

$\mathbf{a}_0$

$$= t^{12} y^{24} + 497664t^{12} y^{20} + 648t^{10} y^{20}$$
$$+ 103195607040t^{12} y^{16} - 6635520t^{10} y^{18}$$
$$- 447897600t^{10} y^{16}$$
$$+ 114126085737676 80t^{12} y^{12}$$
$$- 1651129712640t^{10} y^{14} - 291600t^8 y^{16}$$
$$- 113927950172160t^{10} y^{12}$$
$$- 2149908480t^8 y^{14}$$
$$+ 70995555415693983744 0t^{12} y^8$$
$$- 1369513028852121 60t^{10} y^{10}$$

$$\begin{aligned}
&+ 935210188800t^8 y^{12} \\
&- 6162808629834547200t^{10} y^8 \\
&+ 594406696550400t^8 y^{10} + 125971200t^6 y^{12} \\
&+ 2355462139359728715064 9344t^{12} y^4 \\
&- 3786429622170345799680t^{10} y^6 \\
&+ 304663152316907520t^8 y^8 \\
&+ 386983526400t^6 y^{10} \\
&+ 1533503996978990048870 40t^{10} y^4 \\
&+ 27445041098196516864 0t^8 y^6 \\
&+ 727916013158400t^6 y^8 \\
&+ 325619086145088897570576531456t^{12} \\
&- 2622398328167196524544 0t^8 y^4 \\
&- 47790298402652160t^6 y^6 \\
&+ 23808556800t^4 y^8 \\
&+ 1385011737943520484458181 4272t^{10} \\
&+ 1751585765260787712 00t^6 y^4 \\
&- 208971104256000t^4 y^6 \\
&+ 4508501751118230743678 9760t^8 \\
&+ 68145477097881600t^4 y^4 \\
&+ 732141665224344207360000t^6 \\
&- 1714216089600t^2 y^4 \\
&+ 5839742646192439296000t^4 \\
&- 629671731344179200t^2 - 30855889612800
\end{aligned} \tag{A.3}$$

$$\mathbf{b}_{12} = 0,$$

$$\mathbf{b}_{11} = 0,$$

$$\mathbf{b}_{10} = 213986410758144ty,$$

$$\mathbf{b}_9 = 178322008965120t^2 y^3 + 51356738581954560t^2 y,$$

$$\begin{aligned}
\mathbf{b}_8 &= 66870753361920t^3 y^5 \\
&+ 34237825721303040t^3 y^3 \\
&+ 5546527766851092480t^3 y \\
&+ 802449040343040ty,
\end{aligned}$$

$$\begin{aligned}
\mathbf{b}_7 &= 14860167413760t^4 y^7 \\
&+ 9986032502046720t^4 y^5 \\
&+ 2875977360589455360t^4 y^3 \\
&+ 354977777078469918720t^4 y \\
&+ 534966026895360t^2 y^3 \\
&+ 154070215745863680t^2 y,
\end{aligned}$$

$$\begin{aligned}
\mathbf{b}_6 &= 2167107747840t^5 y^9 \\
&+ 1664338750341120t^5 y^7 \\
&+ 633399775844106240t^5 y^5 \\
&+ 138046913308293857280t^5 y^3 \\
&+ 156031757844480t^3 y^5 \\
&+ 14909066637295736586240t^5 y \\
&+ 59916195012280320t^3 y^3 \\
&+ 12941898122652549120t^3 y \\
&- 2006122600857600ty,
\end{aligned}$$

$$\begin{aligned}
\mathbf{b}_5 &= 216710774784t^6 y^{11} + 173368619827200t^6 y^9 \\
&+ 77035107872931840t^6 y^7 \\
&+ 22186111067404369920t^6 y^5 \\
&+ 26005292974080t^4 y^7 \\
&+ 4141407399248815718400t^6 y^3 \\
&+ 7489524376535040t^4 y^5 \\
&+ 429381119154117213683712t^6 y \\
&+ 2978690837753364480t^4 y^3 \\
&+ 621211109887322357760t^4 y \\
&- 1003061300428800t^2 y^3 \\
&- 442951870269358080t^2 y,
\end{aligned}$$

$\mathbf{b}_4$

$$= 15049359360t^7y^{13} + 11557907988480t^7y^{11}$$
$$+ 5646863617228800t^7y^9$$
$$+ 1917318240392970240t^7y^7$$
$$+ 2708884684800t^5y^9$$
$$+ 468373455867425587200t^7y^5$$
$$+ 7951502206557726179328 0t^7y^3$$
$$+ 72755379657768960t^5y^5$$
$$+ 8587622383082344273674240t^7y$$
$$+ 9860493807735275520 0t^5y^3$$
$$- 208971104256000t^3y^5$$
$$+ 1863633329661967073280 0t^5y$$
$$- 160489808068608000t^3y^3$$
$$- 17332899271409664000t^3y$$
$$+ 4513775851929600ty$$

$\mathbf{b}_3$

$$= 716636160t^8y^{15} + 481579499520t^8y^{13}$$
$$+ 257576235171840t^8y^{11}$$
$$+ 97007172877025280t^8y^9$$
$$+ 180592312320t^6y^{11}$$
$$+ 27938065788583280640t^8y^7$$
$$- 86684309913600t^6y^9$$
$$+ 6152948136026811924480t^8y^5$$
$$- 32811249649582080t^6y^7$$
$$+ 954180264786927141519360t^8y^3$$
$$+ 4108539086556364 80t^6y^5$$
$$- 23219011584000t^4y^7$$
$$+ 117773106967986435753246720t^8y$$
$$+ 26623333280885243904 00t^6y^3$$
$$- 227360561430528sss00t^4y^5$$
$$+ 35781759929509767806976 0t^6y$$
$$- 7575118940838297600t^4y^3$$
$$+ 332791666011065548800t^4y$$
$$+ 1504591950643200t^2y^3$$
$$+ 8184980211499008 00t^2y$$

$\mathbf{b}_2$

$$= 22394880t^9y^{17} + 11465178560t^9y^{15}$$
$$+ 7430083706880t^9y^{13}$$
$$+ 2853152143441920t^9y^{11}$$
$$+ 7524679680t^7y^{13}$$
$$+ 924421294475182080t^9y^9$$
$$- 8668430991360t^7y^{11}$$
$$+ 236651851385646612480t^9y^7$$
$$- 3358397835509760t^7y^9$$
$$+ 51116799399299668295680t^9y^5$$
$$- 616280862983454720t^7y^7$$
$$- 1451188224000t^5y^9$$
$$+ 6542950387110357541847040t^9y^3$$
$$+ 369768517790072832 00t^7y^5$$
$$- 1560317578444800t^5y^7$$
$$+ 105995796271187792177922 0480t^9y$$
$$+ 5395662211592742764544 0t^7y^3$$
$$- 805658836504412160t^5y^5$$
$$+ 4293811191541172136837120t^7y$$
$$- 80116512187849113600t^5y^3$$
$$+ 188073993830400t^3y^5$$
$$+ 32613583269084423782400t^5y$$
$$+ 120367356051456000t^3y^3$$
$$+ 340880352337723392 00t^3y$$
$$+ 3949553870438400ty,$$

$\mathbf{b}_1$

$$= 414720t^{10}y^{19} + 1194393360t^{10}y^{17}$$
$$+ 137594142720t^{10}y^{15}$$
$$+ 39627113103360t^{10}y^{13} + 179159040t^8y^{15}$$

$$+ 17118912860651520t^{10}y^{11}$$
$$- 361184624640t^{8}y^{13}$$
$$+ 4930246903867637760t^{10}y^{9}$$
$$- 113927950172160t^{8}y^{11}$$
$$+ 946607405542586449920t^{10}y^{7}$$
$$- 32811249649582080t^{8}y^{9}$$
$$- 48372940800t^{6}y^{11}$$
$$+ 27262293279626489757696 0t^{10}y^{5}$$
$$- 287597736058945536 0t^{8}y^{7}$$
$$- 51081825484800t^{6}y^{9}$$
$$+ 19628851161331072625541120t^{10}y^{3}$$
$$+ 1183259256928233062 40t^{8}y^{5}$$
$$- 16583946833756160t^{6}y^{7}$$
$$+ 56531091344633489161558425 60t^{10}y$$
$$+ 6474794653911291317452 80t^{8}y^{3}$$
$$- 19977771308380323840t^{6}y^{5}$$
$$+ 10448555212800t^{4}y^{7}$$
$$+ 294432767419966089383116 80t^{8}y$$
$$+ 20337268478454005760 00t^{6}y^{3}$$
$$+ 3009183901286400t^{4}y^{5}$$
$$+ 6709079986783081463808 00t^{6}y$$
$$- 674057193888153600t^{4}y^{3}$$
$$+ 10630844886464593920 00t^{4}y$$
$$+ 658258978406400t^{2}y^{3}$$
$$- 2437438960041984 00t^{2}y,$$

$\mathbf{b}_0$

$$= 3456t^{11}y^{21} + 1433272320t^{11}y^{17}$$
$$+ 1866240t^{9}y^{17} + 237762678620160t^{11}y^{13}$$
$$- 5733089280t^{9}y^{15} - 1031956070400t^{9}y^{13}$$
$$+ 1972098761547055104 0t^{11}y^{9}$$
$$- 475525357240320t^{9}y^{11} - 671846400t^{7}y^{13}$$
$$- 196867497897492480t^{9}y^{9}$$
$$- 619173642240t^{7}y^{11}$$
$$+ 81786879838879469273088 0t^{11}y^{5}$$
$$+ 3944197523094110208 0t^{9}y^{7}$$
$$+ 664992491765760t^{7}y^{9}$$
$$- 70995555415693983744 00t^{9}y^{5}$$
$$- 17118912860651520 0t^{7}y^{7}$$
$$+ 217678233600t^{5}y^{9}$$
$$+ 1356746192271203739877402 2144t^{11}y$$
$$+ 3271475193555178770923520t^{9}y^{3}$$
$$- 12233175130221576192 0t^{7}y^{5}$$
$$- 111451255603200t^{5}y^{7}$$
$$+ 883298302259898268149350 40t^{9}y$$
$$+ 354977777078469918720 00t^{7}y^{3}$$
$$+ 22227838417502208 0t^{5}y^{5}$$
$$+ 45238367910880206441676 80t^{7}y$$
$$- 77035107872931840000t^{5}y^{3}$$
$$+ 27427457433600t^{3}y^{5}$$
$$+ 156966735801885917184 00t^{5}y$$
$$- 12036735605145600t^{3}y^{3}$$
$$- 16213482860131123200t^{3}y$$
$$- 987388467609600ty \tag{A.4}$$

$$\mathbf{c}_{12} = 8916100448256,$$

$$\mathbf{c}_{11} = 8916100448256ty^{2} + 2567836929097728t,$$

$\mathbf{c}_{10}$

$$= 4086546038784t^{2}y^{4}$$
$$+ 2139864107581440t^{2}y^{2}$$
$$+ 338954474640900096t^{2}$$
$$- 13374150672384,$$

$\mathbf{c}_9$

$$= 1135151677440t^{3}y^{6} + 802449040343040t^{3}y^{4}$$

$$+ 231105323618795520t^3y^2$$
$$+ 271163579712720076 80t^3$$
$$- 11145125560320ty^2 + 1069932053790720t,$$

$\mathbf{c}_8$

$$= 212840939520t^4y^8 + 178322008965120t^4y^6$$
$$+ 69545583496396800t^4y^4$$
$$+ 14790740711602913280t^4y^2$$
$$- 4179422085120t^2y^4$$
$$+ 1464283330448688414720t^4$$
$$+ 577763309046988800t^2$$
$$+ 75229597532160,$$

$\mathbf{c}_7$

$$= 28378791936t^5y^{10} + 26005292974080t^5y^8$$
$$+ 12125896609628160t^5y^6$$
$$+ 3492258223572910080t^5y^4$$
$$- 928760463360t^3y^6$$
$$+ 621211109887322357760t^5y^2$$
$$- 267483013447680t^3y^4$$
$$+ 56228479889229635125248t^5$$
$$+ 231105323618795520t^3y^2$$
$$+ 665583332022131097 60t^3$$
$$+ 50153065021440ty^2$$
$$+ 14444082726174720t,$$

$\mathbf{c}_6$

$$= 2759049216t^6y^{12} + 2600529297408t^6y^{10}$$
$$+ 1352275234652160t^6y^8$$
$$+ 467916951524474880t^6y^6$$
$$- 135444234240t^4y^8$$
$$+ 112163117062988759040t^6y^4$$
$$- 104021171896320t^4y^6$$
$$+ 17890879964754883903488t^6y^2$$
$$+ 28888165452349440t^4y^4$$
$$+ 1574397436898429783506944t^6$$
$$+ 25883796245305098240t^4y^2$$
$$+ 14627977297920t^2y^4$$
$$+ 40378722142675953254 40t^4$$
$$+ 7222041353087360t^2y^2$$
$$+ 1521443380490403840t^2$$
$$- 325994922639360$$

$\mathbf{c}_5$

$$= 197074944t^7y^{14} + 180592312320t^7y^{12}$$
$$+ 100553799499776t^7y^{10}$$
$$+ 39230841972326400t^7y^8$$
$$- 13544423424t^5y^{10}$$
$$+ 11298482488030003200t^7y^6$$
$$- 19503969730560t^5y^8$$
$$+ 2402016291564313116672t^7y^4$$
$$- 5349660268953 60t^5y^6$$
$$+ 357817599295097678069760t^7y^2$$
$$+ 3749041916482682880t^5y^4$$
$$+ 2437996216320t^3y^6$$
$$+ 32387604416196269832142848t^7$$
$$+ 1397724997246475304960t^5y^2$$
$$+ 1504591950643200t^3y^4$$
$$+ 152072479700416513179648t^5$$
$$+ 741462913276968960t^3y^2$$
$$+ 102610763686745210880t^3$$
$$- 162997461319680ty^2$$
$$- 61387351586242560t,$$

$\mathbf{c}_4$

$$= 10264320t^8y^{16} + 8599633920t^8y^{14}$$
$$+ 5056584744960t^8y^{12}$$
$$+ 2139864107581440t^8y^{10} - 940584960t^6y^{12}$$
$$+ 697595699071549440t^8y^8$$

$$- 2167107747840t^{6}y^{10}$$
$$+ 1774888885392349593 60t^{8}y^{6}$$
$$- 501530650214400t^{6}y^{8}$$
$$+ 347878221536900520345 60t^{8}y^{4}$$
$$+ 222545867188469760t^{6}y^{6}$$
$$+ 253957939200t^{4}y^{8}$$
$$+ 4907212790332768156385280t^{8}y^{2}$$
$$+ 1679365351629914112 00t^{6}y^{4}$$
$$+ 167176883404800t^{4}y^{6}$$
$$+ 4858140662429440474821427 20t^{8}$$
$$+ 4472719991188720975 8720t^{6}y^{2}$$
$$+ 138422459459174400t^{4}y^{4}$$
$$+ 3757084792598525619732480t^{6}$$
$$+ 323547453066313728 00t^{4}y^{2}$$
$$- 33957804441600t^{2}y^{4}$$
$$+ 44094895746466185216 00t^{4}$$
$$- 18055103407718400t^{2}y^{2}$$
$$- 570541267683901440 0t^{2}$$
$$+ 117546246144000,$$

$\mathbf{c}_3$

$$= 380160t^{9}y^{18} + 268738560t^{9}y^{16}$$
$$+ 171992678400t^{9}y^{14}$$
$$+ 75951966781440t^{9}y^{12} - 447897 60t^{7}y^{14}$$
$$+ 27104945362698240t^{9}y^{10}$$
$$- 150493593600t^{7}y^{12}$$
$$+ 7806224264457093120t^{9}y^{8}$$
$$- 55725627801600t^{7}y^{10}$$
$$+ 1814330860623290695680t^{9}y^{6}$$
$$+ 10699320537907 20t^{7}y^{8}$$
$$+ 16930529280t^{5}y^{10}$$
$$+ 340778665995331121971200t^{9}y^{4}$$
$$+ 866216990748966912 0t^{7}y^{6}$$
$$+ 10448555212800t^{5}y^{8}$$
$$+ 44164915112994913407467520t^{9}y^{2}$$
$$+ 375684814074713997312 0t^{7}y^{4}$$
$$+ 1225963811635200 0t^{5}y^{6}$$
$$+ 518201670659140317314285568 0t^{9}$$
$$+ 8945439982377441951744 00t^{7}y^{2}$$
$$+ 8666449635704832000t^{5}y^{4}$$
$$- 3773089382400t^{3}y^{6}$$
$$+ 6134015987915960195481600 0t^{7}$$
$$+ 24959374950829916 1600t^{5}y^{2}$$
$$- 1755357275750400t^{3}y^{4}$$
$$+ 1187400664327481878118 40t^{5}$$
$$- 240734712102912 00t^{3}y^{2}$$
$$- 333562017089794867200t^{3}$$
$$+ 39182082048000ty^{2}$$
$$- 12789031580467200t,$$

$\mathbf{c}_2$

$$= 9504t^{10}y^{20} + 4976640t^{10}y^{18}$$
$$+ 3941498880t^{10}y^{16} + 1651129712640t^{10}y^{14}$$
$$- 1399680t^{8}y^{16} + 653847366205440t^{10}y^{12}$$
$$- 6449725440t^{8}y^{14}$$
$$+ 205426954327818240t^{10}y^{10}$$
$$- 2941074800640t^{8}y^{12}$$
$$+ 54232715942544015360t^{10}y^{8}$$
$$- 416084687585280t^{8}y^{10} + 705438720t^{6}y^{12}$$
$$+ 11359288866511037399040t^{10}y^{6}$$
$$+ 181888449144422400t^{8}y^{8}$$
$$+ 348285173760t^{6}y^{10}$$
$$+ 2249139195569185405009920t^{10}y^{4}$$
$$+ 142977160212161495040t^{8}y^{6}$$
$$+ 487599243264000t^{6}y^{8}$$
$$+ 235546213935972871506493440t^{10}y^{2}$$

$$
\begin{aligned}
&+ 42242355472337920327680t^{8}y^{4} \\
&+ 930840886797926400t^{6}y^{6} \\
&- 235818086400t^{4}y^{8} \\
&+ 3731052028745810284662856089 6t^{10} \\
&+ 1104122877824872835186688 0t^{8}y^{2} \\
&+ 256141733677498368000t^{6}y^{4} \\
&- 41794220851200t^{4}y^{6} \\
&+ 64039126913842624440827904 0t^{8} \\
&- 1890256662942852317184 0t^{6}y^{2} \\
&+ 57174494124441600t^{4}y^{4} \\
&+ 19360487961859749366988 80t^{6} \\
&+ 35050974082183987200t^{4}y^{2} \\
&+ 4897760256000t^{2}y^{4} \\
&- 111623871307878236160 00t^{4} \\
&+ 1880739938304000t^{2}y^{2} \\
&- 2193695064037785600t^{2} \\
&- 105791621529600,
\end{aligned}
$$

$\mathbf{c}_1$

$$
\begin{aligned}
&= 144t^{11}y^{22} + 41472t^{11}y^{20} + 59719680t^{11}y^{18} \\
&+ 17199267840t^{11}y^{16} - 25920t^{9}y^{18} \\
&+ 9906778275840t^{11}y^{14} - 156764160t^{9}y^{16} \\
&+ 2853152143441920t^{11}y^{12} \\
&- 77396705280t^{9}y^{14} \\
&+ 821707817311272960t^{11}y^{10} \\
&- 12383472844800t^{9}y^{12} + 16796160t^{7}y^{14} \\
&+ 236651851385646612480t^{11}y^{8} \\
&- 1069932053790720t^{9}y^{10} \\
&+ 4837294080t^{7}y^{12} \\
&+ 340778665995331121971 20t^{11}y^{6} \\
&+ 174612911178645504 0t^{9}y^{8} \\
&+ 4179422085120t^{7}y^{10} \\
&+ 9814425580665536312770560t^{11}y^{4} \\
&+ 13607481454674680021760t^{9}y^{6} \\
&+ 24742178743910400t^{7}y^{8} \\
&- 7860602880t^{5}y^{10} \\
&+ 565310913446334891615584256t^{11}y^{2} \\
&+ 18742826629743211708416 0t^{9}y^{4} \\
&+ 237653307787994726640t^{7}y^{6} \\
&+ 2612138803200t^{5}y^{8} \\
&+ 1628095430725444487852882 65728t^{11} \\
&+ 77288601447741098463068160t^{9}y^{2} \\
&+ 3176311567816725626880t^{7}y^{4} \\
&- 3176360784691200t^{5}y^{6} \\
&+ 3886512529943552379857141760t^{9} \\
&- 501583599011877995151360t^{7}y^{2} \\
&+ 556899633998069760 0t^{5}y^{4} \\
&+ 272097792000t^{3}y^{6} \\
&+ 17481945565560486557122560t^{7} \\
&+ 404434316332892160000t^{5}y^{2} \\
&+ 579894814310400t^{3}y^{4} \\
&- 1886263162950719530598 40t^{5} \\
&- 844076084310835200t^{3}y^{2} \\
&- 113674931054995046400t^{3} \\
&- 17631936921600ty^{2} \\
&- 16362437463244800t,
\end{aligned}
$$

$\mathbf{c}_0$

$$
\begin{aligned}
&= t^{12}y^{24} + 497664t^{12}y^{20} - 216t^{10}y^{20} \\
&+ 103195607040t^{12}y^{16} - 1658880t^{10}y^{18} \\
&- 806215680t^{10}y^{16} \\
&+ 11412608573767680t^{12}y^{12} + 174960t^{8}y^{16} \\
&- 173368619827200t^{10}y^{12} \\
&+ 709955554156939837440t^{12}y^{8} \\
&+ 684756514426060 80t^{10}y^{10} \\
&- 148343685120t^{8}y^{12}
\end{aligned}
$$

$$
\begin{aligned}
&- 1109305553370218496 0t^{10}y^{8} \\
&- 534966026895360t^{8}y^{10} - 109175040t^{6}y^{12} \\
&+ 2355462139359728715064934 4t^{12}y^{4} \\
&+ 7572859244340691599360t^{10}y^{6} \\
&+ 666300186498170880t^{8}y^{8} \\
&+ 116095057920t^{6}y^{10} \\
&- 5111679989929966829568 0t^{10}y^{4} \\
&+ 1216127569620683980 80t^{8}y^{6} \\
&- 400527949824000t^{6}y^{8} \\
&+ 32561908614508889757057653145 6t^{12} \\
&+ 235546213935972871506493440t^{10}y^{2} \\
&+ 1175863886572431605760 0t^{8}y^{4} \\
&+ 164056248247910400t^{6}y^{6} \\
&+ 5668704000t^{4}y^{8} \\
&+ 1045825189875719549488830873 6t^{10} \\
&- 368040959274957611728896 0t^{8}y^{2} \\
&+ 1993283416212111360 0t^{6}y^{4} \\
&+ 27862813900800t^{4}y^{6} \\
&+ 671674750676797641405235 20t^{8} \\
&- 612336665460360609792 0t^{6}y^{2} \\
&+ 25013841179443200t^{4}y^{4} \\
&- 124222447308861054805606 40t^{6} \\
&- 3428062300345466880 0t^{4}y^{2} \\
&- 734664038400t^{2}y^{4} \\
&- 49716532743493386240 0t^{4} \\
&+ 1880739938304000t^{2}y^{2} \\
&- 1126187075056435200t^{2} + 4407984230400
\end{aligned}
\quad (A.5)
$$

Conflicts of Interest

The author declares that there are no conflicts of interest.

References

[1] R. S. Johnson, "Water waves and Kortewegde Vries equations," *Journal of Fluid Mechanics*, vol. 97, no. 4, pp. 701–719, 1980.

[2] R. S. Johnson, *A Modern Introduction to the Mathematical Theory of Water Waves*, Cambridge University Press, Cambridge, UK, 1997.

[3] M. J. Ablowitz, *Nonlinear Dispersive Waves: Asymptotic Analysis and Solitons*, Cambridge University Press, Cambridge, UK, 2011.

[4] V. D. Lipovskii, "On the nonlinear internal wave theory in fluid of finite depth," *Izv. Akad. Nauka, Phys. of Atmosphere and Ocean*, vol. 21, no. 8, pp. 864–871, 1985.

[5] V. I. Golinko, V. S. Dryuma, and Y. A. Stepanyants, "Nonlinear quasicylindrical waves: exact solutions of the cylindrical Kadomtsev-Petviashvili equation," in *Nonlinear and Turbulent Processes in Physics: Proceedings of the Second International Workshop on Nonlinear and Turbulent Processes in Physics, Kiev, USSR, 10–25 October 1983*, pp. 1353–1360, Harwood Academic Publishers; Gordon and Breach, 1984.

[6] V. D. Lipovskii, V. B. Matveev, and A. O. Smirnov, "On a connection between the Kadomtsev-Petviashvili equation and the Johnson equation," *Zapiski Nauchnykh Seminarov LOMI*, vol. 150, pp. 70–75, 1986.

[7] B. B. Kadomtsev and W. I. Petviashvili, "On the stability of solitary waves in weakly dispersing media," *Soviet Physics, Doklady*, vol. 15, no. 6, pp. 539–541, 1970.

[8] K. Klein, V. B. Matveev, and A. O. Smirnov, "Cylindrical Kadomtsev-Petviashvili equation: old and new results," *Theoretical and Mathematical Physics*, vol. 152, no. 2, pp. 1132–1145, 2007.

[9] K. R. Khusnutdinova, C. Klein, V. B. Matveev, and A. O. Smirnov, "On the integrable elliptic cylindrical Kadomtsev-Petviashvili equation," *Chaos*, vol. 23, no. 1, Article ID 013126, 15 pages, 2013.

[10] M. J. Ablowitz and H. Segur, "On the evolution of packets of water waves," *Journal of Fluid Mechanics*, vol. 92, no. 4, pp. 691–715, 1979.

[11] D. E. Pelinovsky, Y. A. Stepanyants, and Y. S. Kivshar, "Self-focusing of plane dark solitons in nonlinear defocusing media," *Physical Review E: Statistical, Nonlinear, and Soft Matter Physics*, vol. 51, no. 5, pp. 5016–5026, 1995.

[12] P. Gaillard, "Families of rational solutions of order 5 to the KPI equation depending on 8 parameters," *New Horizons in Mathematical Physics*, vol. 1, no. 1, pp. 26–31, 2017.

[13] P. Gaillard, "Families of quasi-rational solutions of the NLS equation and multi-rogue waves," *Journal of Physics A: Mathematical and Theoretical*, vol. 44, pp. 1–15, 2010.

[14] P. Gaillard, "Degenerate determinant representation of solution of the NLS equation, higher Peregrine breathers and multi-rogue waves," *Journal of Mathematical Physics*, vol. 54, Article ID 013504, 32 pages, 2013.

2

Adaptive Finite-Time Mixed Interlayer Synchronization of Two-Layer Complex Networks with Time-Varying Coupling Delay

Xiaoyun Tang, Zhanying Yang, and Jie Zhang

School of Mathematics and Statistics, South-Central University for Nationalities, Wuhan, Hubei 430074, China

Correspondence should be addressed to Zhanying Yang; yangzhanying8011@163.com

Academic Editor: Emmanuel Lorin

This paper is concerned with two-layer complex networks with unidirectional interlayer couplings, where the drive and response layer have time-varying coupling delay and different topological structures. An adaptive control scheme is proposed to investigate finite-time mixed interlayer synchronization (FMIS) of two-layer networks. Based on the Lyapunov stability theory, a criterion for realizing FMIS is derived. In addition, several sufficient conditions for realizing mixed interlayer synchronization are given. Finally, some numerical simulations are presented to verify the correctness and effectiveness of theoretical results. Meanwhile, the proposed adaptive control strategy is demonstrated to be nonfragile with the noise perturbation.

1. Introduction

As an important and typical dynamic behavior of complex networks, synchronization has been extensively investigated in many fields, such as physics, mathematics, information science, biology, and sociology. In the literatures, many works have primarily focused on synchronization within a network having no connection with other networks. However, in real-world situations, many systems are often composed of some interacting networks. For example, a transportation system consists of road network, railway network, and air network. A communication system is composed of some subnetworks depending on phone, email, QQ, Wechat, etc. Thus, multiplex networks, proposed by Mucha et al. [1], would be more appropriate for describing systems in the real-world than traditional (single-layer) complex networks. In the past few years, many efforts have been made to investigate various problems of multiplex networks, such as topological structure, dynamic behavior [2], synchronizability [3, 4], spectral property [5], diffusion process [6], and synchronization [7–10].

Interlayer synchronization is also called counterpart synchronization [11] which describes how the nodes in one network behave coherently with the corresponding ones in other connected networks. This concept can be regarded as the development of synchronization for coupled drive-response systems [12]. In a drive-response chaotic system, the response network is driven by signals from the drive network, but the latter is not influenced by the former. These two coupled single-layer networks, whose topological structures may be different, can be viewed as a two-layer network. For example, a two-layer network with unidirectional interlayer couplings is shown in Figure 1, where the layer X and the layer Y represent the drive layer and the response layer, respectively. Synchronization between the layer X and the layer Y exists extensively in the real world. This kind of interlayer synchronization can be also understood as outer synchronization [13] between two coupled networks. In the past decade, it has gained considerable attention—see [11, 13–20] and the references therein. Some control schemes have been proposed to realize all kinds of outer synchronization such as complete outer synchronization [13–17], inverse outer synchronization [18], generalized outer synchronization [19], and finite-time outer synchronization [21]. As a special case of generalized outer synchronization, mixed outer synchronization (MOS) was first proposed in [22]. Wang et al. [22] studied MOS between two complex networks with the same topological structure and time-varying coupling delay by

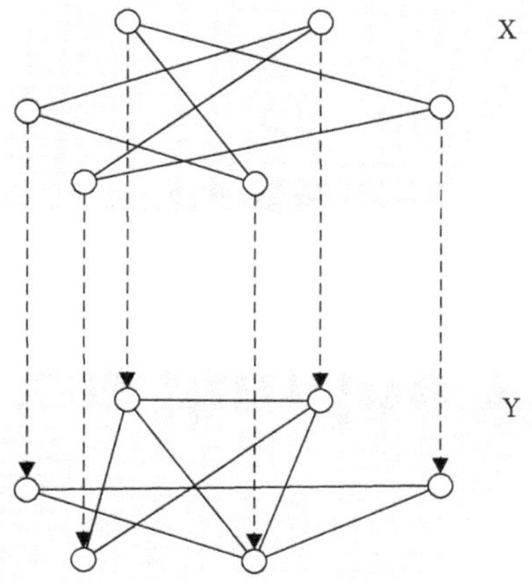

Figure 1: A two-layer network with unidirectional interlayer couplings.

designing robust linear feedback controllers. Later, Zheng and Shao [23] reported MOS between two complex networks with the same topological structure and output couplings via impulsive hybrid control. Sheng et al. [24] considered MOS between two complex networks with time-varying delay coupling and nondelay coupling by using pinning feedback control and impulsive control.

Finite-time mixed outer synchronization (FMOS) [25] between two complex networks is a recently developed MOS. Essentially, FMOS can be regarded as finite-time mixed interlayer synchronization (FMIS). In the state of FMIS, different state variables of the corresponding nodes can attain finite-time synchronization, finite-time antisynchronization, and even finite-time amplitude death simultaneously. FMIS is a kind of finite-time synchronization, in which the synchronization error remains within a prescribed range in a fixed time interval for a given range of initial error. Notice that finite-time synchronization is defined in a fixed finite-time interval. Hence, it has attracted increasing attention [26–31]. In [25], He et al. discussed FMOS between two complex networks with time-varying coupling delay and the same topological structures by designing a simple and robust linear state feedback controller. However, for most two-layer networks in real-world situations, the nodes are not always identical and topological structures of two layers are not always the same. In this paper, we consider FMIS of two-layer networks with time-varying coupling delay and different topological structures by designing an adaptive control scheme. Based on the Lyapunov stability theory, a sufficient condition for realizing FMIS of two-layer networks is derived. In addition, a criterion for realizing MIS is obtained. Finally, some numerical simulations are provided to verify the effectiveness of our theoretical results. Moreover, FMIS is analyzed when the two-layer networks are disturbed by the additive noise.

The rest of this paper is organized as follows. In Section 2, we formulate the two-layer network model and introduce some preliminaries. In Section 3, some sufficient conditions for realizing FMIS or MIS of two-layer networks are given under the proposed adaptive controllers. In Section 4, some numerical simulations are presented. Conclusions are finally drawn in Section 5.

2. Model and Preliminaries

Consider a dynamical two-layer complex network (See Figure 1) which is composed of the drive layer X and the response layer Y. Assume that each layer consists of N nodes. The drive layer X and the response layer Y are described as follows:

$$\begin{aligned}\dot{x}_i(t) &= A_d^i x_i(t) + A_n^i x_i(t) + f_i(x_i(t)) \\ &\quad + \sum_{j=1}^{N} a_{ij} G\left(x_j(t-\tau(t))\right), \quad i = 1, 2, \cdots, N, \\ \dot{y}_i(t) &= A_d^i y_i(t) + A_n^i y_i(t) + f_i(y_i(t)) \\ &\quad + \sum_{j=1}^{N} b_{ij} G\left(y_j(t-\tau(t))\right) + u_i, \\ &\qquad i = 1, 2, \cdots, N,\end{aligned} \tag{1}$$

where $x_i = (x_{i1}, x_{i2}, \cdots, x_{in})^T \in \mathbb{R}^n$ and $y_i = (y_{i1}, y_{i2}, \cdots, y_{in})^T \in \mathbb{R}^n$ are the state vectors, A_d and A_n are the diagonal and nondiagonal matrices representing the linear part of the i-th node dynamics, $f_i : \mathbb{R}^n \longrightarrow \mathbb{R}^n$ is a smooth nonlinear vector function such that $Hf_i(x) = f_i(Hx)$, where H is a scaling matrix defined as $H = \mathrm{diag}(h_1, h_2, \cdots, h_n)$ ($h_i \in \{-1, 0, 1\}$). $G(x_i(t)) = (g(x_{i1}(t), g(x_{i2}(t)), \cdots, g(x_{in}(t)))^T$ is a nonlinear vector-valued function which denotes the inner connection of each node. $A = (a_{ij})_{N\times N} \in \mathbb{R}^{N\times N}$ is the weight configuration matrix of the drive layer X, where a_{ij} is defined as follows: if there is a connection from node i to node j ($j \neq i$), $a_{ij} \neq 0$; otherwise, $a_{ij} = 0$. The diagonal elements of matrix A are defined as $a_{ii} = -\sum_{j=1, j\neq i}^{N} a_{ij}$, $i = 1, 2, \cdots, N$. $B = (a_{ij})_{N\times N} \in \mathbb{R}^{N\times N}$ denotes the weight configuration matrix of the response layer Y and has the same meanings as that of A. $\tau(t) \geq 0$ is the time-varying coupling delay. u_i is a controller for node i to be designed later.

Now, we introduce some assumptions, definitions, and lemmas, which are necessary to the proofs of main results.

Assumption 1. The nonlinear function $f_i(x)$ ($i = 1, 2, \cdots, N$) satisfies the Lipschitz condition; i.e., there exists a positive constant α_i such that

$$\left\| f_i(x) - f_i(y) \right\| \leq \alpha_i \left\| x - y \right\|, \quad \forall x, y \in \mathbf{R}^n. \tag{2}$$

Assumption 2. The nonlinear function $G(x)$ satisfies $HG(x) = G(Hx)$ and the following inequality:

$$\left\| G(x) - G(y) \right\| \leq \sqrt{L} \left\| x - y \right\|, \quad \forall x, y \in \mathbf{R}^n, \tag{3}$$

where L is a known positive constant.

Assumption 3. The time-varying coupling delay $\tau(t)$ is a differential function with

$$0 \le \dot{\tau}(t) \le \varepsilon < 1, \quad 0 \le \tau(t) \le \overline{\tau}. \tag{4}$$

Notice that this assumption is obviously satisfied when the delay $\tau(t)$ is a constant.

For $i = 1, 2, \cdots, N$, let $e_i(t) = y_i(t) - Hx_i(t)$ be the synchronization error of node i.

Definition 4. Suppose that q_1, q_2, and T are three positive constants. Let $q_1 \le q_2$. The two-layer network (1) is said to achieve finite-time mixed interlayer synchronization (FMIS) with respect to (q_1, q_2, T, H) if there exists a controller $u_i(t)$ $(i = 1, 2, \cdots, N)$ for node i such that

$$\sup_{-\overline{\tau}\le v\le 0} \|e(v)\|^2 \le q_1, \tag{5}$$

and one has

$$\|e(t)\|^2 \le q_2, \quad \forall t \in [0, T], \tag{6}$$

where $e(t) = (\|e_1(t)\|, \|e_2(t)\|, \ldots, \|e_N(t)\|)^T$.

Remark 5. In view of Definition 4, when $H = I$ and $H = -I$, FMIS is reduced to the known notions: finite-time outer synchronization and finite-time outer antisynchronization, respectively.

Definition 6. The two-layer network (1) is said to achieve mixed interlayer synchronization (MIS) with respect to the scaling matrix H if there exists a controller $u_i(t)$ $(i = 1, 2, \cdots, N)$ for node i such that

$$\lim_{t\to\infty} \|e_i(t)\| = 0. \tag{7}$$

Lemma 7. *Let $x, y \in R^n$ and $P \in R^{n\times n}$. One has the following inequality:*

$$2x^T P^T y \le \eta x^T P^T P x + \frac{1}{\eta} y^T y \tag{8}$$

for any scalar $\eta > 0$.

3. Main Results

In this section, we will give some sufficient conditions for realizing FMIS or MIS of two-layer networks based on adaptive control. The adaptive controllers are designed as follows:

$$u_i(t) = HA_n^i x_i(t) - A_n^i y_i(t) + \sum_{j=1}^{N} m_{ij} G(y_j(t-\tau(t))) - d_i(t) e_i(t), \quad i = 1, 2, \ldots, N. \tag{9}$$

where

$$\dot{d}_i = k_i e_i^T(t) e_i(t), \quad \dot{m}_{ij} = -e_i^T(t) G(y_j(t-\tau(t))). \tag{10}$$

Here, k_i $(i = 1, 2, \ldots, N)$ is a positive constant.

The error system can be described as

$$\begin{aligned}\dot{e}_i(t) &= A_d^i e_i(t) + f_i(y_i(t)) - Hf_i(x_i(t)) + \sum_{j=1}^{N} b_{ij} G(y_j(t-\tau(t))) \\ &\quad - \sum_{j=1}^{N} a_{ij} HG(x_j(t-\tau(t))) + \sum_{j=1}^{N} m_{ij} G(y_j(t-\tau(t))) - d_i(t) e_i(t), \\ &\quad i = 1, 2, \ldots, N.\end{aligned} \tag{11}$$

Theorem 8. *Suppose that Assumptions 1–3 hold. Let q_1, q_2, and T be three positive constants, and let $q_1 \le q_2$. FMIS of the two-layer network (1) can be realized with respect to (q_1, q_2, T, H) under the adaptive controllers (9) if there exist a diagonal matrix $D^* = \mathrm{diag}(d_1^*, d_2^*, \cdots, d_N^*)$ and a scalar $\theta > 0$ such that*

$$\left(\alpha + \xi + \frac{L}{2(1-\varepsilon)}\right) I_N + \frac{1}{2} AA^T - D^* - \theta I_N < 0, \tag{12}$$

$$e^{\theta T} q_1 \left(1 + \frac{L\overline{\tau}}{1-\varepsilon}\right) + e^{\theta T} \mu \le q_2. \tag{13}$$

Here, $\alpha = \max\{\alpha_1, \alpha_2, \cdots, \alpha_n\}$, $\xi = \max\{\lambda_{max}(A_d^1), \lambda_{max}(A_d^2), \ldots, \lambda_{max}(A_d^N)\}$, and μ is a positive constant which depends on $m_{ij}(0)$ $(i, j = 1, 2, \cdots, N)$ and $d_i(0)$ $(i = 1, 2, \cdots, N)$.

Proof. Consider the following Lyapunov function:

$$\begin{aligned}V(t) &= \frac{1}{2}\sum_{i=1}^{N} e_i^T(t) e_i(t) + \frac{1}{2}\sum_{i=1}^{N} \frac{1}{k_i}(d_i - d_i^*)^2 + \frac{1}{2}\sum_{i=1}^{N}\sum_{j=1}^{N}(b_{ij} + m_{ij} - a_{ij})^2 \\ &\quad + \frac{L}{2(1-\varepsilon)} \int_{t-\tau(t)}^{t} \sum_{i=1}^{N} e_i^T(s) e_i(s)\, ds,\end{aligned} \tag{14}$$

where d_i^* is a undetermined positive constant.

The derivative of $V(t)$ along (11) is

$$\begin{aligned}\dot{V}(t) &= \sum_{i=1}^{N} e_i^T \dot{e}_i + \sum_{i=1}^{N} (d_i - d_i^*) \dot{d}_i + \sum_{i=1}^{N}\sum_{j=1}^{N}(b_{ij} + m_{ij} - a_{ij}) \dot{m}_{ij} + \frac{L}{2(1-\varepsilon)} \\ &\quad \cdot \sum_{i=1}^{N} e_i^T(t) e_i(t) - \frac{L(1-\dot{\tau}(t))}{2(1-\varepsilon)} \cdot \sum_{i=1}^{N} e_i^T(t-\tau(t)) e_i(t-\tau(t)) = \sum_{i=1}^{N} e_i^T A_d^i e_i + \sum_{i=1}^{N} e_i^T\end{aligned}$$

$$\begin{aligned}
&\cdot \left(f_i\left(y_i(t)\right) - Hf_i\left(x_i(t)\right)\right) + \sum_{i=1}^{N}\sum_{j=1}^{N} b_{ij} e_i^T(t) \\
&\cdot G\left(y_j(t-\tau(t))\right) - \sum_{i=1}^{N}\sum_{j=1}^{N} a_{ij} e_i^T HG\left(x_j(t-\tau(t))\right) \\
&+ \sum_{i=1}^{N}\sum_{j=1}^{N} m_{ij} e_i^T G\left(y_j(t-\tau(t))\right) - \sum_{i=1}^{N} d_i(t) e_i^T(t) \\
&\cdot e_i(t) + \sum_{i=1}^{N} \left(d_i - d_i^*\right) e_i^T(t) e_i(t) \\
&- \sum_{i=1}^{N}\sum_{j=1}^{N} \left(b_{ij} + m_{ij} - a_{ij}\right) e_i^T(t) G\left(y_j(t-\tau(t))\right) \\
&+ \frac{L}{2(1-\varepsilon)} \sum_{i=1}^{N} e_i^T(t) e_i(t) - \frac{L(1-\dot{\tau}(t))}{2(1-\varepsilon)} \\
&\cdot \sum_{i=1}^{N} e_i^T(t-\tau(t)) e_i(t-\tau(t)) = \sum_{i=1}^{N} e_i^T A_d^i e_i \\
&+ \sum_{i=1}^{N} e_i^T \left[f_i\left(y_i(t)\right) - f_i\left(Hx_i(t)\right)\right] + \sum_{i=1}^{N}\sum_{j=1}^{N} a_{ij} e_i^T \\
&\cdot \left[G\left(y_j(t-\tau(t))\right) - HG\left(x_j(t-\tau(t))\right)\right] \\
&- \sum_{i=1}^{N} d_i^* e_i^T e_i + \frac{L}{2(1-\varepsilon)} \sum_{i=1}^{N} e_i^T(t) e_i(t) \\
&- \frac{L(1-\dot{\tau}(t))}{2(1-\varepsilon)} \sum_{i=1}^{N} e_i^T(t-\tau(t)) e_i(t-\tau(t)).
\end{aligned} \tag{15}$$

Let $z_j(t-\tau(t)) = Hx_j(t-\tau(t))$. Then we get

$$\begin{aligned}
\dot{V}(t) &= \sum_{i=1}^{N} e_i^T A_d^i e_i + \sum_{i=1}^{N} e_i^T \left[f_i\left(y_i(t)\right) - f_i\left(Hx_i(t)\right)\right] \\
&+ \sum_{m=1}^{n} \widetilde{e}_m^T(t) A(G(\widetilde{y}_m(t-\tau(t))) \\
&- G\left(\widetilde{z}_m(t-\tau(t))\right)) - \sum_{i=1}^{N} d_i^* e_i^T e_i \\
&+ \frac{L}{2(1-\varepsilon)} \sum_{i=1}^{N} e_i^T(t) e_i(t) \\
&- \frac{L(1-\dot{\tau}(t))}{2(1-\varepsilon)} \sum_{i=1}^{N} e_i^T(t-\tau(t)) e_i(t-\tau(t)),
\end{aligned} \tag{16}$$

where $\widetilde{y}_m(t) = (y_{1m}(t), y_{2m}(t), \cdots, y_{Nm}(t))^T$, $\widetilde{z}_m(t) = (z_{1m}(t), z_{2m}(t), \cdots, z_{Nm}(t))^T$, and $\widetilde{e}_m(t) = (e_{1m}(t), e_{2m}(t), \cdots, e_{Nm}(t))^T$.

From Lemma 7, we derive

$$\begin{aligned}
&\sum_{m=1}^{n} \widetilde{e}_m^T(t) A(G(\widetilde{y}_m(t-\tau(t))) - G\left(\widetilde{z}_m(t-\tau(t))\right)). \\
&\le \frac{1}{2} \sum_{m=1}^{n} \widetilde{e}_m^T(t) AA^T \widetilde{e}_m(t) \\
&+ \frac{1}{2} \sum_{m=1}^{n} \left\| G(\widetilde{y}_m(t-\tau(t))) - G\left(\widetilde{z}_m(t-\tau(t))\right) \right\|^2.
\end{aligned} \tag{17}$$

Together with Assumptions 1 and 2, it follows that

$$\begin{aligned}
\dot{V}(t) &\le (\alpha + \xi) \sum_{m=1}^{n} \widetilde{e}_m^T(t) \widetilde{e}_m(t) \\
&+ \frac{1}{2} \sum_{m=1}^{n} \widetilde{e}_m^T(t) \left(AA^T\right) \widetilde{e}_m(t) \\
&+ \frac{L}{2} \sum_{m=1}^{n} \widetilde{e}_m^T(t-\tau(t)) \widetilde{e}_m(t-\tau(t)) - \sum_{i=1}^{N} d_i^* e_i^T e_i \\
&+ \frac{L}{2(1-\varepsilon)} \sum_{m=1}^{n} \widetilde{e}_m^T(t) \widetilde{e}_m(t) \\
&- \frac{L(1-\dot{\tau}(t))}{2(1-\varepsilon)} \sum_{m=1}^{n} \widetilde{e}_m^T(t-\tau(t)) \widetilde{e}_m(t-\tau(t)).
\end{aligned} \tag{18}$$

Here $\alpha = \max\{\alpha_1, \alpha_2, \cdots, \alpha_n\}$ and $\xi = \max\{\lambda_{max}(A_d^1), \lambda_{max}(A_d^2), \ldots, \lambda_{max}(A_d^N)\}$.

Making use of Assumption 3, we have

$$\frac{L}{2}\left(1 - \frac{1-\dot{\tau}(t)}{1-\varepsilon}\right) \le 0. \tag{19}$$

Furthermore, we obtain

$$\begin{aligned}
\dot{V}(t) &\le \sum_{m=1}^{n} \widetilde{e}_m^T(t) \\
&\cdot \left[\left(\alpha + \xi + \frac{L}{2(1-\varepsilon)}\right) I_N + \frac{1}{2} AA^T - D^*\right] \\
&\cdot \widetilde{e}_m(t),
\end{aligned} \tag{20}$$

where $D^* = \mathrm{diag}(d_1^*, d_2^*, \cdots, d_N^*)$. Together with (12), we have

$$\dot{V}(t) \le \theta \sum_{i=1}^{N} e_i^T(t) e_i(t) \le 2\theta V(t). \tag{21}$$

Multiplying (21) by $e^{-\theta t}$, we obtain $(d/dt)(e^{\theta t} V(t)) < 0$. Integrating from 0 to t $(t \in [0, T])$, it follows that

$$e^{-\theta t} V(t) < V(0). \tag{22}$$

By (14), we get

$$\begin{aligned}V(0) &= \frac{1}{2}e^T(0)e(0) + \frac{1}{2}\sum_{i=1}^{N}\frac{1}{k_i}(d_i(0)-d_i^*)^2 \\ &\quad + \frac{1}{2}\sum_{i=1}^{N}\sum_{j=1}^{N}\left(b_{ij}+m_{ij}(0)-a_{ij}\right)^2 \\ &\quad + \frac{L}{2(1-\varepsilon)}\int_{-\tau(t)}^{0}\sum_{i=1}^{N}e_i^T(s)e_i(s)\,ds \\ &\le \frac{1}{2}\|e(0)\|^2 + \frac{L\overline{\tau}}{2(1-\varepsilon)}\sup_{-\overline{\tau}\le v\le 0}\|e(v)\|^2 \\ &\quad + \frac{1}{2}\sum_{i=1}^{N}\frac{1}{k_i}(d_i(0)-d_i^*)^2 \\ &\quad + \frac{1}{2}\sum_{i=1}^{N}\sum_{j=1}^{N}\left(b_{ij}+m_{ij}(0)-a_{ij}\right)^2 \\ &\le \frac{1}{2}q_1\left(1+\frac{L\overline{\tau}}{1-\varepsilon}\right)+\frac{1}{2}\mu,\end{aligned} \tag{23}$$

where $\mu = \sum_{i=1}^{N}(1/k_i)(d_i(0)-d_i^*)^2 + \sum_{i=1}^{N}\sum_{j=1}^{N}(b_{ij}+m_{ij}(0)-a_{ij})^2$. Combining this with (13), we have

$$\begin{aligned}\|e(t)\|^2 &\le 2V(t) \le 2e^{\theta T}V(0) \\ &\le e^{\theta T}q_1\left(1+\frac{L\overline{\tau}}{1-\varepsilon}\right)+e^{\theta T}\mu < q_2.\end{aligned} \tag{24}$$

The proof of Theorem 8 is finished. □

Corollary 9. *Let Assumptions 1–3 be satisfied. Suppose that the configuration matrix of the drive layer is the same as that of the response layer, i.e., $A = B$. Let q_1, q_2, and T be three positive constants and let $q_1 \le q_2$. Under the simplified adaptive controllers*

$$\begin{aligned}u_i(t) &= HA_n^i x_i(t) - A_n^i y_i(t) - d_i(t)e_i(t), \\ \dot{d}_i &= k_i e_i^T(t)e_i(t),\end{aligned} \tag{25}$$

where k_i $(i = 1,2,\ldots,N)$ is any positive constant, FMIS of the two-layer network (1) can be realized with respect to (q_1,q_2,T,H) if there exist a diagonal matrix $D^* = \operatorname{diag}(d_1^*,d_2^*,\cdots,d_N^*)$ and a scalar $\theta > 0$ such that

$$\begin{aligned}&\left(\alpha+\xi+\frac{L}{2(1-\varepsilon)}\right)I_N+\frac{1}{2}AA^T-D^*-\theta I_N<0, \\ &e^{\theta T}q_1\left(1+\frac{L\overline{\tau}}{1-\varepsilon}\right)+e^{\theta T}\mu \le q_2.\end{aligned} \tag{26}$$

Here, $\alpha = \max\{\alpha_1,\alpha_2,\cdots,\alpha_N\}$, $\xi = \max\{\lambda_{max}(A_d^1),\lambda_{max}(A_d^2),\ldots,\lambda_{max}(A_d^N)\}$, and μ is a positive constant which depends on $d_i(0)$ $(i = 1,2,\cdots,N)$.

Theorem 10. *Suppose that Assumptions 1–3 hold. MIS of two-layer network (1) can be realized with the adaptive controllers (9).*

In fact, choose (14) as the Lyapunov function and use the calculation of $\dot{V}(t)$ in the proof of Theorem 8; we have (20). When d_i^* is sufficiently large, we get

$$\left(\alpha+\xi+\frac{L}{2(1-\varepsilon)}\right)I_N+\frac{1}{2}AA^T-D^*<0. \tag{27}$$

This implies $\dot{V}(t) \le 0$. Let

$$E = \left\{(\mathbf{e},\mathbf{d},M)\in\mathbb{R}^{nN}\times\mathbb{R}^N\times\mathbb{R}^{N\times N} : \dot{V}=0\right\}, \tag{28}$$

where $\mathbf{e} = (e_1^T(t),e_2^T(t),\ldots,e_N^T(t))^T$, $\mathbf{d} = (d_1,d_2,\cdots,d_N)^T$, and $M = (m_{ij})_{N\times N}$. The largest invariant set is as follows:

$$\begin{aligned}S = \Bigg\{&(\mathbf{e},\mathbf{d},M)\in\mathbb{R}^{nN}\times\mathbb{R}^N\times\mathbb{R}^{N\times N} : \mathbf{e}=0,\ \dot{\mathbf{d}} \\ &=0,\ \dot{M}=0,\ \sum_{j=1}^{N}\left(b_{ij}+m_{ij}-a_{ij}\right)=0\Bigg\}.\end{aligned} \tag{29}$$

Therefore, by the LaSalle invariance principle, starting from arbitrary initial values, the trajectory asymptotically tends to the largest invariant set S, which implies $\lim_{t\to\infty}e_i(t) = 0$ $(i = 1,2,\cdots,N)$. Thus, MIS in two-layer network (1) can be realized with the adaptive controllers (9).

Remark 11. When the MIS of two-layer network (1) is realized, i.e., $e_i(t) \longrightarrow 0$ $(i = 1,2,\cdots,N)$, it can be further obtained that $\dot{d}_i \longrightarrow 0$ and $\dot{m}_{ij} \longrightarrow 0$ $(i,j = 1,2,\cdots,N)$. This implies that $d_i(t)$ and $m_{ij}(t)$ will tend to be constant when $t \longrightarrow \infty$.

Remark 12. For the two-layer network (1), the nodes in each layer may be nonidentical and topological structures of two layers may be different. For each node, the inner connecting function is nonlinear. In addition, two configuration matrices A and B are not assumed to be symmetric or irreducible. Consequently, Theorems 8 and 10 can be widely applied in practice.

Corollary 13. *Suppose that Assumptions 1–3 hold. MIS of two-layer network (1) can be realized with the following adaptive controllers:*

$$\begin{aligned}u_i(t) &= HA_n^i x_i(t) - A_n^i y_i(t) \\ &\quad + \sum_{j=1}^{N}m_{ij}G\left(y_j(t-\tau(t))\right) - d_i(t)e_i(t), \\ &\qquad i = 1,2,\ldots,N,\end{aligned} \tag{30}$$

where $\dot{d}_i = k_i e_i^T(t)e_i(t)$ and k_i is any positive constant. For $i,j = 1,2,\ldots,N$, $m_{ij}(t)$ is defined as follows:

(i) $\dot{m}_{ij} = -e_i^T(t)G(y_j(t-\tau(t)))$, *when $a_{ij} \ne b_{ij}$.*

(ii) $m_{ij} = 0$, *when* $a_{ij} = b_{ij}$.

Corollary 14. *Suppose that Assumptions 1–3 hold. If the configuration matrix of the drive layer is the same as that of the response layer, i.e., A = B, then MIS of two-layer network (1) can be realized with the following simplified adaptive controllers:*

$$u_i(t) = HA_n^i x_i(t) - A_n^i y_i(t) - d_i(t) e_i(t), \quad i = 1, 2, \ldots, N, \tag{31}$$

where $\dot{d}_i = k_i e_i^T(t) e_i(t)$ *and* k_i *is any positive constant.*

4. Numerical Simulations

In this section, some numerical examples are presented to illustrate the effectiveness of results in Section 3. Each layer of two-layer network (1) is made of 10 nodes. The dynamics of each node is taken as the modified Chua's circuit system, which is described as

$$\begin{aligned} \dot{x}_1 &= p\left(x_2 - \frac{2x_1^3 - x_1}{7}\right), \\ \dot{x}_2 &= x_1 - x_2 + x_3, \\ \dot{x}_3 &= -qx_2, \end{aligned} \tag{32}$$

where $> 0, q > 0$. This system is rewritten into $\dot{x}(t) = A_d x(t) + A_n x(t) + f(x(t))$, where

$$x = \begin{pmatrix} x_1 \\ x_2 \\ x_3 \end{pmatrix}, \quad A_d = \begin{pmatrix} \frac{p}{7} & 0 & 0 \\ 0 & -1 & 0 \\ 0 & 0 & 0 \end{pmatrix}, \quad A_n = \begin{pmatrix} 0 & p & 0 \\ 1 & 0 & 1 \\ 0 & -q & 0 \end{pmatrix}, \quad f(x) = \begin{pmatrix} -\frac{2px_1^3}{7} \\ 0 \\ 0 \end{pmatrix}. \tag{33}$$

Hence, $\xi = \lambda_{max}(A_d) = p/7$. Let $p = 10$ and $q = 100/7$. In view of the calculation in [22], it follows that

$$\|f(x) - f(y)\| \leq \frac{27p}{14} \|x - y\|. \tag{34}$$

This means that Assumption 1 is satisfied with $\alpha_i = 27p/14$ $(i = 1, 2, \cdots, 10)$. Thus, $\alpha = 27p/14$. For the nonlinear function $G(x) = (g(x_1), g(x_2), \ldots, g(x_n))$, $g(x_i)$ $(i = 1, 2, \ldots, n)$ is chosen as $g(x_i) = \sin x_i + x_i$. It can be easily derived that $\|G(x) - G(y)\| \leq 2\|x - y\|$, which implies $L = 4$ in Assumption 2.

Let $\tau(t) = 0.1 - 0.1e^{-t}$. Obviously, it follows that $\varepsilon = 0.2$ and $\overline{\tau} = 0.1$. The scaling matrix is given by $H = \text{diag}(1, 0, -1)$. Then the error variables are written as

$$\begin{aligned} e_{i1}(t) &= y_{i1}(t) - x_{i1}(t), \\ e_{i2}(t) &= y_{i2}(t), \\ e_{i3}(t) &= y_{i3}(t) + x_{i3}(t). \end{aligned} \tag{35}$$

Moreover, it is easily verified that $f(Hx) = Hf(x)$ and $G(Hx) = HG(x)$.

The coupled two-layer network is described as follows:

$$\begin{aligned} \dot{x}_i(t) &= A_d x_i(t) + A_n x_i(t) + f(x_i(t)) + \sum_{j=1}^{10} a_{ij} G(x_j(t - \tau(t))), \quad i = 1, 2, \cdots, 10, \\ \dot{y}_i(t) &= A_d y_i(t) + A_n y_i(t) + f(y_i(t)) + \sum_{j=1}^{10} b_{ij} G(y_j(t - \tau(t))) + u_i, \quad i = 1, 2, \cdots, 10, \end{aligned} \tag{36}$$

where

$$u_i(t) = HA_n x_i(t) - A_n y_i(t) + \sum_{j=1}^{10} m_{ij} G(y_j(t - \tau(t))) - d_i(t) e_i(t) \tag{37}$$

with $\dot{d}_i = k_i e_i^T(t) e_i(t)$, and $\dot{m}_{ij} = -e_i^T(t) G(y_j(t - \tau(t)))$. Here, k_i $(i = 1, 2, \ldots, 10)$ is any positive constant.

Let K_1 and K_2 be two matrices defined as follows:

$$K_1 = \begin{pmatrix} -9 & 1 & 2 & 1 & 1 & 1 & 1 & 0 & 1 & 1 \\ 1 & -7 & 0 & 1 & 0 & 1 & 1 & 0 & 1 & 2 \\ 1 & 1 & -10 & 1 & 1 & 0 & 1 & 1 & 3 & 1 \\ 0 & 1 & 1 & -6 & 1 & 1 & 0 & 1 & 0 & 1 \\ 1 & 0 & 1 & 1 & -8 & 0 & 1 & 2 & 1 & 1 \\ 2 & 1 & 1 & 0 & 1 & -7 & 1 & 0 & 1 & 0 \\ 1 & 1 & 1 & 1 & 1 & 1 & -11 & 1 & 3 & 1 \\ 3 & 1 & 0 & 1 & 1 & 0 & 1 & -9 & 1 & 1 \\ 1 & 1 & 2 & 1 & 3 & 1 & 1 & 1 & -12 & 1 \\ 0 & 2 & 1 & 1 & 1 & 0 & 1 & 3 & 1 & -10 \end{pmatrix} \tag{38}$$

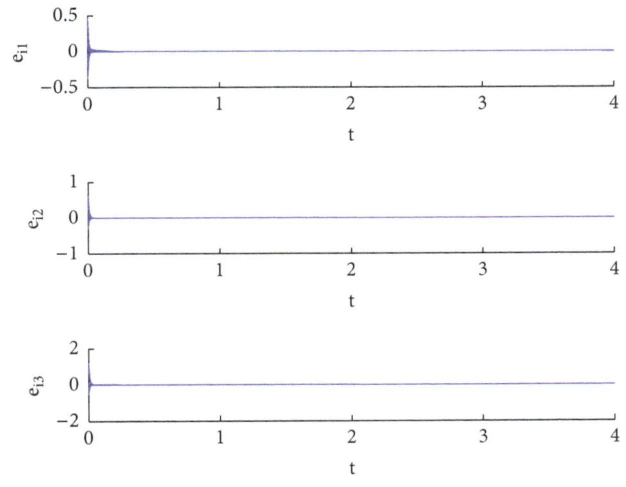

FIGURE 2: Finite-time mixed interlayer synchronization errors e_i $(i = 1, 2, \ldots, 10)$ with $A \neq B$.

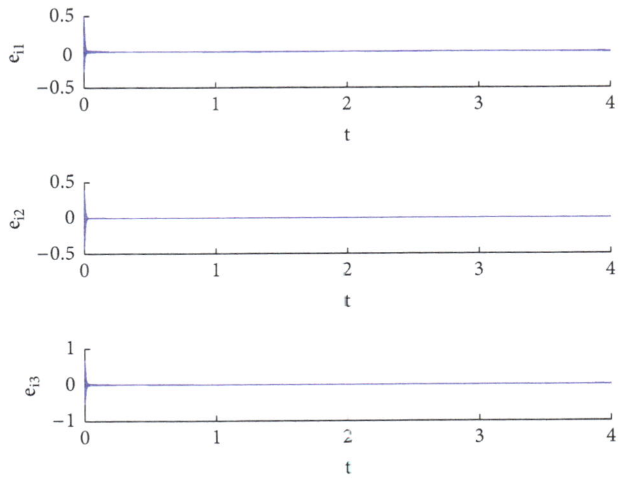

FIGURE 3: Finite-time mixed interlayer synchronization errors e_i $(i = 1, 2, \ldots, 10)$ for Case 1.

and

$$K_2 = \begin{pmatrix} -10 & 2 & 1 & 1 & 0 & 1 & 1 & 1 & 2 & 1 \\ 1 & -9 & 0 & 1 & 1 & 3 & 1 & 0 & 1 & 1 \\ 0 & 1 & -7 & 1 & 1 & 0 & 2 & 1 & 1 & 0 \\ 0 & 1 & 1 & -9 & 1 & 1 & 0 & 1 & 3 & 1 \\ 1 & 0 & 1 & 2 & -8 & 1 & 1 & 1 & 1 & 0 \\ 2 & 1 & 0 & 1 & 0 & -7 & 1 & 0 & 1 & 1 \\ 1 & 0 & 3 & 1 & 1 & 0 & -10 & 1 & 2 & 1 \\ 1 & 2 & 1 & 0 & 1 & 1 & 1 & -9 & 1 & 1 \\ 1 & 1 & 1 & 1 & 3 & 1 & 1 & 1 & -11 & 1 \\ 3 & 1 & 1 & 2 & 1 & 1 & 1 & 1 & 1 & -12 \end{pmatrix}. \quad (39)$$

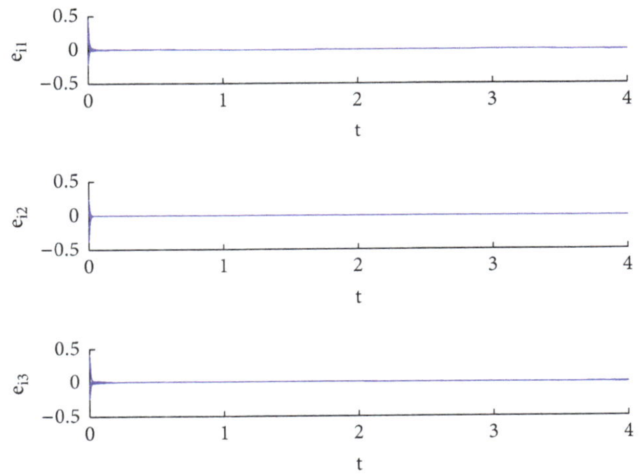

FIGURE 4: Finite-time mixed interlayer synchronization errors e_i $(i = 1, 2, \ldots, 10)$ for Case 2.

4.1. FMIS of Two-Layer Networks with respect to (q_1, q_2, T, H). In the subsection, we consider FMIS of the two-layer network (36) under two cases: $A \neq B$ and $A = B$.

Example 15 ($A \neq B$). Suppose that $A = K_1$ and $B = K_2$. Let $q_1 = 1.5$, $q_2 = 2.4$, and $T = 4$. In view of inequality (12) in Theorem 8, we can get $\theta = 0.01$ and $d_i^* = 128$ $(i = 1, 2, \cdots, N)$. For $v \in [-0.1, 0]$, we arbitrarily take $x_i(v)$ and $y_i(v)$ in the interval $[-1, 1]$ for $i = 1, 2, \cdots, N$. It follows that $\sup_{-\overline{\tau} \le v \le 0} \|e(v)\|^2 \le q_1$. According to inequality (13), we choose k_i, $d_i(0)$ and $m_{ij}(0)$ as $k_i = 10$, $d_i(0) = 128$ and $m_{ij}(0) = b_{ij} - a_{ij}$ for $i, j = 1, 2, \ldots, 10$. Under the adaptive controllers (37), Figure 2 shows that the FMIS of two-layer network (36) is realized with respect to (q_1, q_2, T, H).

As is well-known, the dynamics of networks is always affected by all kinds of noises from the environment. In the following, the FMIS will be further investigated when the network (36) is disturbed by additive noise. Here, two cases are analyzed by numerical simulations. Let $\phi = (0.5(0.25\sin(t) + 0.5\cos(5t)), 0.1\sin(2\pi t), 0.3e^{-t})^T$ and $\varphi(t) = (0.3\cos(2t), 0.2e^{-3t}, 0.1(\cos(t) - \sin(0.1t)))^T$ be two additive noise functions.

Case 1. $\phi(t)$ is added to the right hand of two equations in (36). This means that every node in (36) is affected by the same noise.

Case 2. $\phi(t)$ and $\varphi(t)$ are added to the right hand of two equations in (36), respectively. This indicates that the nodes in the drive layer and the ones in the response layer are affected by different noise.

For these two cases, the steps of above numerical simulation are repeated with the adaptive controllers (37). The synchronization errors for two cases are shown in Figures 3 and 4. Hence, the proposed scheme is demonstrated to be nonfragile with the noise perturbation.

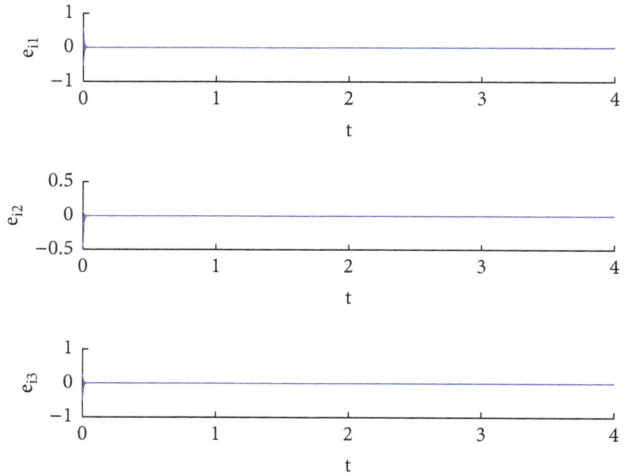

FIGURE 5: Finite-time mixed interlayer synchronization errors e_i $(i = 1, 2, \ldots, 10)$ with $A = B$.

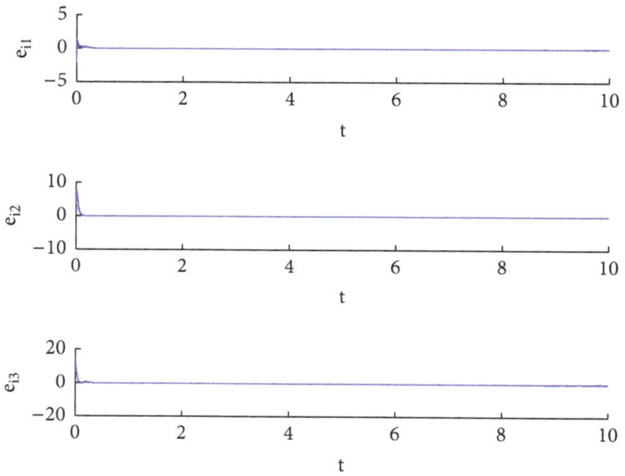

FIGURE 6: Mixed synchronization errors e_i $(i = 1, 2, \ldots, 10)$ with $A \neq B$.

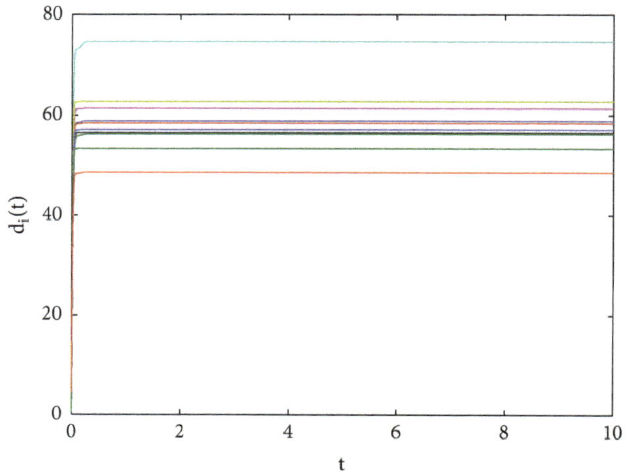

FIGURE 7: Evolution of feedback gains $d_i(t)$ with $k_i = 10$, $i = 1, 2, \ldots, 10$.

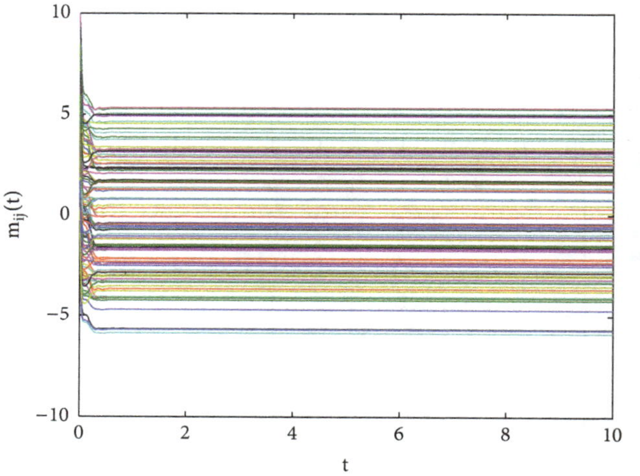

FIGURE 8: Evolution of adaptive parameters $m_{ij}(t)$, $i, j = 1, 2, \ldots, 10$.

Example 16 ($A = B$). Suppose that $A = B = K_1$. Let $q_1 = 1.7$, $q_2 = 2.7$, and $T = 4$. According to Corollary 9, we can get the values of θ and d_i^* $(i = 1, 2, \ldots, 10)$. For $v \in [-0.1, 0]$, we arbitrarily take $x_i(v)$ and $y_i(v)$ in the interval $[-1, 1]$ for $i = 1, 2, \cdots, N$. In view of the calculation, we can take the other parameters and initial values which are the same as those in Example 15. Under the simplified adaptive controllers in Corollary 9, the synchronization errors are shown in Figure 5. The FMIS of two-layer network (36) is realized with respect to (q_1, q_2, T, H).

4.2. MIS of Two-Layer Networks. In the subsection, MIS of the two-layer network (36) is considered under two cases: $A \neq B$ and $A = B$. Let $x_i(0)$, $y_i(0)$, $d_i(0)$, and $m_{ij}(0)$ be arbitrarily given in the interval $[0, 10]$ for $i, j = 1, 2, \ldots, 10$.

Let $A = K_1$ and $B = K_2$. The synchronization error is shown in Figure 6, which demonstrates the effectiveness of Theorem 10. Figures 7 and 8 present the evolution of $d_i(t)$ and $m_{ij}(t)$, respectively. It is obvious that $d_i(t)$ and $m_{ij}(t)$ $(i, j = 1, 2, \ldots, 10)$ tend to be constant.

Suppose that $A = B = K_1$. Figure 9 presents the synchronization error with the simplified adaptive controllers in Corollary 14. Hence, the MIS in two-layer network (36) can be realized.

5. Conclusions

In this paper, we have investigated finite-time mixed interlayer synchronization (FMIS) of two-layer complex network with unidirectional interlayer couplings. For the two-layer network, the nodes in each layer are nonidentical and two single-layer networks have time-varying coupling delay and different topological structures. We have proposed an adaptive control scheme to realize FMIS of two-layer networks. Based on the Lyapunov stability theory, a sufficient condition for realizing FMIS has been derived. In addition, some criteria for realizing MIS have been obtained. Finally, some

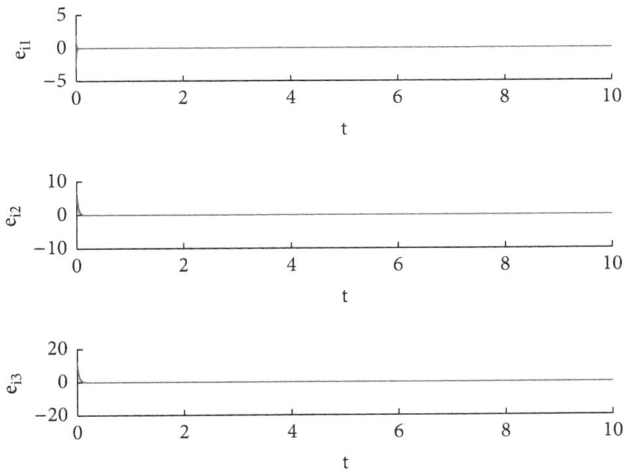

FIGURE 9: Mixed synchronization errors e_i $(i = 1, 2, \ldots, 10)$ with $A = B$.

numerical simulations have been presented to verify the correctness and effectiveness of theoretical results. Moreover, FMIS has been analyzed when the two-layer networks are disturbed by the additive noise. Two numerical simulations show that the proposed adaptive control strategy is nonfragile with the noise perturbation.

Abbreviations

FMIS: Finite-time mixed interlayer synchronization
MOS: Mixed outer synchronization
FMOS: Finite-time mixed outer synchronization
MIS: Mixed interlayer synchronization.

Conflicts of Interest

The authors declare no conflicts of interest.

Acknowledgments

This research was funded by the National Natural Science Foundation of China (no. 11401595).

References

[1] P. J. Mucha, T. Richardson, K. Macon, M. A. Porter, and J.-P. Onnela, "Community structure in time-dependent, multiscale, and multiplex networks," *Science*, vol. 328, no. 5980, pp. 876–878, 2010.

[2] N. Bastas, F. Lazaridis, P. Argyrakis, and M. Maragakis, "Static and dynamic behavior of multiplex networks under interlink strength variation," *EPL (Europhysics Letters)*, vol. 109, no. 3, 2015.

[3] Y. Li, X. Wu, J.-A. Lu, and J. Lu, "Synchronizability of Duplex Networks," *IEEE Transactions on Circuits and Systems II: Express Briefs*, vol. 63, no. 2, pp. 206–210, 2016.

[4] S. K. Dwivedi, M. S. Baptista, and S. Jalan, "Optimization of synchronizability in multiplex networks by rewiring one layer," *Physical Review E: Statistical, Nonlinear, and Soft Matter Physics*, vol. 95, no. 4, 2017.

[5] A. Solé-Ribalta, M. De Domenico, N. E. Kouvaris, A. Díaz-Guilera, S. Gómez, and A. Arenas, "Spectral properties of the Laplacian of multiplex networks," *Physical Review E: Statistical, Nonlinear, and Soft Matter Physics*, vol. 88, no. 3, 2013.

[6] Q. Liu, W. Wang, M. Tang, and H. Zhang, "Impacts of complex behavioral responses on asymmetric interacting spreading dynamics in multiplex networks," *Scientific Reports*, vol. 6, no. 1, 2016.

[7] L. V. Gambuzza, M. Frasca, and J. Gómez-Gardeñes, "Intra-layer synchronization in multiplex networks," *EPL (Europhysics Letters)*, vol. 110, no. 2, p. 20010, 2015.

[8] R. Sevilla-Escoboza, I. Sendiña-Nadal, I. Leyva, R. Gutiérrez, J. M. Buldú, and S. Boccaletti, "Inter-layer synchronization in multiplex networks of identical layers," *Chaos: An Interdisciplinary Journal of Nonlinear Science*, vol. 26, no. 6, p. 065304, 2016.

[9] S. Rakshit, S. Majhi, B. K. Bera, S. Sinha, and D. Ghosh, "Time-varying multiplex network: Intralayer and interlayer synchronization," *Physical Review E: Statistical, Nonlinear, and Soft Matter Physics*, vol. 96, no. 6, 2017.

[10] I. Leyva, R. Sevilla-Escoboza, I. Sendiña-Nadal, R. Gutiérrez, J. Buldú, and S. Boccaletti, "Inter-layer synchronization in non-identical multi-layer networks," *Scientific Reports*, vol. 7, p. 45475, 2017.

[11] X. Wei, X. Wu, J. Lu, and J. Zhao, "Counterpart synchronization of duplex networks with delayed nodes and noise perturbation," *Journal of Statistical Mechanics: Theory and Experiment*, vol. 2015, no. 11, p. P11021, 2015.

[12] L. M. Pecora and T. L. Carroll, "Synchronization in chaotic systems," *Physical Review Letters*, vol. 64, no. 8, pp. 821–824, 1990.

[13] C. Li, W. Sun, and J. Kurths, "Synchronization between two coupled complex networks," *Physical Review E: Statistical, Nonlinear, and Soft Matter Physics*, vol. 76, no. 4, Article ID 046204, 2007.

[14] C. Li, C. Xu, W. Sun, J. Xu, and J. Kurths, "Outer synchronization of coupled discrete-time networks," *Chaos: An Interdisciplinary Journal of Nonlinear Science*, vol. 19, no. 1, Article ID 013106, 2009.

[15] H. Tang, L. Chen, J. Lu, and C. K. Tse, "Adaptive synchronization between two complex networks with nonidentical topological structures," *Physica A: Statistical Mechanics and its Applications*, vol. 387, no. 22, pp. 5623–5630, 2008.

[16] Y. Wu and L. Liu, "Exponential outer synchronization between two uncertain time-varying complex networks with nonlinear coupling," *Entropy*, vol. 17, no. 5, pp. 3097–3109, 2015.

[17] W. Zhang, J. Cao, D. Chen, and F. Alsaadi, "Synchronization in Fractional-Order Complex-Valued Delayed Neural Networks," *Entropy*, vol. 20, no. 1, p. 54, 2018.

[18] W. Sun, Z. Chen, J. Lü, and S. Chen, "Outer synchronization of complex networks with delay via impulse," *Nonlinear Dynamics*, vol. 69, no. 4, pp. 1751–1764, 2012.

[19] X. Wu, W. X. Zheng, and J. Zhou, "Generalized outer synchronization between complex dynamical networks," *Chaos: An Interdisciplinary Journal of Nonlinear Science*, vol. 19, no. 1, Article ID 013109, 9 pages, 2009.

[20] C. Yang, J. Qiu, T. Li, A. Zhang, and X. Chen, "Projective exponential synchronization for a class of complex PDDE networks

with multiple time delays," *Entropy*, vol. 17, no. 11, pp. 7298–7309, 2015.

[21] X. Lin, H. Du, and S. Li, "Finite-time boundedness and L2-gain analysis for switched delay systems with norm-bounded disturbance," *Applied Mathematics and Computation*, vol. 217, no. 12, pp. 5982–5993, 2011.

[22] J. W. Wang, Q. H. Ma, Z. Li, and S. A. E. Mohammed, "Mixed outer synchronization of coupled complex networks with time-varying coupling delay," *Chaos: An Interdisciplinary Journal of Nonlinear Science*, vol. 21, Article ID 013121, pp. 1–8, 2011.

[23] S. Zheng and W. Shao, "Mixed outer synchronization of dynamical networks with nonidentical nodes and output coupling," *Nonlinear Dynamics*, vol. 73, no. 4, pp. 2343–2352, 2013.

[24] S. Sheng, J. Feng, Z. Tang, and Y. Zhao, "Mixed outer synchronization of two coupled complex networks with time-varying delay coupling and non-delay coupling," *Nonlinear Dynamics*, vol. 80, no. 1-2, pp. 803–815, 2015.

[25] P. He, S.-H. Ma, and T. Fan, "Finite-time mixed outer synchronization of complex networks with coupling time-varying delay," *Chaos: An Interdisciplinary Journal of Nonlinear Science*, vol. 22, no. 4, 043151, 11 pages, 2012.

[26] F. Amato, M. Ariola, and P. Dorato, "Finite-time control of linear systems subject to parametric uncertainties and disturbances," *Automatica*, vol. 37, no. 9, pp. 1459–1463, 2001.

[27] R. Wang, J. Xing, C. Zhou, P. Wang, and Q. Yang, "Finite-time asynchronously switched control of switched systems with sampled-data feedback," *Circuits, Systems and Signal Processing*, vol. 33, no. 12, pp. 3713–3738, 2014.

[28] Z. Yu, S. Wang, M. Zeng, and Y. Liu, "Finite-time boundedness and weighted L2-gain analysis for uncertain switched linear systems with both stable and unstable time-delay subsystems," in *Proceedings of the 27th Chinese Control and Decision Conference, CCDC 2015*, pp. 2406–2411, China, May 2015.

[29] B. Li and H. Xu, "Finite-time outer synchronization of two general complex dynamical networks with hybrid coupling based on pinning control," in *Proceedings of the 2016 35th Chinese Control Conference (CCC)*, pp. 7189–7193, Chengdu, China, July 2016.

[30] C. Zheng, M. Hu, and L. Guo, "Finite-time boundedness analysis for a new multi-layer switched system with time-delay," *Neurocomputing*, vol. 171, pp. 277–282, 2016.

[31] X. Lin, X. Li, S. Li, and Y. Zou, "Finite-time boundedness for switched systems with sector bounded nonlinearity and constant time delay," *Applied Mathematics and Computation*, vol. 274, pp. 25–40, 2016.

3

Classical and Quantum Exact Solutions for a FRW Multiscalar Field Cosmology with an Exponential Potential Driven Inflation

J. Socorro ,[1] Omar E. Núñez,[1] and Rafael Hernández-Jiménez[2]

[1]*Departamento de Física, DCeI, Universidad de Guanajuato-Campus León, 37150 León, GTO, Mexico*
[2]*School of Physics and Astronomy, University of Edinburgh, Edinburgh EH9 3FD, UK*

Correspondence should be addressed to J. Socorro; socorro@fisica.ugto.mx

Academic Editor: Eugen Radu

A flat Friedmann-Robertson-Walker (FRW) multiscalar field cosmology is studied with a particular potential of the form $V(\phi,\sigma) = V_0 e^{-\lambda_1\phi-\lambda_2\sigma}$, which emerges as a relation between the time derivatives of the scalars field momenta. Classically, by employing the Hamiltonian formalism of two scalar fields (ϕ,σ) with standard kinetic energy, exact solutions are found for the Einstein-Klein-Gordon (EKG) system for different scenarios specified by the parameter $\lambda^2 = \lambda_1^2 + \lambda_2^2$, as well as the e-folding function N_e which is also computed. For the quantum scheme of this model, the corresponding Wheeler-DeWitt (WDW) equation is solved by applying an appropriate change of variables.

1. Introduction

The inflation paradigm is considered the most accepted mechanism to explain many of the fundamental problems of the early stages in the evolution of our universe [1–4], such as the flatness, homogeneity, and isotropy observed in the present universe. Another important aspect of inflation is its ability to correlate cosmological scales that would otherwise be disconnected. Fluctuations generated during this early phase of inflation yield a primordial spectrum of density perturbation [5–8], which is nearly scale invariant, adiabatic, and Gaussian, which is in agreement with cosmological observations [9].

The single-field scalar models have been broadly used to describe the primordial expansion, the most phenomenological successful are those with a quintessence scalar field and slow-roll inflation [10–19]. However, if another component is included, i.e., a multiscalar field theory, it is also possible to produce an inflationary scenario [20, 21], even if the fields are noninteracting [16]. Even more the dynamical possibilities in multifield inflationary scenarios are considerably richer than in single-field models, such as in the primordial inflation perturbations analysis [22, 23] or the assisted inflation as discussed in [24], furthermore, the general assisted inflation as in [21]. In this sense the multiscalar fields' cosmology is an attractive candidate to explain such phenomenon.

Recent works have shown that multiscalar field models are very fruitful when studying the early stages of the universe, such as the case in [25], where the authors perform a semianalytic study of preheating in inflationary models comprised of multiple scalar fields coupled nonminimally to gravity. In [26] the authors show the sensitivity of the cosmological observables to the reheating phase following inflation driven by many scalar fields, where they find that, for certain decay rate, reheating following multifield inflation can have a significant impact on the prediction of cosmological observables.

Indeed the multiscalar field models for inflation are of interest even on most recent studies, such as the above-mentioned cases; however, one of the most important features in such models is the potential associated with the scalar fields, and in many cases, the employed potentials are simple polynomial powers of the scalar fields or in other cases the employed potential is a series of lineally summed exponentials; however, it has been shown that a potential of the form $V(\phi,\sigma) = V_0 e^{-\lambda_1\phi-\lambda_2\sigma}$ is a good candidate to model the inflation phenomenon for multiscalar field theory, as discussed in previous work [27], and might provide a richer postinflation scenario.

Generally, in the studies of inflationary cosmology one employs the usual slow-roll approximation with the objective to extract simple expression for basics observable, such as the scalar and tensor spectral indices, the running of the scalar spectral index, and the tensor-to-scalar ratio. Moreover, in the slow-roll regime the set of EKG equations reduces in such a way that one can quickly obtain the solution of the scale factor. Nevertheless, there is an alternative approach which allows for an easy derivation of many inflation results. It is called Hamilton's formulation, widely used in analytical mechanics. Using this method we obtain the exact solutions of the complete set of EKG equations without using the aforementioned approximation.

On the other hand, we implement a basic formulation in quantum cosmology by means of the Wheeler-DeWitt (WDW) equation. The WDW equation has been analyzed with different approaches in order to solve it, and there are several papers on the subject, such is the case in [28], where they debate what a typical wave function for the universe is. Reference [29] has a review on quantum cosmology where the problem of how the universe emerged from big bang singularity can no longer be neglected in the GUT epoch. Moreover, the best candidates for quantum solutions are those that have a damping behavior with respect to the scale factor, since only such wave functions allow for good classical solutions when using a Wentzel-Kramers-Brillouin (WKB) approximation for any scenario in the evolution of our universe [30, 31]. Furthermore, in the context of a single scalar field a family of scalar potentials is obtained in the Bohmian formalism [27, 32], where among others a general potential of the form $V(\phi) = V_0 e^{-\lambda\phi}$ is examined. Given this insight, for a two scalar field scenario we consider a potential of the form $V(\phi,\sigma) = V_0 e^{-\lambda_1\phi-\lambda_2\sigma}$ in order to solve the WDW equation.

This work is arranged as follows. In Section 2 we present the model with the action and the corresponding EKG equations for our cosmological model and the associated Hamiltonian density. In Section 3 general equations for the classical solutions of scale factor, scalar fields, and their associated momenta are derived in terms of the free parameters of the model. In Sections 3.1, 3.2, 3.3, and 3.4 the particular solutions and their number of e-folds are computed for different cases of the λ parameter. in Section 4 we use the Hamiltonian density to compute the corresponding WDW equation, which is solved by using a change of variables; an ansatz for the wave function is employed in terms of a generic function and parameters which are to be determined. In Sections 4.1 and 4.2 the corresponding wave function and their constants relations are presented for different cases of the δ parameter, which in turn is related to the λ parameter of the classical solutions. Finally, in Section 5 we present our conclusions for this work.

2. The Model

We begin with the construction of two scalar fields cosmological paradigm, which requires canonical scalar fields ϕ, σ. The action of a universe with the constitution of such fields is

$$\mathcal{L} = \sqrt{-g}\left(R + \frac{1}{2}g^{\mu\nu}\nabla_\mu\phi\nabla_\nu\phi + \frac{1}{2}g^{\mu\nu}\nabla_\mu\sigma\nabla_\nu\sigma - V(\phi,\sigma)\right), \quad (1)$$

where R is the Ricci scalar, $V(\phi,\sigma)$ is the corresponding scalar field potential, and the reduced Planck mass $M_P^2 = 1/8\pi G = 1$. The corresponding variations of (1), with respect to the metric and the scalar fields give the Einstein-Klein-Gordon field equations

$$\begin{aligned} G_{\alpha\beta} = {} & \frac{1}{2}\left(\nabla_\alpha\phi\nabla_\beta\phi - \frac{1}{2}g_{\alpha\beta}g^{\mu\nu}\nabla_\mu\phi\nabla_\nu\phi\right) \\ & + \frac{1}{2}\left(\nabla_\alpha\sigma\nabla_\beta\sigma - \frac{1}{2}g_{\alpha\beta}g^{\mu\nu}\nabla_\mu\sigma\nabla_\nu\sigma\right) \\ & - \frac{1}{2}g_{\alpha\beta}V(\phi,\sigma), \end{aligned} \quad (2)$$

$$\Box\phi - \frac{\partial V}{\partial\phi} = g^{\mu\nu}\phi_{,\mu\nu} - g^{\alpha\beta}\Gamma^\nu_{\alpha\beta}\nabla_\nu\phi - \left(\frac{\partial V}{\partial\phi}\right)_\sigma = 0, \quad (3)$$

$$\Box\sigma - \frac{\partial V}{\partial\sigma} = g^{\mu\nu}\sigma_{,\mu\nu} - g^{\alpha\beta}\Gamma^\nu_{\alpha\beta}\nabla_\nu\sigma - \left(\frac{\partial V}{\partial\sigma}\right)_\phi = 0. \quad (4)$$

The line element to be considered in this work is the flat FRW

$$ds^2 = -N(t)^2 dt^2 + e^{2\Omega(t)}\left[dr^2 + r^2\left(d\theta^2 + \sin^2\theta d\phi^2\right)\right], \quad (5)$$

where N is the lapse function, which in a special gauge one can directly recover the cosmic time t_{phys} ($Ndt = dt_{phys}$), the scale factor $A(t) = e^{\Omega(t)}$ is in the Misner's parametrization, and the scalar function has an interval, $\Omega \in (-\infty,\infty)$. Consequently the field equations are

$$\frac{3\dot\Omega^2}{N^2} - \frac{\dot\phi^2}{4N^2} - \frac{\dot\sigma^2}{4N^2} - \frac{1}{2}V(\phi,\sigma) = 0, \quad (6)$$

$$\frac{2\ddot\Omega}{N^2} + \frac{3\dot\Omega^2}{N^2} - \frac{2\dot\Omega\dot N}{N^3} + \frac{\dot\phi^2}{4N^2} + \frac{\dot\sigma^2}{4N^2} - \frac{1}{2}V(\phi,\sigma) = 0, \quad (7)$$

$$\frac{\ddot\phi\dot\phi}{N^2} + \frac{3\dot\Omega\dot\phi^2}{N^2} - \frac{\dot N\dot\phi^2}{N^3} + (\dot V)_\sigma = 0, \quad (8)$$

$$\frac{\ddot\sigma\dot\sigma}{N^2} + \frac{3\dot\Omega\dot\sigma^2}{N^2} - \frac{\dot N\dot\sigma^2}{N^3} + (\dot V)_\phi = 0. \quad (9)$$

By building the corresponding Lagrangian and Hamiltonian densities for this cosmological model, classical solutions to Einstein-Klein-Gordon equations ((2)-(4)) can be found using the Hamilton's approach, and the quantum formalism can be determined and solved. In that sense, we use the metric equation (5) into (1) having

$$\mathcal{L} = e^{3\Omega}\left(\frac{6\dot\Omega^2}{N} - \frac{\dot\phi^2}{2N^2} - \frac{\dot\sigma^2}{2N^2} + NV(\phi,\sigma)\right), \quad (10)$$

where upper "•" represents the first time derivative and the corresponding momenta are defined in the usual way $\Pi_q = \partial\mathscr{L}/\partial\dot{q}$. We obtain

$$\Pi_\Omega = 12\frac{e^{3\Omega}}{N}\dot{\Omega},$$
$$\dot{\Omega} = \frac{Ne^{-3\Omega}}{12}\Pi_\Omega,$$
$$\Pi_\phi = -\frac{e^{3\Omega}}{N}\dot{\phi},$$
$$\dot{\phi} = -Ne^{-3\Omega}\Pi_\phi, \tag{11}$$
$$\Pi_\sigma = -\frac{e^{3\Omega}}{N}\dot{\sigma},$$
$$\dot{\sigma} = -Ne^{-3\Omega}\Pi_\sigma.$$

By performing the variation of the canonical Lagrangian with respect to N, i.e., $\delta\mathscr{L}_{\text{canonical}}/\delta N = 0$, where $\mathscr{L}_{\text{canonical}} = \Pi_q\dot{q} - N\mathscr{H}$, it implies the constraint $\mathscr{H} = 0$. Hence the Hamiltonian density is

$$\mathscr{H} = \frac{e^{-3\Omega}}{24}\left[\Pi_\Omega^2 - 12\Pi_\phi^2 - 12\Pi_\sigma^2 - 24V(\phi,\sigma)e^{6\Omega}\right]. \tag{12}$$

In the gauge $N = 24e^{3\Omega}$ and using the Hamilton equations $\dot{q} = \partial\mathscr{H}/\partial\Pi_q$ and $\dot{\Pi}_q = -\partial\mathscr{H}/\partial q$, we have the following set of equations:

$$\dot{\Omega} = 2\Pi_\Omega,$$
$$\dot{\phi} = -24\Pi_\phi,$$
$$\dot{\sigma} = -24\Pi_\sigma,$$
$$\dot{\Pi}_\Omega = 6U, \tag{13}$$
$$\dot{\Pi}_\phi = \frac{\partial U}{\partial\phi},$$
$$\dot{\Pi}_\sigma = \frac{\partial U}{\partial\sigma},$$

where $U = 24V(\phi,\sigma)e^{6\Omega}$. Given a particular form of the potential $V(\phi,\sigma)$ one can derive a relation between the time derivative of the momenta such as $\dot{\Pi}_\phi \propto \dot{\Pi}_\sigma$, provided that $\partial V/\partial\phi = \alpha\partial V/\partial\sigma$, where α is a constant. Such connection can be obtained considering two different configurations of the potential: $V(\phi,\sigma) = f[\pm(\alpha_1\phi + \alpha_2\sigma)]$ or $V(\phi,\sigma) = V_1 f[\pm(\alpha_1\phi)] + V_2 f[\pm(\alpha_1\sigma)]$, where V_1 and V_2 are constants, and $f(\phi,\sigma)$ is an arbitrary function. We select the simplest form of the potential $V(\phi,\sigma) = A(\phi)B(\sigma)$:

$$V = V_0 e^{-\lambda_1\phi-\lambda_2\sigma}, \tag{14}$$

where V_0 is a constants and λ_1 and λ_2 are distinguishing parameters. This class of potential has been obtained by other methods; see, for instance, [27, 32–35]. Therefore the time derivative of the momenta is simply $\dot{\Pi}_\phi = -\lambda_1 U$ and $\dot{\Pi}_\sigma = -\lambda_2 U$, of which solutions are

$$\Pi_\phi = -\frac{\lambda_1}{6}\Pi_\Omega + p_\phi,$$
$$\Pi_\sigma = -\frac{\lambda_2}{6}\Pi_\Omega + p_\sigma, \tag{15}$$

where p_ϕ and p_σ are integration constants. Henceforth we will employ this scheme in order to find analytic classic and quantum solutions.

3. Classical Solutions

We start from the Hamilton equations (13) in order to find relations between the scale factor and the scalar fields, such as

$$\dot{\phi} = -24\Pi_\phi = 4\lambda_1\Pi_\Omega - 24p_\phi = 2\lambda_1\dot{\Omega} - 24p_\phi, \tag{16}$$

$$\dot{\sigma} = -24\Pi_\sigma = 4\lambda_2\Pi_\Omega - 24p_\sigma = 2\lambda_2\dot{\Omega} - 24p_\sigma, \tag{17}$$

of which solutions are

$$\phi = \phi_1 + 2\lambda_1\Omega - 24p_\phi t, \tag{18}$$

$$\sigma = \sigma_1 + 2\lambda_2\Omega - 24p_\sigma t, \tag{19}$$

where ϕ_1 and σ_1 are integration constants, and they can be determined by suitable conditions. These expressions are indeed general relations, since they satisfy the Einstein-Klein-Gordon equations ((6)-(9)). Then by taking into account the constraint $\mathscr{H} = 0$, we obtain the temporal dependence for $\Pi_\Omega(t)$ which allows us to construct a master equation:

$$\frac{d\Pi_\Omega}{m_1\Pi_\Omega^2 + m_2\Pi_\Omega - m_3} = dt, \tag{20}$$

where the parameters m_i, $i = 1, 2, 3$, are

$$m_1 = 2\left(3 - \lambda_1^2 - \lambda_2^2\right) = 2\left(3 - \lambda^2\right),$$
$$m_2 = 24\left[\lambda_1 p_\phi + \lambda_2 p_\sigma\right], \tag{21}$$
$$m_3 = 72\left[p_\phi^2 + p_\sigma^2\right].$$

Subsequently by analyzing the parameter $\lambda^2 = \lambda_1^2 + \lambda_2^2$ we will obtain three different solutions.

3.1. Solution for $\lambda^2 = 3$. Having $\lambda^2 = 3$ implies that $m_1 = 0$, so the integral to solve becomes

$$\int\frac{d\Pi_\Omega}{m_2\Pi_\Omega - m_3} = \int dt, \tag{22}$$

then we parameterize λ_i such as $\lambda_1 = \sqrt{3(1-\epsilon)}$, $\lambda_2 = \sqrt{3\epsilon}$, $\lambda^2 = 3$, where $\epsilon \in (0,1)$ measures the corresponding weight for each scalar field during inflation; so the constants (21) become

$$m_2 = 24\sqrt{3(1-\epsilon)}\left[p_\phi + \sqrt{\frac{\epsilon}{(1-\epsilon)}}p_\sigma\right],$$
$$m_3 = 72\left[p_\phi^2 + p_\sigma^2\right]. \tag{23}$$

Thus $\Pi_\Omega(t)$ becomes

$$\Pi_\Omega(t) = \frac{m_3}{m_2} + c_1 e^{bt}, \tag{24}$$

where c_1 is an integration constant. Using the relations from (13) and after some algebra, the solutions of the set of variables (Ω, ϕ, σ) and (Π_ϕ, Π_σ) are

$$\Omega = \Omega_0 + \frac{2m_3}{m_2}t + \frac{2c_1}{m_2}e^{m_2 t}, \tag{25}$$

$$\phi = \phi_0 + 4\sqrt{3(1-\epsilon)}\frac{m_3}{m_2}t - 24p_\phi t + \frac{4c_1\sqrt{3(1-\epsilon)}}{m_2}e^{m_2 t}, \tag{26}$$

$$\sigma = \sigma_0 + 4\sqrt{3\epsilon}\frac{m_3}{m_2}t - 24p_\sigma t + \frac{4c_1\sqrt{3\epsilon}}{m_2}e^{m_2 t}, \tag{27}$$

$$\Pi_\phi = -\frac{\sqrt{3(1-\epsilon)}}{6}\left(\frac{m_3}{m_2} + c_1 e^{m_2 t}\right) + p_\phi, \tag{28}$$

$$\Pi_\sigma = -\frac{\sqrt{3\epsilon}}{6}\left(\frac{m_3}{m_2} + c_1 e^{m_2 t}\right) + p_\sigma, \tag{29}$$

where $(\Omega_0, \phi_0, \sigma_0)$ are all integration constants. In order to make the above results fulfill the EKG equations ((6)-(9)), all constants must satisfy that $144V_0 = m_2 c_1 e^{-6\Omega_0+\lambda_1\phi_0+\lambda_2\sigma_0}$. Finally the scale factor A(t) for this case is

$$A = A_0 e^{(2m_3/m_2)t}\exp\left[\frac{2c_1}{m_2}e^{m_2 t}\right], \tag{30}$$

where $A_0 = e^{\Omega_0}$. Given that the scale factor is an exponential of an exponential function, it might exhibit a highly substantial growth.

3.2. Solution for $\lambda^2 > 3$. For this case $m_1 = 2(3-\lambda^2) < 0$, so the integral to solve becomes

$$\frac{d\Pi_\Omega}{-m_1\Pi_\Omega^2 + m_2\Pi_\Omega - m_3} = dt, \tag{31}$$

where we include the minus sign in this equation, such the constant $m_1 = 2(\lambda^2 - 3) = 2\beta > 0$. Then we define $\omega^2 = m_2^2 - 8\beta m_3$, so we change variable as $z = 4\beta\Pi_\Omega - m_2$ in order to integrate (31). Thus the solution to the momenta $\Pi_\Omega(t)$ becomes

$$\Pi_\Omega = \frac{m_2}{4\beta} + \frac{\omega}{4\beta}\tanh\left(\frac{\omega}{2}t\right). \tag{32}$$

Using the relations from (13) and after some algebra, the solutions of the set of variables (Ω, ϕ, σ) and (Π_ϕ, Π_σ) are

$$\Omega = \Omega_0 + \frac{m_2}{2\beta}t + \frac{1}{\beta}\ln\left[\cosh\left(\frac{\omega}{2}t\right)\right], \tag{33}$$

$$\phi = \phi_0 + \left(\lambda_1\frac{m_2}{\beta} - 24p_\phi\right)t - \frac{2\lambda_1}{\beta}\ln\left[\cosh\left(\frac{\omega}{2}t\right)\right], \tag{34}$$

$$\sigma = \sigma_0 + \left(\lambda_2\frac{m_2}{\beta} - 24p_\sigma\right)t - \frac{2\lambda_2}{\beta}\ln\left[\cosh\left(\frac{\omega}{2}t\right)\right], \tag{35}$$

$$\Pi_\phi = -\frac{\lambda_1}{6}\left(\frac{m_2}{4\beta} + \frac{\omega}{4\beta}\tanh\left(\frac{\omega}{2}t\right)\right) + p_\phi, \tag{36}$$

$$\Pi_\sigma = -\frac{\lambda_2}{6}\left(\frac{m_2}{4\beta} + \frac{\omega}{4\beta}\tanh\left(\frac{\omega}{2}t\right)\right) + p_\sigma, \tag{37}$$

where $(\Omega_0, \phi_0, \sigma_0)$ are all integration constants. In order to make the above results fulfill the EKG equations ((6)-(9)), all constants must satisfy that $1152\beta V_0 = \omega^2 e^{\lambda_1\phi_0+\lambda_2\sigma_0+2\beta\Omega_0}$. Finally the scale factor becomes

$$A = A_0 e^{(m_2/2\beta)t}\left[\cosh\left(\frac{\omega}{2}t\right)\right]^{1/\beta}, \tag{38}$$

where $A_0 = e^{\Omega_0}$. For this case, given that $\beta > 0$, one would expect that the scale factor grows slower than the previous case $\lambda^2 = 3$.

3.3. Solution When $\lambda^2 < 3$. For this case we modify the relation between the momenta equation (15), by changing the sign in the constants, $(p_\phi, p_\sigma, m_2) \longrightarrow (-p_\phi, -p_\sigma, -m_2)$, and $m_1 = 2(3-\lambda^2) = 2\eta > 0$; therefore the integral to solve becomes

$$\frac{d\Pi_\Omega}{2\eta\Pi_\Omega^2 - m_2\Pi_\Omega - m_3} = dt. \tag{39}$$

Thus $\Pi_\Omega(t)$ is

$$\Pi_\Omega = \frac{1}{4\eta}\left[m_2 - \alpha\coth\left(\frac{\alpha}{2}t\right)\right], \tag{40}$$

where $\alpha^2 = m_2^2 + 8\eta m_3$. Using the relations from (13) and after some algebra, the solutions of the set of variables (Ω, ϕ, σ) and (Π_ϕ, Π_σ) are

$$\Omega = \Omega_0 + \frac{m_2}{2\eta} t + \ln\left[\operatorname{csch}\left(\frac{\alpha}{2} t\right)\right]^{1/\eta}, \quad (41)$$

$$\phi = \phi_0 + \left(\lambda_1 \frac{m_2}{\eta} + 24 p_\phi\right) t - \ln\left[\sinh\left(\frac{\alpha}{2} t\right)\right]^{2\lambda_1/\eta}, \quad (42)$$

$$\sigma = \sigma_0 + \left(\lambda_2 \frac{m_2}{\eta} + 24 p_\sigma\right) t - \ln\left[\sinh\left(\frac{\alpha}{2} t\right)\right]^{2\lambda_2/\eta}, \quad (43)$$

$$\Pi_\phi = -\frac{1}{24}\left[\frac{\lambda_1 m_2}{\eta} + 24 p_\phi\right] + \frac{\lambda_1 \alpha}{24\eta} \coth\left(\frac{\alpha}{2} t\right), \quad (44)$$

$$\Pi_\sigma = -\frac{1}{24}\left[\frac{\lambda_2 m_2}{\eta} + 24 p\phi\right] + \frac{\lambda_2 \alpha}{24\eta} \coth\left(\frac{\alpha}{2} t\right), \quad (45)$$

where $(\Omega_0, \phi_0, \sigma_0)$ are integration constants. In order to fulfill the EKG equations ((6)-(9)), all constants must satisfy that $1152\eta V_0 = \alpha^2 e^{\lambda_1\phi_0 + \lambda_2\sigma_0 - 2\eta\Omega_0}$. Finally the scale factor $A = e^{\Omega}$ becomes

$$A = A_0 e^{(m_2/2\eta)t} \operatorname{csch}^{1/\eta}\left(\frac{\alpha}{2} t\right), \quad (46)$$

where $A_0 = e^{\Omega_0}$. For this case, given that $\eta > 0$, one would expect the scale factor to grow in a similar way as the previous case $\lambda^2 > 3$.

3.4. Number of e-Folds. Inflation is characterised by the number of e-folds it expands during such period that corresponds to $A''_{phys} > 0$, where the primes represent the derivatives with respect to the cosmic time t_{phys}. The e-folding function $N_e = \int dt_{phys} H(t_{phys})$ is described by t_{phys}: computing the integral from $t_{phys}*$ to $t_{phys\ end}$, where $t_{phys}*$ represents the time when the relevant cosmic microwave background (CMB) modes become superhorizon at 50-60 e-folds before inflation ends at $t_{phys\ end}$; and $H(t_{phys}) = H_{phys} = A'_{phys}/A_{phys}$ is the Hubble parameter. Although, in our prescription we use a proper time t, we can evaluate the Hubble function in the corresponding gauge as $H_{phys} = \dot{\Omega}/N$.

At the end of inflation the expansion rate of the scale factor must be null which translates to $-H'_{phys} = H^2_{phys}$ or $\ddot{\Omega} = 2\dot{\Omega}^2$. From here we can compute the time when inflation ends (t_{end}) given each particular case. Table 1 shows the computation of the e-folding function N_e and t_{end} for each case given by the λ parameter.

4. Quantum Solutions

The Wheeler-DeWitt equation for this model is acquired by replacing $\Pi_{q^\mu} = -i\hbar\partial_{q^\mu}$ in (12). The factor $e^{-3\Omega}$ may be factor ordered with $\widehat{\Pi}_\Omega$ in several forms. Hartle and Hawking [30] have suggested what might be called a semigeneral factor ordering, which in this case would order $e^{-3\Omega}\widehat{\Pi}^2_\Omega$ as

$$-e^{-(3-p)\Omega}\partial_\Omega e^{-p\Omega}\partial_\Omega = -e^{-3\Omega}\partial^2_\Omega + p e^{-3\Omega}\partial_\Omega, \quad (47)$$

where p is any real constant that measures the ambiguity in the factor ordering for the variable Ω; in the following we will assume such factor ordering for the Wheeler-DeWitt equation, which becomes

$$\hbar^2 \Box\Psi + \hbar^2 p \frac{\partial\Psi}{\partial\Omega} - U(\Omega, \phi, \sigma)\Psi = 0, \quad (48)$$

where $\Box = -\partial^2/\partial\Omega^2 + (1/12)(\partial^2/\partial\phi^2) + (1/12)(\partial^2/\partial\sigma^2)$ is the d'Alambertian in the coordinates $q^\mu = (\Omega, \phi, \sigma)$ and the potential is $U = +24V_0 e^{6\Omega - \lambda_1\phi - \lambda_2\sigma}$. Then we transform the coordinates to obtain a potential that only depends on a single variable, employing the following transformation:

$$\begin{aligned} \zeta &= 6\Omega - \lambda_1\phi - \lambda_2\sigma, \\ \kappa &= \phi + \sigma, \\ \eta &= \phi - \sigma. \end{aligned} \quad (49)$$

Now we find the partial derivatives of ψ with respect to the old coordinates (a, ϕ, σ) but in terms of the new variables (ζ, κ, η),

$$\begin{aligned} \frac{\partial\Psi}{\partial\Omega} &= \frac{\partial\Psi}{\partial\zeta}\frac{\partial\zeta}{\partial\Omega} + \frac{\partial\Psi}{\partial\kappa}\frac{\partial\kappa}{\partial\Omega} + \frac{\partial\Psi}{\partial\eta}\frac{\partial\eta}{\partial\Omega} = 6\frac{\partial\Psi}{\partial\zeta}, \\ \frac{\partial\Psi}{\partial\phi} &= \frac{\partial\Psi}{\partial\zeta}\frac{\partial\zeta}{\partial\phi} + \frac{\partial\Psi}{\partial\kappa}\frac{\partial\kappa}{\partial\phi} + \frac{\partial\Psi}{\partial\eta}\frac{\partial\eta}{\partial\phi} \\ &= -\lambda_1\frac{\partial\Psi}{\partial\zeta} + \frac{\partial\Psi}{\partial\kappa} + \frac{\partial\Psi}{\partial\eta}, \\ \frac{\partial\Psi}{\partial\sigma} &= \frac{\partial\Psi}{\partial\zeta}\frac{\partial\zeta}{\partial\sigma} + \frac{\partial\Psi}{\partial\kappa}\frac{\partial\kappa}{\partial\sigma} + \frac{\partial\Psi}{\partial\eta}\frac{\partial\eta}{\partial\sigma} \\ &= -\lambda_2\frac{\partial\Psi}{\partial\zeta} + \frac{\partial\Psi}{\partial\kappa} - \frac{\partial\Psi}{\partial\eta}; \end{aligned} \quad (50)$$

from here we use these new relations in the quantum Hamiltonian density, obtaining

$$\begin{aligned} &12\hbar^2\left(\lambda_1^2 + \lambda_2^2 - 3\right)\frac{\partial^2\Psi}{\partial\zeta^2} + 24\hbar^2\left(-\lambda_1 + \lambda_2\right)\frac{\partial^2\Psi}{\partial\eta\partial\zeta} \\ &\quad + 24\hbar^2\frac{\partial^2\Psi}{\partial\eta^2} - 24\hbar^2\left(\lambda_1 + \lambda_2\right)\frac{\partial^2\Psi}{\partial\kappa\partial\zeta} + 24\hbar^2\frac{\partial^2\Psi}{\partial\kappa^2} \\ &\quad - 6\hbar^2 p\frac{\partial\Psi}{\partial\zeta} - 24V_0 e^{\zeta}\Psi = 0, \end{aligned} \quad (51)$$

At this point, we propose the following ansatz, $\Psi = e^{(1/\hbar)(c_2\kappa + c_3\eta)} G(\zeta)$, where the parameters c_i are constants and $G(\zeta)$ is a function to be determined. By introducing the aforementioned into (51) we obtain the following differential equation of the function G:

$$\delta_0\frac{d^2G}{d\zeta^2} + \alpha_0\frac{dG}{d\zeta} + \left(\beta_0 e^{\zeta} + \rho_0\right)G = 0, \quad (52)$$

TABLE 1: Computation of the number of e-folds N_e and t_{end} for each case provided by the λ parameter. Note that $M_{\pm}^{(1)} > 1$, $0 < M_{\pm}^{(2)} < 1$, and $M_{\pm}^{(3)} > 1$ in order to have that $t_{end} > 0$.

	t_{end}	$M_{\pm}^{(n)}$	N_e
$\lambda = 3$	$\frac{1}{m_2} \ln \left[M_{\pm}^{(1)} \right]$	$\frac{1}{8c_1} \left[\frac{m_2^2 - 8m_3}{m_2} \pm \sqrt{m_2^2 - 16m_3} \right]$	$\frac{2}{m_2} \left[\frac{m_3}{m_2} \ln \left[M_{\pm}^{(1)} \right] + c_1 M_{\pm}^{(1)} - m_3 t_* - c_1 e^{m_2 t_*} \right]$
$\lambda > 3$	$\frac{2}{\omega} \tanh^{-1} \left[M_{\pm}^{(2)} \right]$	$\frac{-2m_2 \pm \sqrt{\beta \left[\omega^2 (2 - \beta) + 2m_2^2 \right]}}{\omega(2 - \beta)}$	$\frac{1}{2\beta} \left[\left(\frac{m_2 + \omega}{\omega} \right) \ln \left[1 + M_{\pm}^{(2)} \right] - \left(\frac{m_2 - \omega}{\omega} \right) \ln \left[1 - M_{\pm}^{(2)} \right] - m_2 t_* - 2 \ln \left[\cosh \left(\frac{\omega}{2} t_* \right) \right] \right]$
$\lambda < 3$	$\frac{2}{\alpha} \coth^{-1} \left[M_{\pm}^{(3)} \right]$	$\frac{-2m_2 \pm \sqrt{\eta \left[\alpha^2 (\eta - 2) + 2m_2^2 \right]}}{\alpha(\eta - 2)}$	$\frac{1}{2\eta} \left[\left(\frac{m_2 + \alpha}{\alpha} \right) \ln \left[M_{\pm}^{(3)} + 1 \right] - \left(\frac{m_2 - \alpha}{\alpha} \right) \ln \left[M_{\pm}^{(3)} - 1 \right] - m_2 t_* + 2 \ln \left[\sinh \left(\frac{\alpha}{2} t_* \right) \right] \right]$

where the constants are

$$\begin{aligned}
\delta_0 &= 12\hbar^2\left(\lambda_1^2+\lambda_2^2-3\right),\\
\alpha_0 &= -6\hbar\left[4\lambda_1\left(c_2+c_3\right)+4\lambda_2\left(c_2-c_3\right)+\hbar p\right],\\
\beta_0 &= -24V_0,\\
\rho_0 &= 24\left(c_2^2+c_3^2\right).
\end{aligned} \tag{53}$$

The solution of (52) is dependant on the value of constant δ_0, which turns in three different cases, (I) $\delta = 0$ implying that $\lambda_1^2+\lambda_2^2 = 3$, (II) $\delta < 0$ implying that $\lambda_1^2+\lambda_2^2 < 3$, and (III) $\delta > 0$ implying that $\lambda_1^2+\lambda_2^2 > 3$, which can be analyzed in two different cases.

4.1. Case $\delta = 0$. For this case, (52) becomes

$$\alpha_0\frac{dG}{d\zeta}+\left(\beta_0 e^{\zeta}+\rho_0\right)G = 0, \tag{54}$$

of which solution is

$$\begin{aligned}
G(\zeta) &= e^{c_1\zeta}\\
&\cdot\exp\left[-\frac{4V_0}{\hbar\left[4\lambda_1\left(c_2+c_3\right)+4\lambda_2\left(c_2-c_3\right)+\hbar p\right]}e^{\zeta}\right],\\
c_1 &= \frac{4\left(c_2^2+c_3^2\right)}{\hbar\left[4\lambda_1\left(c_2+c_3\right)+4\lambda_2\left(c_2-c_3\right)+\hbar p\right]},
\end{aligned} \tag{55}$$

and therefore, the corresponding wave function for this case becomes

$$\begin{aligned}
\Psi &= \psi_0 e^{(1/\hbar)(c_1\zeta+c_2\kappa+c_3\eta)}\exp\left[-c_4 e^{\zeta}\right],\\
c_4 &= \frac{4V_0}{\hbar\left[4\lambda_1\left(c_2+c_3\right)+4\lambda_2\left(c_2-c_3\right)+\hbar p\right]}.
\end{aligned} \tag{56}$$

Note that wave function has a damping behavior with respect to the scale factor, which is a required feature.

4.2. Case $\delta \neq 0$. For this case, (52) becomes, which is similar to that in [36],

$$\begin{aligned}
y''+ay'+\left(be^{\kappa x}+c\right)y &= 0,\\
y &= e^{-ax/2}Z_\nu\left(\frac{2\sqrt{b}}{\kappa}e^{\kappa x/2}\right),
\end{aligned} \tag{57}$$

where Z_ν is the Bessel function and $\nu = \sqrt{a^2-4c}/\kappa$ is the corresponding order, and its relations are

$$\begin{aligned}
a &= \frac{\alpha_0}{\delta_0} = -\frac{4\lambda_1\left(c_2+c_3\right)+4\lambda_2\left(c_2-c_3\right)+\hbar p}{\hbar\left(\lambda_1^2+\lambda_2^2-3\right)},\\
b &= \frac{\beta_0}{\alpha_0}\\
&= \begin{cases} -\dfrac{2V_0}{\hbar^2\left(\lambda_1^2+\lambda_2^2-3\right)}, & \text{when } \lambda_1^2+\lambda_2^2 > 3\\ \dfrac{2V_0}{\hbar^2\left(\lambda_1^2+\lambda_2^2-3\right)}, & \text{when } \lambda_1^2+\lambda_2^2 < 3\end{cases}\\
c &= \frac{\rho_0}{\alpha_0} = \frac{2\left(c_2^2+c_3^2\right)}{\hbar^2\left(\lambda_1^2+\lambda_2^2-3\right)},\\
\kappa &= 1,
\end{aligned} \tag{58}$$

of which, according to the constant b, the solution to the function G becomes

$$\begin{aligned}
G(\zeta) &= e^{((4\lambda_1(c_2+c_3)+4\lambda_2(c_2-c_3)+\hbar p)/2\hbar(\lambda_1^2+\lambda_2^2-3))\zeta}K_\nu\left(\frac{2}{\hbar}\right.\\
&\left.\cdot\sqrt{\frac{2V_0}{\lambda_1^2+\lambda_2^2-3}e^{\zeta/2}}\right),\quad \lambda_1^2-\lambda_2^2 > 3
\end{aligned} \tag{59}$$

$$\begin{aligned}
G(\zeta) &= e^{-((4\lambda_1(c_2+c_3)+4\lambda_2(c_2-c_3)+\hbar p)/2\hbar(3-\lambda_1^2+\lambda_2^2))\zeta}J_\nu\left(\frac{2}{\hbar}\right.\\
&\left.\cdot\sqrt{\frac{2V_0}{3-\lambda_1^2+\lambda_2^2}e^{\zeta/2}}\right),\quad \lambda_1^2+\lambda_2^2 < 3.
\end{aligned} \tag{60}$$

and the resulting wave functions are

$$\begin{aligned}
\Psi &= e^{c_5\zeta+c_2\kappa+c_3\eta}K_\nu\left(\frac{2}{\hbar}\sqrt{\frac{2V_0}{\lambda_1^2+\lambda_2^2-3}e^{\zeta/2}}\right),\\
&\qquad \lambda_1^2+\lambda_2^2 > 3
\end{aligned} \tag{61}$$

$$\begin{aligned}
\Psi &= e^{-c_6\zeta+c_2\kappa+c_3\eta}J_\nu\left(\frac{2}{\hbar}\sqrt{\frac{2V_0}{3-\lambda_1^2+\lambda_2^2}e^{\zeta/2}}\right),\\
&\qquad \lambda_1^2+\lambda_2^2 < 3.
\end{aligned} \tag{62}$$

where the constants are

$$\begin{aligned}
c_5 &= \frac{4\lambda_1\left(c_2+c_3\right)+4\lambda_2\left(c_2-c_3\right)+\hbar p}{2\hbar\left(\lambda_1^2+\lambda_2^2-3\right)},\\
c_6 &= \frac{4\lambda_1\left(c_2+c_3\right)+4\lambda_2\left(c_2-c_3\right)+\hbar p}{2\hbar\left(3-\lambda_1^2+\lambda_2^2\right)}.
\end{aligned} \tag{63}$$

Whilst $c_5 < 0$ and $c_6 > 0$, the wave functions ((61), (62)) will remain suppressed by the growth of the scale factor, yielding an expected damped wave function.

5. Conclusions

We studied a flat Friedmann-Robertson-Walker (FRW) multiscalar field cosmological model. We introduce the corresponding Einstein-Klein-Gordon (EKG) system of equations and the associated Hamiltonian density. Exact solutions to the EKG system are derived by means of Hamilton's approach where a particular scalar potential of the form $V = V_0 e^{-\lambda_1\phi-\lambda_2\sigma}$ was utilized, which gave rise to different cases dependant of the free parameter λ, for which the scalar fields, the scale factor, and the e-folding function were found. The Hamiltonian density was employed in order to compute the Wheeler-DeWitt (WDW) equation, which was solved by means of a change of variables. An ansatz for the wave function was proposed which in turn allowed us to find the exact form of the generic function and its constants which was composed by the aforementioned in terms of the free parameter λ. We found the model to be rather simple and its solutions to be quite interesting for a model building inflation.

Conflicts of Interest

The authors declare that they have no conflicts of interest.

Acknowledgments

This work was partially supported by CONACYT 167335 and 179881 grants and PROMEP grants UGTO-CA-3. Rafael Hernández-Jiménez acknowledges CONACYT for financial support. This work is part of the collaboration within the Instituto Avanzado de Cosmología. Many calculations were done by Symbolic Program REDUCE 3.8.

References

[1] A. H. Guth, "Inflationary universe: a possible solution to the horizon and flatness problems," *Physical Review D: Particles, Fields, Gravitation and Cosmology*, vol. 23, no. 2, pp. 347–356, 1981.

[2] A. D. Linde, "A new inflationary universe scenario: a possible solution of the horizon, flatness, homogeneity, isotropy and primordial monopole problems," *Physics Letters B*, vol. 108, no. 6, pp. 389–393, 1982.

[3] J. D. Barrow and M. S. Turner, "Inflation in the Universe," *Nature*, vol. 292, no. 5818, pp. 35–38, 1981.

[4] A. A. Starobinsky, "A new type of isotropic cosmological models without singularity," *Physics Letters B*, vol. 91, no. 1, pp. 99–102, 1980.

[5] A. A. Starobinsky, "Spectrum of relict gravitational radiation and the early state of the universe," *JETP Lett*, vol. 30, no. 682, 1979.

[6] V. F. Mukhanov and G. V. Chibisov, "Quantum Fluctuations and a Nonsingular Universe," *JETP Lett*, vol. 33, no. 532, 1981.

[7] H. Kodama and M. Sasaki, "Cosmological perturbation theory," *Progress of Theoretical and Experimental Physics Supplement*, vol. 78, pp. 1–166, 1984.

[8] B. A. Bassett, S. Tsujikawa, and D. Wands, "Inflation dynamics and reheating," *Reviews of Modern Physics*, vol. 78, no. 2, pp. 537–589, 2006.

[9] P. A. R. Ade and Planck Collaboration, "Planck 2018 results. X. Constraints on inflation," 2018, https://arxiv.org/abs/1807.06211.

[10] J. D. Barrow, "Slow-roll inflation in scalar-tensor theories," *Physical Review D: Particles, Fields, Gravitation and Cosmology*, vol. 51, p. 2729, 1995.

[11] A. R. Liddle and R. J. Scherrer, "Classification of scalar field potentials with cosmological scaling solutions," *Physical Review D: Particles, Fields, Gravitation and Cosmology*, vol. 59, no. 2, Article ID 023509, 7 pages, 1998.

[12] P. G. Ferreira and M Joyce, "Cosmology with a primordial scaling field," *Phys. Rev. D*, vol. 58, Article ID 023503, 1998.

[13] E. J. Copeland, M. Sami, and S. Tsujikawa, "Dynamics of dark energy," *International Journal of Modern Physics D: Gravitation, Astrophysics, Cosmology*, vol. 15, no. 11, pp. 1753–1935, 2006.

[14] E. J. Copeland, A. R. Liddle, and D. Wands, "Exponential potentials and cosmological scaling solutions," *Physical Review D: Particles, Fields, Gravitation and Cosmology*, vol. 57, no. 8, pp. 4686–4690, 1998.

[15] T. Barreiro, E. J. Copeland, and N. J. Nunes, "Quintessence arising from exponential potentials," *Physical Review D: Particles, Fields, Gravitation and Cosmology*, vol. 61, no. 12, 2000.

[16] G. Calcagni and A. R. Liddle, "Stability of multifield cosmological solutions," *Physical Review D: Particles, Fields, Gravitation and Cosmology*, vol. 77, no. 2, 023522, 6 pages, 2008.

[17] D. Sáez-Gómez, *Scalar-Tensor theories and current Cosmology, Problems of Modern Cosmology*, 2008, https://arxiv.org/abs/0812.1980.

[18] M. Capone, C. Rubano, and P. Scudellaro, "Slow rolling, inflation, and quintessence," *EPL (Europhysics Letters)*, vol. 73, no. 1, pp. 149–155, 2006.

[19] E. W. Kolb and M. S. Turner, *The Early Universe*, Addison-Wesley Publishing co., Illinois, USA, 1998.

[20] A. A. Coley and R. J. van den Hoogen, "Dynamics of multi-scalar-field cosmological models and assisted inflation," *Physical Review D: Particles, Fields, Gravitation and Cosmology*, vol. 62, no. 2, 023517, 11 pages, 2000.

[21] E. J. Copeland, A. Mazumdar, and N. J. Nunes, "Generalized assisted inflation," *Physical Review D: Particles, Fields, Gravitation and Cosmology*, vol. 60, no. 8, 1999.

[22] S. Yokoyama, T. Suyama, and T. Tanaka, "Primordial non-Gaussianity in multiscalar inflation," *Physical Review D: Particles, Fields, Gravitation and Cosmology*, vol. 77, no. 8, 2008.

[23] T. Chiba and M. Yamaguchi, "Extended slow-roll conditions and primordial fluctuations: multiple scalar fields and generalized gravity," *Journal of Cosmology and Astroparticle Physics*, vol. 2009, no. 01, pp. 019-019, 2009.

[24] A. R. Liddle, A. Mazumdar, and F. E. Schunck, "Assisted inflation," *Physical Review D: Particles, Fields, Gravitation and Cosmology*, vol. 58, no. 6, 1998.

[25] M. P. DeCross, D. I. Kaiser, A. Prabhu, C. Prescod-Weinstein, and E. I. Sfakianakis, "Preheating after multifield inflation with nonminimal couplings. I. Covariant formalism and attractor behavior," *Physical Review D: Particles, Fields, Gravitation and Cosmology*, vol. 97, no. 2, 2018.

[26] S. C. Hotinli, J. Frazer, A. H. Jaffe, J. Meyers, L. C. Price, and E. R. Tarrant, "Effect of reheating on predictions following multiple-field inflation," *Physical Review D: Particles, Fields, Gravitation and Cosmology*, vol. 97, no. 2, 2018.

[27] J. Socorro and O. E. Nuñez, "Scalar potentials with multi-scalar fields from quantum cosmology and supersymmetric quantum mechanics," *The European Physical Journal Plus*, vol. 132, no. 168, 2017, https://arxiv.org/abs/1702.00478.

[28] G. W. Gibbons and L. P. Grishchuk, "What is a typical wave function for the universe?" *Nuclear Physics B*, vol. 313, no. 3, pp. 736–748, 1989.

[29] L. Z. Fang and R. Ruffini, Eds., *Quantum cosmology*, vol. 3 of *Advanced Series in Astrophysics and Cosmology*, World Scientific Publishing Co., Inc., Teaneck, NJ, USA, 1987.

[30] J. B. Hartle and S. W. Hawking, "Wave function of the Universe," *Physical Review D*, vol. 28, no. 2960, 1983.

[31] S. W. Hawking, "The quantum state of the universe," *Nuclear Physics. B. Theoretical, Phenomenological, and Experimental High Energy Physics. Quantum Field Theory and Statistical Systems*, vol. 239, no. 1, pp. 257–276, 1984.

[32] W. Guzmán, M. Sabido, J. Socorro et al., "Scalar potentials out of canonical quantum cosmology," *International Journal of Modern Physics D: Gravitation, Astrophysics, Cosmology*, vol. 16, no. 4, pp. 641–653, 2007.

[33] J. Socorro and M. D'oleire, "Inflation from supersymmetric quantum cosmology," *Physical Review D*, vol. 82, no. 4, Article ID 044008, pp. 1–7, 2010.

[34] J. Socorro, P. Romero, L. O. Pimentel, and M. Aguero, "Quintom potentials from quantum cosmology using the FRW cosmological model," *International Journal of Theoretical Physics*, vol. 52, no. 8, pp. 2722–2734, 2013.

[35] J. Socorro, M. Sabido, W. Ramírez et al., "Inflación cosmológica vista desde la mecánica cuántica supersimétrica," *Notabilis Scientia*, 2013.

[36] V. F. Zaitsev and A. D. Polyanin, *Handbook of Exact Solutions for Ordinary Differential Equations*, Taylor & Francis Editorial, 2002.

4

The Perturbed Riemann Problem with Delta Shock for a Hyperbolic System

Xinli Han[1] **and Lijun Pan**[2]

[1]School of Science, Nanjing University of Posts and Telecommunications, Nanjing 210023, China
[2]Department of Mathematics, Nanjing University of Aeronautics and Astronautics, Nanjing 211106, China

Correspondence should be addressed to Lijun Pan; 98010149@163.com

Academic Editor: Carlo Bianca

In this paper, we study the perturbed Riemann problem with delta shock for a hyperbolic system. The problem is different from the previous perturbed Riemann problems which have no delta shock. The solutions to the problem are obtained constructively. From the solutions, we see that a delta shock in the corresponding Riemann solution may turn into a shock and a contact discontinuity under a perturbation of the Riemann initial data. This shows the instability and the internal mechanism of a delta shock. Furthermore, we find that the Riemann solution of the hyperbolic system is instable under this perturbation, which is also quite different from the previous perturbed Riemann problems.

1. Introduction

In this paper, we are concerned with the following nonstrictly hyperbolic system of conservation laws in the form

$$\begin{aligned} u_t + \left(u^2\right)_x &= 0, \\ v_t + (uv)_x &= 0. \end{aligned} \tag{1}$$

Equations (1) can be derived from a two-dimensional hyperbolic system of conservation laws

$$\begin{aligned} u_t + \left(u^2\right)_x + (uv)_y &= 0, \\ v_t + (uv)_x + \left(v^2\right)_y &= 0. \end{aligned} \tag{2}$$

The above system is the mathematical simplification of Euler equations of gas dynamics. In 1991 and 1994, Yang and Zhang [1] and Tan and Zhang [2] obtained both numerical and analytical solutions to the Riemann problem of system (2). The form of Dirac delta functions supported on shocks was found necessary and was used as parts in the Riemann solutions. We call it a delta shock. A delta shock is the generalization of an ordinary shock. It is more compressive than an ordinary shock in the sense that more characteristics enter the discontinuity line. Mathematically, the delta shocks are new type singular solutions such that their components contain delta functions and their derivatives. Physically, they are interpreted as the process of formation of the galaxies in the universe, or the process of concentration of particles [3].

To investigate the validity of delta shock, in 1994, Tan et al. [4] considered the Riemann problem for the one-dimensional model (1). They found that there exist delta shock as the limit of vanishing viscosity for system (1). As for delta shock, there are numerous excellent papers. We refer readers to [3–14] and the references cited therein. There are still many open and complicated problems in the delta shock theories. Study of this area gives a new perspective in the theory of conservation law systems.

In this paper, we are interested in the internal mechanism and instability of a delta shock. For this purpose, we study system (1) with the following initial data:

$$(u, v)|_{t=0} = (u_0(x), v_0(x)) = \begin{cases} \left(u_0^-(x), v_0^-(x)\right), & x < 0, \\ \left(u_0^+(x), v_0^+(x)\right), & x > 0, \end{cases} \tag{3}$$

where $u_0^\pm(x)$ and $v_0^\pm(x)$ are all bounded C^1 functions with the following property:

$$\left(u_0^\pm(0\pm), v_0^\pm(0\pm)\right) = \left(\widehat{u}^\pm, \widehat{v}^\pm\right). \tag{4}$$

Here $\widehat{u}^\pm$ and $\widehat{v}^\pm$ are constants with $(\widehat{u}^-, \widehat{v}^-) \neq (\widehat{u}^+, \widehat{v}^+)$. The initial value (3) is a perturbation of Riemann initial value (5) at the neighborhood of the origin in the $x-t$ plane. The perturbation on the Riemann initial data is reasonable. For example, error is unavoidable in computation and the error forms a perturbation of the initial data.

We divide our work into two parts according to the presence of delta shock or not. When the delta shock is not involved, the perturbed Riemann problem (1) and (3) is classical. More importantly, there is no delta shock in the corresponding Riemann solutions for the previous work on the perturbed Riemann problem. Therefore, we only pay attention to the perturbed Riemann problem (1) and (3) when the delta shock is involved. To overcome the difficulty caused by delta shock, we adopt the method of characteristic analysis and the local existence and uniqueness theorem proposed by Li Ta-tsien and Yu Wen-ci [15]. We construct the solution to the perturbed problem (1) and (3) locally in time.

Our result shows that a perturbation of initial data may bring essential change when a delta shock appears in the corresponding Riemann solution. A delta shock may turn into a shock and a contact discontinuity. This shows the instability of the delta shock, which allows us to better investigate the internal mechanism of a delta shock. Furthermore, the previous works [15, 16] about the perturbed Riemann problem pay more attention to the stability of the corresponding Riemann solution. In other words, the Riemann solution has a local structure stability with respect to the perturbation of Riemann initial data. A distinctive feature for this paper is that, for some initial data (3), the preceding local structure stability fails. We pay more attention to the differences between Riemann solution and perturbed Riemann solution.

The paper is organized as follows. In Section 2, we present some preliminary knowledge about the hyperbolic system (1). Then, the construction and proof of the solution to the perturbed Riemann problem (1) and (3) with delta shock are presented in Section 3.

2. Preliminaries

In this section, we recall the main properties of system (1) with Riemann initial data

$$(u, v)(x, 0) = \left(\widehat{u}^\pm, \widehat{v}^\pm\right), \quad \pm x > 0, \tag{5}$$

where $\widehat{u}^\pm$ and $\widehat{v}^\pm$ are constants with $(\widehat{u}^-, \widehat{v}^-) \neq (\widehat{u}^+, \widehat{v}^+)$ (see [4, 17, 18] for a more detailed study of the model).

The eigenvalues of the hyperbolic system (1) are

$$\begin{aligned} \lambda_1(u, v) &= u, \\ \lambda_2(u, v) &= 2u, \end{aligned} \tag{6}$$

with the corresponding left eigenvectors,

$$\begin{aligned} l_1(u, v) &= (v, -u), \\ l_2(u, v) &= (1, 0), \end{aligned} \tag{7}$$

and the corresponding right eigenvectors,

$$\begin{aligned} r_1(u, v) &= (0, 1)^T, \\ r_2(u, v) &= (u, v)^T. \end{aligned} \tag{8}$$

By a direct calculation,

$$\begin{aligned} \nabla\lambda_1 \cdot r_1 &\equiv 0, \\ \nabla\lambda_2 \cdot r_2 &\equiv 2u. \end{aligned} \tag{9}$$

Therefore, λ_1 is always linearly degenerate; λ_2 is genuinely nonlinear if $u \neq 0$ and linearly degenerate if $u = 0$.

The Riemann invariants of system (1) along the characteristic fields are

$$\begin{aligned} \zeta(u, v) &= u, \\ \varsigma(u, v) &= \frac{u}{v}. \end{aligned} \tag{10}$$

Definition 1 (see [3, 19]). A pair of (u, v) is called a generalized delta shock solution to (1) with the initial data (3) on local time $[0, T)$, if there exists a smooth curve $l = \{(x_\delta(t), t) : 0 \leq t < T\}$ and a weight $\omega(x, t)$ such that u and v are represented in the following form:

$$\begin{aligned} u &= U(x, t), \\ v &= V(x, t) + \omega(x, t)\,\delta(l), \end{aligned} \tag{11}$$

in which $\delta(x)$ is the delta function, $\omega \in C^1(l)$, $U, V \in L^\infty(R \times [0, T); R)$ and satisfy

$$\begin{aligned} &\int_0^T \int_{-\infty}^{+\infty} \left(U\phi_t + U^2\phi_x\right) dx\,dt \\ &\quad + \int_{-\infty}^{+\infty} u_0(x)\,\phi(x, 0)\,dx = 0, \end{aligned} \tag{12}$$

$$\begin{aligned} &\int_0^T \int_{-\infty}^{+\infty} \left(V\phi_t - UV\phi_x\right) dx\,dt \\ &\quad + \int_0^T \omega(x_\delta(t), t)\,\frac{\partial\phi(x, t)}{\partial l}\sqrt{1 + \sigma_\delta^2}\,dt \\ &\quad + \int_{-\infty}^{+\infty} v_0(x)\,\phi(x, 0)\,dx = 0 \end{aligned} \tag{13}$$

for all the test functions $\phi \in C_0^\infty((-\infty, +\infty) \times [0, T))$. Here σ_δ is the tangential derivative of the curve l, and $\partial\phi(x, t)/\partial l$ stands for the tangential derivative of the function ϕ on the curve l.

Definition 2. For an $n \times n$ matrix $A = (a_{ij})$, define

$$\|A\| = \max_{i=1,\cdots,n} \sum_{j=1}^{n} \left|a_{ij}\right| \tag{14}$$

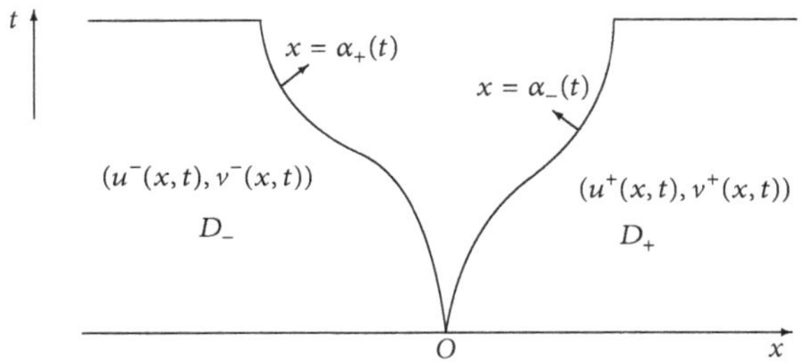

FIGURE 1: The local time.

and

$$\|A\|_{\min} = \inf\left\{\left\|\gamma A\gamma^{-1}\right\|; \gamma = \operatorname{diag}\{\gamma_i\}, \gamma_i \neq 0, i = 1, \cdots, n\right\}. \tag{15}$$

3. The Perturbed Riemann Problem with Delta Shock

In this section, we construct the perturbed Riemann solutions of hyperbolic system (1) with initial data (3) for local time and investigate the internal mechanism and instability of delta shock. About the perturbed Riemann problem, we have six cases according to the different constructions of the solutions to the corresponding Riemann problem (1) and (5) as follows:

(1) When $\widehat{u}^+ < \widehat{u}^- < 0$, the Riemann solution is $\overleftarrow{S} + J$

(2) When $\widehat{u}^+ \leq 0 \leq \widehat{u}^-$, the Riemann solution is delta shock δS

(3) When $\widehat{u}^- < \widehat{u}^+ \leq 0$, the Riemann solution is $\overleftarrow{R} + J$

(4) When $\widehat{u}^- < 0 < \widehat{u}^+$, the Riemann solution is $\overleftarrow{R}_1 + \overrightarrow{R}_2$

(5) When $\widehat{u}^+ > \widehat{u}^- > 0$, the Riemann solution is $J + \overrightarrow{R}$

(6) When $\widehat{u}^- > \widehat{u}^+ > 0$, the Riemann solution is $J + \overrightarrow{S}$

Here "+" means "followed by"; the capitals S, J, and R denote shock, contact discontinuity, and rarefaction wave, respectively.

Since the perturbed Riemann problem (1) and (3) with no delta shock is classical and well known, it will not be pursued here. In this section, we mainly study the differences between the perturbed Riemann solution and the corresponding Riemann solution. Thus we only consider the perturbed Riemann problem (1) and (3) with delta shock, a.e., $\widehat{u}^+ \leq 0 \leq \widehat{u}^-$.

It is known from classical theory that the classical solution $(u^-(x,t), v^-(x,t))$ and $(u^+(x,t), v^+(x,t))$ can be defined in a strip domains D_- and D_+ for local time, respectively (see Figure 1). Here $(u^-(x,t), v^-(x,t))$ and $(u^+(x,t), v^+(x,t))$ are local smooth solutions to the initial problem (1) with corresponding initial data $(u_0^-(x), v_0^-(x))$ and $(u_0^+(x), v_0^+(x))$ on both sides of $x = 0$, respectively. The right boundary of domain D_- is a $I-$characteristic : $x = \alpha_-(t)$; namely,

$$\frac{u^-(\alpha_-(t), t)}{v^-(\alpha_-(t), t)} = \frac{\widehat{u}^-}{\widehat{v}^-}, \qquad \frac{d\alpha_-(t)}{dt} = u^-(\alpha_-(t), t). \tag{16}$$

The left boundary of domain D_+ is $I-$characteristic : $x = \alpha_+(t)$; namely,

$$\frac{u^+(\alpha_+(t), t)}{v^+(\alpha_+(t), t)} = \frac{\widehat{u}^+}{\widehat{v}^+}, \qquad \frac{d\alpha_+(t)}{dt} = u^+(\alpha_+(t), t). \tag{17}$$

Now we turn our attention to the solution of (1) and (3) between the right boundary of domain D_- and the left boundary of domain D_+. We note that the corresponding Riemann solution is a delta shock δS with speed $\widehat{u}^- + \widehat{u}^+$ separating two states $(\widehat{u}^-, \widehat{v}^-)$ and $(\widehat{u}^+, \widehat{v}^+)$ (see Figure 2(a)). If the perturbed Riemann problem (1) and (3) has a solution by using a delta shock δS connecting two states (u^-, v^-) and (u^+, v^+) on local time $[0, T)$, we must choose u and v to be

$$u = u^+ + [u] H(-x + x_\delta(t)), \qquad v = v^+ + [v] H(-x + x_\delta(t)) + \omega(x_\delta(t), t)\,\delta(l), \tag{18}$$

Where, here and below, we use the usual notation $[u] = u^- - u^+$ with u^- and u^+ the values of the function u on the left-hand and right-hand sides of the discontinuity $x = x_\delta(t)(x_\delta(0) = 0)$, etc.; $H(x)$ is the Heaviside function, that is, 0 when $x < 0$ and 1 when $x > 0$; $\omega(x_\delta(t), t)$ and σ_δ are the weight and the tangential derivative of curve $l \triangleq \{(x_\delta(t), t) : 0 \leqslant t < T\}$, which can be defined by

$$0 = -\sigma_\delta [u] + [u^2], \qquad \frac{d\sqrt{1+\sigma_\delta^2}\,\omega(x_\delta(t), t)}{dt} = -\sigma_\delta [v] + [uv], \qquad \omega(0,0) = 0. \tag{19}$$

From (19), one can get that the propagating speed of the delta shock

$$\sigma_\delta = \frac{dx_\delta(t)}{dt} = u^-(x_\delta(t), t) + u^+(x_\delta(t), t). \tag{20}$$

We will prove that the delta shock solution constructed in (18) is a solution of the initial value problem (1) and (3) in the sense of distributions on $[0, T)$.

Proposition 3. *The delta shock solution constructed in (18) satisfies (1) and (3) in the sense of distributions on a domain*

$$R(\epsilon) = \{(x,t) \mid -\infty < x < \infty, 0 \leq t < T\}, \tag{21}$$

where $T > 0$ is a finite time.

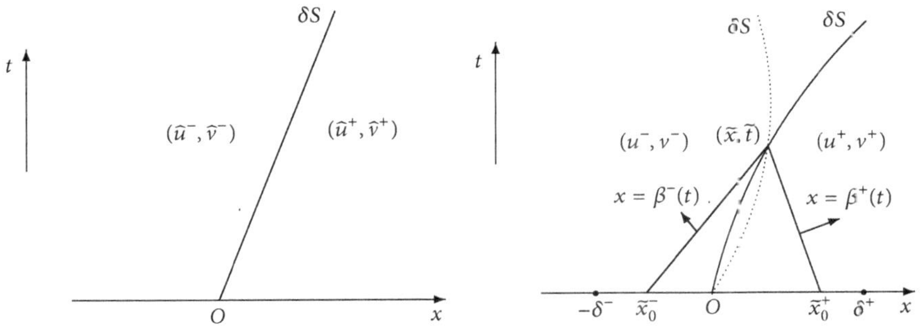

(a) The Riemann solution (b) The perturbed Riemann solution

FIGURE 2: Case 1, Case 2 with $\dot{u}_0^-(0) < 0$, and Case 3 with $\dot{u}_0^+(0) < 0$.

Proof. Let

$$\begin{aligned} U(x,t) &= u^+ + [u]\,H(-x + x_\delta(t)), \\ V(x,t) &= v^+ + [v]\,H(-x + x_\delta(t)). \end{aligned} \tag{22}$$

Then the delta shock solution (18) can be reduced to

$$\begin{aligned} u &= U(x,t), \\ v &= V(x,t) + \omega(x_\delta(t), t)\,\delta(l). \end{aligned} \tag{23}$$

We need to check that (u, v) satisfies (1), which is (12) and (13).

For any test function $\phi \in C_0^\infty((-\infty,+\infty)\times[0,T))$, we plug (23) into the left-hand side of equation (12) and get

$$\begin{aligned} &\int_0^T \int_{-\infty}^{+\infty} \left(U\phi_t + U^2\phi_x\right) dx\,dt + \int_{-\infty}^{+\infty} u_0(x)\,\phi(x,0)\,dx \\ &= \int_0^T \int_{-\infty}^{x_\delta(t)} \left(u^-\phi_t + (u^-)^2\phi_x\right) dx\,dt \\ &+ \int_0^T \int_{x_\delta(t)}^{+\infty} \left(u^+\phi_t + (u^+)^2\phi_x\right) dx\,dt \\ &+ \int_{-\infty}^{+\infty} u_0(x)\,\phi(x,0)\,dx. \end{aligned} \tag{24}$$

Now using the fact that (u^-, v^-) and (u^+, v^+) are C^1 solutions to problem (1) with initial data $(u_0^-(x), v_0^-(x))$ and $(u_0^+(x), v_0^+(x))$ in the domains D_- and D_+, respectively, the divergence theorem gives

$$\begin{aligned} &\int_0^T \int_{-\infty}^{+\infty} \left(U\phi_t + U^2\phi_x\right) dx\,dt + \int_{-\infty}^{+\infty} u_0(x)\,\phi(x,0)\,dx \\ &= \int_0^T \int_{-\infty}^{x_\delta(t)} \left((u^-\phi)_t + \left((u^-)^2\phi\right)_x\right) dx\,dt \\ &+ \int_0^T \int_{x_\delta(t)}^{+\infty} \left((u^+\phi)_t + \left((u^+)^2\phi\right)_x\right) dx\,dt \\ &+ \int_{-\infty}^{+\infty} u_0(x)\,\phi(x,0)\,dx = -\int_{-\infty}^{0} u_0^-(x) \\ &\cdot \phi(x,0)\,dx + \int_0^T \phi(x_\delta(t), t) \\ &\cdot \left(-u^-(x_\delta(t), t)\,\sigma_\delta(x_\delta(t), t)\right. \\ &\left. + \left(u^-(x_\delta(t), t)\right)^2\right) dt - \int_0^{+\infty} u_0^+(x) \\ &\cdot \phi(x,0)\,dx - \int_0^T \phi(x_\delta(t), t) \\ &\cdot \left(-u^+(x_\delta(t), t)\,\sigma_\delta(x_\delta(t), t)\right. \\ &\left. + \left(u^+(x_\delta(t), t)\right)^2\right) dt + \int_{-\infty}^{+\infty} u_0(x) \\ &\cdot \phi(x,0)\,dx = -\int_{-\infty}^{+\infty} u_0(x)\,\phi(x,0)\,dx \\ &+ \int_0^T \phi(x_\delta(t), t)\left(\left[u^2\right] - [u]\,\sigma_\delta\right) dt \\ &+ \int_{-\infty}^{+\infty} u_0(x)\,\phi(x,0)\,dx = 0. \end{aligned} \tag{25}$$

Similarly, we plug (23) into the left-hand side of (13) and get

$$\begin{aligned} &\int_0^T \int_{-\infty}^{+\infty} \left(V\phi_t + UV\phi_x\right) dx\,dt + \int_0^T \omega(x_\delta(t), \\ &t)\,\frac{\partial\phi(x,t)}{\partial l}\sqrt{1+\sigma_\delta^2}\,dt \\ &+ \int_{-\infty}^{+\infty} v_0(x)\,\phi(x,0)\,dx \\ &= \int_0^T \int_{-\infty}^{x_\delta(t)} \left(v^-\phi_t + u^-v^-\phi_x\right) dx\,dt \\ &+ \int_0^T \int_{x_\delta(t)}^{+\infty} \left(v^+\phi_t + u^+v^+\phi_x\right) dx\,dt + \int_0^T \omega(x_\delta(t), \end{aligned}$$

$$t)\frac{\partial\phi(x,t)}{\partial l}\sqrt{1+\sigma_\delta^2}dt$$

$$+\int_{-\infty}^{+\infty}v_0(x)\phi(x,0)\,dx$$

$$=\int_0^T\int_{-\infty}^{x_\delta(t)}\left((v^-\phi)_t+(u^-v^-\phi)_x\right)dxdt$$

$$+\int_0^T\int_{x_\delta(t)}^{+\infty}\left((v^+\phi)_t+(u^+v^+\phi)_x\right)dxdt$$

$$+\int_0^T\omega(x_\delta(t),$$

$$t)\frac{\partial\phi(x,t)}{\partial l}\sqrt{1+\sigma_\delta^2}dt$$

$$+\int_{-\infty}^{+\infty}v_0(x)\phi(x,0)\,dx=-\int_{-\infty}^{0}v_0^-(x)\phi(x,$$

$$0)\,dx+\int_0^T\phi(x_\delta(t),t)\left(-v^-(x_\delta(t),t)\right.$$

$$\left.\cdot\sigma_\delta(x_\delta(t),t)+u^-(x_\delta(t),t)v^-(x_\delta(t),t)\right)dt$$

$$-\int_0^{+\infty}v_0^+(x)\phi(x,0)\,dx-\int_0^T\phi(x_\delta(t),t)$$

$$\cdot\left(-v^+(x_\delta(t),t)\sigma_\delta(x_\delta(t),t)+u^+(x_\delta(t),t)\right.$$

$$\left.\cdot v^+(x_\delta(t),t)\right)dt+\int_0^T\omega(x_\delta(t),$$

$$t)\frac{\partial\phi(x,t)}{\partial l}\sqrt{1+\sigma_\delta^2}dt$$

$$+\int_{-\infty}^{+\infty}v_0(x)\phi(x,0)\,dx=-\int_{-\infty}^{+\infty}v_0(x)\phi(x,$$

$$0)\,dx+\int_0^T\phi(x_\delta(t),t)\left([uv]-[v]\sigma_\delta\right)dt$$

$$+\int_0^T\omega(x_\delta(t),$$

$$t)\frac{\partial\phi(x,t)}{\partial l}\sqrt{1+\sigma_\delta^2}dt$$

$$+\int_{-\infty}^{+\infty}v_0(x)\phi(x,0)\,dx=\int_0^T\phi(x_\delta(t),t)\left([uv]\right.$$

$$\left.-[v]\sigma_\delta\right)dt+\int_0^T\omega(x_\delta(t),t)$$

$$\cdot\sqrt{1+\sigma_\delta^2}d\phi(x_\delta(t),t)=\int_0^T\phi(x_\delta(t),t)\left([uv]-[v]\right.$$

$$\left.\cdot\sigma_\delta\right)dt+\omega(x_\delta(t),t)\sqrt{1+\sigma_\delta^2}\phi(x_\delta(t),t)\Big|_{t=0}^{t=T}$$

$$-\int_0^T\phi(x_\delta(t),$$

$$t)\,d\left(\omega(x_\delta(t),t)\sqrt{1+\sigma_\delta^2}\right)$$

$$=0, \tag{26}$$

which gets equality (13). Then we complete the proof of the proposition. ☐

Furthermore, to guarantee uniqueness of the solution, the delta shock solution constructed in (18) should satisfy the entropy condition (27) on the discontinuity $x=x_\delta(t)$.

Definition 4. The delta shock solution constructed in (18) is an admissible solution of the initial value problem (1) and (3) in the sense of distributions on $[0,T)$, if (u,v) satisfies Definition 1 and the entropy condition

$$\begin{aligned}\lambda_2(u^+,v^+)\le\lambda_1(u^+,v^+)\le\frac{dx_\delta(t)}{dt}\le\lambda_1(u^-,v^-)\\ \le\lambda_2(u^-,v^-)\end{aligned} \tag{27}$$

on the discontinuity $x=x_\delta(t)$.

We now turn to check the entropy condition (27) on $x=x_\delta(t)$. The discussion is divided into three cases: $\widehat{u}^+<0<\widehat{u}^-$, $\widehat{u}^+<0=\widehat{u}^-$, and $\widehat{u}^+=0<\widehat{u}^-$. We start with the following case.

Case 1 ($\widehat{u}^+<0<\widehat{u}^-$). Due to $\widehat{u}^+<0<\widehat{u}^-$, there exist constants $\delta^->0$ and $\delta^+>0$ so small that, for any $x_0^-\in(-\delta^-,0)$ and $x_0^+\in(0,\delta^+)$, the C^1 functions $u_0^-(x)$ and $u_0^+(x)$ satisfy the following condition:

$$u_0^+(x_0^+)<0<u_0^-(x_0^-). \tag{28}$$

Let $x=\beta^-(t)$ (resp., $x=\beta^+(t)$) be the downwards left (resp., right) $II-$characteristic from any point $(\widetilde{x},\widetilde{t})$ on the delta shock curve $x=x_\delta(t)$ (see Figure 2(b)). When the time $\widetilde{t}$ is small enough, the $II-$characteristic $x=\beta^-(t)$ (resp., $x=\beta^+(t)$) starting at $(\widetilde{x},\widetilde{t})$ will intersect at a point $(\widetilde{x}_0^-,0)$ (resp., $(\widetilde{x}_0^+,0)$) on the initial axis $t=0$ with $0>\widetilde{x}_0^->-\delta^-$ (resp., $0<\widetilde{x}_0^+<\delta^+$). Since the Riemann invariant $\zeta(u,v)=u$ must be a constant along $II-$characteristic $x=\beta^-(t)$ and $x=\beta^+(t)$, we have

$$\begin{aligned}\zeta\left(u^-(\beta^-(t),t),v^-(\beta^-(t),t)\right)=u^-(\beta^-(t),t)\\ =u^-(\widetilde{x},\widetilde{t})=\zeta\left(u_0^-(\widetilde{x}_0^-),v_0^-(\widetilde{x}_0^-)\right)=u_0^-(\widetilde{x}_0^-),\end{aligned} \tag{29}$$

and

$$\begin{aligned}\zeta\left(u^+(\beta^+(t),t),v^+(\beta^+(t),t)\right)=u^+(\beta^+(t),t)\\ =u^+(\widetilde{x},\widetilde{t})=\zeta\left(u_0^+(\widetilde{x}_0^+),v_0^+(\widetilde{x}_0^+)\right)=u_0^+(\widetilde{x}_0^+).\end{aligned} \tag{30}$$

Using Rankine-Hugoniot condition and the above two expressions, we obtain the propagation speed of the delta shock $x=x_\delta(t)$ at the point $(\widetilde{x},\widetilde{t})$

$$\begin{aligned}\sigma_\delta(\widetilde{x},\widetilde{t})=\left.\frac{dx_\delta(t)}{dt}\right|_{t=\widetilde{t}}=u^-(\widetilde{x},\widetilde{t})+u^+(\widetilde{x},\widetilde{t})\\ =u_0^-(\widetilde{x}_0^-)+u_0^+(\widetilde{x}_0^+).\end{aligned} \tag{31}$$

Together with (28) and (31), we get the entropy condition

$$\begin{aligned}\lambda_2 (u^+ (\tilde{x},\tilde{t}), v^+ (\tilde{x},\tilde{t})) &= 2u^+ (\tilde{x},\tilde{t}) = 2u_0^+ (\tilde{x}_0^+) \\ &< \lambda_1 (u^+ (\tilde{x},\tilde{t}), v^+ (\tilde{x},\tilde{t})) = u_0^+ (\tilde{x}_0^+) < \sigma_\delta (\tilde{x},\tilde{t}) \\ &= u_0^- (\tilde{x}_0^-) + u_0^+ (\tilde{x}_0^+) < \lambda_1 (u^- (\tilde{x},\tilde{t}), v^- (\tilde{x},\tilde{t})) \\ &= u_0^- (\tilde{x}_0^-) = u^- (\tilde{x},\tilde{t}) < \lambda_2 (u^- (\tilde{x},\tilde{t}), v^- (\tilde{x},\tilde{t})) \\ &= 2u_0^- (\tilde{x}_0^-) = 2u^- (\tilde{x},\tilde{t}),\end{aligned} \tag{32}$$

which is valid at any point $(\tilde{x},\tilde{t})$ on the curve $x = x_\delta(t)$ locally in time. Then we have the following.

Theorem 5. *In case of $\hat{u}^+ < 0 < \hat{u}^-$, the perturbed Riemann problem (1) and (3) has a delta shock solution constructed in (18) for local time. The admissible solution has a structure similar to that of the corresponding Riemann problem (1) and (5) (see Figure 2). Furthermore, the delta shock curve $x = x_\delta(t)$ possesses the following property:*

(a) *If $\dot{u}_0^-(0) < \dot{u}_0^+(0)$, then the curve is convex*

(b) *On the other hand, if $\dot{u}_0^-(0) > \dot{u}_0^+(0)$, then the curve is concave*

Proof. By Proposition 3 and inequality (32), we know that the delta shock solution constructed in (18) satisfies (1) and (3) in the sense of distributions and the entropy condition (27) for local time, respectively. Obviously, the delta shock curve $x = x_\delta(t)$ retains its form in a neighborhood of the origin $(0,0)$. Namely, the solution of the perturbed Riemann problem (1) and (3) has a structure similar to the corresponding Riemann solution of (1) and (5) for local time.

In order to show the behavior of the delta shock curve $x = x_\delta(t)$ near the origin, we need a certain a priori estimate on the value of $\ddot{x}_\delta(0)$. From (20), it can be easily checked that

$$\frac{dx_\delta (\tilde{t})}{d\tilde{t}} = u^- (x_\delta (\tilde{t}),\tilde{t}) + u^+ (x_\delta (\tilde{t}),\tilde{t}). \tag{33}$$

Differentiating the above equality with respect to $\tilde{t}$ and letting $\tilde{t} = 0$, one obtains

$$\ddot{x}_\delta (0) = \left.\frac{du^- (x_\delta (\tilde{t}),\tilde{t})}{d\tilde{t}}\right|_{\tilde{t}=0} + \left.\frac{du^+ (x_\delta (\tilde{t}),\tilde{t})}{d\tilde{t}}\right|_{\tilde{t}=0}. \tag{34}$$

Next, we compute the values of $(du^-(x_\delta(\tilde{t}),\tilde{t})/d\tilde{t})|_{\tilde{t}=0}$ and $(du^+(x_\delta(\tilde{t}),\tilde{t})/d\tilde{t})|_{\tilde{t}=0}$, respectively. From (1), it can be checked that

$$\begin{aligned}\frac{du}{dt} &= \frac{\partial u}{\partial t} + \frac{\partial u}{\partial x}\frac{dx}{dt} = -\frac{\partial u^2}{\partial x} + \frac{\partial u}{\partial x}\frac{dx}{dt} \\ &= -2u\frac{\partial u}{\partial x} + \frac{\partial u}{\partial x}\frac{dx}{dt} = \left(-2u + \frac{dx}{dt}\right)\frac{\partial u}{\partial x}.\end{aligned} \tag{35}$$

Along the delta shock curve $x = x_\delta(t)$, by using the fact that $\dot{x}_\delta(0) = \hat{u}^- + \hat{u}^+$ and condition (35), we have

$$\begin{aligned}\left.\frac{du^- (x_\delta (\tilde{t}),\tilde{t})}{d\tilde{t}}\right|_{\tilde{t}=0} &= (-2\hat{u}^- + \dot{x}_\delta (0)) \cdot \dot{u}_0^- (0) \\ &= (\hat{u}^+ - \hat{u}^-) \cdot \dot{u}_0^- (0)\end{aligned} \tag{36}$$

and

$$\begin{aligned}\left.\frac{du^+ (x_\delta (\tilde{t}),\tilde{t})}{d\tilde{t}}\right|_{\tilde{t}=0} &= (-2\hat{u}^+ + \dot{x}_\delta (0)) \cdot \dot{u}_0^+ (0) \\ &= (\hat{u}^- - \hat{u}^+) \cdot \dot{u}_0^+ (0).\end{aligned} \tag{37}$$

Substituting (36) and (37) into (34), we transform (34) into the form

$$\begin{aligned}\ddot{x}_\delta (0) &= (\hat{u}^+ - \hat{u}^-) \cdot \dot{u}_0^- (0) + (\hat{u}^- - \hat{u}^+) \cdot \dot{u}_0^+ (0) \\ &= (\hat{u}^+ - \hat{u}^-) \cdot (\dot{u}_0^- (0) - \dot{u}_0^+ (0)).\end{aligned} \tag{38}$$

With $\hat{u}^+ - \hat{u}^- < 0$ in mind, (38) shows the second derivative of the delta shock curve at origin $\ddot{x}(0) > 0$ when $\dot{u}_0^-(0) < \dot{u}_0^+(0)$; otherwise, $\ddot{x}(0) < 0$ when $\dot{u}_0^-(0) > \dot{u}_0^+(0)$. Thus the proof of Theorem 5 is completed. □

Case 2 $(\hat{u}^+ < 0 = \hat{u}^-)$. When $\dot{u}_0^-(0) < 0$, due to $\hat{u}^+ < 0 = \hat{u}^-$, there exist constants $\delta^- > 0$ and $\delta^+ > 0$ so small that, for any $x_0^- \in (-\delta^-, 0)$ and $x_0^+ \in (0, \delta^+)$, the C^1 functions $u_0^-(x)$ and $u_0^+(x)$ also satisfy the inequality (28). Then a discussion similar to that for Case 1 shows that the delta shock δS solution defined by (18) also satisfies Definition 1 and the entropy condition (27). The delta shock δS can retain its form in a neighborhood of the origin. That is, the perturbed Riemann solution of (1) and (3) has a structure similar to the Riemann solution of (1) and (5) for this case (see Figure 2).

When $\dot{u}_0^-(0) > 0$, we can prove (29)~(31) correspondingly for this subcase. In the same way as Case 1, by virtue of $\hat{u}^- = 0$, there exist constants $\delta^- > 0$ and $\delta^+ > 0$ so small that, for any $x_0^- \in (-\delta^-, 0)$ and $x_0^+ \in (0, \delta^+)$, we have

$$u_0^+ (x_c^-) < u_0^- (x_c^-) < 0. \tag{39}$$

Then, from (39), we get

$$\begin{aligned}\lambda_1 (u^+ (\tilde{x},\tilde{t}), v^+ (\tilde{x},\tilde{t})) &= u_0^+ (\tilde{x}_0^+) > \sigma_\delta (\tilde{x},\tilde{t}) \\ &= u_0^- (\tilde{x}_0^-) + u_0^+ (\tilde{x}_0^+)\end{aligned} \tag{40}$$

on the delta shock curve $x = x_\delta(t)$. Inequality (40) shows that the entropy condition (27) does not hold on the curve $x = x_\delta(t)$. Therefore, the perturbed Riemann solution to (1) and (3) should not be a delta shock solution for this subcase.

Now we prove that the perturbed Riemann solution should be a backward shock $\overleftarrow{S}$ followed by a contact discontinuity J for this subcase. The structure of the perturbed solution can be indicated in Figure 3(b), in which $\overleftarrow{S} : x = x_s(t)$ and $J : x = x_c(t)$ are free boundaries. Furthermore, on OA_1, we have

$$\frac{dx_s (t)}{dt} = u^- + v^*, \tag{41}$$

$$u^- v^* = u^* v^-. \tag{42}$$

On OA_2, we have

$$\frac{dx_c (t)}{dt} = u^*, \tag{43}$$

$$u^* = u^+. \tag{44}$$

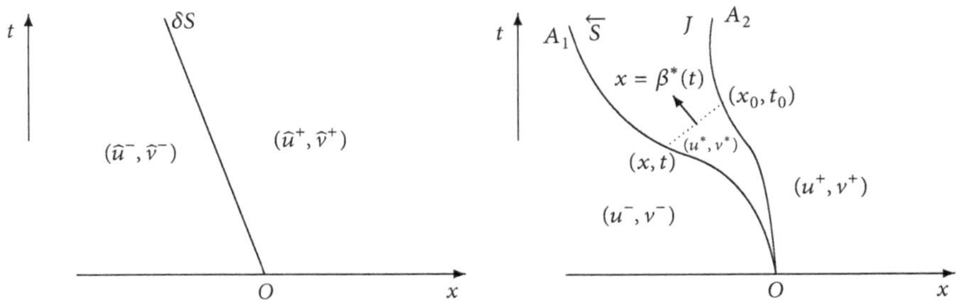

(a) The Riemann solution

(b) The perturbed Riemann solution

FIGURE 3: Case 2 with $\dot{u}_0^-(0) > 0$.

On the domain $\{(x,t) \mid x < x_s(t), 0 \le t < \epsilon\}$ ($\epsilon > 0$ so small), the perturbed Riemann solution to (1) and (3) is $(u^-(x,t), v^-(x,t))$. On the domain $\{(x,t) \mid x > x_c(t), 0 \le t < \epsilon\}$, the perturbed Riemann solution is $(u^+(x,t), v^+(x,t))$. On the domain $\{(x,t) \mid x_s(t) < x < x_c(t), 0 \le t < \epsilon\}$, the perturbed Riemann solution is denoted by $(u^*(x,t), v^*(x,t))$, which is an unknown regular solution to problem (1) and (3); moreover,

$$\lim_{(x,t)\longrightarrow(0,0)} u^*(x,t) = \widehat{u}^* = \widehat{u}^+, \qquad \lim_{(x,t)\longrightarrow(0,0)} v^*(x,t) = \widehat{v}^* = +\infty. \tag{45}$$

Noticing the corresponding results in [15, 16], the above perturbed Riemann problem is equivalent to the free boundary problem (1) with boundary conditions (41)~(44) on the fan-shaped domain $\{(x,t) \mid x_s(t) < x < x_c(t), 0 \le t < \epsilon\}$ ($\epsilon > 0$ so small).

Set

$$\begin{pmatrix} \mu^*(x,t) \\ \psi^*(x,t) \end{pmatrix} = \begin{pmatrix} l_2(\widehat{u}^*,\widehat{v}^*) \\ l_1(\widehat{u}^*,\widehat{v}^*) \end{pmatrix} \begin{pmatrix} u^*(x,t) \\ v^*(x,t) \end{pmatrix} = \begin{pmatrix} 1 & 0 \\ \widehat{v}^* & -\widehat{u}^* \end{pmatrix} \begin{pmatrix} u^*(x,t) \\ v^*(x,t) \end{pmatrix}. \tag{46}$$

Boundary condition (42) on $x = x_s(t)$ then reduces to

$$\psi^* = \frac{(u^-\widehat{v}^* - v^-\widehat{u}^*)\cdot \mu^*}{u^-}. \tag{47}$$

Boundary condition (44) on $x = x_c(t)$ can be written as

$$\mu^* = u^+. \tag{48}$$

Thus, the characterizing matrix H of this problem is of the form [15]

$$H = \begin{pmatrix} 0 & 0 \\ \dfrac{u^-\widehat{v}^* - v^-\widehat{u}^*}{u^-} & 0 \end{pmatrix}. \tag{49}$$

According to the local existence and uniqueness theorem (c.f. Chapter 6 in Li Ta-tsien and Yu Wen-ci [15]), if the minimal characterizing number

$$\|H\|_{\min} < 1, \tag{50}$$

then there exists a unique C^1 solution on the fan-shaped domain $\{(x,t) \mid x_s(t) < x < x_c(t), 0 \le t < \epsilon\}$, where $\epsilon > 0$ is suitably small.

By Remark 4.4 in the introduction of [15], it is not hard to prove that the minimal characterizing number of this problem for this case is

$$\|H\|_{\min} = 0 < 1. \tag{51}$$

Then the free boundary problem under consideration admits a unique piecewise smooth solution on the fan-shaped domain $\{(x,t) \mid x_s(t) < x < x_c(t), 0 \le t < \epsilon\}$. Hence we have the following.

Theorem 6. *In case of $\widehat{u}^+ < 0 = \widehat{u}^-$ and $\dot{u}_0^-(0) > 0$, the solution to the perturbed Riemann problem (1) and (3) is composed of a backward shock $x = x_s(t)$ followed by a contact discontinuity $x = x_c(t)$ for local time. The perturbed solution is dramatically different from the corresponding Riemann solution of (1) and (5), which is a delta shock (see Figure 3).*

Next, we proceed to prove that the perturbed Riemann problem (1) and (3) for this subcase admits a solution that contains a backward shock $\overleftarrow{S}$ and a contact discontinuity J near the origin in another way.

Let $x = \beta^*(t)$ be the upwards right II−characteristic from any point (x,t) on the shock curve $x = x_s(t)$ (see Figure 3(b)). The point (x_0,t_0) is the intersection point of the II−characteristic curve $x = \beta^*(t)$ and the contact discontinuity $x = x_c(t)$. Since the Riemann invariant $\zeta(u,v) = -u + v$ must be a constant along II−characteristic, we have

$$\zeta(u^*,v^*) = u^*(x,t) = u^*(\beta^*(t),t) = u^*(x_0,t_0) = u^*(x_s(t),t) = u^*(x_c(t_0),t_0). \tag{52}$$

From (52), it is easy to see that the propagating speed of II−characteristic

$$\frac{d\beta^*(t)}{dt} = 2u^*(\beta^*(t),t) = 2u^*(x_0,t_0) \tag{53}$$

is a constant. That is, the characteristic $x = \beta^*(t)$ is a straight line. Using the above equality, we arrive at

$$\frac{x_s(t) - x_c(t_0)}{t - t_0} = 2u^*(x_s(t),t) = 2u^*(x_c(t_0),t_0). \tag{54}$$

Noting (54), we get

$$x_s(t) - x_c(t_0) = 2(t - t_0) \cdot u^*(x_s(t), t). \quad (55)$$

Differentiate the above equation with respect to t and let $t = 0$; then one obtains

$$\dot{x}_s(0) - \dot{x}_c(0) \left.\frac{dt_0}{dt}\right|_{t=0} = 2\left(1 - \left.\frac{dt_0}{dt}\right|_{t=0}\right) \cdot \widehat{u}^*. \quad (56)$$

Substituting

$$\dot{x}_s(0) = \dot{x}_c(0) = \widehat{u}^+ = \widehat{u}^* \quad (57)$$

into (56), with $\widehat{u}^* = \widehat{u}^+ < 0$, we get

$$\left.\frac{dt_0}{dt}\right|_{t=0} = 1. \quad (58)$$

Moreover, differentiating the last equality in (52) with respect to t and letting $t = 0$, we obtain

$$\left.\frac{du^*(x_s(t), t)}{dt}\right|_{t=0} = \left.\frac{du^*(x_c(t_0), t_0)}{dt_0}\right|_{t_0=0} \cdot \left.\frac{dt_0}{dt}\right|_{t=0}. \quad (59)$$

In view of (58) and (59), it is easy to see that

$$\left.\frac{du^*(x_s(t), t)}{dt}\right|_{t=0} = \left.\frac{du^*(x_c(t_0), t_0)}{dt_0}\right|_{t_0=0}. \quad (60)$$

In the following, firstly, we will accumulate the second derivative of the shock at the origin $\ddot{x}_s(0)$. Along $x = x_s(t)$, differentiating the above equality (41) with respect to t and letting $t = 0$, by (44) and (52), we have

$$\ddot{x}_s(0) = \left.\frac{du^-(x_s(t), t)}{dt}\right|_{t=0} + \left.\frac{du^*(x_s(t), t)}{dt}\right|_{t=0}. \quad (61)$$

On the one hand, from (35), we have

$$\frac{du^-(x_s(t), t)}{dt} = \left(-2u^- + \frac{dx_s(t)}{dt}\right) \cdot \frac{\partial u^-}{\partial x}. \quad (62)$$

Noting (57), (62), and $\widehat{u}^- = 0$, one can get that

$$\left.\frac{du^-(x_s(t), t)}{dt}\right|_{t=0} = \widehat{u}^+ \cdot \dot{u}_0^-(0). \quad (63)$$

On the other hand, by (60) and (44), we have

$$\begin{aligned} \left.\frac{du^*(x_s(t), t)}{dt}\right|_{t=0} &= \left.\frac{du^*(x_c(t_0), t_0)}{dt_0}\right|_{t_0=0} \\ &= \left.\frac{du^+(x_c(t_0), t_0)}{dt_0}\right|_{t_0=0}. \end{aligned} \quad (64)$$

Along the contact discontinuity wave curve $x = x_c(t)$, from (35), (57), and (64), it follows that

$$\begin{aligned} \left.\frac{du^*(x_s(t), t)}{dt}\right|_{t=0} &= \left.\frac{du^*(x_c(t_0), t_0)}{dt_0}\right|_{t_0=0} \\ &= -\widehat{u}^+ \cdot \dot{u}_0^+(0). \end{aligned} \quad (65)$$

Then, substituting (63) and (65) into (61), we get

$$\ddot{x}_s(0) = \widehat{u}^+ \cdot ((\dot{u}_0^-(0) - \dot{u}_0^+(0)). \quad (66)$$

Secondly, we now have an estimate of the second derivative of the contact discontinuity wave curve at the origin, a.e. $\ddot{x}_c(0)$. Along $x = x_c(t)$, with (43), it holds that

$$\frac{dx_c(t_0)}{dt_C} = u^*(x_c(t_0), t_0). \quad (67)$$

Differentiating (67) with respect to t_0 and letting $t_0 = 0$ give

$$\ddot{x}_c(0) = \left.\frac{du^*(x_c(t_0), t_0)}{dt_0}\right|_{t_0=0} = -\widehat{u}^+ \cdot \dot{u}_0^+(0). \quad (68)$$

Thirdly, combining (66) and (68), it follows that

$$\ddot{x}_s(0) - \ddot{x}_c(0) = \widehat{u}^+ \cdot \dot{u}_0^-(0) < 0. \quad (69)$$

Finally, due to $x_s(0) = x_c(0) = 0$, (57), and (69), the perturbed Riemann solution to (1) and (3) in this subcase clearly consists of a backward shock $\overleftarrow{S}$ connecting (u^-, v^-) to (u^*, v^*), followed by a contact discontinuity J connecting (u^*, v^*) to (u^+, v^+) near the origin (see Figure 3(b)). Thus we have completed the construction and proof of the perturbed Riemann solution for this subcase.

Case 3 ($\widehat{u}^+ = 0 < \widehat{u}^-$). When $\dot{u}_0^+(0) < 0$, there exist constants $\delta^- > 0$ and $\delta^+ > 0$ so small that, for any $x_0^- \in (-\delta^-, 0)$ and $x_0^+ \in (0, \delta^+)$, the C^1 functions $u_0^-(x)$ and $u_0^+(x)$ also satisfy inequality (28). By a similar argument used in Case 1, we get that the delta shock solution defined by (18) satisfies the entropy condition (27). That is, in a neighborhood of the origin, the perturbed Riemann solution of (1) and (3) has a structure similar to the Riemann solution to (1) and (5) for this subcase (see Figure 2).

When $\dot{u}_0^+(0) > 0$, by virtue of $\widehat{u}^- \equiv 0$, there exist constants $\delta^- > 0$ and $\delta^+ > 0$ so small that, for any $x_0^- \in (-\delta^-, 0)$ and $x_0^+ \in (0, \delta^+)$, we have

$$0 < u_0^+(x_0^+) < u_0^-(x_0^-). \quad (70)$$

Then in the same way as Case 1, we can prove (29)~(31) corresponding for this subcase. Finally, from (70), we get

$$\begin{aligned} u_0^-(\overline{x}_0^-) + u_0^+(\overline{x}_0^+) &= \sigma_\delta(\overline{x}, \overline{t}) \\ &> \lambda_1(u^-(\overline{x}, \overline{t}), v^-(\overline{x}, \overline{t})) \\ &= u_0^-(\overline{x}_0^-) \end{aligned} \quad (71)$$

on the delta shock curve $x = x_\delta(t)$. Equality (53) shows that the entropy condition (27) does not hold for $x = x_\delta(t)$. Therefore, we should not expect the perturbed Riemann solution to (1) and (3) to be a delta shock solution for this subcase.

We will prove that the perturbed Riemann solution is a contact discontinuity J followed by a forward shock $\overrightarrow{S}$

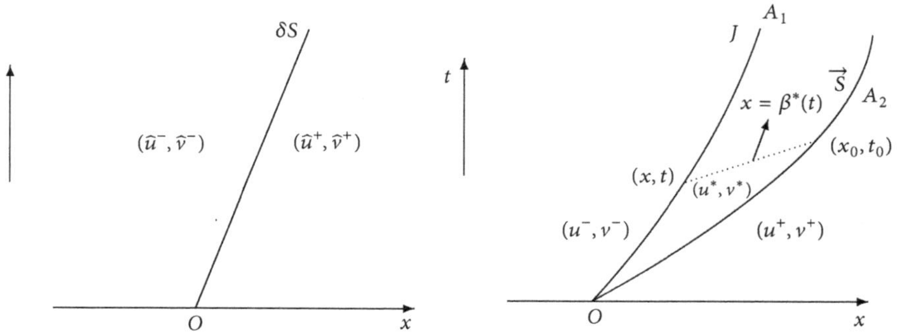

(a) The Riemann solution

(b) The perturbed Riemann solution

FIGURE 4: Case 3 with $\dot{u}_0^+(0) > 0$.

locally in time for this subcase. The structure of the perturbed solution to (1) and (5) can be indicated in Figure 4(b). As shown in Figure 4(b),

$$OA_1 : x = x_c(t) \quad (x_c(0) = 0) \tag{72}$$

and

$$OA_2 : x = x_s(t) \quad (x_s(0) = 0) \tag{73}$$

are free boundaries. Furthermore, OA_1 is a contact discontinuity, on which we have

$$\frac{dx_c(t)}{dt} = u^-, \tag{74}$$

$$u^- = u^*. \tag{75}$$

OA_2 is a forward shock, on which we have

$$\frac{dx_s(t)}{dt} = u^* + u^+, \tag{76}$$

$$u^* v^+ = v^* u^+. \tag{77}$$

On the domain $\{(x,t) \mid x < x_c(t), 0 \le t < \epsilon\}$ ($\epsilon > 0$ so small), the perturbed Riemann solution to (1) and (3) is $(u^-(x,t), v^-(x,t))$. On the domain $\{(x,t) \mid x > x_s(t), 0 \le t < \epsilon\}$, the perturbed Riemann solution is $(u^+(x,t), v^+(x,t))$. On the domain $\{(x,t) \mid x_c(t) < x < x_s(t), 0 \le t < \epsilon\}$, the perturbed Riemann solution is denoted by $(u^*(x,t), v^*(x,t))$, which is an unknown regular solution to problem (1) with (3) and satisfies condition (45).

Since $(u^-(x,t), v^-(x,t))$ and $(u^+(x,t), v^+(x,t))$ are known, in order to get the perturbed Riemann solution, we have to solve the free boundary problem (1) and (74)~(77) on the fan-shaped domain $\{(x,t) \mid x_c(t) < x < x_s(t), 0 \le t < \epsilon\}$ ($\epsilon > 0$ so small).

We introduce the change of variables

$$\begin{aligned}\begin{pmatrix} \mu^*(x,t) \\ \psi^*(x,t) \end{pmatrix} &= \begin{pmatrix} l_1(\hat{u}^*, \hat{v}^*) \\ l_2(\hat{u}^*, \hat{v}^*) \end{pmatrix} \begin{pmatrix} u^*(x,t) \\ v^*(x,t) \end{pmatrix} \\ &= \begin{pmatrix} \hat{v}^* & -\hat{u}^* \\ 1 & 0 \end{pmatrix} \begin{pmatrix} u^*(x,t) \\ v^*(x,t) \end{pmatrix}.\end{aligned} \tag{78}$$

Boundary condition (75) on $x = x_c(t)$ then reduces to

$$\psi^* = u^-. \tag{79}$$

Boundary condition (77) on $x = x_s(t)$ can be written as

$$\mu^* = \frac{(u^+\hat{v}^* - v^+\hat{u}^*) \cdot \psi^*}{u^+}. \tag{80}$$

Hence, the characterizing matrix H of this problem is of the form [15]

$$H = \begin{pmatrix} \dfrac{u^+\hat{v}^* - v^+\hat{u}^*}{u^+} & 0 \\ 0 & 0 \end{pmatrix}. \tag{81}$$

According to the local existence and uniqueness theorem, what remains is to prove that $\|H\|_{\min} < 1$.

By Remark 4.4 in the introduction of [15], it is not hard to prove that for this subcase

$$\|H\|_{\min} = 0 < 1. \tag{82}$$

Then the free boundary problem under consideration admits a unique piecewise C^1 solution on the fan-shaped domain $\{(x,t) \mid x_c(t) < x < x_s(t), 0 \le t < \epsilon\}$ ($\epsilon > 0$ so small). We can conclude the following.

Theorem 7. *In case of $\hat{u}^+ = 0 < \hat{u}^-$ and $\dot{u}_0^+(0) > 0$, the solution of the perturbed Riemann problem (1) and (3) is composed of a contact discontinuity $x = x_c(t)$ followed by a forward shock $x = x_s(t)$ for local time. The perturbed solution is different from the corresponding Riemann solution of (1) and (5), which is a delta shock (see Figure 4).*

Next, we proceed to prove that the perturbed Riemann problem (1) and (3) for this subcase admits a solution which contains a contact discontinuity $x = x_c(t)$ followed by a forward shock $x = x_s(t)$ near the origin. Let $x = \beta^*(t)$ be the upwards right II−characteristic from any point (x,t) on the contact discontinuity curve $x = x_c(t)$ (see Figure 4). The point (x_0, t_0) is the intersection point of the II−characteristic curve $x = \beta^*(t)$ and the shock curve $x = x_s(t)$. Since the

Riemann invariant $\zeta(u,v) = u$ must be a constant along II–characteristic, we have

$$\begin{aligned}\zeta(u^*,v^*) &= u^*(x,t) = u^*(\beta^*(t),t) = u^*(x_0,t_0)\\ &= u^*(x_c(t),t) = u^*(x_s(t_0),t_0).\end{aligned} \tag{83}$$

From (83), it is easy to see that the propagating speed of the II–characteristic

$$\frac{d\beta^*(t)}{dt} = 2u^*(\beta^*(t),t) = 2u^*(x_0,t_0) \tag{84}$$

is a constant. That is, the characteristic $x = \beta^*(t)$ is a straight line. Using the above equality, we arrive at

$$\frac{x_c(t) - x_s(t_0)}{t - t_0} = 2u^*(x_c(t),t) = 2u^*(x_s(t_0),t_0). \tag{85}$$

Noting (85), we get

$$x_c(t) - x_s(t_0) = (t - t_0)\cdot 2u^*(x_c(t),t). \tag{86}$$

Differentiate the above equation with respect to t and let $t = 0$; then one obtains

$$\dot{x}_c(0) - \dot{x}_s(0)\left.\frac{dt_0}{dt}\right|_{t=0} = \left(1 - \left.\frac{dt_0}{dt}\right|_{t=0}\right)\cdot 2\widehat{u}^-. \tag{87}$$

Substituting

$$\dot{x}_s(0) = \dot{x}_c(0) = \widehat{u}^- \tag{88}$$

into (87), with $\widehat{u}^- > 0$, we get

$$\left.\frac{dt_0}{dt}\right|_{t=0} = 1. \tag{89}$$

Moreover, differentiating the last equality in (83) with respect to t and letting $t = 0$, we obtain

$$\left.\frac{du^*(x_c(t),t)}{dt}\right|_{t=0} = \left.\frac{du^*(x_s(t_0),t_0)}{dt_0}\right|_{t_0=0}\cdot\left.\frac{dt_0}{dt}\right|_{t=0}. \tag{90}$$

From (89) and (90), we derive

$$\left.\frac{du^*(x_c(t),t)}{dt}\right|_{t=0} = \left.\frac{du^*(x_s(t_0),t_0)}{dt_0}\right|_{t_0=0}. \tag{91}$$

In the following, firstly, we will accumulate the second derivative of the contact discontinuity curve at the origin $\ddot{x}_c(0)$. Along $x = x_c(t)$, differentiating (74) with respect to t and letting $t = 0$, it yields

$$\ddot{x}_c(0) = \left.\frac{du^-(x_c(t),t)}{dt}\right|_{t=0}. \tag{92}$$

From (35), we have

$$\frac{du^-(x_c(t),t)}{dt} = \left(-2u^-(x_c(t),t) + \frac{dx_c(t)}{dt}\right)\cdot\frac{\partial u^-}{\partial x}. \tag{93}$$

Noting (88) and (93), one can get that

$$\ddot{x}_c(0) = \left.\frac{du^-(x_c(t),t)}{dt}\right|_{t=0} = -\widehat{u}^-\cdot\dot{u}_0^-(0). \tag{94}$$

Secondly, we now estimate the second derivative of the shock curve at the origin $\ddot{x}_s(0)$. Along $x = x_s(t)$, with (76), it holds that

$$\frac{dx_s(t_0)}{dt_0} = u^*(x_s(t_0),t_0) + u^+(x_s(t_0),t_0). \tag{95}$$

Differentiating (95) with respect to t_0 and letting $t_0 = 0$ give

$$\begin{aligned}\ddot{x}_s(0) = &\left.\frac{du^*(x_s(t_0),t_0)}{dt_0}\right|_{t_0=0}\\ &+ \left.\frac{du^+(x_s(t_0),t_0)}{dt_0}\right|_{t_0=0}.\end{aligned} \tag{96}$$

On the one hand, by (75) and (91), we get

$$\begin{aligned}\left.\frac{du^*(x_s(t_0),t_0)}{dt_0}\right|_{t_0=0} &= \left.\frac{du^*(x_c(t),t)}{dt}\right|_{t=0}\\ &= \left.\frac{du^-(x_c(t),t)}{dt}\right|_{t=0}.\end{aligned} \tag{97}$$

Along the delta shock curve $x = x_s(t)$, it follows from (35), (88), and (97) that

$$\begin{aligned}\left.\frac{du^*(x_s(t_0),t_0)}{dt_0}\right|_{t_0=0} &= \left.\frac{du^*(x_c(t),t)}{dt}\right|_{t=0}\\ &= -\widehat{u}^-\cdot\dot{u}_0^-(0).\end{aligned} \tag{98}$$

On the other hand, based on (35) and (88), we obtain

$$\left.\frac{du^+(x_s(t_0),t_0)}{dt_0}\right|_{t_0=0} = \widehat{u}^-\cdot\dot{u}_0^+(0). \tag{99}$$

We substitute (98) and (99) into (96) and get

$$\ddot{x}_s(0) = \widehat{u}^-\cdot\left(\dot{u}_0^+(0) - \dot{u}_0^-(0)\right). \tag{100}$$

Thirdly, combining (94) and (100), it follows that

$$\ddot{x}_s(0) - \ddot{x}_c(0) = \widehat{u}^-\cdot\dot{u}_0^+(0) > 0. \tag{101}$$

Finally, by virtue of $x_c(0) = x_s(0) = 0$, (88), and (101), the perturbed Riemann solution to (1) and (3) in this subcase clearly consists of a contact discontinuity J connecting (u^-,v^-) to (u^*,v^*), followed by a forward shock $\overrightarrow{S}$ connecting (u^*,v^*) to (u^+,v^+) near the origin. Thus we have completed the construction and proof of the perturbed Riemann solution for this subcase (see Figure 4).

4. Conclusion

So far, we have locally constructed the solution to the perturbed Riemann problem (1) and (3) with delta shock. From the above discussion, we discover a delta shock in the corresponding Riemann solution may turn into a shock and a contact discontinuity after the perturbation of Riemann initial data. This shows the internal mechanism and instability of a delta shock. Then we can obtain the main theorem of this paper.

Theorem 8. *There is a unique local solution to the perturbed Riemann problem (1) and (3) with delta shock. The Riemann solution of (1) and (5) is instable under a perturbation of the Riemann initial data, which is quite different from the previous perturbed Riemann problems.*

Conflicts of Interest

The authors declare that they have no conflicts of interest.

Acknowledgments

This paper was supported by NSF of China (11301264), China Scholarship Council, China Postdoctoral Science Foundation (2013M531343), the Fundamental Research Funds for the Central Universities (NZ2014107), Jiangsu Overseas Research & Training Program for University Prominent Young & Middle-aged Teachers and Presidents, and the NSF of Jiangsu Province of China (BK20130779).

References

[1] S. L. Yang and T. Zhang, "The MmB difference sliutions of 2-D Riemann problems for a 2×2 hyperbolic system of conservation laws," *Impact of Computing in Science and Engineering*, vol. 3, no. 2, pp. 146–180, 1991.

[2] D. C. Tan and T. Zhang, "Two-dimensional Riemann problem for a hyperbolic system of nonlinear conservation laws(I): four J's cases, (II):initial data consists of some rarefaction," *Journal of Differential Equations*, vol. 111, no. 2, pp. 203–254, 1994.

[3] H. Yang and Y. Zhang, "New developments of delta shock waves and its applications in systems of conservation laws," *Journal of Differential Equations*, vol. 252, no. 11, pp. 5951–5993, 2012.

[4] D. C. Tan, T. Zhang, and Y. X. Zheng, "Delta-shock waves as limits of vanishing viscosity for hyperbolic systems of conservation laws," *Journal of Differential Equations*, vol. 112, no. 1, pp. 1–32, 1994.

[5] G.-Q. Chen and H. Liu, "Formation of delta-shocks and vacuum states in the vanishing pressure limit of solutions to the isentropic Euler equations," *SIAM Journal on Mathematical Analysis*, vol. 34, no. 4, pp. 925–938, 2003.

[6] D. Korchinski, *Solutions of a Riemann Problem for a System of Conservation Laws Possessing No Classical Weak Solution, Thesis*, Adelphi University, 1977.

[7] J. Li, T. Zhang, and S. Yang, "The two-dimensional Riemann problem in gas dynamics," in *Pitman Monographs and Surveys in Pure and Applied Mathematics*, vol. 98, Longman, Harlow, England, 1998.

[8] M. Nedeljkov, "Delta and singular delta locus for one-dimensional systems of conservation laws," *Mathematical Methods in the Applied Sciences*, vol. 27, no. 8, pp. 931–955, 2004.

[9] V. M. Shelkovich, "The Riemann problem admitting δ- and δ'-shocks and vaccum states (the vanishing viscosity approach)," *Journal of Differential Equations*, vol. 231, no. 2, pp. 459–500, 2006.

[10] C. Shen and M. Sun, "Formation of delta shocks and vacuum states in the vanishing pressure limit of Riemann solutions to the perturbed Aw-RAScle model," *Journal of Differential Equations*, vol. 249, no. 12, pp. 3024–3051, 2010.

[11] W. Sheng and T. Zhang, "The Riemann problem for the transportation equations in gas dynamics," *Memoirs of the American Mathematical Society*, vol. 137, no. 654, pp. 1–77, 1999.

[12] J. Smoller, *Shock Waves and Reaction-Diffusion Equations*, Springer, New York, NY, USA, 1994.

[13] M. Sun, "Delta shock waves for the chromatography equations as self-similar viscosity limits," *Quarterly of Applied Mathematics*, vol. 69, no. 3, pp. 425–443, 2011.

[14] M. Sun, "Interactions of delta shock waves for the chromatography equations," *Applied Mathematics Letters*, vol. 26, no. 6, pp. 631–637, 2013.

[15] T. T. Li and W. C. Yu, *Boundary Value Problems for Quasilinear Hyperbolic Systems*, Duke University Mathematics, Durham, NC, England, 1985.

[16] T.-T. Li, D. Serre, and H. Zhang, "The generalized Riemann problem for the motion of elastic strings," *SIAM Journal on Mathematical Analysis*, vol. 23, no. 5, pp. 1189–1203, 1992.

[17] C. Shen and M. Sun, "Stability of the Riemann solutions for a nonstrictly hyperbolic system of conservation laws," *Nonlinear Analysis. Theory, Methods and Applications. An International Multidisciplinary Journal*, vol. 73, no. 10, pp. 3284–3294, 2010.

[18] A. Yao and W. Sheng, "Interaction of elementary waves on a boundary for a hyperbolic system of conservation laws," *Mathematical Methods in the Applied Sciences*, vol. 31, no. 11, pp. 1369–1381, 2008.

[19] C. Shen, "The Riemann problem for the Chaplygin gas equations with a source term," *ZAMM - Journal of Applied Mathematics and Mechanics/Zeitschrift für Angewandte Mathematik und Mechanik*, vol. 96, no. 6, pp. 681–695, 2016.

5

Higher-Spin Symmetries and Deformed Schrödinger Algebra in Conformal Mechanics

Francesco Toppan[1] **and Mauricio Valenzuela**[2]

[1]*CBPF, Rua Dr. Xavier Sigaud 150, Urca, cep 22290-180, Rio de Janeiro (RJ), Brazil*
[2]*Facultad de Ingeniería y Tecnología, Universidad San Sebastián, General Lagos 1163, Valdivia 5 110693, Chile*

Correspondence should be addressed to Francesco Toppan; toppan@cbpf.br

Academic Editor: Ziemowit Popowicz

The dynamical symmetries of 1 + 1-dimensional Matrix Partial Differential Equations with a Calogero potential (with/without the presence of an extra oscillatorial de Alfaro-Fubini-Furlan, DFF, term) are investigated. The first-order invariant differential operators induce several invariant algebras and superalgebras. Besides the $sl(2) \oplus u(1)$ invariance of the Calogero Conformal Mechanics, an $osp(2 \mid 2)$ invariant superalgebra, realized by first-order and second-order differential operators, is obtained. The invariant algebras with an infinite tower of generators are given by the universal enveloping algebra of the deformed Heisenberg algebra, which is shown to be equivalent to a deformed version of the Schrödinger algebra. This vector space also gives rise to a higher-spin (gravity) superalgebra. We furthermore prove that the pure and DFF Matrix Calogero PDEs possess isomorphic dynamical symmetries, being related by a similarity transformation and a redefinition of the time variable.

1. Introduction

In this paper we investigate the dynamical symmetries of 1 + 1-dimensional Matrix Partial Differential Equations with a Calogero [1] potential (with/without the presence of an extra oscillatorial de Alfaro-Fubini-Furlan, DFF, [2] term). The Matrix Partial Differential Equations correspond to Schrödinger Equations. The pure Calogero Hamiltonian is $\mathcal{N} = 2$ supersymmetric, while the DFF Hamiltonian is, as explained later, only "softly" supersymmetric. The first-order invariant differential operators induce several invariant algebras and superalgebras. Besides the $sl(2) \oplus u(1)$ invariance of the Calogero Conformal Mechanics, an $osp(2 \mid 2)$ invariant superalgebra, realized by first-order and second-order differential operators, is obtained. The invariant (super) algebras with an infinite tower of generators are given by the universal enveloping algebra of a deformed Heisenberg algebra [3–6]. It also contains a deformed version of the Schrödinger algebra (in both cases the deformation stems from the presence of a Klein operator). As a vector space the universal enveloping algebra carries the $\mathbb{Z}_2$-graded higher-spin superalgebra $q(2; \nu)$.

The higher-spin superalgebra $q(2; \nu)$ was introduced in [6] and applied to the construction of a Chern-Simons higher-spin gravity. In [6] it was also shown that $q(2; \nu)$ is provided by the universal enveloping algebra of the $osp(2 \mid 2)$ superalgebra. The $q(2; \nu)$ algebra was also employed to extend the Chern-Simons higher-spin gravity model [6] with fractional spin fields [7, 8].

We will show that these algebraic structures appear in the different context of nonrelativistic quantum mechanics in the presence of a Calogero potential, either with or without an extra harmonic potential term. We highlight the main features and results of our approach.

We use the standard tools (see, e.g., [9]) of dynamical symmetries of (matrix) PDEs. The operators under consideration are *local* (matrix) differential operators, in contrast to the nonlocal realizations given in [3–5] (see also [10]).

The construction based on an invariant PDE allows us to detect an invariant deformed Schrödinger algebra which has not been discussed in the existing literature. This deformation can be regarded as a deformation of the mass-central-charge of the algebra of Galileo boosts and translations by a term proportional to the Klein operator. This algebra is connected with Vasiliev's $q(2;\nu)$ superalgebra. As a vector space it is spanned by the same infinite set of differential operators which, on the other hand, are endowed with different brackets (ordinary commutators versus $\mathbb{Z}_2$-graded commutators). We furthermore prove that the models with pure Calogero potentials and those with the extra DFF term possess isomorphic dynamical symmetry (super) algebras. This results from the fact that a triplet of matrix PDEs close an $sl(2)$ algebra. In a given canonical form, the Matrix PDE associated with the positive root of $sl(2)$ corresponds to the Schrödinger equation of the pure Matrix Calogero system (it would have been tantamount to work with the negative root of $sl(2)$), while a canonical form exists such that the $sl(2)$ Cartan generator is associated with the Schrödinger equation for the DFF Matrix Calogero system. The two canonical forms are related by a similarity transformation coupled with a redefinition of the time variable.

The pure Matrix Calogero Hamiltonian is an $\mathcal{N} = 2$ Supersymmetric Quantum Mechanical System. On the other hand the DFF Matrix Calogero Hamiltonian is not supersymmetric (this is a common feature of superconformal models possessing an oscillatorial term; see [11]); it is, nevertheless, related to supersymmetric Hamiltonians via a shift realized by a diagonal operator; following [12] we can state that, in this case, the system enjoys a "soft" supersymmetry.

In the last part of the paper we discuss the appearance of the higher-spin superalgebra, which is now interpreted as a dynamical symmetry of the models under consideration (the higher-spin symmetry of the free nonrelativistic particle case was pointed out in [13]).

The scheme of the paper is as follows. In Section 2 the Matrix Calogero PDEs (with/without the DFF oscillatorial potential term) are introduced. It is explained why, in the presence of a Calogero potential, a Matrix PDE is needed to get (deformed) Heisenberg symmetry generators. The dynamical symmetry generators of these systems are computed. In Section 3 their dynamical symmetry subalgebras are investigated. In particular the $osp(2 \mid 2)$ superalgebra (realized by first-order and second-order differential symmetry operators) is derived, as well as the deformed Heisenberg algebra. The existence of an $\mathcal{N} = 2$ supersymmetry for the pure Matrix Calogero Hamiltonian and a "soft" supersymmetry for the DFF Matrix Calogero Hamiltonian is pointed out. The connection between pure and DFF Matrix Calogero PDEs is presented in Section 4. In Section 5 it is shown that both a deformed Schrödinger algebra and a higher-spin superalgebra appear as dynamical symmetries of (pure and DFF) Matrix Calogero PDEs. In the Conclusions we comment our results and sketch some directions for future investigations.

2. Matrix PDEs with Calogero Potentials

It is convenient to systematically review the arising of higher-spin (super) algebras in Calogero systems by analyzing the symmetry, realized by first-order differential operators, of the (matrix) Partial Differential Equations containing Calogero potentials. This analysis uses well-known methods.

To fix our conventions, if $\Omega = \Omega^\dagger$ is a Hermitian (matrix) PDE, a first-order differential operator Σ is a symmetry operator if it satisfies the equation

$$[\Sigma, \Omega] = \Phi_\Sigma \cdot \Omega, \tag{1}$$

for a given matrix-valued function Φ_Σ.

In our analysis we do not need to introduce symmetry operators which, unlike (1), are defined by a "right" convention; i.e., $[\Sigma, \Omega] = \Omega \cdot \Phi'_\Sigma$. One should note, on the other hand, that the symmetry operator Σ is not necessarily Hermitian (in that case its Hermitian conjugate $\Sigma^\dagger$ satisfies the "right" equation with $\Phi'_{\Sigma^\dagger} = -\Phi^*_\Sigma$).

Before investigating matrix PDEs (our results require a 2×2 matrix whose entries are differential operators) we recall some basic features of the standard Partial Differential Equations. The Schrödinger equation in $1+1$ dimension induces the maximal 6-generator Schrödinger algebra of invariant operators for three choices (up to normalization and trivial shift) of the potential $V(x)$ [14–17]. $V(x)$ is either constant (corresponding to the free equation), linear (giving rise to Airy's functions), or quadratic (corresponding to the harmonic oscillator). For all other choices of the potential the symmetry algebra possesses fewer generators. This is particularly the case for the Calogero potential.

2.1. The Ordinary Calogero PDEs. The pure Calogero PDE is defined by the operator

$$\Omega_{Cal} = i\partial_t + \frac{1}{2}\partial_x^2 - \frac{g}{x^2}, \tag{2}$$

where g is the coupling constant.

We mention that the operator (2) can be applied to describe the dynamics of a two-particle Calogero model with omitted center of mass degree of freedom. The operator also appears in the dynamics of a test particle in the proximity of the horizon of a Reissner-Nordström black hole; see [18]. The symmetry algebra for the Calogero potential is $sl(2) \oplus (1)$. It contains four generators. Two extra generators, corresponding to the Heisenberg subalgebra of the Schrödinger algebra, are no longer encountered in the presence of a nonvanishing, $g \neq 0$, Calogero potential. We have in this case, as suitably normalized symmetry generators,

$$\begin{aligned} c &= 1, \\ z_+ &= \partial_t, \\ z_0 &= -2t\partial_t - x\partial_x - \frac{1}{2}, \\ z_- &= -t^2\partial_t - tx\partial_x + i\frac{x^2}{2} - \frac{t}{2}. \end{aligned} \tag{3}$$

They satisfy the nonvanishing commutators

$$\begin{aligned} [z_0, z_\pm] &= \pm 2z_\pm, \\ [z_+, z_-] &= z_0, \end{aligned} \tag{4}$$

while c is the identity operator.

The nonvanishing commutators involving Ω_{Cal} are

$$\begin{aligned}[z_0, \Omega_{Cal}] &= 2\Omega_{Cal},\\ [z_-, \Omega_{Cal}] &= 2t\Omega_{Cal}.\end{aligned} \tag{5}$$

Based on (1), this means that

$$\begin{aligned}\Phi_{z_0} &= 2,\\ \Phi_{z_-} &= 2t.\end{aligned} \tag{6}$$

The conveniently chosen symmetry operators,

$$\begin{aligned}\overline{z}_+ &= iz_+,\\ \overline{z}_0 &= i(z_0 - c) \equiv i(z_0 - 1),\end{aligned} \tag{7}$$

are Hermitian.

One should note that the four (3) symmetry operators do not depend on the Calogero coupling constant g.

In the case of the Calogero potential with the extra addition (see [2]) of the DFF quadratic term, we can introduce without loss of generality (by suitably normalizing the frequency of the harmonic oscillator), the Ω_{DFF} operator given by

$$\Omega_{DFF} = i\partial_t + \frac{1}{2}\partial_x^2 - \frac{g}{x^2} - \frac{1}{2}x^2. \tag{8}$$

In this case as well we obtain a $sl(2) \oplus u(1)$ symmetry algebra. Its four generators are

$$\begin{aligned}\widehat{c} &= 1,\\ \widehat{z}_+ &= \frac{1}{2}e^{-2it}\left(\partial_t - ix\partial_x + ix^2 - \frac{i}{2}\right),\\ \widehat{z}_0 &= i\partial_t,\\ \widehat{z}_- &= \frac{1}{2}e^{2it}\left(\partial_t + ix\partial_x + ix^2 + \frac{i}{2}\right).\end{aligned} \tag{9}$$

The nonvanishing commutators are

$$\begin{aligned}[\widehat{z}_0, \widehat{z}_\pm] &= \pm 2\widehat{z}_\pm,\\ [\widehat{z}_+, \widehat{z}_-] &= \widehat{z}_0.\end{aligned} \tag{10}$$

It is worth pointing out that, when $g \neq 0$, for no other nonconstant $w(x)$, the potential $g/x^2 + w(x)$ produces a 4-generator symmetry algebra. In particular, for a linear $w(x)$ ($w(x) = kx$, $k \neq 0$), we only recover a 2-generator symmetry algebra.

The arising of a (generalized) Heisenberg and of higher-spin (super) algebras requires a Matrix Calogero PDE. For this purpose a 2×2 matrix system is sufficient.

2.2. The 2×2-Matrix Calogero PDEs. We check under which conditions we can obtain off-diagonal symmetry generators for a 2×2-matrix differential operator containing Calogero potentials.

Let us denote with e_{ij} $(i, j = 1, 2)$ the matrix with entry $+1$ at the i-th column and j-th row and 0 otherwise. We are looking for symmetry generators of the matrix PDE defined by the diagonal operator

$$\Omega = (e_{11} + e_{22})\left(i\partial_t + \frac{1}{2}\partial_x^2\right) - v_1(x)e_{11} - v_2(x)e_{22}. \tag{11}$$

Since Ω is a diagonal operator, from either (3) or (9), we obtain $4 + 4$ diagonal symmetry generators (4 of them are associated with the upper diagonal element e_{11}; the remaining 4 generators are associated with the lower diagonal element e_{22}).

Concerning the nondiagonal symmetry operators, they should be expressed as

$$\begin{aligned}\Sigma_{up} &= (f(x,t)\partial_t + g(x,t)\partial_x + h(x,t))e_{12},\\ \Sigma_{down} &= \left(\overline{f}(x,t)\partial_t + \overline{g}(x,t)\partial_x + \overline{h}(x,t)\right)e_{21},\end{aligned} \tag{12}$$

for appropriate functions $f(x,t)$, $g(x,t)$, $h(x,t)$, $\overline{f}(x,t)$, $\overline{g}(x,t)$, and $\overline{h}(x,t)$.

We assume the potentials $v_1(x)$, $v_2(x)$ to be

$$v_i(x) = \frac{a_i}{x^2} + b_i x^2 + c_i, \tag{13}$$

with $a_2 \neq a_1$. Without loss of generality we can set, via similarity transformations, $c_1 = c_2 = 0$.

It is easily realized that the requirement

$$b_2 = b_1 \tag{14}$$

is implied by the existence of a nonvanishing solution Σ_{up} (Σ_{down}) for the respective equations

$$\begin{aligned}[\Sigma_{up}, \Omega] &= 0,\\ [\Sigma_{down}, \Omega] &= 0.\end{aligned} \tag{15}$$

The constraint on the functions entering Σ_{up}, Σ_{down} is derived with standard techniques. The solutions of constraint (15) for the functions entering (12) are presented here. We get, for Σ_{up},

$$\begin{aligned}&\partial_x\partial_t : f_x = 0,\\ &\partial_x^2 : g_x = 0,\\ &\partial_t : -i\dot{f} - \frac{1}{2}f_{xx} + (v_1 - v_2)f = 0 \Longrightarrow f = 0 \quad \text{for } v_2 \neq v_1,\\ &\partial_x : -i\dot{g} - h_x + (v_1 - v_2) = 0,\\ &1 : -i\dot{h} - \frac{1}{2}h_{xx} + (v_1 - v_2)h - gv_{2x} = 0.\end{aligned} \tag{16}$$

Similar equations are obtained for $\overline{f}(x,t)$, $\overline{g}(x,t)$, $\overline{h}(x,t)$ entering Σ_{down}.

The above system of equations tells us that, in particular, the condition

$$a_1 + a_2 - a_1^2 - a_2^2 + 2a_1a_2 = 0 \tag{17}$$

has to be fulfilled. After setting $\nu = 2(a_2 - a_1)$, it is solved by

$$a_1 = \frac{1}{8}\nu(\nu - 2), \quad a_2 = \frac{1}{8}\nu(\nu + 2). \tag{18}$$

We can therefore write the most general (11) operator admitting off-diagonal symmetry operators. It depends on the parameter ϵ, whose values are either $\epsilon = 0$ for the pure Calogero system, or $\epsilon = 1$ for the Calogero system with the extra oscillatorial term. Due to (14), the oscillatorial term is the same in both upper and lower diagonal entries. We can write, with a proper normalization,

$$\Omega_\epsilon = (e_{11} + e_{22})\left(i\partial_t + \frac{1}{2}\partial_x^2 - \frac{1}{2}\epsilon x^2 - \frac{\nu^2}{8x^2}\right) + (e_{11} - e_{22})\frac{\nu}{4x^2}. \tag{19}$$

The operator K,

$$K = e_{11} - e_{22}, \tag{20}$$

plays different roles, depending on the context. It is either the Fermion Parity Operator (in Supersymmetric Quantum Mechanics) or the Klein operator (entering the definition of the deformed oscillators). We note that, in particular, $K^2 = \mathbb{I}_2$.

In both $\epsilon = 0, 1$ cases we get two upper triangular and two lower triangular symmetry operators.

At $\epsilon = 0$ we have, up to normalization,

$$\begin{aligned}\Sigma_1 &= e_{12}\left(\partial_x + \frac{\nu}{2x}\right),\\ \Sigma_2 &= e_{12}\left(t\partial_x + t\frac{\nu}{2x} - ix\right),\\ \Sigma_3 &= e_{21}\left(\partial_x - \frac{\nu}{2x}\right),\\ \Sigma_4 &= e_{21}\left(t\partial_x - t\frac{\nu}{2x} - ix\right).\end{aligned} \tag{21}$$

At $\epsilon = 1$ we have, up to normalization,

$$\begin{aligned}\Xi_1 &= e_{12}e^{it}\left(\partial_x + \frac{\nu}{2x} + x\right),\\ \Xi_2 &= e_{12}e^{-it}\left(\partial_x + \frac{\nu}{2x} - x\right),\\ \Xi_3 &= e_{21}e^{it}\left(\partial_x - \frac{\nu}{2x} + x\right),\\ \Xi_4 &= e_{21}e^{-it}\left(\partial_x - \frac{\nu}{2x} - x\right).\end{aligned} \tag{22}$$

3. Symmetry Superalgebra of the Pure Calogero Matrix System

For the 2×2 pure Calogero Matrix (19) $\Omega_{\epsilon=0}$ operator we can introduce the basis of 4 off-diagonal Hermitian operators. They are given by

$$Q_1 = \frac{i}{\sqrt{2}}(\Sigma_1 + \Sigma_3), \tag{23}$$

$$Q_2 = iKQ_1 = \frac{1}{\sqrt{2}}(\Sigma_3 - \Sigma_1), \tag{24}$$

$$\widetilde{Q}_1 = \frac{i}{\sqrt{2}}(\Sigma_2 + \Sigma_4), \tag{25}$$

$$\widetilde{Q}_2 = iK\widetilde{Q}_1 = \frac{1}{\sqrt{2}}(\Sigma_4 - \Sigma_2). \tag{26}$$

By construction

$$Q_i^\dagger = Q_i, \quad \widetilde{Q}_i^\dagger = \widetilde{Q}_i \quad (i = 1, 2). \tag{27}$$

Since, for K given by (20), we have

$$\{K, Q_i\} = \{K, \widetilde{Q}_i\} = 0, \tag{28}$$

the quantum mechanical system defined by $\Omega_{\epsilon=0}$ is $\mathcal{N} = 2$ supersymmetric. Indeed,

$$\{Q_i, Q_j\} = 2\delta_{ij}\overline{\mathbf{H}}, \tag{29}$$

where

$$\overline{\mathbf{H}} = \mathbb{I}_2\left(-\frac{1}{2}\partial_x^2 + \frac{\nu^2}{8x^2}\right) + \frac{\nu}{4x^2}K. \tag{30}$$

The following nonvanishing anticommutators are recovered:

$$\begin{aligned}\{Q_i, Q_j\} &= 2\delta_{ij}\overline{\mathbf{H}},\\ \{\widetilde{Q}_i, \widetilde{Q}_j\} &= 2\delta_{ij}\overline{\mathbf{K}},\\ \{Q_i, \widetilde{Q}_j\} &= \delta_{ij}\overline{\mathbf{D}} + \epsilon_{ij}\overline{\mathbf{J}},\end{aligned} \tag{31}$$

where, besides $\overline{\mathbf{H}}$ given in (30), we have

$$\overline{\mathbf{D}} = \mathbb{I}_2\left(-t\partial_x^2 + ix\partial_x + \frac{i}{2} + t\frac{\nu^2}{4x^2}\right) - Kt\frac{\nu}{2x^2}, \tag{32}$$

$$\overline{\mathbf{K}} = \frac{1}{2}\mathbb{I}_2\left(-t^2\partial_x^2 + 2itx\partial_x + x^2 + it + t^2\frac{\nu^2}{4x^2}\right) - \frac{1}{4}Kt^2\frac{\nu}{x^2}, \tag{33}$$

$$\overline{\mathbf{J}} = \frac{1}{2}(K + \nu\mathbb{I}_2). \tag{34}$$

The four odd (fermionic) operators $Q_i, \widetilde{Q}_i$, together with the four even (bosonic) operators $\overline{\mathbf{H}}, \overline{\mathbf{D}}, \overline{\mathbf{K}}, \overline{\mathbf{J}}$, close the finite simple Lie superalgebra $osp(2 \mid 2)$, as it can be easily checked. By construction $\overline{\mathbf{H}}, \overline{\mathbf{D}}, \overline{\mathbf{K}}, \overline{\mathbf{J}}$ commute with $\Omega_{\epsilon=0}$. Therefore, $osp(2 \mid 2)$ is an invariant subalgebra of the 2×2 pure Matrix Calogero system.

We have indeed, for any generator $\mathfrak{g} \in osp(2 \mid 2)$,

$$[\mathfrak{g}, \Omega_{\epsilon=0}] = 0. \tag{35}$$

The set of generators $Q_1, \widetilde{Q}_1, \overline{\mathbf{H}}, \overline{\mathbf{D}}, \overline{\mathbf{K}}$ (and, similarly, $Q_2, \widetilde{Q}_2, \overline{\mathbf{H}}, \overline{\mathbf{D}}, \overline{\mathbf{K}}$) close an $osp(1 \mid 2)$ superalgebra. $Q_1, \widetilde{Q}_1$ (respectively, $Q_2, \widetilde{Q}_2$) are odd generators. If, nevertheless, we compute their commutators, we obtain

$$[Q_1, \widetilde{Q}_1] = [Q_2, \widetilde{Q}_2] = \frac{i}{2}(\mathbb{I}_2 + \nu K), \tag{36}$$

therefore recovering the (ν-)deformed Heisenberg algebra.

For future convenience we also consider the non-Hermitian linear combinations

$$\begin{aligned} \overline{A} &= i(Q_1 - \widetilde{Q}_1), \\ \overline{A}^\dagger &= Q_1 + \widetilde{Q}_1, \end{aligned} \tag{37}$$

which satisfy

$$[\overline{A}, \overline{A}^\dagger] = \mathbb{I}_2 + \nu K. \tag{38}$$

4. Connection between Pure and DFF Calogero Matrix Systems

It is convenient to normalize the $sl(2)$ root symmetry generators of the pure Calogero system Ω_{Cal}, given in (2), according to $z'_\pm = \mp i z_\pm$, with $z_\pm$ entering (3). The corresponding symmetry operators for the pure Calogero Matrix system $\Omega_{\epsilon=0}$, given in (19), are $z'_\pm \mathbb{I}_2$.

It is convenient to redefine $\Omega_{\epsilon=0}$ as

$$\Omega_{+1} := \Omega_{\epsilon=0}. \tag{39}$$

Indeed, a triplet of operators, $\Omega_{\pm 1}, \Omega_0$, carrying a $sl(2)$ representation generated by $z'_\pm \mathbb{I}_2, z_0 \mathbb{I}_2$, is encountered. We have

$$\begin{aligned} [z'_- \mathbb{I}_2, \Omega_{+1}] &= \Omega_0 = 2it\Omega_{+1}, \\ [z'_- \mathbb{I}_2, \Omega_0] &= \Omega_{-1} = -2t^2 \Omega_{+1}. \end{aligned} \tag{40}$$

One should note that the commutator $[z'_- \mathbb{I}_2, \Omega_{-1}] = 0$ is vanishing, so that no further operator is generated.

The operators $\Omega_{\pm 1}, \Omega_0$ close an $sl(2)$ algebra. We have, indeed,

$$\begin{aligned} [\Omega_0, \Omega_{\pm 1}] &= \pm 2\Omega_{\pm 1}, \\ [\Omega_{+1}, \Omega_{-1}] &= -2\Omega_0. \end{aligned} \tag{41}$$

The connection between the pure 2×2 Matrix Calogero system obtained from (19) at $\epsilon = 0$ and the 2×2 Matrix Calogero-DFF system obtained from (19) at $\epsilon = 1$ can now be made explicit, by adapting to the present case the construction discussed in [19]. Under redefinition of the time coordinate and a similarity transformation, the suitably normalized Cartan operator $N\Omega_C$ (N is a constant) is mapped into the DFF $\Omega_{\epsilon=1}$ operator. As a corollary, the symmetry operators of the pure Calogero system ($\epsilon = 0$) are mapped into the symmetry operators of the Calogero system with the oscillatorial ($\epsilon = 1$) term.

The new time variable τ is introduced from the equation

$$t = Ce^{2i\tau}, \tag{42}$$

(C is here a suitable constant), such that the time-derivative operator entering $N\Omega_0$ can be expressed as $2iNt\partial_t = i\partial_\tau$. This is accomplished by choosing $N = i, C = -i/2$.

We point out that for our purposes the time t can be complexified and the isomorphism established; a reality condition imposed on the new parameter τ entering (42) allows interpreting it as the time for the Hamiltonian with oscillatorial term.

The further similarity transformation which maps the symmetry generators of $\Omega_{\epsilon=0}$ into the symmetry generators of $\Omega_{\epsilon=1}$ is given by

$$\mathfrak{g} \longmapsto \check{\mathfrak{g}} = e^{S_2} e^{S_1} \mathfrak{g} e^{-S_1} e^{-S_2}, \tag{43}$$

where

$$\begin{aligned} S_1 &= i\tau\left(x\partial_x - \frac{1}{2}\right), \\ S_2 &= \frac{1}{2}x^2. \end{aligned} \tag{44}$$

It is rather straightforward to check that, in particular, we get

$$\begin{aligned} i\Omega_0 &\longmapsto i\check{\Omega}_0 \\ &= \mathbb{I}_2\left(i\partial_\tau + \frac{1}{2}\partial_x^2 - \frac{1}{8}\frac{\nu^2}{x^2} - \frac{1}{2}x^2\right) + K\frac{1}{4}\frac{\nu}{x^2}. \end{aligned} \tag{45}$$

The similarity transformation (43) and the time redefinition (42) apply to all 2×2 matrix symmetry generators. We have, in particular,

$$\begin{aligned} \Sigma_1 &\longmapsto \Xi_2 = e_{12} e^{-i\tau}\left(\partial_x + \frac{\nu}{2x} - x\right), \\ \Sigma_3 &\longmapsto \Xi_4, \\ \Sigma_2 &\longmapsto -\frac{i}{2}\Xi_1, \\ \Sigma_4 &\longmapsto -\frac{i}{2}\Xi_3. \end{aligned} \tag{46}$$

The operators in the r.h.s. coincide, for the time variable τ, with the corresponding operators given in (22). It follows, in particular, that $osp(2 \mid 2)$ is a symmetry superalgebra of the matrix system with the oscillatorial term. Since the $osp(2 \mid 2)$ generators are connected by similarity transformations their (anti)commutation relations are preserved.

It is convenient to define the Hermitian generators $\widehat{Q}_1, \widehat{Q}_2, \widehat{Q}'_1, \widehat{Q}'_2$ through the positions

$$\begin{aligned}\widehat{Q}_1 &= \frac{i}{\sqrt{2}}\left(\Xi_1+\Xi_4\right),\\ \widehat{Q}_2 &= \frac{1}{\sqrt{2}}\left(\Xi_4-\Xi_1\right) = iK\widehat{Q}_1,\\ \widehat{Q}'_1 &= \frac{i}{\sqrt{2}}\left(\Xi_2+\Xi_3\right),\\ \widehat{Q}'_2 &= \frac{1}{\sqrt{2}}\left(\Xi_3-\Xi_2\right) = iK\widehat{Q}'_1.\end{aligned} \tag{47}$$

In the presence of the oscillatorial term we obtain two $\mathcal{N}=2$ supersymmetric quantum mechanics, since

$$\begin{aligned}\left\{\widehat{Q}_i, \widehat{Q}_j\right\} &= 2\delta_{ij}\widehat{H},\\ \left\{\widehat{Q}'_i, \widehat{Q}'_j\right\} &= 2\delta_{ij}\widehat{H}'.\end{aligned} \tag{48}$$

The supersymmetric Hamiltonians are, respectively, given by

$$\begin{aligned}\widehat{H} &= \mathbb{I}_2\left(-\frac{1}{2}\partial_x^2+\frac{1}{2}x^2+\frac{1}{8}\frac{\nu^2}{x^2}+\frac{\nu}{2}\right)\\ &\quad+\frac{1}{2}K\left(1-\frac{\nu}{2x^2}\right),\\ \widehat{H}' &= \mathbb{I}_2\left(-\frac{1}{2}\partial_x^2+\frac{1}{2}x^2+\frac{1}{8}\frac{\nu^2}{x^2}-\frac{\nu}{2}\right)\\ &\quad-\frac{1}{2}K\left(1+\frac{\nu}{2x^2}\right).\end{aligned} \tag{49}$$

The supersymmetric Hamiltonians correspond to a shift of the original H_{osc} Hamiltonian with oscillatorial term, whose expression is

$$H_{osc} = \mathbb{I}_2\left(-\frac{1}{2}\partial_x^2+\frac{1}{2}x^2+\frac{1}{8}\frac{\nu^2}{x^2}\right)-\frac{1}{4}K\frac{\nu}{x^2}. \tag{50}$$

Indeed, we have

$$\begin{aligned}\widehat{H} &= H_{osc}+\frac{1}{2}\left(K+\mathbb{I}_2\nu\right),\\ \widehat{H}' &= H_{osc}-\frac{1}{2}\left(K+\mathbb{I}_2\nu\right).\end{aligned} \tag{51}$$

The shift operators $\pm(1/2)(K+\mathbb{I}_2\nu)$ commute with H_{osc}; we have $[H_{osc}, (K+\mathbb{I}_2\nu)] = 0$.

An $osp(1 \mid 2)$ symmetry superalgebra can be expressed in terms of the operators A, $A^\dagger$ and their anticommutators, where

$$\begin{aligned}A &= \frac{i}{\sqrt{2}}\left(\Xi_1+\Xi_3\right),\\ A^\dagger &= \frac{i}{\sqrt{2}}\left(\Xi_2+\Xi_4\right).\end{aligned} \tag{52}$$

The Hamiltonian H_{osc} is recovered from the anticommutator

$$\left\{A, A^\dagger\right\} = 2H_{osc}. \tag{53}$$

The remaining two operators $(E_\pm)$ that, together with $A, A^\dagger, H_{osc}$, close the $osp(1 \mid 2)$ superalgebra are given by

$$\begin{aligned}\{A, A\} &:= E_+ = e^{2it}\left[\mathbb{I}_2\left(-\partial_x^2-2x\partial_x-x^2+\frac{\nu^2}{4x^2}-1\right)\right.\\ &\quad\left.-K\frac{\nu}{2x^2}\right]\\ \left\{A^\dagger, A^\dagger\right\} &:= E_-\\ &= e^{-2it}\left[\mathbb{I}_2\left(-\partial_x^2+2x\partial_x-x^2+\frac{\nu^2}{4x^2}+1\right)\right.\\ &\quad\left.-K\frac{\nu}{2x^2}\right].\end{aligned} \tag{54}$$

The operators $A, A^\dagger$ induce the deformed Heisenberg algebra since their commutator is given by

$$\left[A, A^\dagger\right] = \mathbb{I}_2+\nu K. \tag{55}$$

We point out that, following [12], the Hilbert space can be taken to be given by a pair of square-integrable functions on the real line. It is then given by a direct sum of lowest-weight representations of the $osp(2 \mid 2)$ spectrum-generating superalgebra, with lowest weights determined by ν. Unlike the $(1+2)$-dimensional theories discussed in [12, 20], ν is not quantized to be an integer or a half-integer value. One should further note that, for $(1+1)$-dimensional 4×4 matrix PDEs, the spectrum-generating superalgebra itself, not just its representations, is determined by ν, being given $D(2,1;\alpha)$, with a linear relation between α and ν (see [12]).

5. Deformed Schrödinger and Higher-Spin (Super) Algebras

We are now in the position to discuss the arising of infinitely generated deformed Schrödinger algebra and higher-spin superalgebra as dynamical symmetries of the pure and DFF Matrix Calogero PDEs. Pure and DFF Matrix Calogero dynamical symmetries are isomorphic, since they are related by the time redefinition (42) and the similarity transformation (43). It is therefore sufficient to present the results for the pure Matrix Calogero case.

The introduction of the deformed Schrödinger algebra requires to couple an $sl(2)\oplus u(1)$ invariant subalgebra to a pair of deformed Heisenberg oscillators. Several invariant deformed Schrödinger algebras can be recovered as a consequence. Indeed we can pick, e.g., as $sl(2)\oplus u(1)$ subalgebra, either the first-order differential operators $g\cdot\mathbb{I}_2$, with g denoting one of the four operators entering (3), or the second-order differential operators entering the even sector of the $osp(2 \mid 2)$ superalgebra and recovered from (30) and (32)-(34). Similarly, up to normalization, the deformed

Heisenberg operators can be given by several possible pairs. We have, e.g., $Q_1, \widetilde{Q}_1$, $Q_2, \widetilde{Q}_2$, or $\overline{A}, \overline{A}^\dagger$, whose respective deformed Heisenberg commutators are given in formulas (36) and (38). The choice $\mathcal{Q}_1 = \sqrt{2}\widetilde{Q}_1$, $\mathcal{Q}_2 = -\sqrt{2}Q_1$, so that

$$\begin{aligned} \mathcal{Q}_1 &= (e_{12} + e_{21})(it\partial_x + x) + (e_{12} - e_{21})\, it\frac{\nu}{2x}, \\ \mathcal{Q}_2 &= -(e_{12} + e_{21})\, i\partial_x - (e_{12} - e_{21})\, i\frac{\nu}{2x}, \end{aligned} \tag{56}$$

with

$$[\mathcal{Q}_1, \mathcal{Q}_2] = i(\mathbb{I}_2 + \nu K), \quad (K = e_{11} - e_{22}), \tag{57}$$

is particularly convenient. The limit $\nu \longrightarrow 0$ exists. At $\nu = 0$ (in the absence of the Calogero potential) they coincide, respectively, with the ordinary Galileo boost and space translation multiplied by $e_{12} + e_{21}$ (namely, the operator exchanging the fields entering a 2-component multiplet). The closure, at $\nu = 0$, of the ordinary $1+1$-dimensional Schrödinger algebra, with the extra $sl(2) \oplus u(1)$ first-order differential operators $g \cdot \mathbb{I}_2$ introduced above, immediately follows.

We should recall that, for the original Calogero and de Alfaro-Fubini-Furlan models, the wavefunctions $\Psi(x)$ in [1, 2] (see also [21]) are defined on the half-line and respect the Dirichlet boundary condition $\Psi(0) = 0$ at the origin. It was shown in [22, 23] that, in a certain range of the Calogero coupling constant g, a more general class of self-adjoint domain for the one-dimensional Calogero Hamiltonian is allowed. It consists of square-integrable functions defined on the real line (see [22, 23] for details). The construction of the Hilbert space for the Matrix Calogero and de Alfaro-Fubini-Furlan oscillators induced by a spectrum-generating superalgebra was presented in [12, 24]. We summarize here the main properties. The Hilbert space is given by an n-ple of square-integrable functions on the line. It is determined by a direct sum of the superalgebra lowest-weight representations. The lowest-weight vectors are defined by imposing the condition $A\Psi_{lw} = 0$, A being an annihilation operator (in the present work A is the annihilation operator defined in (52)). A simple inductive proof shows that

(i) if Ψ_{lw} is normalized as a n-ple of square-integrable functions, all created states $(A^\dagger)^N \Psi_{lw}$ are normalized as n-ple of square-integrable functions for any positive integer N

(ii) if $\Psi_{lw}(x) \longrightarrow 0$ for $x \longrightarrow 0$, for any positive integer N all created states $(A^\dagger)^N \Psi_{lw}(x) \longrightarrow 0$ in the $x \longrightarrow 0$ limit (this lowest-weight representation can be interpreted to be formed by states belonging to the half-line and satisfying the Dirichlet condition at the origin)

The analysis of [12, 24], applied to the (19) operator, proves that

(i) for $\nu > -1$ the Hilbert space can be selected to be a lowest-weight representation with bosonic lowest-weight

(ii) for $\nu < 1$ the Hilbert space can be selected to be a lowest-weight representation with fermionic lowest-weight

(iii) in the range $-1 < \nu < 1$ an alternative choice for the Hilbert space consists in selecting it to be the direct sum of the two (the bosonic and the fermionic) lowest-weight representations

At $\nu \neq 0$ the appearance in the r.h.s. of (57) of the Klein operator K implies that (an infinite tower of) new generators must be introduced to close a Lie algebra. Indeed, for $\alpha = 1, 2$, we have $[\mathcal{Q}_\alpha, K] = 2\mathcal{Q}_\alpha K$, so that $[\mathcal{Q}_\alpha, \mathcal{Q}_\alpha K] = 2\mathcal{Q}_\alpha^2 K$, $[\mathcal{Q}_\alpha, \mathcal{Q}_\alpha^2 K] = 2\mathcal{Q}_\alpha^3 K$, and so on.

We can now introduce the associative algebra of Weyl ordered (Hermitian) monomials

$$\mathcal{Q}_{\alpha(n)} = \mathcal{Q}_{\alpha_1\alpha_2\cdots\alpha_n} := \sum_{\alpha=1,2;\sigma} \frac{1}{n!} \mathcal{Q}_{\sigma(\alpha_1)} \mathcal{Q}_{\sigma(\alpha_2)} \cdots \mathcal{Q}_{\sigma(\alpha_n)}, \quad n = 0, 1, \ldots, \tag{58}$$

(the σ's denote the members of the permutation group acting on n elements), together with their product with the Klein operator K. We have, at a fixed value ν,

$$Aq(2;\nu) := \left\{(-iK)^b\, \mathcal{Q}_{\alpha(n)},\ n = 0, 1, \ldots,\ b = 1, 2\right\}. \tag{59}$$

The associative algebra $Aq(2;\nu)$ defines the universal enveloping algebra of the deformed Heisenberg algebra (57). This algebra was first introduced in [6] in the context of the quantization of observables on a hyperboloid.

As a vector space $Aq(2;\nu)$ can be endowed with either of two different types of brackets:

(i) ordinary brackets realized by commutators

(ii) $\mathbb{Z}_2$-graded brackets realized by (anti) commutators

In the first case $Aq(2;\nu)$ is an infinite-dimensional Lie algebra (the Jacobi identities being satisfied by construction) that we will denote as $q(2;\nu)_{[,]}$:

$$q(2;\nu)_{[,]} := \{Aq(2;\nu) \mid [\mathfrak{a}, \mathfrak{b}] \in Aq(2;\nu),\ \forall \mathfrak{a}, \mathfrak{b}\}. \tag{60}$$

The infinite-dimensional Lie algebra $q(2;\nu)_{[,]}$ is another realization of the deformed Schrödinger algebra. Since the operators $\mathcal{Q}_\alpha$ commute with the (19) 2×2 matrix operator $\Omega_{\epsilon=0}$, by construction, $q(2;\nu)_{[,]}$ is a dynamical symmetry of the Pure Calogero Matrix PDE; indeed we have

$$\left[(-iK)^b\, \mathcal{Q}_{\alpha_1\alpha_2\cdots\alpha_n}, \Omega_{\epsilon=0}\right] = 0. \tag{61}$$

In the second case $Aq(2;\nu)$ is an infinite-dimensional Lie superalgebra satisfying the graded Jacobi identities and whose consistent $\mathbb{Z}_2$-graded brackets, the (anti) commutators, are introduced through the position

$$[\mathfrak{a}, \mathfrak{b}\} := \mathfrak{a}\mathfrak{b} - (-1)^{\epsilon_\mathfrak{a}\epsilon_\mathfrak{b}}\, \mathfrak{b}\mathfrak{a}. \tag{62}$$

The $\mathbb{Z}_2$-grading of the generators, $\epsilon_\bullet = 0, 1$, is determined by, respectively, the even or odd power of the $\mathcal{Q}_\alpha$'s operators entering any given monomial (the presence of the Klein operator K does not affect the $\mathbb{Z}_2$-grading). We denote the superalgebra, with the given $\mathbb{Z}_2$-graded brackets, as

$$q(2;\nu)_{[,\}} := \{Aq(2;\nu) \mid [\mathfrak{a}, \mathfrak{b}\} \in Aq(2;\nu),\ \forall \mathfrak{a}, \mathfrak{b}\}. \tag{63}$$

This ν-dependent class of superalgebras was introduced in [6] and, together with a Lorentz invariant supertrace formula, was used in the construction of higher-spin Chern-Simons supergravity (see also [25]). In [6] these superalgebras were simply denoted as "$q(2;\nu)$". We introduced here the "[, }" label to distinguish them from the "[,]"-labeled deformed Schrödinger algebras. We keep, however, the simplified notation when no confusion arises.

The infinite-dimensional, ν-dependent, $q(2;\nu)$ superalgebra is a higher-spin superalgebra satisfying the correct spin-statistics connection. In order to understand its relation with higher-spin, two main features have to be recalled. The first one consists in the fact that $q(2;\nu)$ contains the finite dimensional Lie superalgebra $osp(2 \mid 2)$ as subalgebra, so that $osp(2 \mid 2) \subset q(2;\nu)$; the second feature is due to the fact that $osp(2 \mid 2)$ admits a covariant description in terms of three types of indices. They are the vector indices $\mu, \lambda, \ldots = 0, 1, 2$ labeling a three-dimensional Minkowski space, the (Majorana) spinorial indices $\alpha, \beta, \ldots = 1, 2$ labeling the associated real 2-component spinors, and the scalar indices $A, B, \ldots = 1, 2$ labeling an internal space.

In the covariant notation the $osp(2 \mid 2)$ fermions are expressed as $\mathcal{Q}_\alpha^A$, where one can set

$$\mathcal{Q}_\alpha^1 := \mathcal{Q}_\alpha, \qquad \mathcal{Q}_\alpha^2 := iK\mathcal{Q}_\alpha. \tag{64}$$

We note that the $\mathcal{Q}_\alpha^A$'s operators are Hermitian.

The even sector is expressed by the 2×2 diagonal operators J^μ and R ("R" stands for the R-symmetry $u(1)$-generator) which can be recovered from the saturated (due to the presence of both spinorial α, β and internal A, B indices) generalized supersymmetry (see [26])

$$\left\{\mathcal{Q}_\alpha^A, \mathcal{Q}_\beta^B\right\} = \delta^{AB}\left(C\gamma_\mu\right)_{\alpha\beta} J^\mu + \epsilon^{AB} C_{\alpha\beta} R. \tag{65}$$

The Charge Conjugation matrix $C_{\alpha\beta}$ ($C = e_{12} - e_{21}$) and its inverse are used to raise/lower the spinorial indices; the diagonal metric $\eta_{\mu\nu}$ ($\eta_{\mu\nu} = \mathrm{diag}(-1, +1, +1)$) and its inverse raise/lower the vector indices; the three gamma matrices $(\gamma_\mu)_\alpha{}^\beta$ ($\gamma_0 = e_{12} - e_{21}, \gamma_1 = e_{12} + e_{21}, \gamma_2 = e_{11} - e_{22}$) satisfy the composition law of the split-quaternions given by $\gamma_\mu\gamma_\nu = \eta_{\mu\nu}\mathbb{I}_2 + \epsilon_{\mu\nu\lambda}\gamma^\lambda$, with the totally antisymmetric Levi-Civita pseudo-tensor $\epsilon_{\mu\nu\lambda}$ normalized so that $\epsilon_{012} = 1$; by definition $(C\gamma_\mu)_{\alpha\beta} = C_{\alpha\alpha'}(\gamma_\mu)_\beta{}^{\alpha'}$. It is worth pointing out that the Charge Conjugation matrix and the gamma matrices act on the spinorial indices of the algebra and not on the 2-component wave function of the original PDE. The normalization of the antisymmetric tensor ϵ^{AB} is $\epsilon^{12} = 1$.

Straightforward computations show that the 2×2 matrix differential operators J^μ, R are connected with the $\overline{\mathbf{H}}, \overline{\mathbf{D}}, \overline{\mathbf{K}}, \overline{\mathbf{J}}$ operators of formulas (30) and (32)-(34) through the positions

$$\begin{aligned} J^0 &= -2\left(\overline{\mathbf{K}} + \overline{\mathbf{H}}\right), \\ J^1 &= 2\left(\overline{\mathbf{K}} - \overline{\mathbf{H}}\right), \\ J^2 &= 2\overline{\mathbf{D}}, \\ R &= 2\overline{\mathbf{J}}. \end{aligned} \tag{66}$$

The closure of the $osp(2 \mid 2)$ superalgebra is guaranteed by the nonvanishing commutators

$$\begin{aligned} \left[J_\mu, J_\nu\right] &= 4i\epsilon_{\mu\nu\lambda}J^\lambda, \\ \left[J_\mu, \mathcal{Q}_\alpha^A\right] &= 2i\left(\gamma_\mu\right)_\alpha{}^\beta \mathcal{Q}_\beta^A, \\ \left[R, \mathcal{Q}_\alpha^A\right] &= -2iS^A{}_B\mathcal{Q}_\alpha^B. \end{aligned} \tag{67}$$

In the last equation the matrix $S^A{}_B$ is given by $S = e_{12} - e_{21}$.

We point out that the construction of the above superalgebra in terms of a Klein operator was discussed in [10]. Due to the presence of the covariant $osp(2 \mid 2)$ subalgebra, the remaining generators of $q(2;\nu)$ are accommodated into higher-spin representations.

As mentioned before an associative algebra, called $\widehat{Aq}(2;\nu)$, of invariant operators for the DFF Matrix Calogero PDE is recovered from $Aq(2;\nu)$ by applying the time redefinition (42) and the similarity transformation (43). From $\widehat{Aq}(2;\nu)$ one recovers the deformed Schrödinger algebra $\widehat{q}_{[,]}(2;\nu)$ and the higher-spin superalgebra $\widehat{q}_{[,\}}(2;\nu)$. They are isomorphic realizations of $q_{[,]}(2;\nu)$ and $\widehat{q}_{[,\}}(2;\nu)$, respectively.

6. Conclusions

As we have seen, the Matrix Calogero models under consideration possess, in particular, a deformed Schrödinger algebra and the higher-spin superalgebra $q_{[,\}}(2;\nu)$ as dynamical symmetries. The deformed Schrödinger algebra is obtained by computing, similarly to [14], the class of linear operators which leave invariant the space of solutions of the given PDEs. A whole universal enveloping algebra, $Aq(2;\nu)$, is generated by requiring the closure of the Lie algebra induced by the commutators. When the Calogero potential coupling parameter ν takes the 0 value, the infinite-dimensional algebra can be truncated to the standard Schrödinger algebra.

We further observed that the Matrix Calogero models and (at $\nu = 0$) the free particle multiplet exhibit (relativistic) higher-spin symmetries. These results suggest that, extending the conjecture of [13], the nonrelativistic Calogero system studied so far may be dual, in a certain regime and in a certain asymptotic nonrelativistic geometry (see [27]), to the vacuum of the (bulk) Vasiliev's higher-spin gravity. As a matter of fact, the ν-parameter is related to the vacuum expectation value of the scalar field in the Vasiliev-Prokuskin higher-spin gravity [28] in 2 + 1 dimensions.

We should mention that another class of (1 + 1)-dimensional nonrelativistic PDEs, unlike the Matrix Calogero models discussed here, inherently contain higher-spin operators [19, 29]. These PDEs are invariant under the centrally extended ℓ-Conformal Galilei algebras, with $\ell = 1/2 + \mathbb{N}_0 \geq 3/2$. In these models the higher-spin generators span a spin-ℓ representation of an $sl(2)$ subalgebra constructed from $\ell + 1/2$ Heisenberg algebras ($\ell = 1/2$ for the free particle in one dimension). An open problem consists in understanding whether, for these systems, a deformation of the centrally extended ℓ-Conformal Galilei algebra, induced by Matrix Calogero potentials, is allowed.

We expect, more generally, that systems possessing conformal Galileo invariance could be related to higher-spin gravities. This type of nonrelativistic dualities, if they indeed exist, would complement the dualities conjectured in [30–32] (see also [33]).

Conflicts of Interest

The authors declare that they have no conflicts of interest.

Acknowledgments

Francesco Toppan received support from CNPq (PQ Grant no. 306333/2013-9). He thanks the Universidad San Sebastián of Valdivia, where this work was elaborated, for hospitality. Mauricio Valenzuela thanks as coinvestigator the projects Fondecyt Regular no. 1140296 and Conicyt DPI 20140115.

References

[1] F. Calogero, "Solution of a three-body problem in one dimension," *Journal of Mathematical Physics*, vol. 10, p. 2191, 1969.

[2] V. de Alfaro, S. Fubini, and G. Furlan, "Conformal invariance in quantum mechanics," *Il Nuovo Cimento A*, vol. 34, no. 4, pp. 569–612, 1976.

[3] E. P. Wigner, "Do the equations of motion determine the quantum mechanical commutation relations?" *Physical Review A: Atomic, Molecular and Optical Physics*, vol. 77, no. 5, pp. 711-712, 1950.

[4] L. M. Yang, "A note on the quantum rule of the harmonic oscillator," *Physical Review A: Atomic, Molecular and Optical Physics*, vol. 84, no. 4, pp. 788–790, 1951.

[5] M. S. Plyushchay, "Deformed Heisenberg algebra with reflection," *Nuclear Physics B*, vol. 491, p. 619, 1997.

[6] I. A. Batalin and I. V. Tyutin, "Existence theorem for the effective gauge algebra in the generalized canonical formalism with abelian conversion of second-class constraints," *International Journal of Modern Physics A*, vol. 6, no. 18, p. 3255, 1991.

[7] N. Boulanger, P. Sundell, and M. Valenzuela, "Three-dimensional fractional-spin gravity," *Journal of High Energy Physics*, vol. 2014, no. 2, 2014.

[8] N. Boulanger, P. Sundell, and M. Valenzuela, "Gravitational and gauge couplings in Chern-Simons fractional spin gravity," *Journal of High Energy Physics*, vol. 1601, p. 173, 2016.

[9] P. J. Olver, *Applications of Lie Groups to Differential Equations*, Springer, New York, NY, USA, 2nd edition, 1991.

[10] M. S. Plyushchay, "Deformed Heisenberg algebra, fractional spin fields and supersymmetry without fermions," *Annals of Physics*, vol. 245, Article ID 9601116, p. 339, 1996.

[11] N. L. Holanda and F. Toppan, "Four types of (super)conformal mechanics: D-module reps and invariant actions," *Journal of Mathematical Physics*, vol. 55, no. 8, 2014.

[12] I. E. Cunha, N. L. Holanda, and F. Toppan, "High Energy Physics - TheoryFrom worldline to quantum superconformal mechanics with/without oscillatorial terms: $D(2,1;\alpha)$ and $sl(2—1)$ models," *Physical Review D*, vol. 96, no. 6, Article ID 065014, 2017.

[13] M. Valenzuela, "Higher-spin symmetries of the free Schrodinger equation," *Advances in Mathematical Physics*, vol. 2016, 2016.

[14] U. Niederer, "The maximal kinematical invariance group of the free Schrödinger equation," *Helvetica Physica Acta*, vol. 45, p. 802, 1972.

[15] U. Niederer, "The maximal kinematical invariance group of the harmonic oscillator," *Helvetica Physica Acta*, vol. 46, p. 191, 1973.

[16] U. Niederer, "The maximal kinematical invariance groups of Schrödinger equations with arbitrary potentials," *Helvetica Physica Acta*, vol. 47, p. 167, 1974.

[17] F. Toppan, "Symmetries of the Schrödinger equation and algebra/superalgebra duality," *Journal of Physics: Conference Series*, vol. 597, p. 012071, 2015.

[18] R. Britto-Pacumio, J. Michelson, A. Strominger, and A. Volovich, "Lectures on superconformal quantum mechanics and multi-black hole moduli spaces," *Nato Science Series C*, vol. 556, p. 255, 2000.

[19] N. Aizawa, Z. Kuznetsova, and F. Toppan, "Invariant PDEs of conformal galilei algebra as deformations: cryptohermiticity and contractions," *Progress of Theoretical and Experimental Physics*, 2016.

[20] P. A. Horvathy, M. S. Plyushchay, and M. Valenzuela, "Bosons, fermions and anyons in the plane, and supersymmetry," *Annals of Physics*, vol. 325, p. 1931, 2010.

[21] J. M. Guilarte and M. S. Plyushchay, "Perfectly invisible PT-symmetric zero-gap systems, conformal field theoretical kinks, and exotic nonlinear supersymmetry," *Journal of High Energy Physics*, vol. 1712, p. 061, 2017.

[22] H. Miyazaki and I. Tsutsui, "Quantum tunneling and caustics under inverse square potential," *Annals of Physics*, vol. 299, p. 78, 2002.

[23] L. Fehér, I. Tsutsui, and T. Fülöp, "Inequivalent quantizations of the three-particle Calogero model constructed by separation of variables," *Nuclear Physics B*, vol. 715, Article ID 0412095, p. 713, 2005.

[24] N. Aizawa, Z. Kuznetsova, and F. Toppan, "The quasi-nonassociative exceptional $F(4)$ deformed quantum oscillator," *Journal of Mathematical Physics*, vol. 59, 2018.

[25] M. P. Blencowe, "A consistent interacting massless higher-spin field theory in D = 2 + 1," *Classical and Quantum Gravity*, vol. 6, no. 4, pp. 443–452, 1989.

[26] R. D'Auria and P. Fré, "Geometric supergravity in d=11 and its hidden supergroup," *Nuclear Physics B*, vol. 201, p. 101, 1982.

[27] C. Duval, M. Hassaïne, and P. A. Horvathy, "The geometry of Schrödinger symmetry in non-relativistic CFT," *Annals of Physics*, vol. 324, no. 5, pp. 1158–1167, 2009.

[28] S. F. Prokushkin and M. A. Vasiliev, "Higher-spin gauge interactions for massive matter fields in 3D AdS space-time," *Nuclear Physics B*, vol. 545, Article ID 9806236, p. 385, 1999.

[29] N. Aizawa, Z. Kuznetsova, and F. Toppan, "ℓ-oscillators from second-order invariant PDEs of the centrally extended Conformal Galilei Algebras," *Journal of Mathematical Physics*, vol. 56, 2015.

[30] E. Sezgin and P. Sundell, "Massless higher spins and holography," *Nuclear Physics B*, vol. 644, p. 303, 2002.

[31] E. Sezgin and P. Sundell, "Erratum to: "Massless higher spins and holography"," *Nuclear Physics B*, vol. 660, no. 1-2, p. 403, 2003.

[32] I. Klebanov and A. Polyakov, "AdS dual of the critical O(N) vector model," *Physics Letters B*, vol. 550, no. 3-4, pp. 213–219, 2002.

[33] J. Maldacena, D. Martelli, and Y. Tachikawa, "Comments on string theory backgrounds with non-relativistic conformal symmetry," *Journal of High Energy Physics*, vol. 2008, no. 10, article 072, 39 pages, 2008.

Lump Solutions to a (2+1)-Dimensional Fifth-Order KdV-Like Equation

Sumayah Batwa[1,2] and Wen-Xiu Ma[1,3,4,5,6]

[1]Department of Mathematics and Statistics, University of South Florida, Tampa, FL 33620-5700, USA
[2]Mathematics Department, King Abdulaziz University, Jeddah 21589, Saudi Arabia
[3]Department of Mathematics, Zhejiang Normal University, Jinhua 321004, China
[4]College of Mathematics and Systems Science, Shandong University of Science and Technology, Qingdao, Shandong 266590, China
[5]College of Mathematics and Physics, Shanghai University of Electric Power, Shanghai 200090, China
[6]International Institute for Symmetry Analysis and Mathematical Modelling, Department of Mathematical Sciences, North-West University, Mafikeng Campus, Private Bag X2046, Mmabatho 2735, South Africa

Correspondence should be addressed to Sumayah Batwa; sbatwa@mail.usf.edu

Academic Editor: Guozhen Lu

A (2+1)-dimensional fifth-order KdV-like equation is introduced through a generalized bilinear equation with the prime number $p = 5$. The new equation possesses the same bilinear form as the standard (2+1)-dimensional fifth-order KdV equation. By Maple symbolic computation, classes of lump solutions are constructed from a search for quadratic function solutions to the corresponding generalized bilinear equation. We get a set of free parameters in the resulting lump solutions, of which we can get a nonzero determinant condition ensuring analyticity and rational localization of the solutions. Particular classes of lump solutions with special choices of the free parameters are generated and plotted as illustrative examples.

1. Introduction

The (2+1)-dimensional fifth-order KdV equation [1] is

$$\begin{aligned} 36u_t + u_{5x} + 15u_x u_{xx} + 15uu_{3x} + 45u^2 u_x - 5u_{xxy} \\ - 15uu_y - 15u_x \int u_y dx - 5 \int u_{yy} dx = 0, \end{aligned} \quad (1)$$

which is the (2+1)-dimensional analogue of the Caudrey-Dodd-Gibbon-Kotera-Sawada (CDGKS) equation [2]. When $u_y = 0$, (1) reduces to the Sawada-Kotera equation

$$u_t + u_{5x} + 15u_x u_{xx} + 15uu_{3x} + 45u^2 u_x = 0. \quad (2)$$

Konopelchenko and Dubovsky [3] were the first to come up with (1). Lv *et al.* [4] obtained the symmetry transformations for (1) by using its Lax pair. Lü [5] constructed four sets of bilinear Bäcklund transformations in order to obtain multisoliton solutions. Wazwaz [6] derived multiple soliton solutions and multiple singular soliton solutions for (1). Equation (1) has a widespread adoption in many physical branches, such as conserved current of Liouville equation, two-dimensional quantum gravity gauge field, and conformal field theory [7–13].

In recent years, there has been a growing interest in finding exact solutions of nonlinear evolution equations, such as the rational solutions and the rogue wave solutions, which are exponentially localized in certain directions. Lump solutions are a type of rational function solutions, localized in all directions in the space. Lump solutions have been studied for many nonlinear partial differential equations such as the KPI equation [14, 15], the three-dimensional three-wave resonant interaction equation [16], and the B-KP equation [17]. Through Hirota bilinear equations, one of the authors (Ma) [18] introduced a new method to construct lump solutions to the KP equation. Following Ma's method, the lump solutions for more nonlinear evolution equations have been found, for instance, the dimensionally reduced p-gKP and p-gBKP [19], the (2+1)-dimensional Boussinesq equation [20], the BKP equation [21], the (3+1)-dimensional Jambo-Miwa [22], and the KdV equation [23]. In addition to Hirota bilinear forms, generalized bilinear derivatives [24] are used to find rational function solutions to the generalized KdV, KP, and Boussinesq equations [25–27].

In this paper, we investigate lump solutions for a fifth-order KdV-like equation. The organization of the paper is as follows: In Section 2, we formulate a new fifth-order KdV-like equation from generalized bilinear differential equations of KdV type. In Section 3 with the help of Maple, we obtain lump solutions for the constructed equation and analyze their dynamics. Then we draw some figures for a particular classes of lump solutions to show some properties. In the last section, conclusions and some remarks are given.

2. A New (2+1)-Dimensional Fifth-Order KdV-Like Equation

Under the dependent variable transformation

$$u = 2\left(\ln f\right)_{xx} \tag{3}$$

with $f = f(x, y, t)$, the (2+1)-dimensional fifth-order KdV equation (1) becomes the following (2+1)-dimensional Hirota bilinear equation:

$$\begin{aligned} B_{5thKdV} &:= \left(D_x^6 - 5D_x^3 D_y + 36 D_x D_t - 5D_y^2\right) f.f \\ &= 72 f_x f_t - 72 f_{xt} f + 2 f_{6x} f - 12 f_{5x} f_x \\ &\quad + 30 f_{4x} f_{xx} - 20 f_{xxx}^2 + 10 f_{xxx} f_y \\ &\quad - 30 f_{xx} f_{xy} + 30 f_{xxy} f_x - 10 f_{xxxy} f \\ &\quad - 10 f_{yy} f + 10 f_y^2 = 0, \end{aligned} \tag{4}$$

where the Hirota derivatives D_x, D_y, and D_t are defined in [28].

Based on a prime number p, a kind of generalized bilinear operators is introduced as [24, 29]

$$\begin{aligned} \left(D_{p,x}^m D_{p,t}^n\right) f.f &= \left(\partial_x + \alpha_p \partial_{x'}\right)^m \left(\partial_t + \alpha_p \partial_{t'}\right)^n f(x,t) \\ &\quad \cdot f\left(x',t'\right)\Big|_{x'=x,t'=t} = \sum_{i=0}^{m}\sum_{j=0}^{n} \binom{m}{i}\binom{n}{j} \alpha_p^i \alpha_p^j \frac{\partial^{m-i}}{\partial x^{m-i}} \\ &\quad \cdot \frac{\partial^i}{\partial x'^{(i)}} \frac{\partial^{n-j}}{\partial t^{n-j}} \frac{\partial^j}{\partial t'^{(j)}} f(x,t) f\left(x',t'\right)\Big|_{x'=x,t'=t} \\ &= \sum_{i=0}^{m}\sum_{j=0}^{n} \binom{m}{i}\binom{n}{j} \alpha_p^i \alpha_p^j \frac{\partial^{m+n-i-j} f(x,t)}{\partial x^{m-i} t^{n-j}} \frac{\partial^{i+j} f(x,t)}{\partial x^i t^j} \end{aligned} \tag{5}$$

where $m, n \geq 0$, $\alpha_p^s = (-1)^{r_p(s)}$, $s \equiv r_p(s) \bmod p$.

For example, if we assume $p = 5$, we have

$$\begin{aligned} \alpha_5 &= -1, \\ \alpha_5^2 &= 1, \\ \alpha_5^3 &= -1, \\ \alpha_5^4 &= \alpha_5^5 = 1, \\ \alpha_5^6 &= -1, \ldots \end{aligned} \tag{6}$$

With $p = 5$, we can generalize (4) into

$$\begin{aligned} &GB_{5thKdV} \\ &:= \left(D_{5,x}^6 - 5D_{5,x}^3 D_{5,y} + 36 D_{5,x} D_{5,t} - 5D_{5,y}^2\right) f.f \\ &= 30 f_{4x} f_{xx} - 20 f_{xxx}^2 - 10 f_{xxxy} f + 10 f_{xxx} f_y \\ &\quad - 30 f_{xx} f_{xy} + 30 f_{xxy} f_x + 72 f_{xt} f - 72 f_x f_t \\ &\quad - 10 f_{yy} f + 10 f_y^2 = 0. \end{aligned} \tag{7}$$

Equation (7) is a generalized bilinear fifth-order KdV equation. Under the transformations

$$\begin{aligned} u &= 6\left(\ln f\right)_x, \\ v &= 6\left(\ln f\right)_y \end{aligned} \tag{8}$$

which were suggested by the Bell polynomial theories [29–31], (7) is transformed into the following fifth-order KdV-like nonlinear differential equation:

$$\begin{aligned} GP_{5thKdV}(u) &= u_t + \frac{11}{279936} u^6 + \frac{25}{15552} u^4 u_x \\ &\quad + \frac{5}{972} u^3 u_{xx} + \frac{5}{288} u^2 u_x^2 - \frac{5}{2592} u^3 v \\ &\quad + \frac{5}{54} u u_{xx} u_x + \frac{5}{432} u_x^3 - \frac{5}{432} u v u_x \\ &\quad - \frac{5}{432} u_y u^2 + \frac{5}{432} u^2 u_{xxx} + \frac{5}{54} u_{xx}^2 \\ &\quad - \frac{5}{72} u_y u_x + \frac{5}{72} u_x u_{xxx} + \frac{1}{36} u u_{4x} \\ &\quad + \frac{1}{36} u_{5x} - \frac{5}{36} v_y = 0 \end{aligned} \tag{9}$$

Therefore, if f solves the bilinear equation (4) or (7), then $u = 6(\ln f)_{xx}$ or $u = 6(\ln f)_x$ will solve the nonlinear equation (1) or (9).

3. Lump Solutions to the Fifth-Order KdV-Like Equation

In this section, we are going to generate lump solutions to (9) by searching for quadratic function solutions to (7) with the assumption

$$\begin{aligned} f &= g^2 + h^2 + a_9, \\ g &= a_1 x + a_2 y + a_3 t + a_4, \\ h &= a_5 x + a_6 y + a_7 t + a_8 \end{aligned} \tag{10}$$

where a_i, $1 \le i \le 9$, are real constants to be determined later. Note that using a sum involving one square, in the two-dimensional space, will not generate exact solutions which are rationally localized in all directions in the space.

Substituting (10) into (7) and equating all the coefficients of different polynomials of x, y, and t to zero using Maple symbolic computation, we obtain a set of algebraic equations in a_i ($1 \le i \le 9$); solving the set of algebraic equations with the aid of Maple, we attain the following two classes of solutions.

Case 1.

$$\begin{aligned} a_1 &= 0, \\ a_3 &= -\frac{5a_2^3 a_9}{54a_5^4}, \\ a_6 &= -\frac{a_2^2 a_9}{3a_5^3}, \\ a_7 &= \frac{5a_2^2\left(-9a_5^6 + a_2^2 a_9^2\right)}{324a_5^7}, \end{aligned} \tag{11}$$

and $a_i = a_i$ ($i = 2, 4, 5, 8, 9$) are real free parameters which need to satisfy $a_5 \neq 0$ to make the corresponding solutions f to be well defined and $a_9 > 0$ to guarantee the positiveness of f.

The parameters in the sets (11) generate a class of positive quadratic function solutions to (7):

$$\begin{aligned} f &= \left(-\frac{5a_2^3 a_9 t}{54a_5^4} + a_2 y + a_4\right)^2 \\ &+ \left(\frac{5a_2^2\left(-9a_5^6 + a_2^2 a_9^2\right)t}{324a_5^7} + a_5 x - \frac{a_2^2 a_9 y}{3a_5^3} + a_8\right)^2 \\ &+ a_9 \end{aligned} \tag{12}$$

and the resulting class of quadratic function solutions, in turn, yields a few classes of lump solutions to the (2+1)-dimensional fifth-order KdV-like equation (9) through the dependent variable transformation:

$$\begin{aligned} u &= 6(\ln f)_{xx} = \frac{6\left(f_{xx} f - f_x^2\right)}{f^2} \\ &= \frac{12\left(a_1^2 - a_5^2\right)\left(-g^2 + h^2\right) - 48a_1 a_5 gh + 12\left(a_1^2 + a_5^2\right)a_9}{\left(g^2 + h^2 + a_9\right)^2} \end{aligned} \tag{13}$$

where the function f is defined by (10), and the functions g and h are given as follows:

$$\begin{aligned} g &= -\frac{5a_2^3 a_9}{54a_5^4} t + a_2 y + a_4, \\ h &= \frac{5a_2^2\left(-9a_5^6 + a_2^2 a_9^2\right)}{324a_5^7} t + a_5 x - \frac{a_2^2 a_9}{3a_5^3} y + a_8. \end{aligned} \tag{14}$$

Case 2.

$$\begin{aligned} a_3 &= \frac{5\left(a_1 a_2^2 - a_1 a_6^2 + 2a_2 a_5 a_6\right)}{36\left(a_1^2 + a_5^2\right)}, \\ a_7 &= \frac{5\left(2a_1 a_2 a_6 - a_2^2 a_5 + a_5 a_6^2\right)}{36\left(a_1^2 + a_5^2\right)}, \\ a_9 &= \frac{-3\left(a_1 a_2 + a_5 a_6\right)\left(a_1^2 + a_5^2\right)^2}{\left(a_1 a_6 - a_2 a_5\right)^2}, \end{aligned} \tag{15}$$

where a_1, a_2, a_4, a_5, a_6, a_8 are arbitrary constants to be determined with the following restricted conditions:

$$\begin{aligned} \Delta_1 &:= a_1^2 + a_5^2 = \begin{vmatrix} a_1 & -a_5 \\ a_5 & a_1 \end{vmatrix} \neq 0, \\ \Delta_2 &:= a_1 a_2 + a_5 a_6 = \begin{vmatrix} a_1 & -a_5 \\ a_6 & a_2 \end{vmatrix} < 0, \\ \Delta_3 &:= a_1 a_6 - a_2 a_5 = \begin{vmatrix} a_1 & a_2 \\ a_5 & a_6 \end{vmatrix} \neq 0. \end{aligned} \tag{16}$$

Δ_1 makes the corresponding solutions f well defined and Δ_2 assures that the solution f is positive, while Δ_3 guarantees the localization of the solutions u in all directions in the (x, y)-plane.

Since these parameters are arbitrary, the solutions of (9) are more general. The parameters a_1, a_5 indicate that the wave velocity in the x direction is arbitrary and a_2, a_6 illustrate the arbitrariness of the wave velocity in the y direction. The parameters a_4, a_8 represent the invariance of variables and a_3, a_7 show the wave frequency which are represented by other quantities.

This set of parameters, in turn, generates positive quadratic function solutions to (7):

$$f = \left(a_1 x + a_2 y + \frac{5\left(a_1 a_2^2 - a_1 a_6^2 + 2a_2 a_5 a_6\right)}{36\left(a_1^2 + a_5^2\right)} t + a_4 \right)^2 + \left(a_5 x + a_6 y + \frac{5\left(2a_1 a_2 a_6 - a_2^2 a_5 + a_5 a_6^2\right)}{36\left(a_1^2 + a_5^2\right)} t + a_8 \right)^2 + \frac{-3\left(a_1 a_2 + a_5 a_6\right)\left(a_1^2 + a_5^2\right)^2}{\left(a_1 a_6 - a_2 a_5\right)^2} \tag{17}$$

Consequently, a kind of lump solutions to (9) through the transformation $u = 6(\ln f)_{xx}$ and (10) is achieved as follows:

$$u = 6\left(\ln f\right)_{xx} = \frac{6\left(f_{xx} f - f_x^2\right)}{f^2} = \frac{12\left(a_1^2 - a_5^2\right)\left(-g^2 + h^2\right) - 48 a_1 a_5 g h + 12\left(a_1^2 + a_5^2\right) a_9}{\left(g^2 + h^2 + a_9\right)^2} \tag{18}$$

where the functions g and h are given by

$$g = a_1 x + a_2 y + \frac{5\left(a_1 a_2^2 - a_1 a_6^2 + 2a_2 a_5 a_6\right)}{36\left(a_1^2 + a_5^2\right)} t + a_4,$$
$$h = a_5 x + a_6 y + \frac{5\left(2a_1 a_2 a_6 - a_2^2 a_5 + a_5 a_6^2\right)}{36\left(a_1^2 + a_5^2\right)} t + a_8. \tag{19}$$

Choosing a special value for the free parameters in Cases 1 and 2, we construct specific lump solutions u of (9). One special pair of positive quadratic function solutions and lump solutions with specific values of the parameters in Case 1 is given as follows. First, the selection of the parameters,

$$a_2 = 4, \quad a_4 = 0, \quad a_5 = 2, \quad a_8 = 0, \quad a_9 = 1, \tag{20}$$

leads to

$$f = \frac{34225}{26244} t^2 - \frac{370}{243} t y + \frac{148}{9} y^2 - \frac{350}{81} t x + 4x^2 - \frac{8}{3} x y + 1 \tag{21}$$

and lump solution

$$u = -\frac{1259712\left(27025 t^2 - 113400 t x + 115560 t y + 104976 x^2 - 69984 x y - 408240 y^2 - 26244\right)}{\left(34225 t^2 - 113400 t x - 39960 t y + 104976 x^2 - 69984 x y + 431568 y^2 + 26244\right)^2} \tag{22}$$

If we take a particular choice of the parameters in Case 2 as

$$a_1 = 1, \quad a_2 = -\frac{1}{2}, \quad a_4 = 0, \quad a_5 = 0, \quad a_6 = 3, \quad a_8 = 0, \tag{23}$$

then we have

$$f = \frac{34225}{26244} t^2 - \frac{370}{243} t y + \frac{148}{9} y^2 - \frac{350}{81} t x + 4x^2 - \frac{8}{3} x y + 1 \tag{24}$$

and lump solution

$$u = -\frac{248832\left(27025 t^2 - 113400 t x + 115560 t y + 104976 x^2 - 69984 x y - 408240 y^2 - 26244\right)}{\left(34225 t^2 - 113400 t x - 39960 t y + 104976 x^2 - 69984 x y + 431568 y^2 + 26244\right)^2} \tag{25}$$

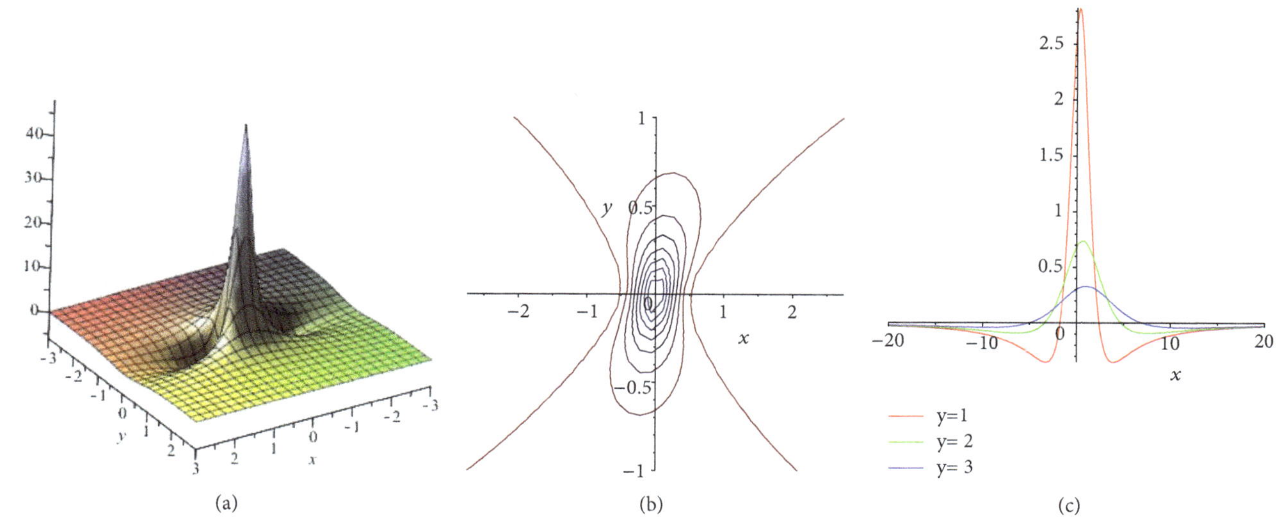

FIGURE 1: Plots of lump solution in Case 1 with $a_2 = 4, a_4 = 0, a_5 = 2, a_8 = 0, a_9 = 1$ at $t = 0$: (a) 3D plot, (b) density plot, and (c) x-curves.

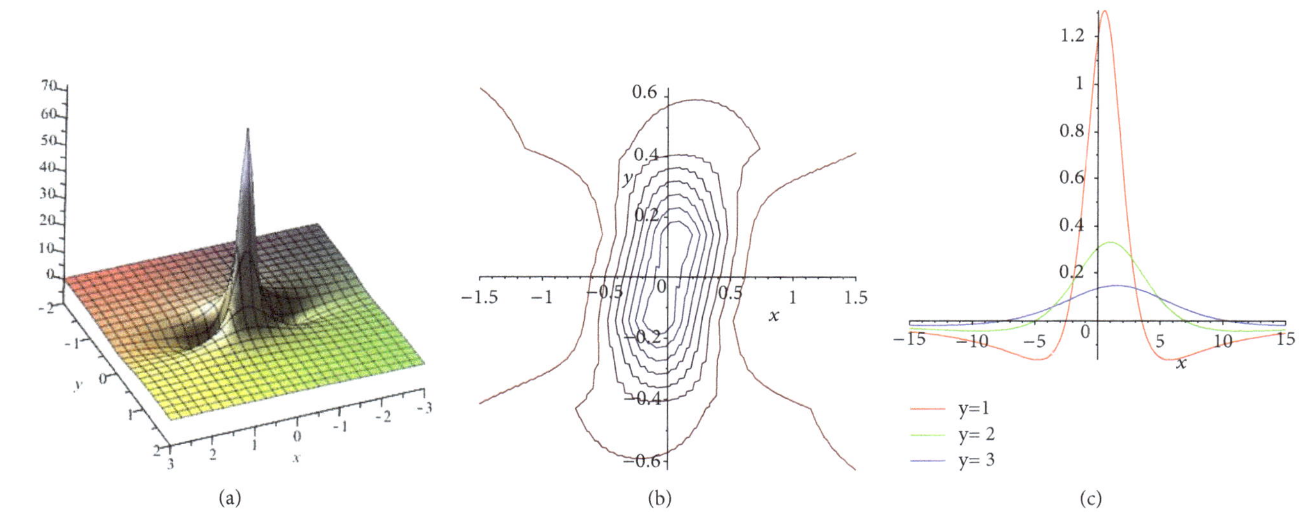

FIGURE 2: Plots of lump solution in Case 2 with $a_1 = 1, a_2 = -1/2, a_4 = 0, a_5 = 0, a_6 = 3, a_8 = 0$ at $t = 0$: (a) 3D plot, (b) density plot, and (c) x-curves.

Figure 1 shows the profile of lump solutions in Case 1 with the special choice of the parameters (20) at $t = 0$, while Figure 2 shows the profile of lump solutions in Case 2 with the special choice of the parameters (23) at $t = 0$.

4. Conclusions

In this paper, we studied a new (2+1)-dimensional fifth-order KdV-like equation, obtained by using the generalized Hirota bilinear formulation with $p = 5$. Through symbolic computation with Maple we constructed a few classes of lump solutions. The analyticity and localization of the resulting lump solutions are guaranteed by a nonzero determinant condition and a positivity condition. A subclass of lump solutions under special choices of the parameters involved covers the lump solutions. Contour plots with small determinant values are sequentially made to exhibit that the corresponding lump solution tends to zero when the determinant tends to zero. Recently, there have been some systematical studies on lump solutions [32] and interaction solutions between lumps and solitons for many integrable equations in (2+1)-dimensions. We refer the reader to [33] for lump-kink interaction solutions and [34, 35] for lump-soliton interaction solutions.

Conflicts of Interest

The authors declare that they have no conflicts of interest.

References

[1] X. Meng, "The Periodic Solitary Wave Solutions for the (2+1)-Dimensional Fifth-Order KdV Equation," *Journal of Applied Mathematics and Physics*, vol. 02, no. 07, pp. 639–643, 2014.

[2] X.-G. Xu, X.-H. Meng, C.-Y. Zhang, and Y.-T. Gao, "Analytical investigation of the Caudrey-Dodd-Gibbon-Kotera-Sawada equation using symbolic computation," *International Journal of Modern Physics B*, vol. 27, no. 6, 2013.

[3] B. G. Konopelchenko and V. G. Dubrovsky, "Some new integrable nonlinear evolution equations in (2+1) dimensions," *Physics Letters A*, vol. 102, no. 1-2, pp. 15–17, 1984.

[4] N. Lü, J.-Q. Mei, and H.-Q. Zhang, "Symmetry reductions and group-invariant solutions of (2+1)-dimensional Caudrey–Dodd–Gibbon–Kotera–Sawada equation," *Communications in Theoretical Physics*, vol. 53, no. 4, pp. 591–595, 2010.

[5] X. Lü, "New bilinear Bäcklund transformation with multisoliton solutions for the (2+1)-dimensional Sawada–Kotera model," *Nonlinear Dynamics*, vol. 76, no. 1, pp. 161–168, 2014.

[6] A.-M. Wazwaz, "Multiple soliton solutions for (2+1)-dimensional Sawada–Kotera and Caudrey–Dodd–Gibbon equations," *Mathematical Methods in the Applied Sciences*, vol. 34, no. 13, pp. 1580–1586, 2011.

[7] Z. Xu, H. Chen, and W. Chen, "The multisoliton solutions for the (2+1)-dimensional Sawada-Kotera equation," *Abstract and Applied Analysis*, vol. 2013, 5 pages, 2013.

[8] J. Satsuma and D. J. Kaup, "A Bäcklund transformation for a higher order Korteweg-de Vries equation," *Journal of the Physical Society of Japan*, vol. 43, no. 2, pp. 692–697, 1977.

[9] D. J. Kaup, "On the inverse scattering problem for cubic eigenvalue problems of the class $\psi_{xxx} + 6Q\psi_x + 6R_\psi = \lambda\psi$," *Studies in Applied Mathematics*, vol. 62, no. 3, pp. 189–216, 1980.

[10] J. Weiss, "On classes of integrable systems and the Painlevé property," *Journal of Mathematical Physics*, vol. 25, no. 1, pp. 13–24, 1984.

[11] C. Rogers and S. Carillo, "On reciprocal properties of the Caudrey-Dodd-Gibbon and Kaup-KUPershmidt hierarchies," *Physica Scripta. An International Journal for Experimental and Theoretical Physics*, vol. 36, no. 6, pp. 865–869, 1987.

[12] Y. Ma and X. Geng, "Darboux and Bäcklund transformations of the bidirectional Sawada-Kotera equation," *Applied Mathematics and Computation*, vol. 218, no. 12, pp. 6963–6965, 2012.

[13] S. Y. Lou, "Symmetries of the KDV Equation and four hierarchies of the integrodifferential KDV Equations," *Journal of Mathematical Physics*, vol. 35, no. 5, pp. 2390–2396, 1994.

[14] S. V. Manakov, V. E. Zakharov, L. A. Bordag, A. R. Its, and V. B. Matveev, "Two-dimensional solitons of the Kadomtsev-Petviashvili equation and their interaction," *Physics Letters A*, vol. 63, no. 3, pp. 205-206, 1977.

[15] J. Satsuma and M. J. Ablowitz, "Two-dimensional lumps in nonlinear dispersive systems," *Journal of Mathematical Physics*, vol. 20, no. 7, pp. 1496–1503, 1979.

[16] D. J. Kaup, "The lump solutions and the Bäcklund transformation for the three-dimensional three-wave resonant interaction," *Physical Review A*, vol. 90, no. 5, 1980.

[17] C. R. Gilson and J. J. Nimmo, "Lump solutions of the BKP equation," *Physics Letters A*, vol. 147, no. 8-9, pp. 472–476, 1990.

[18] W.-X. Ma, "Lump solutions to the Kadomtsev-Petviashvili equation," *Physics Letters A*, vol. 379, no. 36, pp. 1975–1978, 2015.

[19] W. X. Ma, Z. Qin, and X. Lü, "Lump solutions to dimensionally reduced p -gKP and p -gBKP equations," *Nonlinear Dynamics*, vol. 84, no. 2, pp. 923–931, 2016.

[20] H.-C. Ma and A.-P. Deng, "Lump solution of (2+1)-dimensional Boussinesq equation," *Communications in Theoretical Physics*, vol. 65, no. 5, pp. 546–552, 2016.

[21] J.-Y. Yang and W.-X. Ma, "Lump solutions to the BKP equation by symbolic computation," *International Journal of Modern Physics B*, vol. 30, no. 28-29, 2016.

[22] H.-Q. Sun and A.-H. Chen, "Lump and lump-kink solutions of the (3+1)-dimensional Jimbo-Miwa and two extended Jimbo–Miwa equations," *Applied Mathematics Letters*, vol. 68, pp. 55–61, 2017.

[23] C. Wang, "Spatiotemporal deformation of lump solution to (2+1)-dimensional KdV equation," *Nonlinear Dynamics*, vol. 84, no. 2, pp. 697–702, 2016.

[24] W. X. Ma, "Generalized bilinear differential equations," *Studies in Nonlinear Sciences*, vol. 2, no. 4, pp. 140–144, 2011.

[25] Y. Zhang and W.-X. Ma, "Rational solutions to a KdV-like equation," *Applied Mathematics and Computation*, vol. 256, pp. 252–256, 2015.

[26] Y. F. Zhang and W. X. Ma, "A study on rational solutions to a KP-like equation," *Zeitschrift für Naturforschung A*, vol. 70, no. 4, p. 263, 2015.

[27] C.-G. Shi, B.-Z. Zhao, and W.-X. Ma, "Exact rational solutions to a Boussinesq-like equation in 1+1-dimensions," *Applied Mathematics Letters*, vol. 48, pp. 170–176, 2015.

[28] R. Hirota, *The Direct Method in Soliton Theory*, Cambridge University Press, Cambridge, UK, 2004.

[29] W. X. Ma, "Bilinear equations, Bell polynomials and linear superposition principal," vol. 411, p. 12021, 2013.

[30] Y.-H. Wang and Y. Chen, "Binary Bell polynomials, bilinear approach to exact periodic wave solutions of (2+1)-dimensional nonlinear evolution equations," *Communications in Theoretical Physics*, vol. 56, no. 4, p. 1, 2011.

[31] C. Gilson, F. Lambert, J. Nimmo, and R. Willox, "On the combinatorics of the Hirota D-Operators," *Proceedings of the Royal Society A Mathematical, Physical and Engineering Sciences*, vol. 452, no. 1945, pp. 223–234, 1996.

[32] W.-X. Ma and Y. Zhou, "Lump solutions to nonlinear partial differential equations via Hirota bilinear forms," *Journal of Differential Equations*, vol. 264, no. 4, pp. 2633–2659, 2018.

[33] J.-b. Zhang and W.-X. Ma, "Mixed lump-kink solutions to the BKP equation," *Computers and Mathematics with Applications*, 2017.

[34] J. Y. Yang, W. X. Ma, and Z. Qin, "Lump and lump-soliton solutions to the (2+1)-dimensional Ito equation," *Analysis and Mathematical Physics*, vol. 95, no. 1, p. 1, 2017.

[35] W.-X. Ma, X. Yong, and H.-Q. Zhang, "Diversity of interaction solutions to the (2+1)-dimensional Ito equation," *Computers and Mathematics with Applications*, 2017.

7

Self-Similarity Analysis of the Nonlinear Schrödinger Equation in the Madelung Form

Imre F. Barna ,[1,2] Mihály A. Pocsai,[1,3] and L. Mátyás[4]

[1] *Wigner Research Center for Physics of the Hungarian Academy of Sciences, Konkoly-Thege Miklós út 29-33, 1121 Budapest, Hungary*
[2] *ELI-HU Nonprofit Kft., Dugonics Tér 13, H-6720 Szeged, Hungary*
[3] *University of Pécs, Institute of Physics, Ifjúság Útja 6, H-7624 Pécs, Hungary*
[4] *Sapientia University, Department of Bioengineering, Libertătii Sq. 1, 530104 Miercurea Ciuc, Romania*

Correspondence should be addressed to Imre F. Barna; barna.imre@wigner.mta.hu

Academic Editor: Emmanuel Lorin

In the present study a particular case of Gross-Pitaevskii or nonlinear Schrödinger equation is rewritten to a form similar to a hydrodynamic Euler equation using the Madelung transformation. The obtained system of differential equations is highly nonlinear. Regarding the solutions, a larger coefficient of the nonlinear term yields stronger deviation of the solution from the linear case.

1. Introduction

In 1926, at the advent of the quantum mechanics Madelung [1, 2] gave a hydrodynamical foundation of the Schrödinger equation and raised interpretation questions which are open till today. Later, people realized that the Madelung Anzatz is just the complex Cole-Hopf transformation [3, 4] which is sometimes applied to linearize nonlinear partial-differential equations (PDEs). The Madelung form of the Schrödinger equation has a remarkable property that Planck's constant (representing quantum mechanics) appears only once, namely, as the coefficient of the quantum potential. This quantity also appears in the nonmainstream attempt of quantum mechanics (the de Broglie-Bohm pilot wave theory [5] (in other context)) which interprets quantum mechanics as a deterministic nonlocal theory. The Madelung transformation may create a common language between scientist of fluid dynamics and quantum mechanics and can be fruitful to give an additional interpretation of quantum mechanics. In the last decades, hydrodynamical description of quantum mechanical systems become a popular technical tool in numerical simulations. Detailed reviews in this field can be found in a booklet [6] or in Wyatt's monograph [7].

The application of the original Madelung (or to say Cole-Hopf) transformation grew out the linear quantum theory and were applied to more sophisticated field theoretical problems. Recently, Chavanis and Matos [8] applied the Madelung Ansatz to the Klein-Gordon-Maxwell-Einstein equation in curved space-time possibly describing self-gravitating Bose-Einstein condensates, coupled to an electromagnetic field. This study clearly presents the wide applicability of the original Madelung Ansatz to refined quantum field theories with the possible hydrodynamic interpretation picture.

As an analytical tool of investigation we use the self-similar Ansatz of [9]. It is a powerful method to study the global properties of the dispersive solutions of various nonlinear partial-differential equations (PDEs) In other words, the intermediate asymptotic of the problem is described. Therefore this solution is valid when the precise initial conditions are no longer important; however, the final steady state is not yet reached. For real physical systems this region attracts the most attention. In contrast to the full solution it is much simpler and easier to understand for the entire parameter space. The second reason for studying them is that they are solutions of a system of ordinary differential equations (ODEs) (not a PDE) and thus do not suffer the full inherent numerical problems of the full PDEs. Most of the cases this tool helps to understand some of the global properties of the solution, e.g., shock waves or the existence of compact supports. First, we applied this Ansatz to the modified Cattaneo-Venotte heat-conduction problem [10] and obtained a compact solution with finite propagation speed. Later, we successfully generalized the Ansatz to multispace dimensions and studied numerous viscous fluid equations [11, 12] ending up with a book chapter of [13]. As main results we are now able to interpret the birth of Benard cells in heated fluids [14].

In our former study [15] we investigated the Schrödinger equation in the Madelung form with the self-similar Ansatz and presented analytic solutions with discussion. Unlike to classical fluids the obtained density of the Madelung fluid has infinite number of zeroes which is a fingerprint of the quantum mechanical origin. Now we analyze the Madelung form of nonlinear Schrödinger (NLS) equation with the same Ansatz and compare the obtained results to the previous linear one. As far as we know, there are no direct clear-cut analytic solutions existing for the Madelung form of the NLS equation by now. It is important to mention that celebrated standard numerical algorithms exist for solving the NLS [16, 17]. We hope that our present study helps to attract attention to this question. Parallel to the well-known quantum mechanical picture there is a looming hydrodynamical world hidden behind the Schrödinger equation from the very beginning which is poorly understood.

2. Theory and Results

In a weakly interacting Bose system, the macroscopic wave function Ψ appears as the order parameter of the Bose-Einstein condensation, obeying the Gross-Pitaevskii (GP) equation [18–21]:

$$i\hbar\frac{\partial\Psi}{\partial t} = \left(-\frac{\hbar^2\nabla^2}{2m} + n|\Psi|^2 - \mu\right)\Psi. \tag{1}$$

Here, $n = 4\pi\hbar^2 a/m$ represents the coupling constant characterized by the s-wave scattering length a. The particle mass and the chemical potential are m and μ, respectively. Due to the second (nonlinear) term of the right hand side of this equation, it is definitely different from the linear Schrödinger equation [15]; therefore the following analysis with the presented results are independent as well.

The role and the properties of the GP system has been investigated from various viewpoints. Without completeness we mention some of them. The question of collective excitations of the trapped Bose gases was analyzed by [22]. Later, it is considered in the study of self-organization of the Bose-Einstein (BE) condensates [23].

Certain generalized versions of (1) have been also applied to the study of BE condensate of polaritons [24–32].

Considering the Madelung Ansatz for the order parameter with the amplitude and phase $\Psi = \sqrt{\rho}e^{i\phi}$ where the density of the condensate is ρ and the superfluid velocity is $\mathbf{v} = (\hbar/m)\nabla\phi$. (The expression "superfluid velocity" is taken from Tsubota et al. [33] and means, in this content, that the fluid is inviscid.)

The obtained hydrodynamical equations, the continuity and the Euler, are as follows:

$$\begin{aligned} &\frac{\partial\rho}{\partial t} + \nabla\cdot(\rho\mathbf{v}) = 0,\\ &\frac{\partial\mathbf{v}}{\partial t} + \mathbf{v}\cdot\nabla\mathbf{v} = \frac{\hbar^2}{2m^2}\nabla\left(\frac{\Delta\sqrt{\rho}}{\sqrt{\rho}}\right) - \frac{\nabla}{m\rho}\left(\frac{n\rho^2}{2}\right), \end{aligned} \tag{2}$$

where we consider free motion (no extra potential energy term U($\mathbf{r}$) is used) and the chemical potential μ was set to zero as well. It is easy to show that an additional chemical potential term μ gives no contribution to the final hydrodynamical equations. The role of the external potential energy term $U(\mathbf{r})$ will be mentioned later.

The first term of the right hand side of the second equation is the quantum pressure term. The second term makes the difference to the original Madelung equations; this is a kind of *second quantum pressure* due to the nonlinear origin of the GP equation. Not, that these are most general vectorial equation for the velocity field $\mathbf{v}$ which means that one-, two-, or three-dimensional motions can be investigated as well. In the following we will consider the two-dimensional flow motion $\mathbf{v} = (u, v)$ in Cartesian coordinates. The functional form of the three- and one-dimensional equation system will be mentioned briefly, too.

We search the solution of (2) in the form of the self-similar Ansatz [9]:

$$\begin{aligned} &\rho(x,y,t) = t^{-\alpha}f\left(\frac{x+y}{t^{\beta}}\right) := t^{-\alpha}f(\eta),\\ &u(x,y,t) = t^{-\delta}g(\eta),\\ &v(x,y,t) = t^{-\epsilon}h(\eta), \end{aligned} \tag{3}$$

where $f, g,$ and h are the shape functions of the density and the velocity field, respectively. The similarity exponents $\alpha, \beta, \delta, \epsilon$ are of primary physical importance since α, δ, ϵ represent the damping of the magnitude of the shape function. The exponent β represents the spreading of the solution. Additional technical details about the general features of the Ansatz can be found in our former papers [12, 14]. Except some extraordinary cases only all positive similarity exponents correspond to physically relevant dispersive solutions which mean decaying features at $x, y, t \longrightarrow \infty$. Usually negative exponents describe exploding solutions. Substituting the Ansatz (3) into (2) and executing additional algebraic manipulation the following ODE system can written down for the shape functions:

$$\begin{aligned} &-\alpha f - \frac{\alpha}{2}f'\eta + f'g + fg' + f'h + fh' = 0,\\ &-\left(\frac{3}{2}\alpha - 1\right)g - \frac{\alpha}{2}g'\eta + gg' + hg'\\ &= \frac{\hbar^2}{2m^2}\left(\frac{f'^3}{2f^3} - \frac{f'f''}{f^2} + \frac{f'''}{2f}\right) - \frac{n}{m}f', \end{aligned}$$

$$-\left(\frac{3}{2}\alpha-1\right)h-\frac{\alpha}{2}h'\eta+gh'+hh' = \frac{\hbar^2}{2m^2}\left(\frac{f'^3}{2f^3}-\frac{f'f''}{f^2}+\frac{f'''}{2f}\right)-\frac{n}{m}f', \tag{4}$$

where prime means derivation with respect to the new variable η. The relations among the similarity exponents are the following: α can be taken arbitrary, $\beta = \alpha/2$ and $\epsilon = \delta = 3\beta - 1 = 3\alpha/2 - 1$. At this point we have to emphasize, that a particular form of the external potential $U(x, y, t)$ can be added to (1) without leading to contradiction among the similarity exponents. The form in 2D is the following: $U(x, y, t) = 1/t^{2\beta}\cdot v(\eta)$ where $v(\eta)$ can be any kind of function with existing first derivative.

(Of course, there is no evidence that the obtained modified (4) could be solved with analytical means.)

Note, that the ODE system of the linear Schrödinger equation [15] has a much simpler form of

$$\begin{aligned}
&-\frac{1}{2}f-\frac{1}{2}f'\eta+f'g+fg'+f'h+fh'=0,\\
&-\frac{1}{2}g-\frac{1}{2}g'\eta+gg'+hg' = \frac{\hbar^2}{2m^2}\left(\frac{f'^3}{2f^3}-\frac{f'f''}{f^2}+\frac{f'''}{2f}\right),\\
&-\frac{1}{2}h-\frac{1}{2}h'\eta+gh'+hh' = \frac{\hbar^2}{2m^2}\left(\frac{f'^3}{2f^3}-\frac{f'f''}{f^2}+\frac{f'''}{2f}\right),
\end{aligned} \tag{5}$$

where all the similarity exponents have the fixed value of $+1/2$. Our experience tells that this is usual for regular heat conduction, diffusion or for noncompressible Navier-Stokes equations [13]. For the linear case after some additional trivial algebraic steps a decoupled ODE can be derived for the shape function of the density:

$$2f''f-\left(f'\right)^2+\frac{m^2\eta^2 f^2}{2\hbar^2}=0. \tag{6}$$

For one, two, or three dimensions the multiplicative factor of $\hbar$ in the last term is different. It is 1, 2, or 3, respectively.

With the help of Bessel functions [34] the solution of (6) can be expressed:

$$f(\eta) = \frac{\pi\eta}{64}\left[c_1 J_{1/4}\left(\frac{\sqrt{2}m\eta^2}{8}\right)-c_2 Y_{1/4}\left(\frac{\sqrt{2}m\eta^2}{8}\right)\right]^2, \tag{7}$$

where c_1 and c_2 are the usual integration constants. All the additional properties of this solution can be found in our former paper [15]. (There is a misprint in the final [15] at Eq. 7, the parenthesis is missing in front of the exponent 2. This is the correct form.)

The most general equation of (4) cannot be fully integrated by analytical means. The first ODE of such systems [13] explicitly means mass conservation. Furthermore in the most cases it can be integrated analytically. However, the first equation of (4) is not a total derivative for any α. This is highly unusual among our investigated systems till now. As a second case let us consider $(3/2\alpha - 1) = \alpha/2$ this gives $\alpha = 1$, $\beta = \delta = \epsilon = 1/2$. Now the sum of the second and third equation of (5) can be integrated leading to

$$\frac{g^2}{2}+g\left(h-\frac{\eta}{2}\right)+\frac{h^2}{2}+\frac{h\eta}{2}+\frac{2nf}{m} - \frac{\hbar^2}{2m^2}\left(-\frac{f'^2}{4f^2}+\frac{f''}{2f}\right)+c_1=0. \tag{8}$$

For the sake of simplicity we set c_1 to zero. Note, that the variables g, f and h are still coupled in this single equation. However, (8) is quadratic in $g(\eta)$ and the solutions can be expressed with the well-know formula of

$$g_{1,2}=\frac{\eta}{2}-h\pm\sqrt{\frac{\eta^2}{4}-\frac{4nf}{m}+\frac{\hbar^2}{2m^2}\left(-\frac{f'^2}{4f^2}+\frac{f''}{2f}\right)}. \tag{9}$$

Even this includes both functions of h and f. To avoid this problem let us fix the discriminant to zero, which means that the velocity shape function will be single valued. The remaining ODE reads as follows:

$$2ff''-\left(f'\right)^2-f^2\left(\frac{\eta^2m^2}{2\hbar^2}-\frac{8nmf}{\hbar}\right)=0 \tag{10}$$

with the additional restraint that $4f^2 \neq 0$ meaning that the physical density of a fluid must be positive. Note, that without the last term we get back (6) with the solution of (7). We tried numerous variable transformations unfortunately, we could not find any closed solution for the full equation (10). (Note that if the coefficient of the second term would be twice as the first term there would be a possible solution for the full equation including the third order term.) It is also clear from (9) that the velocity shape functions have the trivial form of $g+h=\eta/2=(x+y)/2\sqrt{t}$ which is the same as for the linear case.

Figure 1 shows the analytical solution of (7) and numerically solution of Eq. (10) for various nonlinear parameters n with the same initial conditions of $f(0) = 1$ and $f'(0) = 0$. The $\hbar = m = 1$ units are used. All solutions have a strong damping with stronger and stronger oscillations at large arguments. The functions are positive for all values of the argument, (which is physical for a fluid density), but such oscillatory profiles are completely unknown in regular fluid mechanics [13]. The most interesting feature is the infinite number of zero values which cannot be interpreted physically for a classical real fluid. Remembering the restraint $4f^2 \neq 0$ of the calculation the solution function falls into infinite number of finite intervals. We think that this is a clear fingerprint that the obtained Euler equations have a quantum mechanical origin.

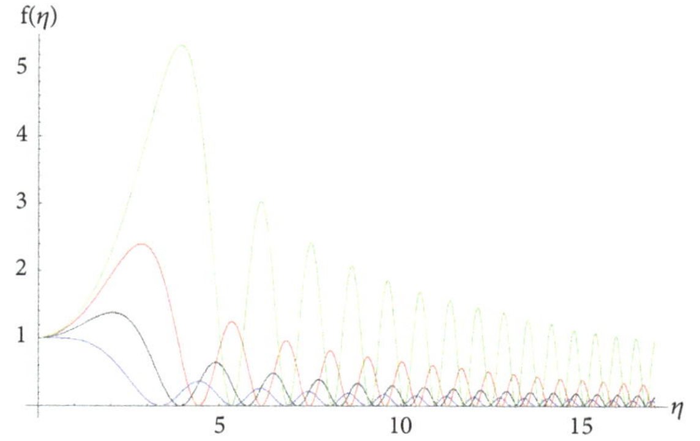

FIGURE 1: Various shape functions of the density. The blue curve is the analytic solution of equation (7) with $n = 0$ nonlinearity parameter. Black, red, and green functions are numerical results of equation (10) for $n = 0.078, 0.11$ and 0.13 nonlinearity parameters. (All curves are for the same initial conditions of $f(0) = 1$ and $f'(0) = 0$.) Note that larger parameters mean larger deviation from the analytic result.

The major difference to the analytic solution is the following the $\int_0^\infty f(\rho)d\rho$ can be evaluated (now of course just numerically up to an arbitrary ρ_{max}) giving us a convergent integral. This clearly shows that the wave function of the corresponding quantum mechanical system (see below) is the element of the L^2 space. We investigated numerous initial conditions all giving the same result of a finite integral. Of course, this is just an empirical statement and not a rigorous mathematical proof.

The $\sqrt{\rho}$ function could be interpreted as the fluid mechanical analogue of the real part of the wave function of the free GP quantum mechanical particle. For the linear case the square root of (7) is a kind of far analogue of the original free Gaussian wave packet. Unfortunately, we cannot find any direct transformation between these two functions.

To obtain the complete original wave function, the imaginary part has to be evaluated as well. It is simple from $\eta = (x+y)/t^{1/2} = 2(g+h)$ that

$$S = \frac{m}{\hbar}\int_{\mathbf{r}_0}^{\mathbf{r}_1} \mathbf{v}d\mathbf{r} = \frac{m}{\hbar}\frac{(x+y)^2}{4t}. \tag{11}$$

Now

$$\begin{aligned}\Psi(x,y,t) &= \sqrt{\rho(x,y,t)}e^{iS(x,y,t)} \\ &= \sqrt{t^{-1}\cdot f\left(\frac{[x+y]}{t^{1/2}}\right)}e^{(im/\hbar)((x+y)^2/4t)},\end{aligned} \tag{12}$$

where $f(\eta)$ is now a numerical function. Figure 2 shows the projection of the real part of the wave function to the x, t subspace. At small times there are clear oscillations, but at larger times the strong damping is evident. If the time is fixed ($t = t_0$ so to say) there are infinite number of spatial oscillations which can be interpreted as a quantum mechanical heritage of the system.

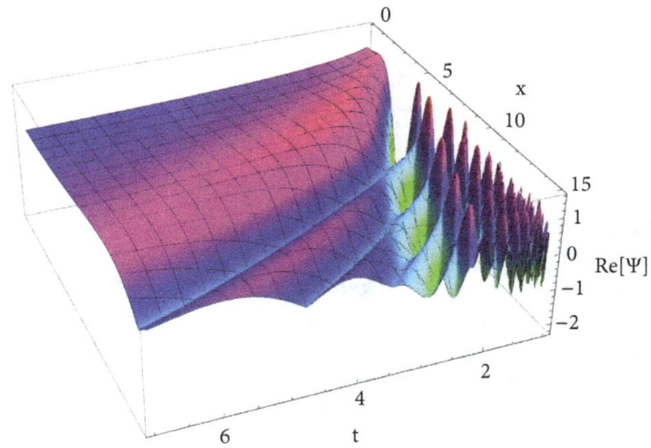

FIGURE 2: The projection of the real part of the wave function (12) along the x-t plane.

3. Summary and Outlook

We gave a short historical review of the interpretation and development of the Madelung equation. As second point we introduced (the not so well known) self-similar Ansatz which is a robust tool to investigate the local and global properties of nonlinear PDEs. Finally, the Madelung form of the free NLS equation was investigated and the results were compared to the ones obtained from the linear Schrödinger equation. We think that the highly oscillatory fluid density (with the infinite number of zeroes) are a clear fingerprint of the quantum mechanical origin of the system. We also believe that the noticeable quadratic argument of the solutions might get an interpretation in the future. However, the presented approach is still evolving and works are in progress. We plan to apply our method to the Madelung equations of more general nonlinear Schrödinger equations like [35].

Conflicts of Interest

The authors declare that they have no conflicts of interest.

Acknowledgments

The ELI-ALPS project (GINOP-2.3.6-15-2015-00001) is supported by the European Union and cofinanced by the European Regional Development Fund.

References

[1] E. Madelung, "Eine anschauliche Deutung der Gleichung von Schrödinger," *Naturwissenschaften*, vol. 14, no. 45, p. 1004, 1926.

[2] E. Madelung and Z. Phys, "Quantentheorie in hydrodynamischer form," *Zeitschrift für Physik*, vol. 40, p. 332, 1927.

[3] E. Hopf, "The partial differential equation $U_x + UU_t = U_{xx}$," *Communications on Pure and Applied Mathematics*, vol. 3, no. 3, pp. 201–230, 1950.

[4] J. D. Cole, "On a quasi-linear parabolic equation occurring in aerodynamics," *Quarterly of Applied Mathematics*, vol. 9, no. 3, pp. 225–236, 1951.

[5] D. Bohm, "A suggested interpretation of the quantum theory in terms of hidden variables, I and II," *Physical Review A: Atomic, Molecular and Optical Physics*, vol. 85, pp. 166–193, 1952.

[6] K. H. Hughes and G. Parlant, *Quantum Trajectories*, CCP6, Daresbury Laboratory, 2011, http://www.ccp6.ac.uk/booklets/CCP6-2011Q_Quantum_Trajectories.pdf.

[7] R. E. Wyatt, *Quantum Dynamics with Trajectories Introduction to Quantum Hydrodynamics*, vol. 28 of *Interdisciplinary Applied Mathematics*, Springer-Verlag, New York, 2005.

[8] P.-H. Chavanis and T. Matos, "Covariant theory of Bose-Einstein condensates in curved spacetimes with electromagnetic interactions: The hydrodynamic approach," *The European Physical Journal Plus*, vol. 132, p. 30, 2017.

[9] L. I. Sedov, *Similarity and Dimensional Methods in Mechanics*, Mir Publishers, Moscow, Russia, 1993.

[10] I. F. Barna and R. Kersner, "Heat conduction: a telegraph-type model with self-similar behavior of solutions," *Journal of Physics A: Mathematical and General*, vol. 43, no. 37, 375210, 6 pages, 2010.

[11] I. F. Barna, "Self-similar solutions of three-dimensional navier—stokes equation," *Communications in Theoretical Physics*, vol. 56, p. 745, 2011.

[12] I. F. Barna and L. Mátyás, "Analytic solutions for the three-dimensional compressible Navier-Stokes equation," *Fluid Dynamics Research. An International Journal*, vol. 46, no. 5, Article ID 055508, 10 pages, 2014.

[13] D. Campos, *Handbook on Navier-Stokes Equations, Theory and Applied Analysis*, Chapter 16, Nova Publishers, New York, NY, USA, 2017.

[14] I. F. Barna and L. Mátyás, "Analytic self-similar solutions of the Oberbeck-Boussinesq equations," *Chaos, Solitons & Fractals*, vol. 78, pp. 249–255, 2015.

[15] I. F. Barna, M. Pocsai, L. Mátyás, and J. Gen, "Self-similarity analysis of the non-linear Schrödinger equation in the Madelung form," *Journal of Generalized Lie Theory and Applications*, vol. 11, p. 271, 2017.

[16] P. Muruganandam and S. K. Adhikari, "Fortran programs for the time-dependent Gross–Pitaevskii equation in a fully anisotropic trap," *Computer Physics Communications*, vol. 180, no. 10, pp. 1888–1912, 2009.

[17] D. Vudragović, I. Vidanović, A. Balaž, P. Muruganandam, and S. K. Adhikari, "C programs for solving the time-dependent Gross–Pitaevskii equation in a fully anisotropic trap," *Computer Physics Communications*, vol. 183, no. 9, pp. 2021–2025, 2012.

[18] E. P. Groos, "Structure of a quantized vortex in boson systems," *Il Nuovo Cimento (1955-1965)*, vol. 20, pp. 454–477, 1961.

[19] L. P. Pitaevskii, "Vortex lines in an imperfect bose gas," *Soviet Physics JETP*, vol. 13, p. 451, 1961.

[20] F. Dalfovo, S. Giorgini, L. P. Pitaevskii, and S. Stringari, "Theory of Bose-Einstein condensation in trapped gases," *Reviews of Modern Physics*, vol. 71, no. 3, pp. 463–512, 1999.

[21] L. Erdös, B. Schlein, and H.-T. Yau, "Derivation of the Gross-Pitaevskii equation for the dynamics of Bose-Einstein condensate," *Annals of Mathematics: Second Series*, vol. 172, no. 1, pp. 291–370, 2010.

[22] A. Csordás and Z. Adam, "Collective excitations of trapped Fermi or Bose gases," *Physical Review A: Atomic, Molecular and Optical Physics*, vol. 74, Article ID 035602, 2006.

[23] D. Nagy, G. Szirmai, and P. Domokos, "Self-organization of a Bose-Einstein condensate in an optical cavity," *The European Physical Journal D*, vol. 48, p. 127, 2008.

[24] M. Wouters and I. Carusotto, "Excitations in a nonequilibrium bose-einstein condensate of exciton polaritons," *Physical Review Letters*, vol. 99, no. 14, 2007.

[25] K. G. Lagoudakis, M. Wouters, M. Richard et al., "Quantized vortices in an exciton-polariton condensate," *Nature Physics*, vol. 4, no. 9, pp. 706–710, 2008.

[26] G. Roumpos, M. Fraser, A. Löffler, S. Löffling, A. Forchel, and Y. Yamamoto, "Single vortex–antivortex pair in an exciton-polariton condensate," *Nature Physics*, vol. 7, pp. 129–133, 2011.

[27] E. A. Ostrovskaya, J. Abdullaev, A. S. Desyatnikov, M. D. Fraser, and Y. S. Kivshar, "Dissipative solitons and vortices in polariton Bose-Einstein condensates," *Physical Review A: Atomic, Molecular and Optical Physics*, vol. 86, no. 1, 2012.

[28] A. S. Rodrigues, P. G. Kevrekidis, R. Carretero-Gonzales, J. Cuevas-Maraver, and D. J. Frantzskakis, "From nodeless clouds and vortices to gray ring solitons and symmetry-broken states in two-dimensional polariton condensates," *Journal of Physics: Condensed Matter*, vol. 26, no. 15, Article ID 155801, 2014.

[29] V. S. Bagnato, D. J. Frantzeskakis, P. G. Kevrekidis, B. A. Malomed, and D. Mihalache, "Bose-Einstein condensation: twenty years after," *Romanian Reports in Physics*, vol. 67, no. 1, pp. 5–50, 2015.

[30] T. C. H. Liew, O. A. Egorov, M. Matuszewski, O. Kyriienko, X. Ma, and E. A. Ostrovskaya, "Instability-induced formation and nonequilibrium dynamics of phase defects in polariton condensates," *Physical Review B*, vol. 91, 2015.

[31] A. M. Belounis and S. Kessal, "Shock wave occurrence and soliton propagation in polariton condensates," *Canadian Journal of Physics*, vol. 95, pp. 1234–1238, 2017.

[32] C. Schneider, K. Winkler, M. D. Fraser et al., "Exciton-polariton trapping and potential landscape engineering," *Reports on Progress in Physics*, vol. 80, no. 1, Article ID 016503, 2017.

[33] M. Tsubota, M. Kobayashi, and H. Takeuchi, "Quantum hydrodynamics," *Physics Reports*, vol. 522, no. 3, pp. 191–238, 2013.

[34] F. W. J. Olver, D. W. Lozier, R. F. Boisvert, and C. W. Clark, *NIST Handbook of Mathematical Functions*, Cambridge University Press, 2010.

[35] R. Carles, R. Danchin, and J.-C. Saut, "Madelung, gross-pitaevskii and korteweg," *Nonlinearity*, vol. 25, no. 10, pp. 2843–2873, 2012.

Decay Rate and Energy Gap for the Singularity Solution of the Inhomogeneous Landau-Lifshitz Equation on $\mathbb{S}^2$

Penghong Zhong

Department of Mathematics, Guangdong University of Education, Guangzhou 510640, China

Correspondence should be addressed to Penghong Zhong; zhonghong737@163.com

Academic Editor: Andrei D. Mironov

The singularity solution for the inhomogeneous Landau-Lifshitz (ILL) equation without damping term in n-dimensional space was investigated. The implicit singularity solution was obtained for the case where the target space is on $\mathbb{S}^2$. This solution can be classified into four types that cover the global and local solutions. An estimation of the energy density of one of these types indicates its exact decay rate, which allows a global solution with finite initial energy under $n > 3$. Analysis of the four aperiodic solutions indicates that energy gaps that are first contributions to the literature of ILL will occur for particular coefficient settings, and these are shown graphically.

1. Introduction

The nonlinear ferromagnetic chain model (FCM) has attracted the attention of physicists and mathematicians. For physicists, FCM represents the possibility of describing magnetic field density evolution with various external fields, while mathematicians study the various solutions or excitations in geometrically motive nonlinear models. Quantitative models depicting FCM were proposed by Landau and Lifshitz in 1935 [1], and the Landau-Lifshitz equation (LLE) was proposed to the dynamics of the nonequilibrium magnetism system:

$$S_t = \alpha S \wedge (\triangle S + H) - \beta S \wedge (S \wedge (\triangle S + H)), \quad (\vec{x}, t) \in R^3 \times R, \tag{1}$$

where H is the variational sum of magnetic anisotropy energy and magnetic field energy, $S = (S_1, S_2, S_3)$ denotes magnetization intensity, $\wedge$ is the cross product of Euclidean 3-space R^3, and α and β are constants. If $H = 0$ and $\alpha = 0$, then LLE is the harmonic map heat flow (HMF). Otherwise, when $\beta = 0$, the Gilbert damping vanishes, and when $\beta = 0$ and $H = 0$, the LLE degenerates into the Schrödinger map heat flow (SMF), which is an important equation of differential geometry. SMF can be regarded as a nonlinear Schrödinger equation that contains a derivative term. Although the existence, uniqueness, and the blow-up problem of some nonlinear Schrödinger equations [2–4] are clear, the theorem of SMF becomes more complicated and some further work still needs to be done. Similarly, comparing HMF and some general harmonic system (or even biharmonic equation) [5–8], the mapping system is more complicated than the nonmapping system due to the curvature flow of the Riemannian manifolds.

For clarity, we set J and h as the complex structure and metric on the Kähler manifold, respectively. Different manifolds and their covariant derivatives are M and D, respectively, and we define SMF as

$$\frac{\partial S}{\partial t} = J \sum_{l=1}^{3} D^l \partial_l S. \tag{2}$$

Thus, (2) defines a mapping $S : R^3 \times R \longrightarrow (M, h, J)$, which can degenerate into other subcases; for example, when $M = \mathbb{S}^2$ and $J = S\Lambda$, (2) is the isotropic LLE without Gilbert damping, and if the manifold and complex structure are $\mathscr{H}^2$ and $\dot{\wedge}$ ($\vec{a} \dot{\wedge} \vec{b} = (\vec{a} \wedge \vec{b})\mathrm{diag}\{1, 1, -1\}$), respectively, then (2) is the hyperbolic isotropic LLE without Gilbert damping.

If $M = \mathbb{S}^2$ and $J = S\Lambda$, then (2) can transform into a Schrödinger-type equation using the Hashimoto transform.

Under n dimensional cylindrical coordinates, let $r = |\vec{x}|$, and the curvature and torsion are, respectively,

$$\begin{aligned} \kappa &= (S_r \cdot S_r)^{1/2}, \\ \tau &= \frac{S \cdot (S_r \Lambda S_{rr})}{\kappa^2}. \end{aligned} \tag{3}$$

Then, employing the Hashimoto transform,

$$Q = \frac{\kappa}{2} \exp\left[i \int \tau(t, \tilde{r})\, d\tilde{r}\right], \tag{4}$$

and (2) transforms to

$$\begin{aligned} &i\frac{\partial}{\partial t}Q + \frac{\partial^2}{\partial r^2}Q + \frac{1}{r}(n-1)\frac{\partial}{\partial r}Q - \frac{1}{r^2}(n-1)Q \\ &\quad + 2|Q|^2 Q = 0. \end{aligned} \tag{5}$$

If $n = 1$, (5) is the standard cubic Schrödinger equation for which the Lax pair is obtained, and explicit solutions can be constructed using the Bäcklund transformation. Hence, (5) can be regarded as a simple case of (2). Similar to the inhomogeneous Schrödinger equation, (2) (or (1)) can be extended to the inhomogeneous expression. If we add the inhomogeneous term $q(t, \vec{x})$ to (2), SME can be generalized into the inhomogeneous SME (ILL) proposed by Balakrishnan [9]:

$$\frac{\partial}{\partial t}S = q(t, \vec{x})\, S \wedge \Delta S + \nabla q(t, \vec{x})\,(S \wedge \nabla S), \tag{6}$$

where the inhomogeneous term, $q(t, \vec{x})$, is a scalar function. Under radially symmetrical coordinates, (6) can be expressed as

$$\begin{aligned} &\frac{\partial}{\partial t}S - q(t,r)\, S \wedge \left(\frac{\partial^2}{\partial r^2}S + \frac{n-1}{r}\frac{\partial}{\partial r}S\right) \\ &\quad - \frac{\partial}{\partial r}q(t,r)\left(S \wedge \frac{\partial}{\partial r}S\right) = 0. \end{aligned} \tag{7}$$

Using the Hashimoto transform (4), (7) transforms into a nonlinear inhomogeneous Schrödinger equation:

$$\begin{aligned} &i\frac{\partial}{\partial t}Q + q\left(\frac{\partial^2}{\partial r^2}Q + \frac{1}{r}(n-1)\frac{\partial}{\partial r}G - \frac{1}{r^2}(n-1)Q\right. \\ &\quad \left. + 2|Q|^2 Q\right) + Q\left(\frac{d^2}{dr^2}q + \frac{1}{r}(n-1)\frac{d}{dr}q\right. \\ &\quad \left. + 2\int\left(\frac{d}{dr}q\right)|Q|^2\, dr + 4(n-1)\int \frac{1}{r}q|Q|^2\, dr\right) \\ &\quad + 2\left(\frac{d}{dr}q\right)\frac{\partial}{\partial r}Q = 0. \end{aligned} \tag{8}$$

Daniel et al. [10] identified the integrable model (8) by analyzing the singularity structure of its solutions and also discussed Lax pairs, Bäcklund transformation, and soliton-like solutions of (7) (or (8)). They showed that inhomogeneous terms play an important role, and Painlevé analysis indicated that (7) (or (8)) was integrable in arbitrary n dimensions if and only if $q(t,r) = \varepsilon_1 r^{-2(n-1)} + \varepsilon_2 r^{-(n-2)}$.

As all know, the bifurcation structures of the general integrable systems and autonomous differential systems [11–14] are clear due to the properties of Hamiltonian systems [15–18]. However, the structure for the unintegrable system is not so clear. If $q(t,r)$ takes other forms, what would be the solution of (7)? The regularity of the unintegrable case has been somewhat clarified [19, 20] in regard of the singularity behavior of the equation. Two different finite time blowup solutions of (7) were constructed in [20], one being an explicit form and the other being an implicit solution. Careful estimation of the two spatial dimension implicit solutions indicates that the energy density decay rate is $1/r$. However, if the spatial dimension is any integer n, does a global smooth (or blowup) solution of (7) exist? Does (7) also present similar regularity aspects under specific smooth initial data? The regularity of n dimensional ILL remains open.

Many physical phenomena develop singular behavior, for example, boundary layers or blowup solutions. Liu [21] analyzed the concentration set of the stationary weak solutions to LLE for the ferromagnetic spin chain. Based on the Ginzburg-Landau approximation, Wang [22] established the existence of a global weak solution for LLE, $n <= 4$, with respect to smooth initial boundary data, which is smooth from a closed set with locally finite n dimensional parabolic Hausdorff measure. In 2008, Huh [23] constructed infinite energy explicit blowup solutions for the modified LLE. Ding [24] constructed an infinite energy blowup solution for LLE on a hyperbolic target.

The finite energy blowup solution was firstly studied by perturbation methods [25, 26], and it was found that the collapse of a symmetric case with large initial energy formed a singularity where the blowup rate could be estimated by scale invariance. To obtain the required resolution for evolving similar solutions, adaptive mesh refinement (AMR) is required [26–28], which dynamically and locally adds numerical resolution. Although Van Den Berg et al. [26] present a formal analysis and AMR simulation for LLE with the Gilbert term, rigorous proof of blowup for 2-dimensional (2D) LLE remains open. However, Ding et al. [29] proved that $n = 3$ or $n = 4$ dimensional LLE with the Gilbert term will lead to a finite time blowup under specific initial boundary conditions. Some regularity and blowup results for LLE were derived considering the Maxwell field [30, 31].

Although the blowup problem is clear for specific settings (initial state or initial boundary conditions), the blowup problem for ILL is unclear. The one-dimensional ILL will not form a singularity as the inhomogeneous term is periodic [32]. However, if the inhomogeneous term is in some other specific format [33], blowup occurs for the inhomogeneous HMF.

We studied the blowup and energy gap for n dimensional aperiodic ILL on $\mathbb{S}^2$ target and investigated what happens as t tends to infinity when the initial data is smooth and sufficiently large, in particular whether the solution develops distinct behaviors (finite or global time singularity) under the large data. Global smooth (or blowup) theory for ILL was not established, but we discuss some special solutions that form a singularity in finite time and classify these solutions and analyses based on their energy density.

In Section 2, we obtain a blowup solution for the ILL and derive the exact decay rate of the energy density for that solution. In Section 3, we obtain another local blowup solution that contains an energy gap under the inhomogeneous term and classify this solution into four types. In Section 4, we obtain the decay rate and prove the decreasing (or increasing) property of the local blowup solution for one of these types under some specific coefficient settings. Finally, in Section 5, we summarize the paper and present our conclusions.

2. Blowup Solution on a Sphere

2.1. Blowup Solution Derivation. Selecting the appropriate solution to construct the blowup solution is difficult. Group invariance (under $O(3,1)$ or $SL(2,C)$) can be applied to search for exact solutions of LLE [34–36], but ILL will not admit any group invariance. Following [19, 20, 37], we use cylindrical coordinates,

$$\begin{aligned} S_1(t,r) &= \cos(m(t,r))\sin(f(r)), \\ S_2(t,r) &= \sin(m(t,r))\sin(f(r)), \\ S_3(r) &= \cos(f(r)), \end{aligned} \tag{9}$$

where $r = |\vec{x}|$, $m(t,r)$, and $f(r)$ are functions to be determined.

For simplicity, consider the case where $q(t\vec{x})$ does not contain t and denote $q(r)$, $m(t,r)$, and $f(r)$ as q, m, and f, respectively. Then, from (9), (7) will be greatly simplified and maybe is transformed into

$$\begin{aligned} &2\frac{q\cos(f)((d/dr)f)(\partial/\partial r)m}{\sin(f)} + q\frac{\partial^2}{\partial r^2}m \\ &\quad + \left(\frac{d}{dr}q\right)\frac{\partial}{\partial r}m + \frac{(n-1)q(\partial/\partial r)m}{r} = 0 \end{aligned} \tag{10}$$

and

$$\begin{aligned} &q\sin(f)\cos(f)\left(\frac{\partial}{\partial r}m\right)^2 r + \left(\frac{\partial}{\partial t}m\right)\sin(f)r \\ &\quad - q\left(\frac{d^2}{dr^2}f\right)r - \left(\frac{d}{dr}q\right)\left(\frac{d}{dr}f\right)r \\ &\quad - (n-1)q\frac{d}{dr}f = 0, \end{aligned} \tag{11}$$

which are nonlinear partial differential equations. To solve these equations, let us assume that m is a variable separation function:

$$m = g(r)h(t) + C, \tag{12}$$

where $g(r)$ and $h(t)$ are dependent on r and t, respectively; and C is a constant. Since $q(t\vec{x})$ does not contain t, $g(r)$ and $h(t)$ are denoted as g and h, respectively. Substituting (12) into (10) and (11),

$$\begin{aligned} &2\left(\frac{d}{dr}g\right)q\cos(f)\left(\frac{d}{dr}f\right)r \\ &\quad + \left(\frac{d}{dr}g\right)\sin(f)\left(\frac{d}{dr}q\right)r + q\left(\frac{d^2}{dr^2}g\right)\sin(f)r \\ &\quad + (n-1)\left(\frac{d}{dr}g\right)q\sin(f) = 0, \end{aligned} \tag{13}$$

$$q\left(\frac{d^2}{dr^2}f\right)r + \left(\frac{d}{dr}q\right)\left(\frac{d}{dr}f\right)r + (n-1)q\frac{d}{dr}f = 0, \tag{14}$$

$$\cos(f)\left(\frac{d}{dr}g\right)^2 q - g = 0, \tag{15}$$

and

$$h = \frac{1}{t-T}, \tag{16}$$

where $T > 0$.

Solving (13)–(15),

$$f = 2\arctan\left(e^F\right), \tag{17}$$

$$q = \frac{C_1\left(e^{-F} + e^F\right)}{2((d/dr)F)r^{n-1}}, \tag{18}$$

and

$$g = \frac{1}{2}C_3\left(-1 + e^{-2F}\right)e^F, \tag{19}$$

where C_1 and C_2 are constants, F is a function of r, and F satisfies

$$\frac{1}{n}r^n - \frac{1}{2}C_3C_1F - \frac{1}{8}C_3C_1e^{2F} + \frac{1}{8}C_3C_1e^{-2F} + C_4 = 0, \tag{20}$$

where C_4 is a constant.

The F derivation of (20) is

$$-\frac{1}{4}C_3{}^2C_1\left(e^{2F} + 2 + e^{-2F}\right). \tag{21}$$

If C_1 and C_3 are nonzero constants, regardless of the value of F, the left sections of (20) and (21) are continuous functions, and (21) is nonzero. Therefore, according to the implicit function theorem, there exists a function F that satisfies (20), and we have the following conclusion.

Theorem 1. *Assuming that q takes the form of (18) and F satisfies (20), there is a solution of (7):*

$$S_1(t,r) = \frac{2e^F}{1+e^{2F}}\cos\left(\frac{C_3\left(-e^F+e^{-F}\right)}{2(t-T)}+C\right),$$
$$S_2(t,r) = \frac{2e^F}{1+e^{2F}}\sin\left(\frac{C_3\left(-e^F+e^{-F}\right)}{2(t-T)}+C\right), \tag{22}$$
$$S_3(r) = -\frac{-1+e^{2F}}{1+e^{2F}},$$

where C, C_1, C_3, and C_4 are arbitrary constants; and constant $T>0$.

On $\mathbb{S}^2$, from (9),

$$|S_r|^2 = S_1^2+S_2^2+S_3^2 = \left(\frac{d}{dr}f\right)^2 + \sin^2(f)\left(\frac{\partial}{\partial r}m\right)^2, \tag{23}$$

and $((\partial/\partial r)m)^2$ of (23) is

$$\left(\frac{\partial}{\partial r}m\right)^2 = \frac{\left(\cot^2(f)+1\right)^2\left((d/dr)f\right)^2 C_2{}^2}{(-t+T)^2} = \frac{C_2{}^2\left((d/dr)F\right)^2\left(e^{2F}+2+e^{-2F}\right)}{4(-t+T)^2}, \tag{24}$$

which indicates that (22) is a solution of (7) which blows up at T.

The energy density of (22) will also blow up, as can be seen from the energy density:

$$W_E = \frac{C_1(d/dr)f}{r^{n-1}} + \frac{C_1\sin^2(f)\left((\partial/\partial r)m\right)^2}{\left((d/dr)f\right)r^{n-1}}. \tag{25}$$

2.2. Decay Rate of the Blowup Solution. From the solutions and the energy density of the upper section, it is difficult to see the decay behavior of energy density. The decay characteristics of the solution determine whether solution energy can be finite value across the whole space. Therefore, we investigate the decay rate of the solution using the implicit solution from Section 2.1.

If we take the derivative of (20) with respect to r,

$$\frac{d}{dr}F = \frac{4r^n}{C_3C_1\left(e^{2F}+2+e^{-2F}\right)}, \tag{26}$$

and, from (17), (24), and (25),

$$W_E = \frac{8e^F}{C_3\left(e^{2F}+2+e^{-2F}\right)\left(1+e^{2F}\right)} + \frac{2C_3\left(e^{3F}+e^{-F}+2e^F\right)}{\left(e^{2F}+2+e^{-2F}\right)\left(1+e^{2F}\right)(-t+T)^2}. \tag{27}$$

From (26), $(d/dr)F|_0 = 0$, so if $r>0$ and $C_3C_1>0$,

$$\frac{d}{dr}F > 0, \tag{28}$$

where, in this case, F is a strictly increasing function in the interval $[0,+\infty)$.

Substituting $F=(n/2)\ln(Kr)$ $(K>0)$ into the left side of (20),

$$H(r) = \frac{r^n}{n} - \frac{1}{8}C_1C_3(Kr)^n - \frac{1}{4}C_1C_3 n\ln(Kr) + \frac{C_1C_3}{8(Kr)^n} + C_4. \tag{29}$$

If $r\longrightarrow+\infty$ and $(1/8)C_3C_1K^n - 1/n \neq 0$, then, from (20),

$$\lim_{r\longrightarrow+\infty} H(r) = -\operatorname{sign}\left(\frac{1}{8}C_3C_1K^n - \frac{1}{n}\right)\cdot\infty, \tag{30}$$

and if $(1/8)C_3C_1K^n - 1/n = 0$,

$$\lim_{r\longrightarrow+\infty} H(r) = -\frac{n}{4}C_3C_1\cdot\infty = -\infty. \tag{31}$$

Equations (30) and (31) mean that K determines the limit to be a positive or a negative value of $H(r)$. Hence,

$$\frac{n}{2}\ln(K_1r) < F < \frac{n}{2}\ln(K_2r) \quad (0<K_1<K_2). \tag{32}$$

Combining $F=(n/2)\ln(Kr)$ and (27),

$$W_E = \frac{8(Kr)^{3/2n}}{C_3\left((Kr)^{2n}+2(Kr)^n+1\right)\left(1+(Kr)^n\right)} + \frac{2C_3\left((Kr)^{5n/2}+2(Kr)^{3n/2}+(Kr)^{n/2}\right)}{\left((Kr)^{2n}+2(Kr)^n+1\right)\left(1+(Kr)^n\right)(-t+T)^2}. \tag{33}$$

From (33), as $r\longrightarrow+\infty$, the decay rate of the energy density is

$$W_E \sim \frac{\widetilde{C}(t)}{r^{n/2}}, \tag{34}$$

where $\widetilde{C}(t)$ is function of t; and as $t\longrightarrow T$, $\widetilde{C}(t)=\infty$.

Thus, we may obtain the decay rate of the $C_1C_3>0$ case. If $C_1C_3<0$ and $r\longrightarrow+\infty$, then the decay rate of the energy density can be similarly proven. Therefore,

$$F\longrightarrow-\infty, \tag{35}$$

where F satisfies

$$-\frac{n}{2}\ln(K_1r) > F > -\frac{n}{2}\ln(K_2r), \tag{36}$$

and from (27), if $r\longrightarrow+\infty$, we may estimate the solution

$$W_E \sim \widetilde{C}(t)e^F \sim \frac{\widetilde{C}(t)}{r^{n/2}}, \tag{37}$$

where $\widetilde{C}(t)$ is a function of t, where if and only if $t=T$, then $\widetilde{C}(t)=\infty$, which has the limit

$$\lim_{r\longrightarrow+\infty} W_E = 0. \tag{38}$$

Thus, W_E is a continuous function with the limit at $+\infty$ of 0. Hence, whether the energy is a finite value depends on the convergence of W_E at 0. From (20), if $r \longrightarrow 0$,

$$\lim_{r\longrightarrow 0} F = \widehat{C}, \tag{39}$$

where $\widehat{C}$ is constant. Thus,

$$\begin{aligned}\lim_{r\longrightarrow 0} W_E = {} & \frac{8e^{\widehat{C}}}{C_3\left(e^{2\widehat{C}}+2+e^{-2\widehat{C}}\right)\left(1+e^{2\widehat{C}}\right)} \\ & + \frac{2C_3\left(e^{3\widehat{C}}+e^{-\widehat{C}}+2e^{\widehat{C}}\right)}{\left(e^{2\widehat{C}}+2+e^{-2\widehat{C}}\right)\left(1+e^{2\widehat{C}}\right)(-t+T)^2},\end{aligned} \tag{40}$$

and if $\widehat{C} = 0$, then

$$\lim_{r\longrightarrow 0} W_E = \frac{T^2 - 2tT + t^2 + C_3{}^2}{T^2C_3 - 2TtC_3 + t^2C_3}. \tag{41}$$

We can obtain the initial energy density from (41) by setting $t = 0$:

$$\lim_{r\longrightarrow 0} W_{E_0} = \frac{T^2 + C_3{}^2}{T^2C_3}. \tag{42}$$

From (40), the initial energy density is a finite value and if and only if $t = T$, W_E will be in infinite value. Combining this and the decay behavior as $r \longrightarrow +\infty$, the energy of the finite spatial area can be estimated as

$$\int_0^A q(r)\left|S_r(t,r)\right|^2 r^{n-1}\mathrm{d}r < \overline{C}(t), \tag{43}$$

where A is a constant and $\overline{C}(t) = \infty$ if and only if $t = T$. Furthermore, the total energy of the spatial region is

$$\int_0^\infty q(r)\left|S_r(t,r)\right|^2 r^{n-1}\mathrm{d}r = \infty. \tag{44}$$

From the above analysis, we may conclude that, assuming that q takes the form of (18) and F satisfies (20), there is the following solution of (7).

Theorem 2. *If $C_1C_3 \neq 0$, q takes the form of (18) and F satisfies (20), the energy density of (22) is*

$$W_E \sim \frac{\widetilde{C}(t)}{r^{n/2}}, \tag{45}$$

as

$$r \longrightarrow +\infty, \tag{46}$$

and the energy in $r \in [0, A]$ satisfies

$$E[0,A] = \int_0^A q(r)\left|S_r(t,r)\right|^2 r^{n-1}\mathrm{d}r \leq \overline{C}(t), \tag{47}$$

where A is constant; and $\overline{C}(t)$ depends on t, such that as $t \longrightarrow T$, $\overline{C}(t) = \infty$.

Similar to the proof of [20], the decay rate of the norm of gradient can now be estimated. Combining (18) and (26),

$$q = \frac{1}{8}C_1{}^2 r^{-2n+2} e^{-F}\left(1+e^{2F}\right)C_3\left(e^{2F}+2+e^{-2F}\right), \tag{48}$$

and to estimate the decay rate as $r \longrightarrow +\infty$, we substitute $F = (n/2)\ln(Kr)$ $(K > 0)$:

$$\begin{aligned} q = {} & \frac{1}{8}C_1{}^2C_3\left((Kr)^{3n/2} + (Kr)^{-3n/2} + 3(Kr)^{n/2}\right. \\ & \left. + 3(Kr)^{-n/2}\right) r^{-2n+2}. \end{aligned} \tag{49}$$

Hence,

$$q \sim r^{-n/2+2}, \quad r \longrightarrow +\infty. \tag{50}$$

The gradient morn and energy density are related:

$$W_E = q\left|S_r\right|^2, \tag{51}$$

so that, combining Theorem 1 and $q \sim r^{-n/2+2}$, as $r \longrightarrow +\infty$,

$$\left|S_r\right|^2 \sim \frac{\widetilde{C}(t)}{r^2}, \tag{52}$$

where $\widetilde{C}(t)$ is function of t, such that if and only if $t \longrightarrow T$, $\overline{C}(t) = \infty$.

3. Multiple Branches of the Blowup Solutions

We investigate the classification of the blowup solution. Since (7) may contain many solutions different from (22), we extend (22) to a general form to find other solutions. In particular, we use (9) as an undetermined solution and assume that C_i $(i = 1, 2, 3, 4)$, C, and $T > 0$ are constant. We set

$$m = gh + C, \tag{53}$$

where

$$\begin{aligned} g &= C_2 + \frac{1}{2}C_3\left(-1 + e^{-2F}\right)e^F, \\ h &= \frac{1}{t-T}, \end{aligned} \tag{54}$$

and

$$q = \frac{C_1\left(e^{-F}+e^F\right)}{2\left((\mathrm{d}/\mathrm{d}r)F\right)r^{n-1}}. \tag{55}$$

Therefore,

$$\frac{\mathrm{d}}{\mathrm{d}r}F = \frac{4r^{n-1}\left(e^{2F}C_3 - 2e^FC_2 - C_3\right)}{C_1C_3{}^2\left(-e^{-2F} + e^{2F} + e^{4F} - 1\right)}. \tag{56}$$

If we set

$$C_4 = \frac{1}{8}C_1C_3\left(e^{2\overline{B}}+e^{-2\overline{B}}\right)+\frac{1}{2}C_1C_2\left(e^{\overline{B}}+e^{-\overline{B}}\right) + \frac{C_1\overline{B}C_2^2}{C_3}+\frac{1}{2}C_1C_3\overline{B} - \frac{2C_1C_2\sqrt{C_2^2+C_3^2}}{C_3}\operatorname{arctanh}\left(\frac{e^{\overline{B}}C_3-C_2}{\sqrt{C_2^2+C_3^2}}\right), \tag{57}$$

then we can obtain a solution of (56) which satisfies the initial condition $F(0)=\overline{B}$:

$$\frac{1}{n}C_3r^n-\frac{1}{8}C_3^2C_1e^{2F}-\frac{1}{2}C_1C_2C_3e^F-\frac{1}{2}C_3^2C_1F + \frac{1}{8}C_3^2C_1e^{-2F}-\frac{1}{2}C_3C_1C_2e^{-F}-C_1FC_2^2 + C_3C_4 + 2C_1C_2\sqrt{C_2^2+C_3^2}\operatorname{arctanh}\left(\frac{e^FC_3-C_2}{\sqrt{C_2^2+C_3^2}}\right) = 0. \tag{58}$$

The derivative of the left of (58) is

$$-\frac{1}{4}\frac{C_3^3\left(e^{4F}+e^{2F}-e^{-2F}-1\right)C_1}{e^{2F}C_3-2e^FC_2-C_3}. \tag{59}$$

We set C_1, C_2, and $C_3 \neq 0$. Thus, (59) is 0 if and only if $F=0$, and

$$F \neq \ln\left(\frac{C_2\pm\sqrt{C_2^2+C_3^2}}{C_3}\right). \tag{60}$$

Regardless of F value, (59) and the left section of (58) are continuous functions, and (59) is not zero. According to the existence theorem of implicit functions, for any

$$F \in R\setminus\left\{0,\ln\left(\frac{C_2\pm\sqrt{C_2^2+C_3^2}}{C_3}\right)\right\}, \tag{61}$$

there is an F that satisfies (58).

From the above process, we obtain the following conclusion.

Theorem 3. *Assuming that q takes the form of (18) and F satisfies (58), then there is a solution of (7):*

$$\begin{aligned} S_1(t,r) &= \frac{2e^F}{1+e^{2F}}\cos\left(\frac{C_3\left(-e^F+e^{-F}\right)}{2(t-T)}+\frac{C_2}{t-T}+C\right), \\ S_2(t,r) &= \frac{2e^F}{1+e^{2F}}\sin\left(\frac{C_3\left(-e^F+e^{-F}\right)}{2(t-T)}+\frac{C_2}{t-T}+C\right), \\ S_3(r) &= -\frac{-1+e^{2F}}{1+e^{2F}}, \end{aligned} \tag{62}$$

where C, C_1, C_2, C_3, and C_4 are any constants; and $T>0$ is constant.

The energy density of the solution from Theorem 3 is

$$W_E = \frac{2e^{-F}\left(e^{2F}C_3-2e^FC_2-C_3\right)\left(C_3^2\left(1+e^{2F}\right)^2+4e^{2F}(t-T)^2\right)}{C_3^2\left(1+e^{2F}\right)\left(-e^{-2F}+e^{2F}+e^{4F}-1\right)(-t+T)^2}. \tag{63}$$

Although (62) is similar to (22), their respective energies ((63) and (27)) are different. In (62), F satisfies (58) which contains arctanh, and this defines the range of F:

$$-1<\frac{e^FC_3-C_2}{\sqrt{C_2^2+C_3^2}}<1. \tag{64}$$

If

$$\frac{e^FC_3-C_2}{\sqrt{C_2^2+C_3^2}}=\pm 1, \tag{65}$$

then

$$F=\ln\left(\frac{\pm\sqrt{C_2^2+C_3^2}+C_2}{C_3}\right). \tag{66}$$

For (56), if F takes the form of (66), the right of (56) is 0, and F_r is 0.

We must solve

$$\frac{C_3e^{2F}-2C_2e^F-C_3}{\left(-e^{-2F}+e^{2F}+e^{4F}-1\right)C_1}>0, \tag{67}$$

which can be expressed as

$$\frac{C_3}{C_1}\frac{\left(e^F-\left(C_2+\sqrt{C_2^2+C_3^2}\right)/C_3\right)\left(e^F-\left(C_2-\sqrt{C_2^2+C_3^2}\right)/C_3\right)}{\left(e^{4F}-1\right)\left(e^{-2F}+1\right)}>0. \tag{68}$$

Table 1: Evolution of F where $C_3 > 0$ and $r > 0$.

	$C_3 > 0,\ C_2 > 0$	$C_3 > 0,\ C_2 < 0$
$C_1 > 0$	If $0 < F < \ln\left(\frac{\sqrt{C_2^2 + C_3^2} + C_2}{C_3}\right)$, $F_r < 0$ If $-\infty < F < 0$, $F_r > 0$	If $-\infty < F < \ln\left(\frac{\sqrt{C_2^2 + C_3^2} + C_2}{C_3}\right)$, $F_r > 0$
$C_1 < 0$	If $0 < F < \ln\left(\frac{\sqrt{C_2^2 + C_3^2} + C_2}{C_3}\right)$, $F_r > 0$ If $-\infty < F < 0$, $F_r < 0$	If $-\infty < F < \ln\left(\frac{\sqrt{C_2^2 + C_3^2} + C_2}{C_3}\right)$, $F_r < 0$

Table 2: Evolution of F with $C_3 < 0$ and $r > 0$.

	$C_3 < 0, C_2 < 0$	$C_3 < 0, C_2 > 0$
$C_1 > 0$	If $0 < F < \ln\left(\frac{-\sqrt{C_2^2 + C_3^2} + C_2}{C_3}\right)$, $F_r < 0$ If $-\infty < F < 0$, $F_r > 0$	If $-\infty < F < \ln\left(\frac{-\sqrt{C_2^2 + C_3^2} + C_2}{C_3}\right)$, $F_r < 0$
$C_1 < 0$	If $0 < F < \ln\left(\frac{-\sqrt{C_2^2 + C_3^2} + C_2}{C_3}\right)$, $F_r > 0$ If $-\infty < F < 0$, $F_r < 0$	If $-\infty < F < \ln\left(\frac{-\sqrt{C_2^2 + C_3^2} + C_2}{C_3}\right)$, $F_r > 0$

If $C_3 > 0$,

$$\frac{\sqrt{C_2^2 + C_3^2} + C_2}{C_3} > 0, \tag{69}$$

and

$$\frac{-\sqrt{C_2^2 + C_3^2} + C_2}{C_3} < 0. \tag{70}$$

Solving

$$\left(e^{4F} - 1\right)\left(e^{-2F} + 1\right) > 0, \tag{71}$$

we obtain

$$F > 0. \tag{72}$$

Similarly, solving

$$\left(e^{4F} - 1\right)\left(e^{-2F} + 1\right) < 0, \tag{73}$$

we obtain

$$F < 0. \tag{74}$$

If $C_3 > 0$ and $C_2 > 0$, then the inequality is

$$\ln\left(\frac{\sqrt{C_2^2 + C_3^2} + C_2}{C_3}\right) > 0, \tag{75}$$

and if $C_3 > 0$ and $C_2 < 0$, then

$$\ln\left(\frac{\sqrt{C_2^2 + C_3^2} + C_2}{C_3}\right) < 0. \tag{76}$$

Thus, if $C_3 > 0$, $C_2 > 0$, $C_1 < 0$, and

$$0 < F < \ln\left(\frac{\sqrt{C_2^2 + C_3^2} + C_2}{C_3}\right), \tag{77}$$

then (68) holds, and since $r > 0$,

$$F_r > 0. \tag{78}$$

If $r > 0$, $C_3 > 0$, $C_2 < 0$, $C_1 > 0$, and

$$-\infty < F < \ln\left(\frac{\sqrt{C_2^2 + C_3^2} + C_2}{C_3}\right), \tag{79}$$

then

$$F_r > 0. \tag{80}$$

Similarly, if $C_3 < 0$, we can solve F from $F_r > 0$, and we can solve $F_r < 0$ for the $C_3 > 0$ (or $C_3 < 0$) case. From (58) and (56), if $C_3 > 0$, the change of F_r is shown in Table 1.

Similar to Table 1, if $C_3 < 0$, the change of F_r is shown in Table 2.

From Tables 1 and 2, F satisfying (58) can be subdivided into the following categories:

(1) $C_3 > 0$, $C_2 > 0$ and $C_1 > 0$ (or $C_3 < 0$, $C_2 < 0$ and $C_1 > 0$).

F is a continuous curve. If $F = 0$, solving (58),

$$r_0 = \left[\frac{n}{C_3}\left(2\operatorname{arctanh}\left(\frac{-C_3 + C_2}{\sqrt{C_2^2 + C_3^2}}\right)\right.\right. \tag{81}$$
$$\left.\left.\cdot\sqrt{C_2^2 + C_3^2 C_1 C_2 + C_1 C_2 C_3 - C_3 C_4}\right)\right]^{1/n},$$

and (56) can be expressed as

$$r^{n-1}\frac{dr}{dF} = \frac{C_1 C_3^2\left(-e^{-2F} + e^{2F} + e^{4F} - 1\right)}{4\left(e^{2F}C_3 - 2e^F C_2 - C_3\right)}. \tag{82}$$

If the right section of (82) is 0, then $F = 0$. Hence, $r_F = 0$ as $r = r_0$. Thus, the tangent of F is perpendicular to the $r-$ axis at $(r_0, 0)$.

If $r = 0$, then $F_r = 0$ from (56). Hence, the tangent of F is parallel to the r–axis at $(0, F(0))$. Specifically, in the fourth quadrant of the rF coordinate system, F is monotonically increasing and $r \in [0, r_0]$. In the first quadrant, F is monotonically decreasing and $r \in [0, r_0]$.

For example, if $n = 2$, $C_1 = C_2 = C_3 = 1$, and $C_4 = 0$, then (58) is

$$\frac{1}{2}r^2 - \frac{1}{8}e^{2F} - \frac{1}{2}e^F - \frac{3}{2}F - \frac{1}{2}e^{-F} + \frac{1}{8}e^{-2F} + 2\sqrt{2}\operatorname{arctanh}\left(\frac{1}{2}e^F\sqrt{2} - \frac{1}{2}\sqrt{2}\right) = 0, \tag{83}$$

and the evolution of F, determined by (83), is shown in Figure 1.

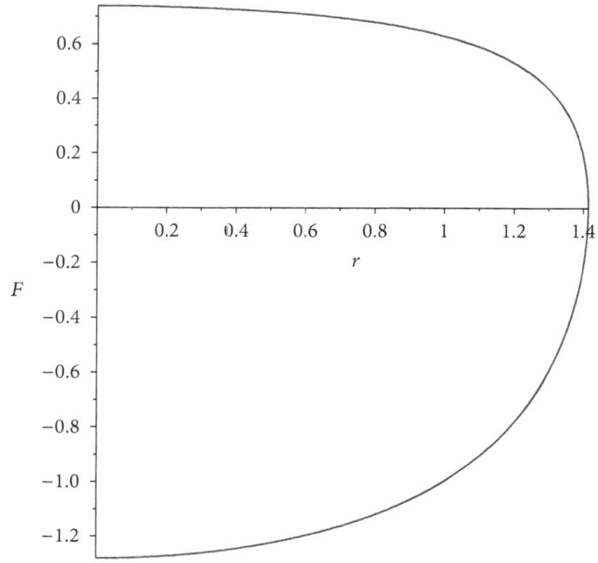

FIGURE 1: Evolution of F from (83) with $r \in [0, 2]$ and $F \in [-3, 3]$.

(2) $C_3 > 0$, $C_2 > 0$ and $C_1 < 0$ (or $C_3 < 0$, $C_2 < 0$ and $C_1 < 0$).

The evolution of F in the first and the fourth quadrant are opposite that of case (1). In the fourth quadrant, F is monotonically decreasing, and in the first quadrant, F is monotonically increasing. The tangent of F is perpendicular to the F–axis at $(0, F(0))$. If we set

$$F_{\max} = \max\left(\ln\left(\frac{\sqrt{C_2^2 + C_3^2} + C_2}{C_3}\right), \ln\left(\frac{-\sqrt{C_2^2 + C_3^2} + C_2}{C_3}\right)\right), \tag{84}$$

the maximum value of F in first quadrant will not exceed $F_{\max}$.

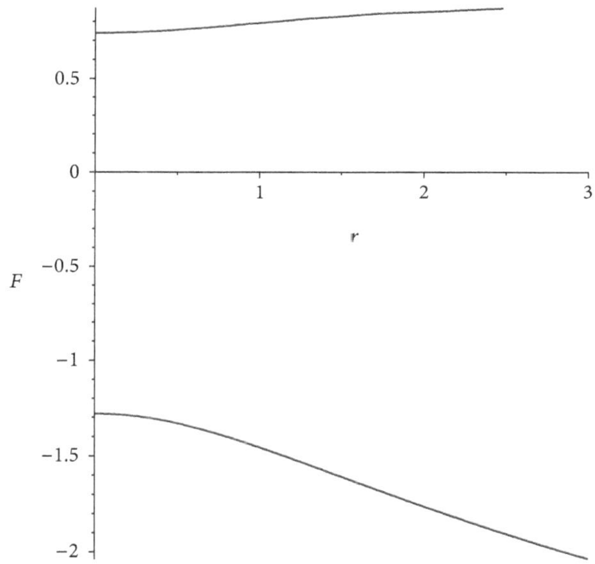

FIGURE 2: Evolution of F from (85) with $r \in [0, 3]$ and $F \in [-3, 2.5]$.

Specifically, if $n = 2$, $C_1 = -1$, $C_2 = C_3 = 1$, and $C_4 = 0$, then (58) is

$$\frac{1}{2}r^2 + \frac{1}{8}e^{2F} + \frac{1}{2}e^F + \frac{3}{2}F + \frac{1}{2}e^{-F} - \frac{1}{8}e^{-2F} + 2\sqrt{2}\operatorname{arctanh}\left(-\frac{1}{2}e^F\sqrt{2} + \frac{1}{2}\sqrt{2}\right) = 0, \tag{85}$$

and the evolution of F from (85) is shown in Figure 2.

(3) $C_3 > 0$, $C_2 < 0$ and $C_1 > 0$ (or $C_3 < 0$, $C_2 > 0$ and $C_1 < 0$).

F is completely in the fourth quadrant, and is monotonically increasing. The maximum value of F will not

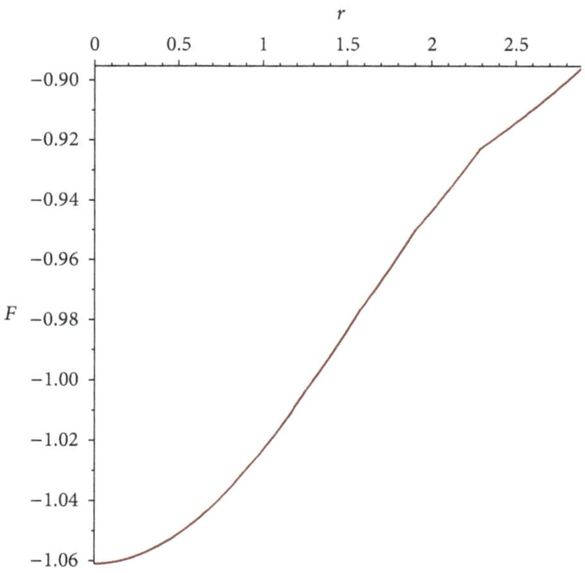

FIGURE 3: Evolution of F from (86) with $r \in [0, 5]$ and $F \in [-4, 4]$.

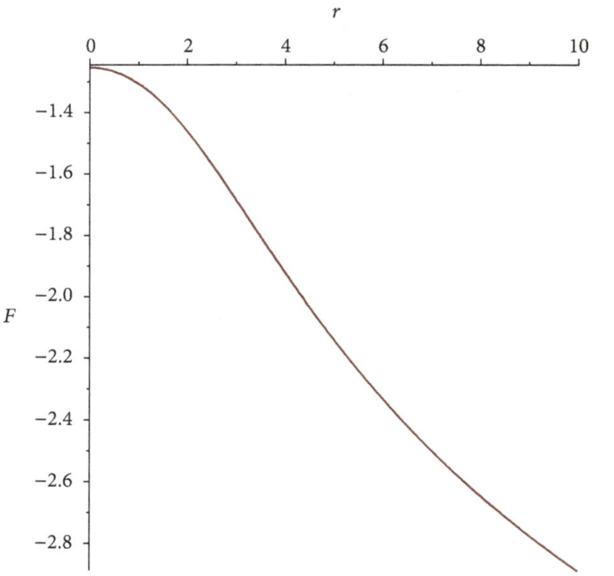

FIGURE 4: Evolution of F from (87) with $r \in [0, 10]$ and $F \in [-4, 4]$.

exceed F_{max}, and the curve in this quadrant has lower bound $F(0)$.

For example, when $n = 2$, $C_1 = 1$, $C_2 = -1$, $C_3 = 1$, and $C_4 = 1$, (58) is

$$\frac{1}{2}r^2 - \frac{1}{8}e^{2F} + \frac{1}{2}e^F - \frac{3}{2}F + \frac{1}{2}e^{-F} + \frac{1}{8}e^{-2F} + 1 + 2\sqrt{2}\,\text{arctanh}\left(-\frac{1}{2}e^F\sqrt{2} - \frac{1}{2}\sqrt{2}\right) = 0, \tag{86}$$

and Figure 3 shows the evolution of F from (86) with $r \in [0, 5]$ and $F \in [-4, 4]$.

(4) $C_3 > 0$, $C_2 < 0$ and $C_1 < 0$ (or $C_3 < 0$, $C_2 > 0$ and $C_1 > 0$).

F is monotonically decreasing in the fourth quadrant, with upper bound $F(0)$.

Cases (1) and (2) are multi-value, i.e., for each r, There are two values of F. We call these multi-branch functions. However, there is only a single monotonic function for cases (3) and (4).

Let us select some specific parameters and investigate the characteristic solutions. If $n = 2$, $C_1 = C_2 = -1$, $C_3 = 1$, and $C_4 = 1$, then (58) is

$$\frac{1}{2}r^2 + \frac{1}{8}e^{2F} - \frac{1}{2}e^F + \frac{3}{2}F - \frac{1}{2}e^{-F} - \frac{1}{8}e^{-2F} + 1 - 2\sqrt{2}\,\text{arctanh}\left(-\frac{1}{2}e^F\sqrt{2} - \frac{1}{2}\sqrt{2}\right) = 0, \tag{87}$$

and the evolution of (87) is shown in Figure 4.

In (56), we may regard F as a function of r and z and decompose (56) as

$$\frac{d}{dz}F = 4r^{n-1}\left(e^{2F}C_3 - 2e^F C_2 - C_3\right), \tag{88}$$

and

$$\frac{d}{dz}r = C_1 C_3^2\left(-e^{-2F} + e^{2F} + e^{4F} - 1\right). \tag{89}$$

$F(z)$ and $r(z)$ cannot be expressed as a Hamiltonian system; that is, there is no function $H(F, r)$ such that

$$\frac{\partial}{\partial F}H(F, r) = \frac{d}{dz}r, \quad \frac{\partial}{\partial r}H(F, r) = \frac{d}{dz}F. \tag{90}$$

Therefore, (90) are noncompatible systems. Solving the first equation of (90),

$$H(F, r) = -C_1 C_3^2\left(-\frac{1}{2}e^{-2F} - \frac{1}{2}e^{2F} - \frac{1}{4}e^{4F}\right) - C_1 C_3^2 F + F_1(r), \tag{91}$$

where F_1 is a function that is only dependent on r; and solving the second equation of (90),

$$H(F, r) = -\frac{4r^n\left(2e^F C_2 - e^{2F}C_3 + C_3\right)}{n} + F_2(F), \tag{92}$$

where F_2 is a function that is only dependent on F.

Since (91) and (92) are contradictory expressions, (90) is an incompatible system.

Equations (88)-(89) contain more solutions than (58). The evolution of these equations can also be classified into four types as follows:

1. Multibranch elliptic case I: $C_3 > 0$, $C_2 > 0$, and $C_1 > 0$ (or $C_3 < 0$, $C_2 < 0$, and $C_1 > 0$).
2. Multibranch hyperbolic case I: $C_3 > 0$, $C_2 > 0$, and $C_1 < 0$ (or $C_3 < 0$, $C_2 < 0$, and $C_1 < 0$).

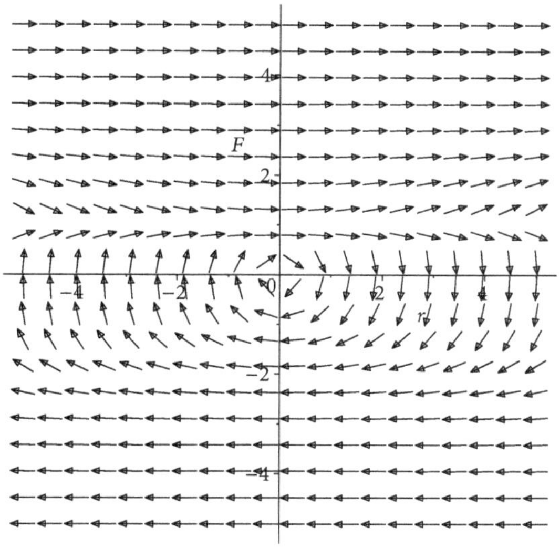

FIGURE 5: Multibranch elliptic case I: evolution of F ($C_1 = C_2 = C_3 = 1$).

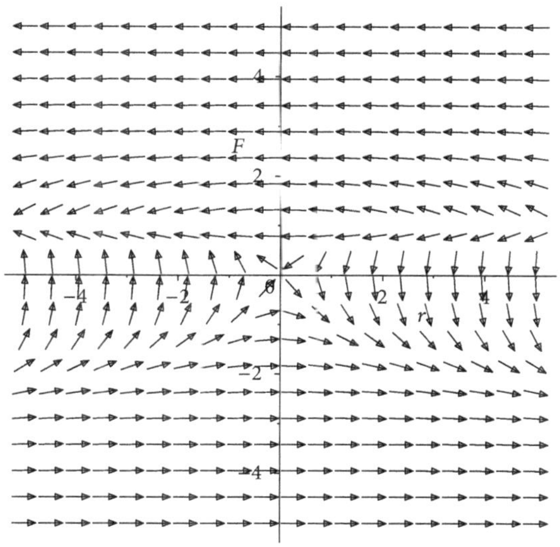

FIGURE 6: Multibranch hyperbolic case I: distribution of F ($C_1 = -1$ and $C_2 = C_3 = 1$).

3. Multibranch hyperbolic case II: $C_3 > 0$, $C_2 < 0$, and $C_1 > 0$ (or $C_3 < 0$, $C_2 > 0$, and $C_1 < 0$).
4. Multibranch elliptic case II: $C_3 > 0$, $C_2 < 0$, and $C_1 < 0$ (or $C_3 < 0$, $C_2 > 0$, and $C_1 > 0$).

To see the evolution of these four solutions directly, we set $n = 2$ and C_1, C_2, and C_3 as follows:

1. $C_1 = C_2 = C_3 = 1$.
2. $C_1 = -1$ and $C_2 = C_3 = 1$.
3. $C_1 = 1$, $C_2 = -1$, and $C_3 = 1$.
4. $C_1 = C_2 = -1$ and $C_3 = 1$.

Figures 5–8 show the orientation fields for these four cases in the rF coordinate, respectively.

The direction of vector field in Figure 5 is opposite to that of Figure 8, and the direction of Figure 6 is opposite to that of Figure 7. Figures 5 and 8 (or Figures 6 and 7) are very similar. However, in Section 4, we show that the energy densities suggest that all of these four multibranch cases are distinct.

4. Blowup Solution and Energy Gap

The solutions presented in Sections 2 and 3 will blow up in finite time, but the energy of the solution presented in Section 2 is infinite on all spatial regions. However, can (7) (or (6)) contain some finite energy solutions on the sphere, since solution (62) presented in Section 3 contains some finite energy subcases? In this section, we investigate the blowup solution (62) and particularly the finite energy subcases.

Let us analyze the properties of the energy density (63) of solution (62). Since (63) contains more parameters than (27), the evolution of F_r is more complex for (56) than (26), which leads to more complex evolutionary behavior of (63).

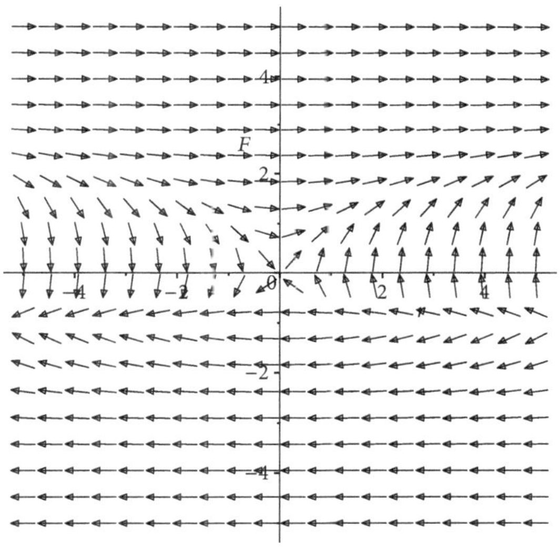

FIGURE 7: Multibranch hyperbolic case II: evolution of F ($C_1 = 1$, $C_2 = -1$, and $C_3 = 1$).

If $C_3 > 0$, $C_2 < 0$, and $C_1 < 0$ (or $C_3 < 0$, $C_2 > 0$, and $C_1 > 0$), then F_r is negative in $[0, +\infty)$, as shown in Figures 4 and 8. Substituting $F = -(n/2)\ln(Lr)$ ($L > 0$) into the left of (58), we obtain a function $J(r)$, with the limit

$$\lim_{r \longrightarrow +\infty} J(r) = \frac{1}{8n} C_3 \operatorname{sign}\left(L^n n C_1 C_3 + 8\right)) \cdot \infty, \tag{93}$$

where $K^n n C_1 C_3 + 8 \neq 0$.

From (93), the positive and negative of $\lim_{r \longrightarrow +\infty} J(r)$ can be determined by L; that is,

$$-\frac{n}{2}\ln(L_1 r) > F > -\frac{n}{2}\ln(L_2 r) \qquad (0 < L_1 < L_2),\ r \longrightarrow +\infty. \tag{94}$$

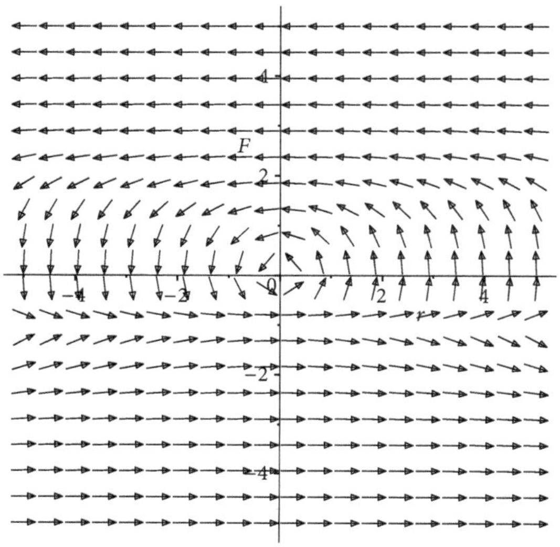

Figure 8: Multibranch elliptic case II: distribution of F ($C_1 = C_2 = -1$ and $C_3 = 1$).

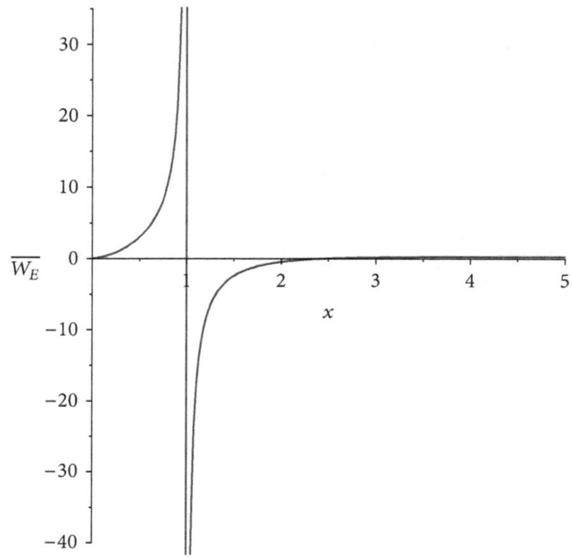

Figure 9: Relationship of $\overline{W}_E$ and x with $C_1 = C_2 = C_3 = C_5 = 1$.

Substituting $F = -(n/2)\ln(Lr)$ into (63), the decay rate of the energy density as $r \longrightarrow +\infty$ is

$$W_E \sim \frac{\widetilde{C}(t)}{r^{n/2}}, \tag{95}$$

where $\widetilde{C}(t)$ is a function of t, such that as $t \longrightarrow T$, $\widetilde{C}(t) = \infty$.

Similar to the proof of Theorem 2, we have the following corollary.

Corollary 4. *If F satisfies (58), $C_3 > 0$, $C_2 < 0$, and $C_1 < 0$ (or $C_3 < 0$, $C_2 > 0$, and $C_1 > 0$), then $r \longrightarrow +\infty$; the energy density of (62) is*

$$W_E \sim \frac{\widetilde{C}(t)}{r^{n/2}}, \tag{96}$$

and the energy in $r \in [0, A]$ satisfies

$$E[0,A] = \int_0^A q(r)\left|S_r(t,r)\right|^2 r^{n-1} dr \le \overline{C}(t), \tag{97}$$

where A is constant; and $\overline{C}(t)$ is only dependent on t, such that as $t \longrightarrow T$, $\overline{C}(t) = \infty$.

In Corollary 4, parameters C_3, C_2, and C_1 belong to one type from Table 1 (or Table 2). The other three parameters types were divided into two types: $C_3C_2 > 0$ and $C_3C_2 < 0$.

If $t - T = C_5$, then (63) may be expressed as

$$\overline{W}_E = \frac{2\left(e^{2F}C_3 - 2e^{F}C_2 - C_3\right)\left(C_3^2\left(1+e^{2F}\right)^2 + 4e^{2F}C_5^2\right)}{C_5^2 e^{F}\left(1+e^{2F}\right)\left(-e^{-2F} + e^{2F} + e^{4F} - 1\right)}, \tag{98}$$

and the monotonicity of (63) and that of (98) are the same under $t - T = C_5$. Substituting $F = \ln(x)$ into (98),

$$\overline{W}_E = \frac{2\left(x^2C_3 - 2xC_2 - C_3\right)\left(C_3^2x^4 + 2C_3^2x^2 + 4x^2C_5^2 + C_3^2\right)x}{C_5^2\left(x^2+1\right)\left(x^6 + x^4 - x^2 - 1\right)}. \tag{99}$$

Let us separate one equation from this, such as

$$x^2C_3 - 2xC_2 - C_3 = 0, \tag{100}$$

and solve to obtain

$$x_{1,2} = \frac{\pm\sqrt{C_2^2 + C_3^2} + C_2}{C_3}. \tag{101}$$

Thus, if $x = C_2/C_3$, the left section of (100) achieves its maximum or minimum value.

The derivative of the left of (99) is

$$\begin{aligned}\left(\overline{W}_E\right)_x = {} & -2C_5^{-2}\left(x^6 + x^4 - x^2 - 1\right)^{-2}\left(x^{10}C_3^3\right. \\ & - 4x^9C_2C_3^2 - x^8C_3^3 + 12x^8C_3C_5^2 - 8x^7C_2C_3^2 \\ & - 32x^7C_2C_5^2 - 2x^6C_3^3 - 36x^6C_3C_5^2 \\ & - 8x^5C_2C_3^2 + 32x^5C_2C_5^2 + 2x^4C_3^3 \\ & + 36x^4C_3C_5^2 - 8x^3C_2C_3^2 - 32x^3C_2C_5^2 + x^2C_3^3 \\ & \left. - 12x^2C_3C_5^2 - 4xC_2C_3^2 - C_3^3\right).\end{aligned} \tag{102}$$

If $C_1 = C_2 = C_3 = C_5 = 1$, then the exact relationship of $\overline{w}_E$ and x (or $\overline{w}_E$ and F) is shown in Figure 9 (Figure 10).

Figure 11 shows the evolution of energy density from (63) for multi ranch elliptic case I with $C_1 = C_2 = C_3 = 1$

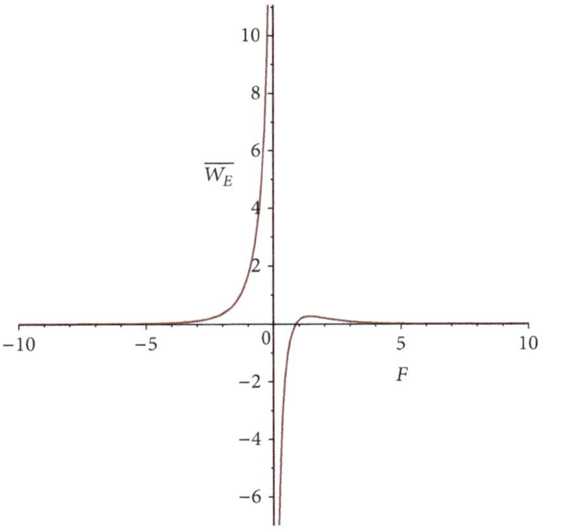

FIGURE 10: Relationship of $\overline{W}_E$ and F with $C_1 = C_2 = C_3 = C_5 = 1$.

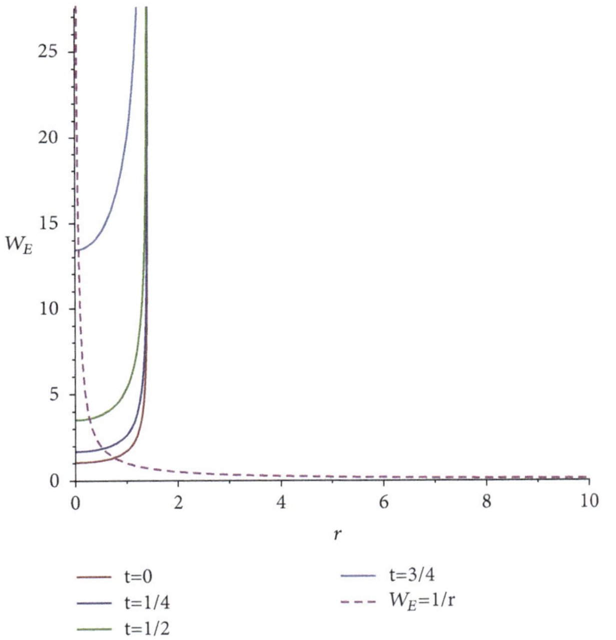

FIGURE 11: W_E at different times with $C_1 = C_2 = C_3 = 1$ and $C_4 = 0$.

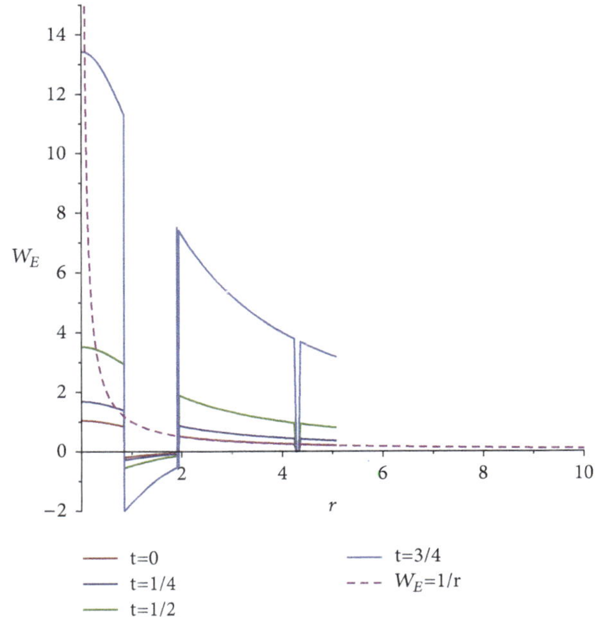

FIGURE 12: W_E in different time under $C_1 = -1$, $C_2 = C_3 = 1$, and $C_4 = 0$.

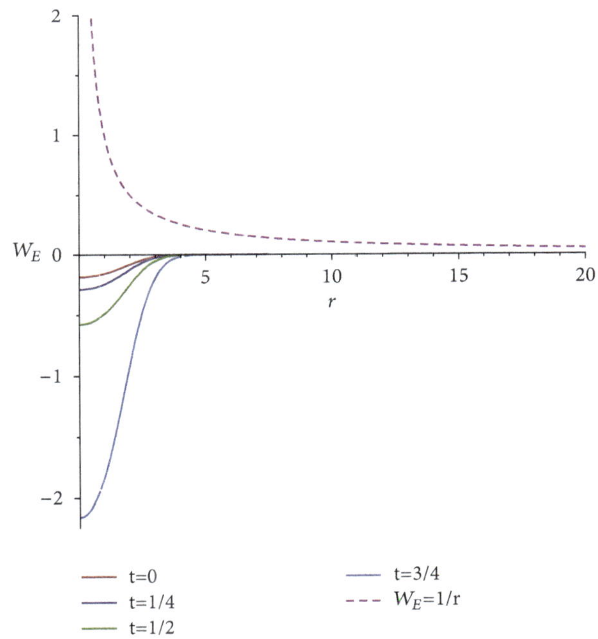

FIGURE 13: W_E for different times with $C_1 = C_3 = -1$ and $C_2 = C_4 = 1$.

and $C_4 = 0$, and Figure 12 shows multibranch hyperbolic case I with $C_1 = -1$, $C_2 = C_3 = 1$, and $C_4 = 0$. In these two cases, $n = 2$ and $T = 1$. The behavior is shown for $t = 0, 1/4, 1/2, 3/4$, and the function $W_E = 1/r$.

The multibranch elliptic case I (Figure 11) is a local solution that terminates at some point of r. The multibranch hyperbolic case I (Figure 12) is also a local solution. However, there are some energy gaps on W_E. Figure 13 shows the evolution of multibranch hyperbolic case II with $C_1 = C_3 = -1$ and $C_2 = C_4 = 1$ is also local with some energy gaps.

The situation for the multibranch elliptic case II is different from the other three cases. As predicted in Corollary 4, this solution diverges, as shown in Figure 14 with $C_1 = C_2 = -1$ and $C_3 = C_4 = 1$. W_E is a global solution and is close to zero as $r \longrightarrow \infty$. Near the W_E peaks, there are also some energy gaps.

We present some connections of ILL with real physics in the ending of this section. Clearly, we return the continuous ILL model back to the discrete Heisenberg model in the real

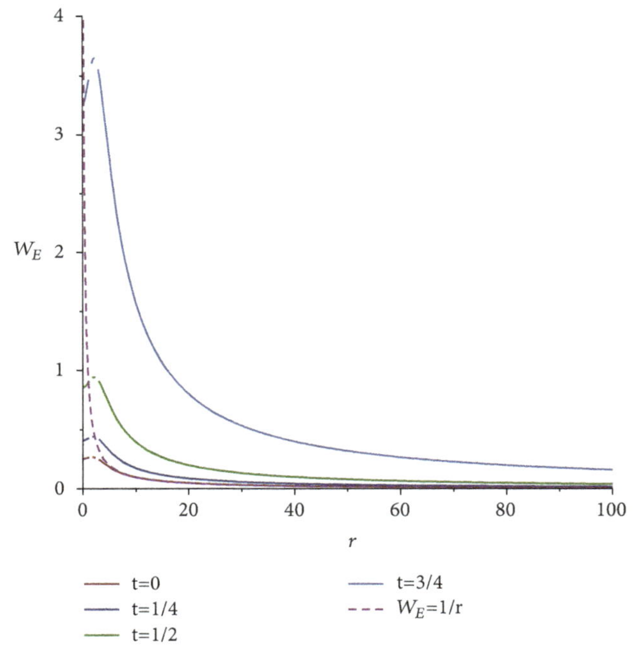

FIGURE 14: W_E for different times with $C_1 = C_2 = -1$ and $C_3 = C_4 = 1$.

physics. We plan to use the inhomogeneous Heisenberg chain model to discuss some physically relevant potential then.

ILL model is the continuum limitation of the discrete inhomogeneous Heisenberg chain, where the corresponding Heisenberg Hamiltonian (representing nearest neighbor interaction between S_i and S_{i+1}) is as follows:

$$H = -\Sigma_i q_i S_i S_{i+1}, \tag{103}$$

where S_i and S_{i+1} are on-site spins and q_i denotes the on-site ferromagnetic parameter which is the inhomogeneous physical interactions at the position i.

Equation (7) is a continuum model that can predicate any value of the site spin vector in a mathematical simulation way. Equation (7) plays a similar role as a Schrödinger equation for the Heisenberg model in quantum physics. As we can see in Section 1, the continuum limit of this system is also a nonlinear inhomogeneous Schrödinger equation (8) under the radial coordinates. For the discrete inhomogeneous Heisenberg chain, H or S_i has been used in the description of the magnetic properties and spin states of conjugated hydrocarbons and a large variety of metallic and organometallic compounds. What does the blowup solution (terminated in a finite time) represent in this discrete lattice system? In the continuum model ((7) or (8)), the blowup solution means that the gradient will be an infinite value, while the time is a finite time T. Accordingly, this finite time means that there are distortion and destruction in every position of i sites of a lattice. In this time T, there is a phase transition in the ferromagnetic solid.

Theorem 2 indicates that the phase transition (distortion and destruction) is possible under some special q_i. More clearly, if the exact quantitative scale of q is $r^{-n/2+2}$ as $r \longrightarrow +\infty$, this breaking down will happen in the ferromagnetic material. From a discrete point of view, i represents a correspondence of r by spatial position. The distribution $|i| \longleftrightarrow r$ means that the value of q_i is the value of $q(r)$. Hence, we expect that the phase transition in real ferromagnetic structure ($n = 3$) will begin in some blowup time T under the setting of $q_i \sim \sqrt{|i|}$ (as $|i| \longrightarrow +\infty$).

Multiple branches solution (see solution (62) in Section 3) allows a finite time blowup too. Accordingly, the discrete inhomogeneous Heisenberg chain of this system will generate a phase transition under some special setting of q_i. In particular, if $n = 3$ and $q_i \sim \sqrt{|i|}$ (as $|i| \longrightarrow +\infty$), the phase transition is a possible case in the multibranch elliptic case II according to Corollary 4. However, although (22) and (62) share a same breakdown in their discrete inhomogeneous Heisenberg chain system, the situation of multiple branches solution (62) is more complex than solution (22). For example, an energy gap can be found in the multibranch elliptic case II, while there is not any gap in solution (22). Why the energy gap will form in solution (62)? The most important reason is that there is a function arctanh in (58). If F touches the edges of the definition domain of arctanh, F jumps discontinuously from one value to another. In fact, if F touches the edges of the definition domain of arctanh, F will go to another track that belongs to the same multibranch. We can see that the difference between (22) and (62) is just a phase shift $C_2/(t-T)$ in the solution. From a point of the discrete inhomogeneous Heisenberg chain, the energy gap is inevitable due to the phase shift. This means that if we force the phase of the site spin vector to do a parallel movement, the jumping of the spin vector can be found in some location $|i|$ in any time before the blowup time. Furthermore, the scale of the jumping enlarges as the time goes on.

In the homogeneous Heisenberg chain ($q_i = 1$), all its lattice sites are equivalent. However, the self-consistent many-body ground states need not preserve all symmetries of the Hamiltonian in the inhomogeneous case. After the balance is broken, some singularity and jumping behaviors are possible according to the examples presented in the preceding sections. As we can see in these solutions, the Hamiltonian itself comprises inequivalent lattice sites that lead to the blowup and the energy gap. The inequivalent lattice sites are the case, for example, the Heisenberg chains with impurities or, more generally, the systems with different types of magnetic atoms. In the real problems, it is very common to find inhomogeneities rather than the homogeneous case. Actually, external nonuniform magnetic fields, anisotropic crystal fields, finite-size effects, boundary conditions, local impurities, lattice defects, and so forth can be induced for generalized inhomogeneities instances.

5. Conclusions

We investigated the evolution of the singular solution of n dimensional ILL on S^2. Two different ansatzes were applied in constructing the exact solutions in radially symmetric coordinates. Accordingly, two singular solutions were obtained ((22) and (62)). If one coefficient of (62) is set to 0, (62) degenerates into (22). All solutions show the same blowup rate $(T-t)^{-1}$. However, (22) is a global solution on the entire spatial domain, whereas (62) contains some local and global subcases.

Energy density plays an important role in the energy analysis of these two solutions. The energy density decay rate (22) scales as $r^{n/2}$, which implies that the initial energy of this solution will be finite for $n > 3$. However, (62) is somewhat more complicated, with four subcases. If the coefficients of the solution (C_1, C_2, and C_3 in (62)) are suitably selected, the solution can be divided into four types: multibranch elliptic cases I and II and multibranch hyperbolic cases I and II. In contrast to (22), these subcases may lead to a new energy gap class of exact solutions.

One of these four cases is the global solution in the spatial domain. In this case, since the decay rate also scales as $r^{n/2}$, the initial energy is finite. The other three subcases can be local solutions, which means that the initial energy will be finite for any spatial dimension. All these multibranch cases lead to energy gaps in the energy density. The evolution of energy density for the cases was shown in Section 3, illustrating the decay property and monotonic increasing and decreasing behaviors. These evolutions were consistent with the theorem and corollary provided in Section 3. Thus, the multibranching of F for the solution plays a decisive role.

Integrability of the physical ILL has been investigated in [10]. The ILL discussed in this paper does not belong to the integrability system due to the inhomogeneous term. While many integrable spin systems are known, the integrability of IIL in the continuum limit has not been investigated. It will be of considerable interest to investigate the underlying singularity structures and energy gaps of such nonintegrable nonlinear evolution equations in detail. It would be also instructive to explore whether similar singular solutions can be obtained for LLE with Gilbert damping.

In conclusion, ILL and LLE display very interesting dynamic structures and further work is required to identify them.

Conflicts of Interest

The author declares that he has no conflicts of interest.

Acknowledgments

This work is supported by NSFC (no. 11161057), National Science Foundation for Young Scientists of China (no. 11301014), TianYuan Special Fund of the National Natural Science Foundation of China (no. 11426068), Project for Young Creative Talents of Ordinary University of Guangdong Province (no. 2014KQNCX228), the PhD Start-up Fund of Natural Science Foundation of Guangdong Province (no. 2014A030310330), Research Award Fund for Outstanding Young Teachers in Guangdong Province (no. Yq20145084601), and Fund for Science and Technology of Guangzhou (no. 201607010352). The author is also grateful for the sponsorship from China Scholarship Council.

References

[1] L. D. Landau and E. M. Lifshitz, "On the theory of the dispersion of magnetic permeability in ferromagnetic bodies," in *Reproduced in Collected Papers of L. D. Landau*, Z. Sowjetunion, Ed., vol. 8, pp. 101–114, Pergamon, NY, USA, 1965.

[2] X. He, A. Qian, and W. Zou, "Existence and concentration of positive solutions for quasi-linear Schrodinger equations with critical growth," *Nonlinearity*, vol. 26, no. 12, pp. 3137–3168, 2013.

[3] Y. Sun, L. Liu, and Y. Wu, "The existence and uniqueness of positive monotone solutions for a class of nonlinear Schrödinger equations on infinite domains," *Journal of Computational and Applied Mathematics*, vol. 321, pp. 478–486, 2017.

[4] X. Zhang, L. Liu, Y. Wu, and Y. Cui, "Entire blow-up solutions for a quasilinear p-Laplacian Schrödinger equation with a non-square diffusion term," *Applied Mathematics Letters*, vol. 74, pp. 85–93, 2017.

[5] X.-N. Ma, P.-H. Wang, and W. Wei, "Constant mean curvature surfaces and mean curvature flow with non-zero Neumann boundary conditions on strictly convex domains," *Journal of Functional Analysis*, vol. 274, no. 1, pp. 252–277, 2018.

[6] F. Sun, L. Liu, and Y. Wu, "Infinitely many sign-changing solutions for a class of biharmonic equation with p-Laplacian and Neumann boundary condition," *Applied Mathematics Letters*, vol. 73, pp. 128–135, 2017.

[7] P. Wang and X. Wang, "The geometric properties of harmonic function on 2-dimensional Riemannian manifolds," *Nonlinear Analysis: Theory, Methods & Applications*, vol. 103, pp. 2–8, 2014.

[8] P. Wang and L. Zhao, "Some geometrical properties of convex level sets of minimal graph on 2-dimensional Riemannian manifolds," *Nonlinear Analysis: Theory, Methods & Applications*, vol. 130, pp. 1–17, 2016.

[9] R. Balakrishnan, "On the inhomogeneous Heisenberg chain," *Journal of Physics C: Solid State Physics*, vol. 15, no. 36, pp. L1305–L1308, 1982.

[10] M. Daniel, K. Porsezian, and M. Lakshmanan, "On the integrability of the inhomogeneous spherically symmetric Heisenberg ferromagnet in arbitrary dimensions," *Journal of Mathematical Physics*, vol. 35, no. 12, pp. 6498–6510, 1994.

[11] M. Han, L. Sheng, and X. Zhang, "Bifurcation theory for finitely smooth planar autonomous differential systems," *Journal of Differential Equations*, vol. 264, no. 5, pp. 3596–3618, 2018.

[12] F. Li and Q. Gao, "Blow-up of solution for a nonlinear Petrovsky type equation with memory," *Applied Mathematics and Computation*, vol. 274, pp. 383–392, 2016.

[13] F. Meng and J. Shao, "Some new Volterra-Fredholm type dynamic integral inequalities on time scales," *Applied Mathematics and Computation*, vol. 223, pp. 444–451, 2013.

[14] H. Tian and M. Han, "Bifurcation of periodic orbits by perturbing high-dimensional piecewise smooth integrable systems," *Journal of Differential Equations*, vol. 263, no. 11, pp. 7448–7474, 2017.

[15] L. Li, F. Meng, and Z. Zheng, "Oscillation results related to integral averaging technique for linear Hamiltonian systems," *Dynamic Systems and Applications*, vol. 18, no. 3-4, pp. 725–736, 2009.

[16] L. Li, F. Meng, and Z. Zheng, "Some new oscillation results for linear Hamiltonian systems," *Applied Mathematics and Computation*, vol. 208, no. 1, pp. 219–224, 2009.

[17] J. Liu and Z. Zhao, "Variational approach to second-order damped Hamiltonian systems with impulsive effects," *Journal of Nonlinear Sciences and Applications. JNSA*, vol. 9, no. 6, pp. 3459–3472, 2016.

[18] Z. Zheng, "Invariance of deficiency indices under perturbation for discrete Hamiltonian systems," *Journal of Difference Equations and Applications*, vol. 19, no. 8, pp. 1243–1250, 2013.

[19] P. Zhong, S. Wang, and S. Chen, "Some periodic and blow-up solutions for Landau-LIFshitz equation," *Modern Physics Letters A*, vol. 26, no. 32, pp. 2437–2452, 2011.

[20] P. Zhong, S. Wang, and M. Zeng, "Two blowup solutions for the inhomogeneous isotropic Landau-LIFshitz equation," *Journal of Mathematical Analysis and Applications*, vol. 409, no. 1, pp. 74–83, 2014.

[21] X.-G. Liu, "Concentration sets of the Landau-LIFshitz system and quasi-mean curvature flows," *Calculus of Variations and Partial Differential Equations*, vol. 27, no. 4, pp. 493–525, 2006.

[22] C. Wang, "On Landau-LIFshitz equation in dimensions at most four," *Indiana University Mathematics Journal*, vol. 55, no. 5, pp. 1615–1644, 2006.

[23] H. Huh, "Blow-up solutions of modified Schrödinger maps," *Communications in Partial Differential Equations*, vol. 33, no. 1-3, pp. 235–243, 2008.

[24] Q. Ding, "Explicit blow-up solutions to the Schrödinger R2 to the hyperbolic 2-space H2," *Journal of Mathematical Physics*, vol. 50, no. 10, 103507, 17 pages, 2009.

[25] F. Merle, P. Raphaël, and I. Rodnianski, "Blowup dynamics for smooth data equivariant solutions to the critical Schrödinger map problem," *Inventiones Mathematicae*, vol. 193, no. 2, pp. 249–365, 2013.

[26] J. B. van den Berg and J. F. Williams, "(In-)stability of singular equivariant solutions to the Landau-LIFshitz-Gilbert equation," *European Journal of Applied Mathematics*, vol. 24, no. 6, pp. 921–948, 2013.

[27] W. Ren and X.-P. Wang, "An iterative grid redistribution method for singular problems in multiple dimensions," *Journal of Computational Physics*, vol. 159, no. 2, pp. 246–273, 2000.

[28] X.-P. Wang and Carlos J., "A Gauss-Seidel projection method for micromagnetics simulations," *Journal of Computational Physics*, vol. 171, no. 1, pp. 357–372, 2001.

[29] S. Ding and C. Wang, "Finite time singularity of the Landau-LIFshitz-Gilbert equation," *International Mathematics Research Notices*, no. 4, Art. ID rnm012, 25 pages, 2007.

[30] S. Ding and J. Lin, "Partially regular solution to Landau-LIFshitz-Maxwell equations in two dimensions," *Journal of Mathematical Analysis and Applications*, vol. 351, no. 1, pp. 291–310, 2009.

[31] J. Lin and S. Ding, "Blow-up solutions to Landau-LIFshitz-Maxwell systems," *Mathematical Methods in the Applied Sciences*, vol. 34, no. 8, pp. 1006–1024, 2011.

[32] J. Lin and S. Ding, "Smooth solution to the one-dimensional inhomogeneous non-automorphic Landau-LIFshitz equation," *Proceedings of the Royal Society A Mathematical, Physical and Engineering Sciences*, vol. 462, no. 2072, pp. 2397–2413, 2006.

[33] Y. Li and Y. Wang, "Bubbling location for F-harmonic maps and inhomogeneous Landau-Lifshitz equations," *Commentarii Mathematici Helvetici*, vol. 81, no. 2, pp. 433–448, 2006.

[34] G. Yang, "The difference between Schrodinger equation derived from Schrödinger map and Landau-Lifshitz equation," *Physics Letters A*, vol. 376, no. 4, pp. 231–235, 2012.

[35] G. YANG, "Spherical cone symmetric families generated by Landau-Lifshitz equation and their evolution," *SCIENCE CHINA Mathematics*, vol. 41, no. 2, p. 181, 2011.

[36] G. Yang and B. Guo, "Some exact solutions to multidimensional Landau-LIFshitz equation with uprush external field and anisotropy field," *Nonlinear Analysis*, vol. 71, no. 9, pp. 3999–4006, 2009.

[37] B. Guo, Y. Han, and G. Yang, "Blow up problem for Landau-LIFshitz equations in two dimensions," *Communications in Nonlinear Science & Numerical Simulation*, vol. 5, no. 1, pp. 43-44, 2000.

The Perturbed Riemann Problem for the Aw-Rascle Model with Modified Chaplygin Gas Pressure

Yujin Liu and Wenhua Sun

School of Mathematics and Statistics, Shandong University of Technology, Zibo, Shandong 255000, China

Correspondence should be addressed to Yujin Liu; yjliu98@126.com

Academic Editor: Antonio Scarfone

This paper is concerned with the perturbed Riemann problem for the Aw-Rascle model with the modified Chaplygin gas pressure. We obtain constructively the solutions when the initial values are three piecewise constant states. The global structure and the large-time asymptotic behaviors of the solutions are discussed case by case. Further, we obtain the stability of the corresponding Riemann solutions as the initial perturbed parameter tends to zero.

1. Introduction

In the present paper we study the Aw-Rascle (AR) macroscopic model of traffic flow which is expressed by

$$\begin{aligned} \rho_t + (\rho u)_x &= 0, \\ (\rho(u+p))_t + (\rho u(u+p))_x &= 0, \end{aligned} \tag{1}$$

where $\rho \geq 0$, $u \geq 0$ are the density and the velocity, respectively, and p is the velocity offset which is called the "pressure" inspired from gas dynamics. The AR model was established to remedy the deficiencies of second order models of car traffic presented by Daganzo [1] and was independently obtained by Zhang [2]. For the related work, we refer to [3–10] and the references cited therein.

In [10], Wang obtained the Riemann solutions to (1) with

$$p = -\frac{B_1}{\rho^\alpha}, \quad 0 < \alpha \leq 1, \ B_1 > 0, \tag{2}$$

which is called the generalized Chaplygin gas. Wang observed that the delta shock wave occurs in the solutions. Let $B_1 = 1$ and $\alpha = 1$, we call it the Chaplygin gas which was introduced by Chaplygin [11]. Chaplygin gas (see [12–15]) is a good candidate for dark energy.

In [16], we constructed the solutions of the elementary wave interactions for the generalized AR model (1) with (2) by resolving two initial value problems case by case and found that the Riemann solutions are globally stable.

In [17], Wang, Liu, and Yang constructed the Riemann solutions for the AR model (1) with

$$p = B_2\rho - \frac{B_1}{\rho^\alpha}, \quad B_2 > 0, \tag{3}$$

which is called the modified Chaplygin gas. The authors analyzed the limit behavior when the pressure tends to zero.

In [18], Cheng and Yang studied the Riemann problem for (1) with

$$p = B_2\rho - \frac{B_1}{\rho}, \tag{4}$$

and they investigated the limits of the solutions when the pressure tends to the Chaplygin gas pressure.

In our present paper, we study the wave interactions for (1) with the state equation (3) which can be regarded as a combination of the perfect fluid and Chaplygin gas. In order to construct the solutions of the all possible wave interactions, we consider (1) and (3) with the following three piecewise constants:

$$(u,\rho)(x,0) = \begin{cases} (u_-,\rho_-), & x < -\zeta, \\ (u_m,\rho_m), & -\zeta < x < \zeta, \\ (u_+,\rho_+), & x > \zeta, \end{cases} \tag{5}$$

where the perturbation parameter $\zeta > 0$ is sufficiently small. The initial data (5) is a small perturbation on the corresponding Riemann initial values

$$(u, \rho)(x, 0) = (u_\pm, \rho_\pm), \quad \pm x > 0, \tag{6}$$

where $u_\pm, \rho_\pm > 0$. We will face the interesting question that whether the solutions of the perturbed Riemann problem (1), (3), and (5) converge to the corresponding Riemann solutions of (1), (3), and (6) as ζ tends to zero.

The paper is organized as follows. In Section 2, for readers' convenience we restate the results of the Riemann problem (1), (3), and (6). In Section 3, the elementary wave interactions are investigated case by case. And we obtain that the perturbed Riemann solutions of the modified AR model (1) and (3) converge to the corresponding Riemann solutions when ζ tends to zero which shows that the corresponding Riemann solutions of (1), (3), and (6) are globally stable.

2. Preliminaries

In this section we first sketch some results on the Riemann solutions to the modified AR model (1) and (3) and the detailed study can be found in [17].

The AR model (1) has two eigenvalues: $\lambda_1 = u - B_2\rho - B_1\alpha/\rho^\alpha$, $\lambda_2 = u$, and the corresponding right eigenvectors are given by

$$\begin{aligned} \vec{r}_1 &= \left(-B_2 - \frac{B_1\alpha}{\rho^{1+\alpha}}, 1\right)^T, \\ \vec{r}_2 &= (0, 1)^T. \end{aligned} \tag{7}$$

By a direct calculation, we have

$$\begin{aligned} \nabla\lambda_1 \cdot \vec{r}_1 &= -2B_2 - \frac{(1-\alpha) B_1\alpha}{\rho^{1+\alpha}} < 0, \\ \nabla\lambda_2 \cdot \vec{r}_2 &\equiv 0. \end{aligned} \tag{8}$$

Therefore, λ_1 is genuinely nonlinear, and λ_2 is always linearly degenerate. And we know that the Riemann solutions of (1), (3), and (6) can be constructed by shock wave, rarefaction wave, or the contact discontinuity connecting the two constant states (u_-, ρ_-) and (u_+, ρ_+).

For the reason that system (1) and the initial values (6) are invariant under stretching of coordinates, i.e., $(x, t) \to (\beta x, \beta t)$ (β is constant), we look for the self-similar solution $(u, \rho)(x, t) = (u, \rho)(\eta)$, $\eta = x/t$. Thus we get the following boundary value problem of the ordinary differential equations:

$$\begin{aligned} &-\eta\rho_\eta + (\rho u)_\eta = 0, \\ &-\eta\left(\rho\left(u + B_2\rho - \frac{B_1}{\rho^\alpha}\right)\right)_\eta + \left(\rho u\left(u + B_2\rho - \frac{B_1}{\rho^\alpha}\right)\right)_\eta \\ &= 0, \end{aligned} \tag{9}$$

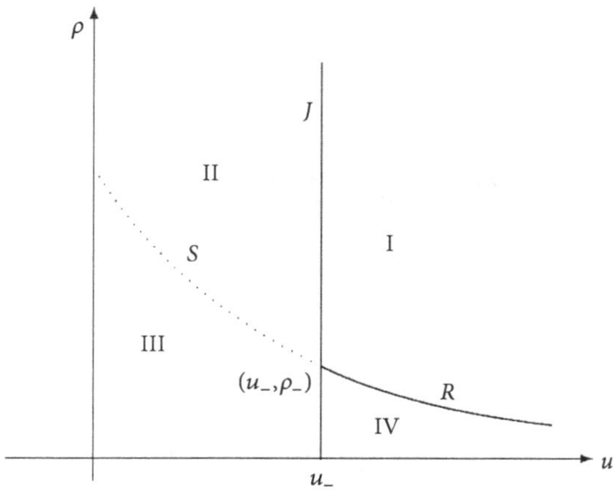

Figure 1: Wave curves in the phase plane (u, ρ).

and $(u, \rho)(\pm\infty) = (u_\pm, \rho_\pm)$. By solving the above problem, we obtain the constant state solution $(u, \rho) = \mathit{constant}$, and the rarefaction wave solution

$$R: \begin{cases} \eta = \lambda_1 = u - B_2\rho - \dfrac{B_1\alpha}{\rho^\alpha}, \\ u = u_- - B_2\rho + \dfrac{B_1}{\rho^\alpha} + B_2\rho_- - \dfrac{B_1}{\rho_-^\alpha}, \\ \rho < \rho_-, \ u > u_-. \end{cases} \tag{10}$$

For a bounded discontinuity at $\eta = \sigma$, the Rankine-Hugoniot conditions are as follows:

$$\begin{aligned} &-\sigma[\rho] + [\rho u] = 0, \\ &-\sigma\left[\rho\left(u + B_2\rho - \frac{B_1}{\rho^\alpha}\right)\right] + \left[\rho u\left(u + B_2\rho - \frac{B_1}{\rho^\alpha}\right)\right] \\ &= 0, \end{aligned} \tag{11}$$

where $[\rho] = \rho_r - \rho_l$, $\rho_l = \rho(\sigma - 0)$, $\rho_r = \rho(\sigma + 0)$, etc.

By solving (11) and using the Lax entropy inequalities, we get the shock wave solution

$$S: \begin{cases} \sigma = u - \dfrac{B_1}{\rho^\alpha} - B_2\rho_- - B_1\dfrac{\rho_-^{1-\alpha} - \rho^{1-\alpha}}{\rho - \rho_-}, \\ u = u_- - B_2\rho + \dfrac{B_1}{\rho^\alpha} + B_2\rho_- - \dfrac{B_1}{\rho_-^\alpha}, \\ \rho > \rho_-, \ u < u_-. \end{cases} \tag{12}$$

Because λ_2 is always linearly degenerate, from (11) we get the contact discontinuity

$$J: \eta = u = u_-, \quad \rho < \rho_- \text{ or } \rho > \rho_-. \tag{13}$$

R, S, and J are called the elementary waves of (1). Notice that system (1) can be written by $Y_t + (uY)_x = 0$ [19]; i.e., it is the Temple type; we can see that the shock curves coincide with the rarefaction curves in the phase plane (u, ρ) (Figure 1). This property simplifies the discussion of the wave interactions.

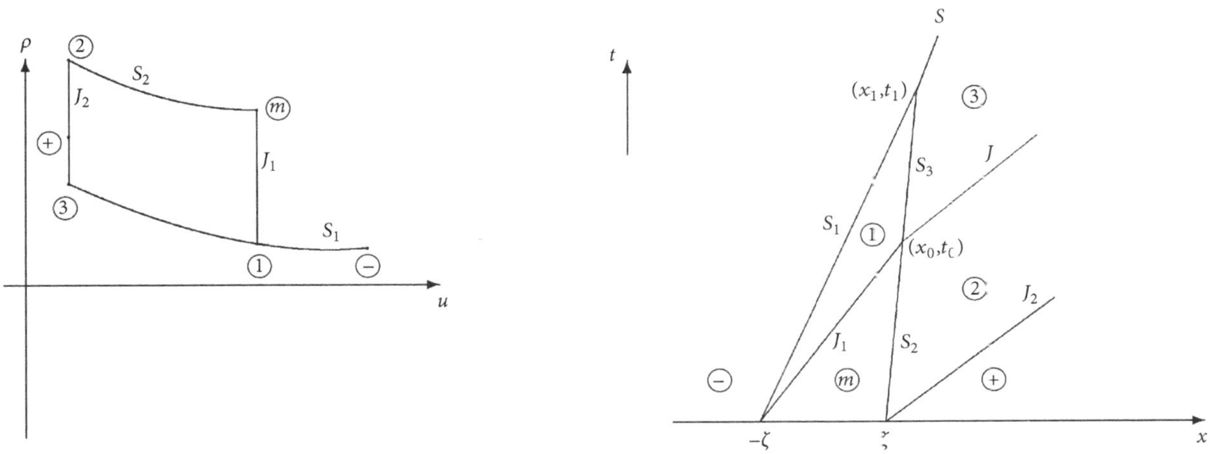

FIGURE 2: Wave interactions as $u_+ < u_m < u_-$.

Since $u_\rho = -B_2 - B_1\alpha/\rho^{\alpha+1} < 0$ and $u_{\rho\rho} = \alpha(\alpha+1)B_1/\rho^{\alpha+2} > 0$, the above two wave curves are monotonic decreasing and convex. And we know that the shock wave curve interacts with ρ-axis, and the rarefaction wave curve has the u-axis as its asymptote.

Based on the above analysis, we have the following result.

Theorem 1. *There exists a unique Riemann solution for (1) and (4) with the initial data (6). When $(u_+, \rho_+) \in I$ or IV, the unique Riemann solution is $R+J$; when $(u_+, \rho_+) \in II$ or III, the unique Riemann solution is $S+J$.*

3. Wave Interactions and Large-Time Behaviors

In this section, we consider the perturbed problem (1) and (3) with initial data (5). The data (5) is a small perturbation to the corresponding Riemann initial values (6). We want to determine whether the solutions of the perturbed Riemann problem (1), (3), and (5) converge to the corresponding Riemann solutions of (1), (3), and (6) as ζ tends to zero, where $(u_\zeta, \rho_\zeta)(x,t)$ are the solutions of (1), (3), and (5).

In order to cover all the cases completely, we divide our discussions into the following four cases: $u_+ < u_m < u_-$, $u_+ > u_m > u_-$, $u_m > u_\pm$, and $u_m < u_\pm$.

Case 1 ($u_+ < u_m < u_-$). In this case, we consider the interaction of $S+J$ emitted from $(-\zeta, 0)$ and $R+J$ emitted from $(\zeta, 0)$ (Figure 2), where "+" means "followed by". The occurrence of this case depends on the condition $u_+ < u_m < u_-$.

The propagation speed of J_1 and S_2 is $\tau_1 = u_m = u_1$ and $\sigma_2 = u_2 - B_1/\rho_2^\alpha - -B_2\rho_m - B_1((\rho_m^{1-\alpha} - \rho_2^{1-\alpha})/(\rho_2 - \rho_m))$, respectively. Since $\sigma_2 < u_2 = u_+ < u_m = \tau_1$, we know that J_1 will overtake S_2 at a point (x_0, t_0) which is determined by

$$\begin{aligned} x_0 + \zeta &= u_m t_0, \\ x_0 - \zeta &= \sigma_2 t_0. \end{aligned} \tag{14}$$

It follows that

$$(x_0, t_0) = \left(\frac{(u_m+\sigma_2)\zeta}{u_m-\sigma_2}, \frac{2\zeta}{u_m-\sigma_2}\right). \tag{15}$$

At the interaction point (x_0, t_0), the new Riemann problem with the initial data (u_1, ρ_1) and (u_2, ρ_2) is formed, which is resolved by a new S_3 and a new J. For the reason that S_3 propagates with the same speed as that of S_2, we obtain that the propagation direction of S_2 is unchanged. From $\tau = \tau_2$ we get that J has the same propagation speed as that of J_2 which implies that J is parallel to J_2.

When $t > t_0$, due to

$$\begin{aligned} \sigma_1 &= u_1 - \frac{B_1}{\rho_1^\alpha} - B_2\rho_- - B_1\frac{\rho_-^{1-\alpha} - \rho_1^{1-\alpha}}{\rho_1 - \rho_-}, \\ \sigma_3 &= u_3 - \frac{B_1}{\rho_3^\alpha} - B_2\rho_1 - B_1\frac{\rho_1^{1-\alpha} - \rho_3^{1-\alpha}}{\rho_3 - \rho_1}, \end{aligned} \tag{16}$$

and $\rho_- < \rho_1 < \rho_3$, by a direct computation it follows that $\sigma_1 - \sigma_3 > 0$. Then we know S_1 will overtake S_3 at (x_1, t_1) which satisfies

$$\begin{aligned} x_1 + \zeta &= \sigma_1 t_1, \\ x_1 - x_C &= \sigma_3 (t_1 - t_0). \end{aligned} \tag{17}$$

From (17) we get $(x_1, t_1) = ((\sigma_1 x_0 - \sigma_1\sigma_3 t_0 + \sigma_3\zeta)/(\sigma_1 - \sigma_3), (x_0 + \zeta - \sigma_3 t_0)/(\sigma_1 - \sigma_3))$. When $t > t_1$, a new shock wave S will occur and its propagating speed is $u_3 - B_1/\rho_3^\alpha - B_2\rho_- - B_1((\rho_-^{1-\alpha} - \rho^{1-\alpha})/(\rho_3 - \rho_-)) < u_3$, which yields that S cannot overtake J_3 forever.

When ζ tends to zero, we get that $(x_0, t_0) \to (0,0)$ and $(x_1, t_1) \to (0,0)$; J and J_2 will coincide with each other. Thus, as the time t is large enough, the solution of the perturbed Riemann problem is $S+J$. For this case, the limit of the perturbed Riemann solution to (1), (3), and (5) is no other than the corresponding Riemann solution to the initial problem (1), (3), and (6) which implies the global stability of the Riemann solution to (1), (3), and (6).

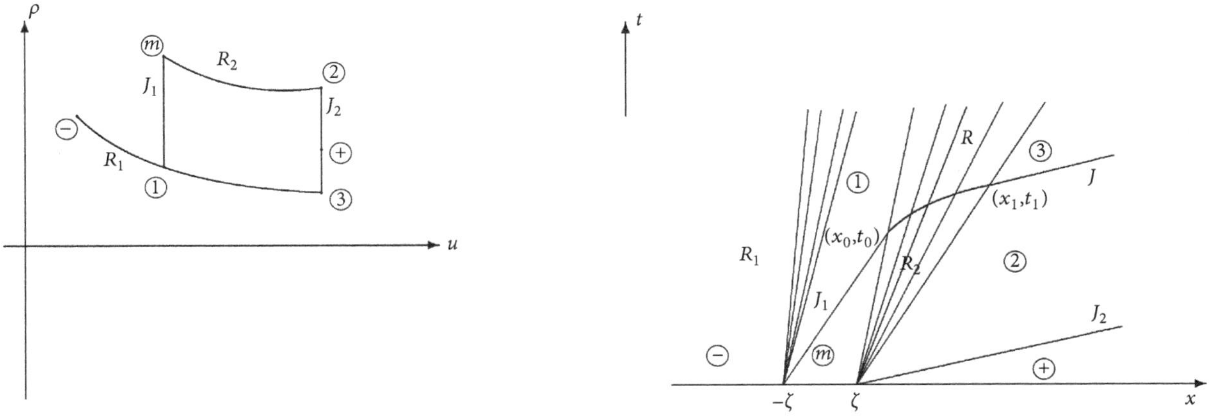

FIGURE 3: Wave interactions as $u_+ > u_m > u_-$.

Case 2 ($u_+ > u_m > u_-$). In this case, we investigate the interaction of $R + J$ emitted from $(-\zeta, 0)$ and $R + J$ emitted from $(\zeta, 0)$ (Figure 3).

Due to $\tau_1 = u_m > \lambda_2 = u_m - B_2\rho_m - B_1\alpha/\rho_m^\alpha$, we know that J_1 will overtake R_2 at the point (x_0, t_0) which satisfies

$$\begin{aligned} x_0 + \zeta &= u_m t_0, \\ x_0 - \zeta &= \lambda_2 t_0. \end{aligned} \tag{18}$$

A direct computation yields that

$$\begin{aligned} x_0 &= \frac{u_m + \lambda_2}{u_m - \lambda_2}\zeta, \\ t_0 &= \frac{2\zeta}{u_m - \lambda_2}. \end{aligned} \tag{19}$$

When $t > t_0$, J_1 begins to penetrate R_2, and during the penetration J_1 satisfies

$$\begin{aligned} &\frac{dx}{dt} = u, \\ &x - \zeta = \left(u - B_2\rho - \frac{B_1\alpha}{\rho^\alpha}\right)t, \\ &u - u_m = -B_2\rho + \frac{B_1}{\rho^\alpha} + B_2\rho_m - \frac{B_1}{\rho_m^\alpha}, \\ &x(t_0) = x_0, \\ &\rho_2 \le \rho \le \rho_m. \end{aligned} \tag{20}$$

Differentiating with respect to t in the second equation of (20), we know that

$$\begin{aligned} \frac{dx}{dt} &= \left(u - B_2\rho - \frac{B_1\alpha}{\rho^\alpha}\right) \\ &+ \left(\frac{du}{dt} - B_2\frac{d\rho}{dt} + B_1\alpha^2\rho^{-\alpha-1}\frac{d\rho}{dt}\right)t. \end{aligned} \tag{21}$$

Differentiating with respect to t in the third equation of (20), we obtain that

$$\frac{du}{dt} = \left(-B_2 - B_1\alpha\rho^{-\alpha-1}\right)\frac{d\rho}{dt}. \tag{22}$$

Substituting the first equation of (20) into the above representations yields that

$$\frac{d\rho}{dt} = \frac{B_2\rho + B_1\alpha}{\left(-2B_2 - B_1\alpha\rho^{-\alpha-1} + B_1\alpha^2\rho^{-\alpha-1}\right)t} < 0, \tag{23}$$

and

$$\begin{aligned} \frac{d^2x}{dt^2} &= \frac{du}{dt} \\ &= \left(B_2\rho + \frac{B_1\alpha}{\rho^\alpha}\right)\frac{1}{t}\frac{-B_2 - B_1\alpha\rho^{-\alpha-1}}{-2B_2 - B_1\alpha\rho^{-\alpha-1} + B_1\alpha^2\rho^{-\alpha-1}} \\ &> 0. \end{aligned} \tag{24}$$

From the above discussions, we conclude that J_1 accelerates during the process of penetration.

From (23) and the condition $\rho_2 < \rho < \rho_m$, we conclude that the contact discontinuity J_1 can penetrate completely the rarefaction wave R_2 in a finite time.

After the penetration, a new R and a new J will occur. Due to $u_1 - B_2\rho_1 - B_1/\rho_1 = u_m - B_2\rho_m - B_1/\rho_m$ and $u_3 - B_2\rho_3 - B_1/\rho_3 = u_2 - B_2\rho_2 - B_1/\rho_2$, the propagation direction of R keeps unchanged during the whole penetration. Notice that J is parallel to J_2 since $\tau = \tau_2$.

When ζ tends to zero, the points (x_0, t_0) and (x_1, t_1) tend to $(0, 0)$ and J coincides with J_2. Since the propagation speed of the wave back in R equals that of the wave front in R_1, it follows that R coincides with R_1 when ζ tends to zero. Thus, we know for this case the result is $R + J$ for large enough time. And the perturbed Riemann solution converges to the corresponding Riemann solution.

Case 3 ($u_m > u_\pm$). In this case, we discuss the interaction of $R + J$ emitted from $(-\zeta, 0)$ and $S + J$ emitted from $(\zeta, 0)$ (Figure 4).

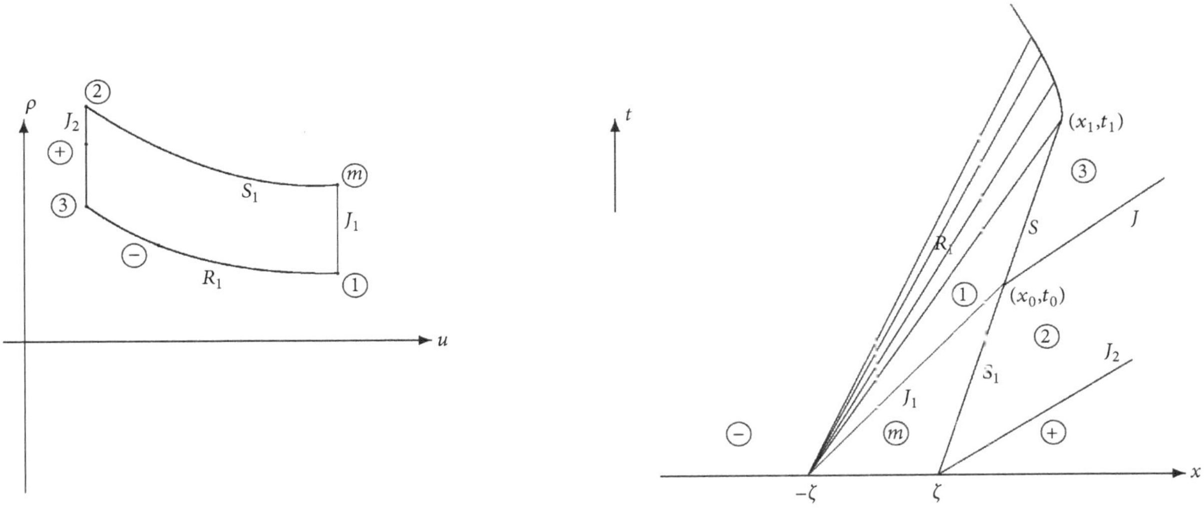

(a) Wave interactions as $u_m > u_- > u_+$

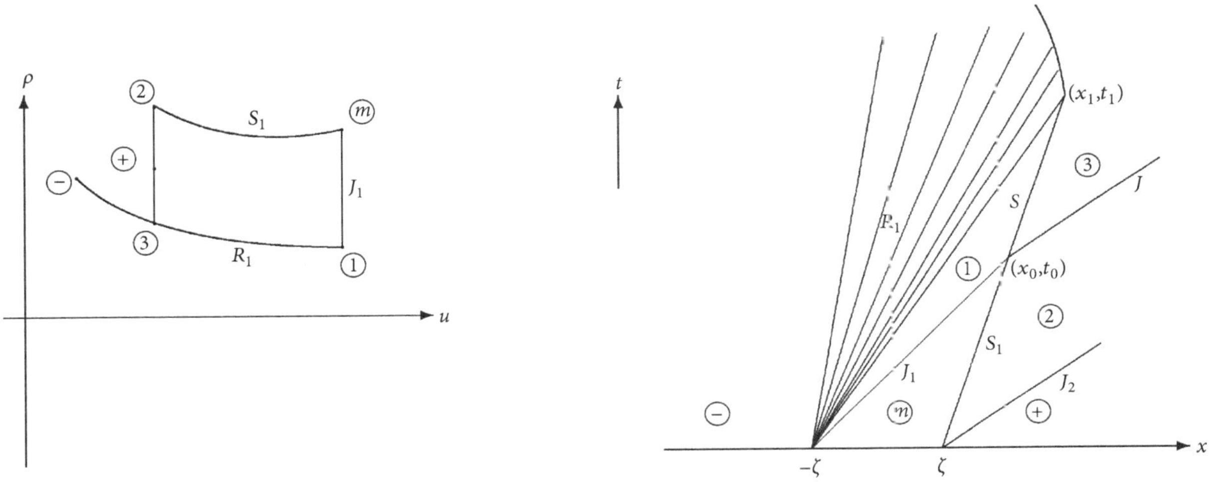

(b) Wave interactions as $u_- \le u_+ < u_m$

FIGURE 4

By similar discussions to that in Case 1, a new shock S and a new J will occur after the interaction of J_1 with S_1. Notice that S propagates with S_1 at the same speed and J is parallel to J_2 due to $\tau = \tau_2$. J_1 intersects with S_1 at (x_0, t_0) which is determined by (14).

Because the propagation speed of the wave front in R_1 is larger than the propagation speed in S, it follows that S will penetrate R_1 at (x_1, t_1) which satisfies

$$x_1 + \zeta = \lambda_1 t_1, \qquad x_1 - x_0 = \sigma (t_1 - t_0), \tag{25}$$

where $\lambda_1 = u_1 - B_2\rho_1 - B_1\alpha/\rho_1^{\alpha}$ and $\sigma = u_3 - B_1/\rho_3^{\alpha} - -B_2\rho_1 - B_1((\rho_1^{1-\alpha} - \rho_3^{1-\alpha})/(\rho_3 - \rho_1))$ are, respectively, the propagation speed of the wave front in R_1 and the propagation speed of S.

By a direct computation, we get $(x_1, t_1) = ((\lambda_1 x_0 - \lambda_1\sigma t_0 + \sigma\zeta)/(\lambda_1 - \sigma), (x_0 + \zeta - \sigma t_0)/(\sigma_1 - \sigma))$. When $t > t_1$, S begins to penetrate R_1 and we have

$$\begin{aligned} \frac{dx}{dt} &= u - \frac{B_1}{\rho^{\alpha}} - B_2\rho_- - B_1\frac{\rho_-^{1-\alpha} - \rho^{1-\alpha}}{\rho - \rho_-}, \\ x + \zeta &= \left(u - B_2\rho - \frac{B_1\alpha}{\rho^{\alpha}}\right)t, \\ u - u_- &= -B_2\rho + \frac{B_1}{\rho^{\alpha}} + B_2\rho_- - \frac{B_1}{\rho_-^{\alpha}}, \\ x(t_1) &= x_1, \\ & u_- \le u \le u_1, \quad \rho_1 \le \rho \le \rho_-. \end{aligned} \tag{26}$$

From the first and the second expression in (26) we know that

$$\frac{du}{dt} + \left(-B_2 + B_1\alpha^2\rho^{-\alpha-1}\right)\frac{d\rho}{dt} = \frac{B_2(\rho - \rho_-) + B_1(\alpha - 1)/\rho^{\alpha}}{t}, \tag{27}$$

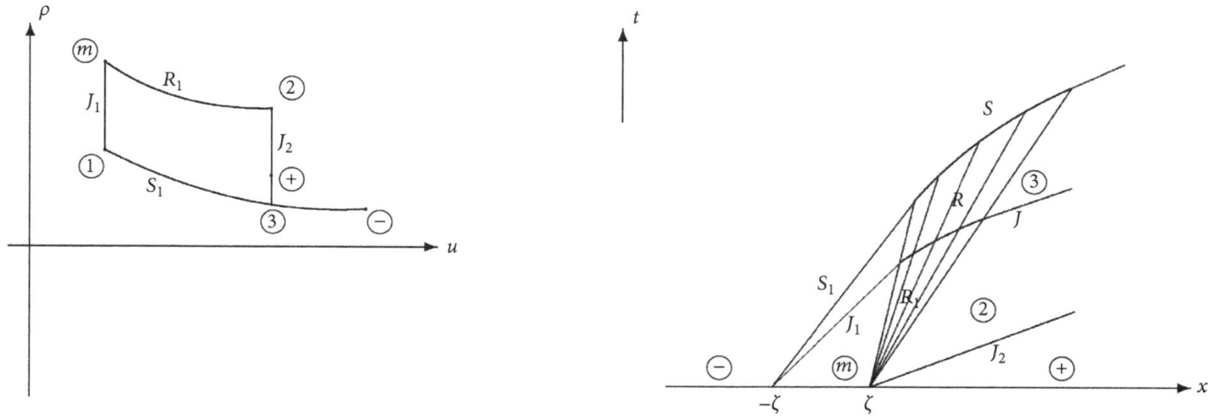

(a) Wave interactions as $u_m < u_+ < u_-$

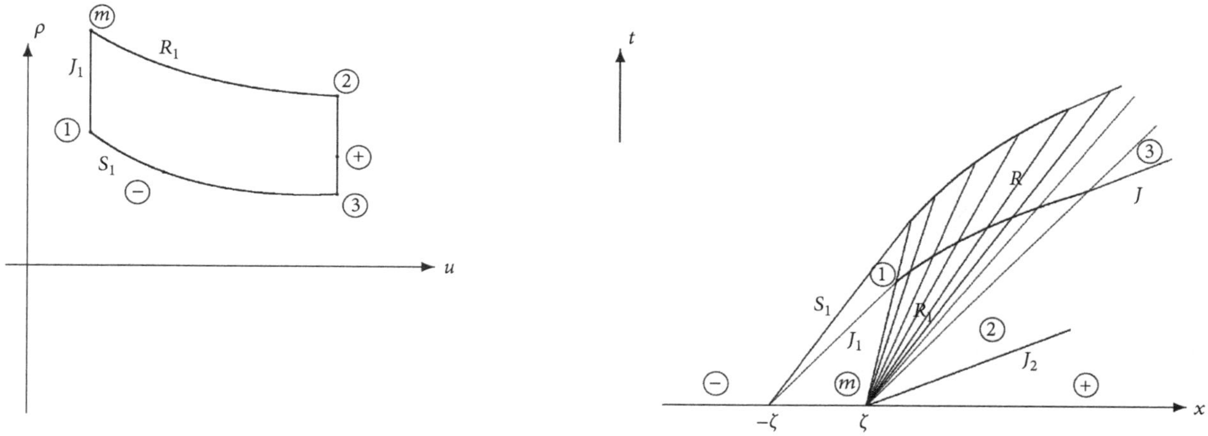

(b) Wave interactions as $u_+ \geq u_- > u_m$

FIGURE 5

and from the third expression we obtain that

$$\frac{du}{dt} = \left(-B_2 - B_1\alpha\rho^{-\alpha-1}\right)\frac{d\rho}{dt}. \tag{28}$$

Substituting (28) into (27), due to $\rho < \rho_-$ we have

$$\frac{d\rho}{dt} = \frac{B_2(\rho - \rho_-) + B_1(\alpha - 1)/\rho^{\alpha}}{\left(-2B_2 - B_1\alpha\rho^{-\alpha-1} + B_1\alpha^2\rho^{-\alpha-1}\right)t} > 0. \tag{29}$$

From (28)

$$\frac{du}{dt} = \left(-B_2 - B_1\alpha\rho^{-\alpha-1}\right)\frac{d\rho}{dt} < 0. \tag{30}$$

From (26), (28), and (29), we have

$$\frac{d^2x}{dt^2} = \left[-B_2 - \frac{B_1(1-\alpha)\rho^{-\alpha}(\rho_- - \rho)}{(\rho - \rho_-)^2} - \frac{B_1\left(\rho^{1-\alpha} - \rho_-^{1-\alpha}\right)}{(\rho - \rho_-)^2}\right]\frac{d\rho}{dt} < 0, \tag{31}$$

telling us that S decelerates during the above penetration. From (29) we conclude that for the case $u_- > u_+$, S will penetrate R_1 completely due to $\rho_3 > \rho_-$ at the finite time. While for the other case $u_- \leq u_+$, S cannot penetrate R_1 forever because $t \to \infty$ when $\rho \to \rho_3$ and S has $x + \zeta = (u_3 - B_2\rho_3 - B_1\alpha/\rho_3^{\alpha})t$ as its asymptote.

For the large time, the solution is $S + J$ (Figure 4(a)) as $u_+ < u_-$, and the solution is $R+J$ (Figure 4(b)) as $u_+ \geq u_-$. For this case, we obtain the stability of the Riemann solution to the initial problem (1), (3), and (6).

Case 4 ($u_m < u_\pm$). In this case, we study the interaction of $S+J$ emitted from $(-\zeta, 0)$ and $R + J$ emitted from $(\zeta, 0)$ (Figure 5).

By similar discussions to that of Case 2, a new R and a new J will occur after the intersection of J_1 and R_1. Notice that R keeps the same propagation speed and J is parallel to J_2 because $\tau = \tau_2$.

Similarly to Case 3, as $u_+ < u_-$, we obtain that S_1 will penetrate R completely in a finite time and a new shock will occur whose propagation speed is less than that of J. It implies that for the large time the solution is $S + J$ (Figure 5(a)). As $u_+ \geq u_-$, S_1 cannot penetrate R forever and it has $x-\zeta = (u_- - B_2\rho_- - B_1\alpha/\rho_-^{\alpha})t$ as its asymptote, which shows that for the large time the solution is $R + J$ (Figure 5(b)).

For this case, we get the global stability of the corresponding Riemann solution.

Now we have discussed all possible wave interactions in the phase plane (u, ρ). Based on the above discussions, we obtain our main result.

Theorem 2. *The limits of the perturbed Riemann solutions of (1), (3), and (5) are exactly the corresponding Riemann solutions of (1), (3), and (6). The asymptotic behavior of the perturbed Riemann solutions is governed completely by the states $(u_{\pm}, \rho_{\pm})$. The Riemann solutions of the initial value problem (1), (3), and (6) are stable under such small perturbation to the initial values.*

Conflicts of Interest

The authors declare that there are no conflicts of interest regarding the publication of this paper.

Acknowledgments

This study is partially supported by NSFC 11326156 and by the Foundation for Young Scholars of Shandong University of Technology (no. 115024).

References

[1] C. F. Daganzo, "Requiem for second-order fluid approximations of traffic flow," *Transportation Research Part B: Methodological*, vol. 29, no. 4, pp. 277–286, 1995.

[2] H. M. Zhang, "A non-equilibrium traffic model devoid of gas-like behavior," *Transportation Research Part B: Methodological*, vol. 36, no. 3, pp. 275–290, 2002.

[3] A. Aw and M. Rascle, "Resurrection of "second order" models of traffic flow," *SIAM Journal on Applied Mathematics*, vol. 60, no. 3, pp. 916–938, 2000.

[4] J. M. Greenberg, "Extensions and amplifications of a traffic model of Aw and Rascle," *SIAM Journal on Applied Mathematics*, vol. 62, no. 3, pp. 729–745, 2001.

[5] M. Garavello and B. Piccoli, "Traffic flow on a road network using the Aw-Rascle model," *Communications in Partial Differential Equations*, vol. 31, no. 1-3, pp. 243–275, 2006.

[6] J. M. Greenberg, A. Klar, and M. Rascle, "Congestion on multilane highways," *SIAM Journal on Applied Mathematics*, vol. 63, no. 3, pp. 818–833, 2003.

[7] S. Moutari and M. Rascle, "A hybrid Lagrangian model based on the Aw-Rascle traffic flow model," *SIAM Journal on Applied Mathematics*, vol. 68, no. 2, pp. 413–436, 2007.

[8] Z. Shao and M. Huang, "Interactions of delta shock waves for the Aw-Rascle traffic model with split delta functions," *Journal of Applied Analysis and Computation*, vol. 7, no. 1, pp. 119–133, 2017.

[9] C. Shen and M. Sun, "Formation of delta shocks and vacuum states in the vanishing pressure limit of Riemann solutions to the perturbed Aw-RAScle model," *Journal of Differential Equations*, vol. 249, no. 12, pp. 3024–3051, 2010.

[10] G. D. Wang, "A Riemann problem for the Aw-RAScle traffic flow model with negative pressure," *Chinese Annals of Mathematics*, vol. 35, no. 1, pp. 73–82, 2014.

[11] S. Chaplygin, "On gas jets," *Scientific Memoirs, Moscow University Mathematic Physics*, vol. 21, pp. 1–121, 1904.

[12] Y. Brenier, "Solutions with concentration to the Riemann problem for the one-dimensional Chaplygin gas equations," *Journal of Mathematical Fluid Mechanics*, vol. 7, no. suppl. 3, pp. S326–S331, 2005.

[13] H. J. Cheng and H. C. Yang, "Riemann problem fot the isentropic relativistic Chaplygin Euler equations," *The Journal of Applied Mathematics and Physics (ZAMP)*, vol. 63, no. 3, pp. 429–440, 2012.

[14] L. Guo, W. Sheng, and T. Zhang, "The two-dimensional Riemann problem for isentropic Chaplygin gas dynamic system," *Communications on Pure and Applied Analysis*, vol. 9, no. 2, pp. 431–458, 2010.

[15] Z. Q. Shao, "The Riemann problem for the relativistic full Euler system with generalized Chaplygin proper energy density-pressure relation," *Journal of Applied Mathematics and Physics*, vol. 69, no. 2, 2018.

[16] Y. Liu and W. Sun, "Wave interactions and stability of Riemann solutions of the Aw-RAScle model for generalized Chaplygin gas," *Acta Applicandae Mathematicae*, vol. 154, pp. 95–109, 2018.

[17] J. H. Wang, J. J. Liu, and H. C. Yang, "Vanishing pressure limit of solutions to the Aw-Rascle model for modified Chaplygin gas," 2014, https://arxiv.org/abs/1410.1110.

[18] H. Cheng and H. Yang "Approaching Chaplygin pressure limit of solutions to the Aw-Rascle model," *Journal of Mathematical Analysis and Applications*, vol. 416, no. 2, pp. 839–854, 2014.

[19] B. Temple, "Systems of conservation laws with coinciding shock and rarefaction curves," *Contemporary Mathematics Journal*, vol. 17, pp. 143–152, 1983.

10

Heat Transport and Majorana Fermions in a Superconducting Dot-Wire System: An Exact Solution

Oscar Bohórquez and A. M. S. Macêdo

Departamento de Física, Laboratório de Física Teórica e Computacional, Universidade Federal de Pernambuco, 50670-901 Recife, Pernambuco, Brazil

Correspondence should be addressed to A. M. S. Macêdo; a.murilos.macedo@gmail.com

Academic Editor: Andrei D. Mironov

We obtain exact expressions for the first three moments of the heat conductance of a quantum chain that crosses over from a superconducting quantum dot to a superconducting disordered quantum wire. Our analytic solution provides exact detailed descriptions of all crossovers that can be observed in the system as a function of its length, which include ballistic-metallic and metallic-insulating crossovers. The two Bogoliubov-de Gennes (BdG) symmetry classes with time-reversal symmetry are accounted for. The striking effect of total suppression of the insulating regime in systems with broken spin-rotation invariance is observed at large length scales. For a single channel system, this anomalous effect can be interpreted as a signature of the presence of the elusive Majorana fermion in a condensed matter system.

1. Introduction

Random-matrix theory (RMT) has been widely used in the study of phase-coherent complex quantum systems and has been particularly successful in uncovering universal properties of quantum transport in chaotic and disordered systems [1]. Much of the success of RMT in quantum transport is due to the strong correspondence between the statistical properties of random-matrix ensembles and the fluctuations of measured observables of complex quantum systems as a function of some control parameter, such as energy or magnetic field. The universality of transport properties, such as the moments of the conductance, lies in their independence of microscopic details of the scattering source. Nevertheless, random-matrix ensembles are sensitive to certain intrinsic symmetries of the system, such as time-reversal (TR), spin-rotation (SR), electron-hole (e-h), and chiral (Ch). It has been established that these symmetries lead to a classification of RMT ensembles into ten universal classes (the tenfold way) [2, 3], which are divided into three categories: (i) Wigner-Dyson (WD, three classes), appropriate to describe normal disordered conductors, (ii) chiral (Ch, three classes), appropriate for systems with a purely off-diagonal disorder, and (iii) Bogoliubov-de Gennes (BdG, four classes), appropriate for normal-superconducting (NS) hybrid systems.

From the perspective of quantum transport, phase-coherent mesoscopic systems can be classified into two types: (I) disordered conductors, in which impurities generate multiple elastic scatterings with an associated mean free path l that is less than the system's dimensions, and (II) ballistic cavities, where l is greater than the system's dimensions and thus the dominant scattering mechanism is reflection at the border of the cavity. Remarkably, RMT can efficiently describe both types of systems since universal transport properties do not depend on the details of either the impurity potential (for disordered systems) or the shape of the cavity (for ballistic systems) [1]. This insensitivity to microscopic details goes as far as to allow an identification, under certain conditions, of a multichannel disordered quantum wire with a chain of ballistic cavities [4].

Besides RMT, there are other well-developed approaches to quantum transport in both disordered wires and chaotic ballistic cavities: the field-theoretic nonlinear sigma model [5] and the trajectory-based semiclassical approach [6] are the most well known. These three approaches have many advantages and pitfalls, but since they are constructed from different and somewhat unrelated statistical hypothesis on the behavior of the underlying degrees of freedom, they may be considered a complementary, albeit equivalent, physical description of the system. Notwithstanding, a full fledged mathematical proof of the equivalence of these three approaches is still missing, in spite of much effort and some successes in particular systems, such as quantum dots with ideal couplings to external leads [7].

In superconducting systems, quantum transport has very striking and different features in comparison with their normal counterpart, which in part is due to what is known as Andreev reflections (AR). The most remarkable phenomenon is probably the possibility of a condensed matter realization of Majorana fermions as protected bound states at the ends of topological superconducting wires [8]. Transport observables in these systems, such as the electrical conductivity, can give information about topological invariants and topological quantum numbers. The thermal conductivity, on the other hand, despite maybe not containing direct information of topological invariants, as can be seen from their random-matrix description [9], can still provide valuable information about topological phase transitions [10]. It was, however, in the study of disordered quantum wires that evidence of the presence of condensed matter Majorana modes emerged most clearly. These can be traced back to the prediction [11] that for quantum wires in the chiral classes (for odd N open scattering channels) and in the superconducting D and DIII classes there is no exponential localization, since, unlike its behavior in the standard classes, the average conductance falls off in the limit of long distances $L \gg Nl$ as $1/\sqrt{L}$, which is a kind of super-ohmic behavior. They also found that in these special classes the average density of states (DOS) diverges logarithmically, $\rho(\varepsilon) \sim 1/|\varepsilon\tau \ln^3(\varepsilon\tau)|$, as energy $\varepsilon \longrightarrow 0$ (where $\tau = N^2 l$ and the Fermi velocity has been set to unity). A similar singularity had already been found by Dyson in the analysis of disordered linear chains [12]. On the other hand, [13] found through a general analysis using strong-disorder renormalization group (RG) that systems of classes D and DIII exhibit localization and an average DOS that vanishes as a power law $\rho(\varepsilon) \sim |\varepsilon\tau|^{\delta-1}$ with $\delta > 0$, as $\varepsilon \longrightarrow 0$. Remarkably, at certain critical points obtained by fine-tuning the disorder, both delocalization and Dyson's divergence can be present. The authors of [13] were also among the first to relate this special type of criticality to transitions between topological phases and also to point out that it could be a signature of the existence of Majorana zero modes. Reference [14] confirmed these results and presented evidence that the delocalization at critical points is well described by the DMPK equation of superconductors. Furthermore, they claimed that there may be "superuniversality" combining the chiral class and the superconducting classes D and DIII, since they have certain similar characteristics in regard to these critical points. Later, the authors of [15] argued that the diverging nature of the average density of states at the band center is a signature of topologically protected zero modes bound to point defects.

In this paper we employ random-matrix theory to study two classes of superconducting dot-wire systems. We obtain, in the continuum limit of a quantum chain with TR symmetry, an exact description of the crossover in thermal conduction between a superconducting chaotic ballistic cavity (a quantum dot) and a disordered multichannel superconducting quantum wire. The calculations were guided by a recent classification scheme of RMT Brownian motion ensembles [16] and were performed by means of a multivariate integral transform method proposed in [16, 17]. More specifically, we obtain exact expressions for the first three moments of the heat conductance of two classes of disordered superconducting quantum wires with time-reversal symmetry and with a crossover to a quantum dot in the small length limit. The analytic solution describes in detail various types of crossovers as a function of the systems' length, which include ballistic-metallic and metallic-insulating crossovers. If the system is realized as a single channel topological superconductor with broken spin-rotation invariance, we can interpret the total suppression of the insulating regime as a signature of the presence of a condensed matter Majorana fermion.

2. The Scattering Problem

Consider a confined quantum system ideally coupled to two electron reservoirs via point contacts with N_1 and N_2 open scattering channels, respectively. According to the Landauer-Büttiker scattering formalism [18], coherent particle transfer through such a device can be efficiently described by its scattering matrix, which can be generically written as

$$S = \begin{bmatrix} r_{N_1\times N_1} & t_{N_1\times N_2} \\ t'_{N_2\times N_1} & r'_{N_2\times N_2} \end{bmatrix}, \tag{1}$$

where r and r' are reflection matrices and t and t' are transmission matrices. The subscripts denote the matrix dimensions and the matrices $tt^\dagger$ and $t't'^\dagger$ are Hermitian and although they have different spectra, the spectrum of the smaller matrix coincides with the nonzero eigenvalues (transmission eigenvalues) of the bigger one $(\tau_1, \ldots, \tau_N) \in [0,1]^N$. Transport observables can be conveniently written in terms of these transmission eigenvalues. For instance, the thermal conductance of a superconducting system, at low temperature T, is given by [19],

$$G = G_0 d \sum_{1}^{N} \tau_i, \tag{2}$$

where $G_0 = \pi^2 k_B^2 T/6h$, $N = \min(N_1, N_2)$, and d is the spin and/or particle-hole degeneracy.

In ballistic chaotic cavities the transmission eigenvalues are strongly correlated random variables, which because of assumption of ergodic dynamics are well described by RMT. According to RMT, the scattering matrix of a ballistic chaotic cavity with ideal contacts is uniformly distributed over its manifold, and thus the probability density is only restricted by the presence or absence of certain symmetries. For the BdG classes the corresponding joint distribution of transmission eigenvalues is given by [19]

$$P(\{\tau\}) = C_N \prod_{i<j} \left|\tau_i - \tau_j\right|^\beta \prod_i \tau_i^{\beta(\mu+1)/2-1} (1-\tau_i)^{\gamma/2}, \tag{3}$$

where $\mu = |N_1 - N_2|$ and C_N is a normalization constant. The values of the parameters β and γ are solely determined

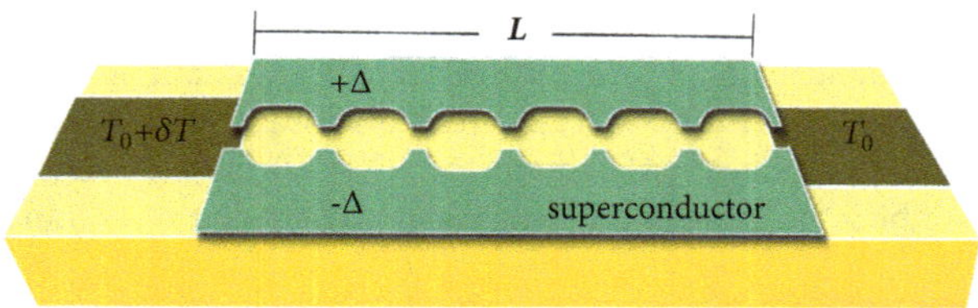

FIGURE 1: By varying the sample's length L, we can go from a quantum dot to a chain of dots, which in the continuous limit is a quantum wire. The conducting device is bounded by superconductors with a phase difference of π (order parameters given by $\pm\Delta$) and connected to electron reservoirs with different temperatures via ideal contacts.

by the symmetries, as shown in Table II of [19]. Here we shall consider only systems in the presence of TR symmetry, which implies $\beta = 2$. Moreover, we must set $\gamma = 1$ for systems in the presence of SR symmetry and $\gamma = -1$ for systems with broken SR symmetry. According to the tenfold way of classifying random-matrix ensembles, these classes are denoted by CI and DIII, respectively. We remark that the system described by (3) differs from a normal ballistic cavity, because, in addition to the two normal contacts coupling to the reservoirs, the cavity is geometrically defined by a normal-superconducting interface, which generates Andreev reflections [1]. As a matter of fact, this type of cavity is also known as an Andreev quantum dot [19]. We remark that if we set $\gamma = 0$ in (3) we recover the Wigner-Dyson A class for normal systems with broken TR symmetry, which turns out to be a useful way to compare our exact expressions for the moments of the thermal conductance with known results of the literature.

We proceed by combining Andreev quantum dots in a chain geometry, as shown in Figure 1. In such a setup the excitation gap induced by the proximity effect in the inner region is closed by adjusting the superconducting boundaries to have a phase difference of π, which also ensures that there is no breaking of TR symmetry [10, 19]. The RMT description of the system can be obtained by appropriately combining the scattering matrices of the Andreev quantum dots, or from the corresponding product of random transfer matrices; see [20] for more details. An equivalent description using the supersymmetric nonlinear sigma model is also possible [21]. Since we want to obtain exact analytical results, we follow [22, 23] and take the continuum limit which leads to a Fokker-Planck equation for the evolution, with the sample's length, of the joint probability distribution of transmission eigenvalues with a zero-length initial condition given by the RMT description of an Andreev quantum dot. See [21, 23] for the corresponding problem with normal quantum dots.

The Fokker-Planck equation of a disordered Andreev quantum wire of length L, with N open scattering channels and localization length ξ, is given by [11]

$$\frac{\partial P}{\partial s} = \sum_i^N \frac{\partial}{\partial q_i} J \frac{\partial}{\partial q_i} \frac{P}{J}, \tag{4}$$

where $s = L/[2l(\beta N - \beta + \alpha + 1)]$ is an adimensional length and

$$J = \prod_{i<j} \left|\cosh(2q_i) - \cosh(2q_j)\right|^{\beta} \prod_i^N \sinh^{\alpha}(2q_i). \tag{5}$$

The variables q_i are related to the transmission eigenvalues through the relation $\tau_i = \operatorname{sech}^2(q_i)$ and since we are considering superconducting TR-symmetric systems, we must set $\beta = 2$. The parameter α can take the values $\alpha = 2$ (in the presence of SR symmetry) and $\alpha = 0$ (in the absence of SR symmetry). The WD class is obtained by setting $\alpha = 1$ and assuming that TR symmetry is broken.

The problem has thus been reduced to solving (4) with an initial condition given by (3), which is the joint probability distribution of transmission eigenvalues of an Andreev quantum dot. Remarkably, an exact analytical solution can be constructed using an integral transform method [16] which provides a complete description of the crossover of the thermal conductance moments as a function of the system's length, covering all transport regimes: ballistic, metallic, and insulating. For a related study of a normal dot-wire system (WD class) with broken TR symmetry, see [22].

3. The Integral Transform Method

A powerful way to represent the probability distributions of the RMT ensembles is to employ a classification scheme based on matrix-valued Brownian motion ensembles [16]. For the BdG classes with TR symmetry (DIII and CI) we may define the following functions:

$$\begin{aligned} J_\beta(\{x\}) &= \prod_{i<j}^N \left|x_i - x_j\right|^{\beta}, \\ \omega_N(\{x\}) &= \prod_i^N \omega(x_i), \end{aligned} \tag{6}$$

where the random variables x_i are related to the transmission eigenvalues τ_i via a simple procedure described in [16]. For a ballistic cavity we may follow [24] and take $x_i = \tau_i$ and write the joint distribution as the stationary solution of the corresponding Brownian motion. We thus get

$$P^{(0)}(\{x\}) = C_N J_\beta(\{x\}) \omega_N(\{x\}), \tag{7}$$

where the function $\omega(x)$ is given in Table 1. Equation (7) will be used as the initial condition of our crossover problem. On the other hand, for a quasi-one-dimensional quantum wire we may follow [25] and take $x_i = \cosh(2q_i) = (2 - \tau_i)/\tau_i$ and write the corresponding Fokker-Planck equation as

$$\frac{\partial P}{\partial s} = \sum_i^N \frac{\partial}{\partial x_i} s(x_i) \omega_N J_\beta \frac{\partial}{\partial x_i} \frac{P}{\omega_N J_\beta}, \tag{8}$$

where the functions $\omega(x)$ and $s(x)$ are also given in Table 1.

Instead of directly solving Fokker-Plank equation (8) with initial condition (7), we employ an integral transform

TABLE 1: Classification of Brownian motion ensembles for the BdG class.

RMT	random variable	$\omega(x)$	$s(x)$	a	b
dot	$x_i = \tau_i$	$x^{\beta(\mu+1)/2-1}(1-x)^{\gamma/2}$	$x(1-x)$	0	1
wire	$x_i = \frac{2}{\tau_i} - 1$	$(x^2-1)^{(\alpha-1)/2}$	x^2-1	1	∞

method that effectively maps (8) onto a much simpler Fokker-Planck problem in an image space of smaller dimension. The multidimensional integral transform is defined as [16]

$$W(\{\vartheta\},s) = \int d^N x \Omega_\beta(\{x\},\{\vartheta\}) P(\{x\},s), \tag{9}$$

where the kernel Ω_β is chosen to have only two $\{\vartheta\}$ variables, which is the minimal number of image variables that allows the exact calculation of the first three heat conductance moments. We remark that the integral transform cannot be inverted in general and thus our method cannot produce the full joint distribution of the transmission eigenvalues $P(\{x\},s)$. However, in [16] a more general choice of the kernel Ω_β is presented and from it one could in principle calculate all moments of the conductance. We proceed by defining

$$\Omega_\beta(\{x\},\{\vartheta\}) = \prod_i^N \frac{x_i - \vartheta_{0,1}}{x_i - \vartheta_{1,1}}. \tag{10}$$

The Fokker-Planck equation in image space $\{\vartheta\}$ is then given by

$$\left(\frac{\partial}{\partial s} - \mathcal{M}_\vartheta^\dagger\right) W(\{\vartheta\},s) = 0, \tag{11}$$

where

$$\mathcal{M}_\vartheta^\dagger = \frac{1}{VB}\sum_{i=0}^{1}(-1)^{1+i}\frac{\partial}{\partial\vartheta_{i,1}}\left(s(\vartheta_{i,1})VB\frac{\partial}{\partial\vartheta_{i,1}}\right) \tag{12}$$

and we defined

$$V = \frac{\omega(\vartheta_{0,1})}{\omega(\vartheta_{1,1})}, \qquad B = \frac{1}{(\vartheta_{0,1}-\vartheta_{1,1})^2}. \tag{13}$$

We can now perform the transformation

$$W(\{\vartheta\},s) = 1 + \omega(\vartheta_{1,1}) B^{-1/2} \Psi(\{\vartheta\},s) \tag{14}$$

that maps (11) onto a Schrödinger equation in imaginary time ($s \longrightarrow it$); thus

$$\frac{\partial \Psi}{\partial t} + \mathcal{H}\Psi = 0, \tag{15}$$

where

$$\mathcal{H} = \sum_{i=0}^{1}(-1)^i \frac{1}{\omega(\vartheta_i)}\frac{\partial}{\partial\vartheta_i}\left(\omega(\vartheta_i)s(\vartheta_i)\frac{\partial}{\partial\vartheta_i}\right) \tag{16}$$

is a free particle Hamiltonian in image space.

We proceed by calculating $W(\{\vartheta\},0)$ using the joint distribution of the Andreev quantum dot (7). Here we may again use the integral transform method with the following modified kernel:

$$\Omega_\beta^{(0)}(\{\tau\},\{\vartheta\}) = \left(\frac{1-\vartheta_0}{1-\vartheta_1}\right)^N \prod_i^N \frac{\tau_i - 2/(1-\vartheta_0)}{\tau_i - 2/(1-\vartheta_1)}, \tag{17}$$

where $\tau_i = x_i$. The choice of kernel is motivated from the connection between the supersymmetric nonlinear sigma model and RMT [22]. From the stationary solution of the corresponding Fokker-Planck equation, we find (Appendix A)

$$W(\{\vartheta\},0) = 1 + (\vartheta_0 - \vartheta_1) \cdot \sum_{l=0}^{N-1}\frac{(1-\vartheta_0)^l}{(1-\vartheta_1)^{l+1}}\left(f_{N-l-1}(\vartheta_0)\, g_{N-l-1}(\vartheta_1) - 1\right), \tag{18}$$

where

$$f_n(\vartheta_0) = F\left[-n, -n-\mu; -2n-\mu-\frac{\gamma}{2}; \frac{1-\vartheta_0}{2}\right],$$
$$g_n(\vartheta_1) = F\left[n+1, n+1+\mu; 2n+\mu+\frac{\gamma}{2}+2; \frac{1-\vartheta_1}{2}\right], \tag{19}$$

and $F[a,b;c;z]$ denotes the hypergeometric function.

We are now in position to unify the notations for the symmetry indices γ and α of the quantum dot and the quantum wire, respectively. For that, we introduce the new index

$$\nu = \frac{\alpha-1}{2} = \frac{\gamma}{2}, \tag{20}$$

which can have two values: $\nu = 1/2$ (system with TR and SR symmetry) or $\nu = -1/2$ (systems with TR symmetry and no SR symmetry). In the tenfold way of S-matrix classification, these classes correspond to CI and DIII, respectively. We remark that $\nu = 0$ corresponds to the Wigner-Dyson A class (systems with no TR symmetry).

4. Exact Solution for the Dot-Wire System

We can now address the full problem and solve (11), which describes a multichannel superconducting quantum wire, with a quantum dot initial condition given by (18). It will prove convenient to use kernel (10) with the choices $\vartheta_{0,1} = -\vartheta_0$ and $\vartheta_{1,1} = -\vartheta_1$, so that we may write

$$\Omega_\beta(\{x\},\{\vartheta\}) = \prod_i^N \frac{x_i + \vartheta_0}{x_i + \vartheta_1}. \tag{21}$$

Note that after using the relation $x_i = 2/\tau_i - 1$ we recover (17), as expected.

Following the procedure introduced in [16, 22] we can find the eigenvalues of the Hamiltonian $\mathscr{H}$ shown in (16). We start by specifying the domains of the variables ϑ_0 and ϑ_1. Inspired by the supersymmetry calculations of [22], we set $-1 \le \vartheta_0 \le 1$ and $1 \le \vartheta_0 \le \infty$. In this domain the eigenvalues of $\mathscr{H}$ are $\varepsilon_{nk} = k^2 + (n + \nu + 1/2)^2$ and the corresponding eigenfunctions are (Appendix B)

$$\varphi_{nk}(\vartheta_0, \vartheta_1) = \frac{A_k^{(-\nu)}}{\left(h_n^{(\nu)}\right)^{1/2}} P_n^{(\nu)}(\vartheta_0) \frac{F_k^{(-\nu)}(\vartheta_1)}{\omega(\vartheta_1)}, \quad (22)$$

where $h_n^{(\nu)} \equiv h_n^{(\nu,\nu)}$ and $A_k^{(\nu)}$ are appropriate normalization constants of the Jacobi polynomials $P_n^{(\nu)}(\vartheta_0) \equiv P_n^{(\nu,\nu)}(\vartheta_0)$ and the hypergeometric function

$$F_k^{(\nu)}(\vartheta_1) = F\left[\nu + \frac{1}{2} + ik, \nu + \frac{1}{2} - ik; \nu + 1; \frac{1-\vartheta_1}{2}\right], \quad (23)$$

respectively (see Appendices A and B).

We can now construct Green's function of the problem by means of its spectral resolution in terms of the eigenfunctions and the eigenvalues of $\mathscr{H}$. We find

$$G(\{\vartheta\}, \{\vartheta'\}, t) = \left(1 - \vartheta_0'^2\right)^\nu \left(\vartheta_1'^2 - 1\right)^\nu \cdot \sum_{n=0}^{\infty} \int_0^\infty dk \varphi_{nk}(\vartheta_0, \vartheta_1) \varphi_{nk}(\vartheta_0', \vartheta_1') e^{-i\varepsilon_{nk}t}. \quad (24)$$

From the completeness of the eigenfunctions, it follows immediately that

$$G(\{\vartheta\}, \{\vartheta'\}, 0) = \delta(\vartheta_0 - \vartheta_0') \delta(\vartheta_1 - \vartheta_1'), \quad (25)$$

as expected for a propagator. Using (24) we can write (14) as

$$W(\{\vartheta\}, s) = 1 + (\vartheta_0 - \vartheta_1) \omega(\vartheta_1) \int_{-1}^{1} d\vartheta_0' \cdot \int_1^\infty d\vartheta_1' G(\{\vartheta\}, \{\vartheta'\}, s) \frac{W(\vartheta', 0) - 1}{(\vartheta_0' - \vartheta_1') \omega(\vartheta_1')}. \quad (26)$$

Inserting (18) and (24) into (26), we get

$$W(\{\vartheta\}, s) = 1 + (\vartheta_0 - \vartheta_1) \cdot \sum_{n=0}^{\infty} \int_0^\infty dk \frac{\left(A_k^{(-\nu)}\right)^2}{h_n^{(\nu)}} e^{-\varepsilon_{nk}s} P_n^{(\nu)}(\vartheta_0) F_k^{(-\nu)}(\vartheta_1) \times \sum_{l=0}^{N-1} \left(I_{nl}^{(1)} J_{kl}^{(1)} - I_{nl}^{(0)} J_{kl}^{(0)}\right), \quad (27)$$

where $I_{nl}^{(i)}$ and $J_{kl}^{(i)}$, $i \in \{0, 1\}$, are obtained from the following integrals:

$$I_{nl}^{(0)} = \int_{-1}^{1} d\vartheta_0 \left(1 - \vartheta_0^2\right)^\nu P_n^{(\nu)}(\vartheta_0) (1 - \vartheta_0)^l,$$
$$J_{kl}^{(0)} = \int_1^\infty d\vartheta_1 \frac{F_k^{(-\nu)}(\vartheta_1)}{(\vartheta_1^2 - 1)^\nu (1 - \vartheta_1)^{l+1}},$$
$$I_{nl}^{(1)} = \int_{-1}^{1} d\vartheta_0 \left(1 - \vartheta_0^2\right)^\nu P_n^{(\nu)}(\vartheta_0) (1 - \vartheta_0)^l f_{N-l-1}(\vartheta_0), \quad (28)$$
$$J_{kl}^{(1)} = \int_1^\infty d\vartheta_1 \frac{F_k^{(-\nu)}(\vartheta_1)}{(\vartheta_1^2 - 1)^\nu (1 - \vartheta_1)^{l+1}} g_{N-l-1}(\vartheta_1).$$

The integrals are calculated in Appendix C for arbitrary N_1 and N_2. The final result is summarized in the following theorem.

Theorem 1. *The solution of Fokker-Planck equation (11) with an initial condition given by (18), in the space of the coordinates $\{\nu\}$, is given by*

$$W(\{\vartheta\}, s) = 1 + 2(\vartheta_0 - \vartheta_1) \sum_{n=0}^{N-1} \frac{P_n^{(\nu)}(\vartheta_0) P_n^{(\nu)}(1)}{h_n^{(\nu)}} \cdot \int_0^\infty d\mu_{nk} c_{nk}^{(\nu)}(N_1) c_{nk}^{(\nu)}(N_2) F_k^{(-\nu)}(\vartheta_1) e^{-\varepsilon_{nk}s}, \quad (29)$$

where

$$d\mu_{nk} = dk \frac{|\Gamma(1/2 - \nu + ik)|^2}{|\Gamma(ik)|^2 \varepsilon_{nk}} \quad (30)$$

and

$$c_{nk}^{(\nu)}(N) = \frac{|\Gamma(N + \nu + 1/2 + ik)|^2}{(N - n - 1)! \Gamma(N + n + 2\nu + 1)}. \quad (31)$$

This theorem is the central result of this work. It can be used as a generating function to find the first three moments of the thermal conductance. A noteworthy feature of the exact result is the separation of the left-right boundary conditions in the form of the product $c_{nk}^{(\nu)}(N_1) c_{nk}^{(\nu)}(N_2)$, which as we will see later allows a simple identification of different conducting regimes.

5. Application: Moments of the Heat Conductance

As an application of (29) we calculate the first three moments of the heat conductance. The Landauer formula for the dimensionless thermal conductance is

$$g = \sum_i^N \tau_i = \sum_i^N \frac{2}{1 + x_i}. \quad (32)$$

From (9), (21), and (32) it is straightforward to verify that the first three moments of thermal conduction can be obtained from the generating function as

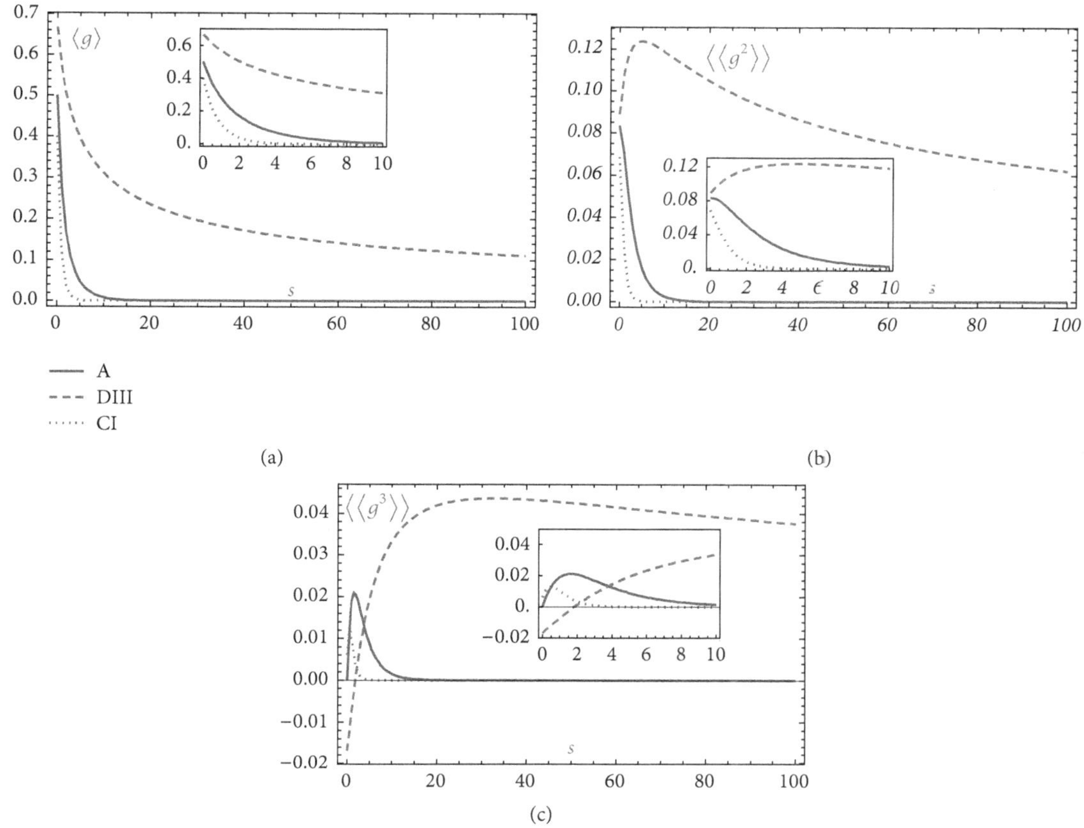

FIGURE 2: The first three cumulants of the thermal conductance for the WD (full line) and BdG classes DIII (dashed line) and CI (dotted line) with a single open channel: $N_1 = N_2 = 1$. The insets show the small s behaviors. Note the qualitative difference of the class DIII BdG system.

$$
\begin{aligned}
\langle g \rangle &= \frac{2\partial W}{\partial \vartheta_0}\bigg|_{\vartheta_0=1=\vartheta_1} \\
\left\langle g^2 \right\rangle &= -\frac{4\partial^2 W}{\partial \vartheta_0 \partial \vartheta_1}\bigg|_{\vartheta_0=1=\vartheta_1}, \\
\left\langle g^3 \right\rangle &= 4\left(\frac{\partial^3 W}{\partial \vartheta_0 \partial \vartheta_1^2} - \frac{\partial^3 W}{\partial \vartheta_1 \partial \vartheta_0^2}\right)\bigg|_{\vartheta_0=1=\vartheta_1}.
\end{aligned}
\tag{33}
$$

From (29) we get

$$
\begin{aligned}
\langle g^m \rangle = 4\sum_{n=0}^{N-1} \frac{\left(P_n^{(\nu)}(1)\right)^2}{h_n^{(\nu)}} \\
\cdot \int_0^\infty d\mu_{nk}\, g_{nk}^{(m)} c_{nk}^{(\nu)}(N_1)\, c_{nk}^{(\nu)}(N_2)\, e^{-\varepsilon_{nk} s},
\end{aligned}
\tag{34}
$$

where

$$
\begin{aligned}
g_{nk}^{(1)} &= 1, \\
g_{nk}^{(2)} &= \frac{k^2 + (1/2-\nu)^2}{(1-\nu)} + \frac{n(n+2\nu+1)}{(1+\nu)}, \\
g_{nk}^{(3)} &= \frac{\left(k^2 + (1/2-\nu)^2\right)\left(k^2 + (3/2-\nu)^2\right)}{2(2-\nu)(1-\nu)} \\
&\quad + \frac{2n(n+2\nu+1)\left(k^2 + (1/2-\nu)^2\right)}{(1+\nu)(1-\nu)} \\
&\quad + \frac{n(n-1)(n+2\nu+1)(n+2\nu+2)}{2(1+\nu)(2+\nu)}.
\end{aligned}
\tag{35}
$$

Equation (34) is a very general exact result. It is valid for any number of channels N_1 and N_2 and for any value of the dimensionless length of the quantum wire s. It is the central application of this paper and as it stands it can serve as a useful tool to compare with results from other nonperturbative approaches, such as the trajectory-based semiclassical technique and field-theoretic methods.

In Figure 2 we show the behavior of the first three cumulants (denoted with double brackets) of the thermal conductance, as a function of the system's length, of both the WD (full line) and BdG classes DIII (dashed line) and CI (dotted line) for the case of a single channel in each lead. This case is particularly interesting, since as shown in [19] class DIII can be realized using Majorana modes of topological superconductors. Therefore, the unusual shape of the heat conductance moments in this class can be interpreted as a signature of the presence of Majorana fermions in the system.

In the long length limit, $s \gg 1$, we may use the saddle-point method in (34) to obtain more explicit analytic expressions for each class. For class A, we get

$$\begin{aligned}\langle g\rangle &= 4\langle g^2\rangle = \frac{64}{9}\langle g^3\rangle \\ &= 2c_{00}^{(0)}(N_1)\,c_{00}^{(0)}(N_2)\left(\frac{\pi}{s}\right)^{3/2} e^{-s/4},\end{aligned} \quad (36)$$

whilst for classes DIII and CI we find, respectively,

$$\langle g\rangle = \frac{3}{2}\langle g^2\rangle = \frac{15}{8}\langle g^3\rangle = \frac{2}{\sqrt{\pi s}}, \quad (37)$$

$$\begin{aligned}\langle g\rangle &= s\langle g^2\rangle = 3s\langle g^3\rangle \\ &= 4c_{00}^{(1/2)}(N_1)\,c_{00}^{(1/2)}(N_2)\frac{e^{-s}}{\sqrt{\pi s}}.\end{aligned} \quad (38)$$

From these equations, we see that the BdG DIII class shows anomalous power law behaviors in all three moments. In contrast, the other two classes (WD and CI) show a rapid exponential decay of the moments as a function of s, which is a common signature of Anderson localization. The significant attenuation in the decay of the moments for the DIII class as a function of the system's length is thus a kind of delocalization effect, which can be interpreted as indicating a type of anomalous metallic behavior. Since a normal disordered quantum wire shows exponential localization when its length exceeds the localization length, we must conclude that the localization length diverges for a class DIII BdG system. A similar conclusion has been reached in [26] for a thick quantum wire in series with a quantum point contact, which can be described by a DMPK Fokker-Planck equation with a delta function initial condition.

Another noteworthy feature comes from the fact that $c_{nk}^{(\nu)}(N) = 1$ when $N \longrightarrow \infty$. Using this property we can recover several previous results of the literature. The most interesting cases are the following: (a) thick wire: we take $N_1 \longrightarrow \infty$ and $N_2 \longrightarrow \infty$, which reproduces the results found in [27], and (b) the disordered quantum wire limit: we take $N_1 \longrightarrow \infty$ with N_2 fixed or $N_2 \longrightarrow \infty$ with N_1 fixed, which reproduces the results shown in [26].

It is also useful to investigate the system's behavior close to the ballistic limit, i.e., when $s \ll 1/g_c$, where $g_c = N_1N_2/(N_1+N_2)$. For that, we expand (34) around $s = 0$ to obtain

$$\langle g\rangle = \frac{N_1N_2}{N_1+N_2+\nu}\left(1 - \frac{(N_1+\nu)(N_2+\nu)(N_1+N_2+2\nu)}{(N_1+N_2+\nu)^2-1}s+\cdots\right) \quad (39)$$

and

$$\langle\langle g^2\rangle\rangle = \frac{N_1(N_1+\nu)N_2(N_2+\nu)}{(N_1+N_2+\nu)^2\left((N_1+N_2+\nu)^2-1\right)}\left(1 + \frac{2\left((N_1-N_2)^2-\nu^2\right)(N_1+N_2+2\nu)}{(N_1+N_2+\nu)^2-4}s+\cdots\right). \quad (40)$$

Note that the zero-order terms reproduce the results of the quantum dot for the three classes [24], as expected. The correction terms are new predictions that may be useful for comparison with alternative techniques, such as the trajectory-based semiclassical approach.

Finally, in Figure 3 we show the behavior of the first three cumulants of the DIII class as a function of the number of channels in each lead. We keep $N_1 = N_2$ and compare the result with the thick-wire limit $N \longrightarrow \infty$. We observe that said limit is reached quickly after an increase of a few channels. The same is also true for the WD and CI classes.

6. Summary and Conclusions

We employed random-matrix theory and matrix-valued Brownian motion models to study two classes of superconducting quantum chains. In the continuum dot-wire limit, we find an exact description of the crossover in thermal conduction between a superconducting chaotic ballistic cavity (a quantum dot) and a disordered multichannel superconducting quantum wire. We obtained exact expressions for the first three moments of the heat conductance of two classes of superconducting dot-wire systems with time-reversal symmetry. The analytic solution describes in detail various types of crossovers as a function of the systems' length, which include ballistic-metallic and metallic-insulating crossovers. Interestingly, in the single channel case, if the system is realized experimentally as a topological superconductor, we can interpret the total suppression of the insulating regime in class DIII as a signature of the presence of condensed matter Majorana fermions.

Appendix

A.

The generating function of the Andreev quantum dot can be calculated by means of (16) and the kernel shown in (17). Using the variables $\eta_0 = (3+\nu_o)/(1-\nu_o)$ and $\eta_1 = (3+\nu_1)/(1-\nu_1)$, we find the effective Hamiltonian

$$\begin{aligned}\mathscr{H} &= \left(1-\eta_0^2\right)\frac{\partial^2}{\partial\eta_0^2} + \left(\mu - \frac{\gamma}{2} - \left(2+\mu+\frac{\gamma}{2}\right)\eta_0\right)\frac{\partial}{\partial\eta_0} \\ &\quad - \left(1-\eta_1^2\right)\frac{\partial^2}{\partial\eta_1^2} \\ &\quad - \left(\mu - \frac{\gamma}{2} - \left(2+\mu+\frac{\gamma}{2}\right)\eta_1\right)\frac{\partial}{\partial\eta_1}.\end{aligned} \quad (A.1)$$

The stationary solution of the corresponding Fokker-Planck equation can be written in terms of the eigenfunctions of (A.1), which are Jacobi polynomials $P_n^{(a,b)}$ and Jacobi

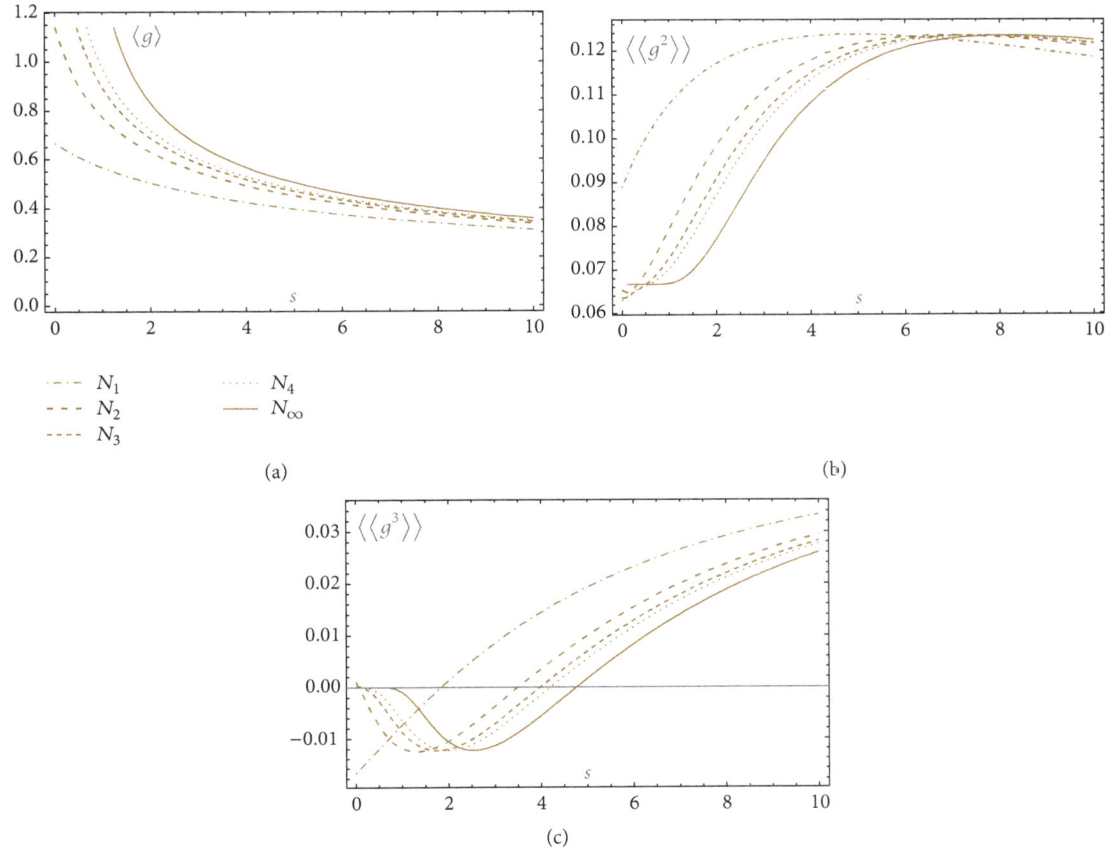

FIGURE 3: Variation of the first three cumulants of the thermal conductance for the DIII class with different numbers of channels N. $N_1 = N_2 = N$.

functions of the second type $Q_n^{(a,b)}$. Inserting this result into (14) we get

$$W(\{\vartheta\},0) = \left(\frac{1-\nu_o}{1-\nu_1}\right)^N \cdot \left(1 + 2^{3+2\mu+\gamma/2} \frac{(\nu_0-\nu_1)(1+\nu_1)^{\gamma/2}}{(1-\nu_0)(1-\nu_1)^{\mu+\gamma/2+1}} R\right), \quad \text{(A.2)}$$

where

$$R = \sum_{n=0}^{N-1} \frac{1}{h_n^{(\gamma/2,\mu)}} P_n^{(\gamma/2,\mu)}\left(\frac{3+\nu_o}{1-\nu_o}\right) Q_n^{(\gamma/2,\mu)}\left(\frac{3+\nu_1}{1-\nu_1}\right), \quad \text{(A.3)}$$

and

$$h_n^{(\alpha,\beta)} = \frac{2^{\alpha+\beta+1}\Gamma(n+\alpha+1)\Gamma(n+\beta+1)}{n!(2n+\alpha+\beta+1)\Gamma(n+\alpha+\beta+1)}, \quad \text{(A.4)}$$

which after some simple algebraic manipulations yields (18).

B.

The generating function of the Andreev quantum wire can be calculated by using (16), which can be written as

$$\mathscr{H} = (\vartheta_0^2 - 1)\frac{\partial^2}{\partial\vartheta_0^2} + 2(\nu+1)\vartheta_0\frac{\partial}{\partial\vartheta_0} - (\vartheta_1^2 - 1)\frac{\partial^2}{\partial\vartheta_1^2} - 2(\nu+1)\vartheta_1\frac{\partial}{\partial\vartheta_1}. \quad \text{(B.1)}$$

Note that the first two terms of this equation match with the operator of the Jacobi differential equation with equal parameters $P_n^{(\nu)}(\vartheta_0) \equiv P_n^{(\nu,\nu)}(\vartheta_0)$

$$\left[(\vartheta_0^2 - 1)\frac{\partial^2}{\partial\vartheta_0^2} + 2(\nu+1)\vartheta_0\frac{\partial}{\partial\vartheta_0}\right] P_n^{(\nu)}(\vartheta_0) = n(n+2\nu+1) P_n^{(\nu)}(\vartheta_0). \quad \text{(B.2)}$$

On the other hand, the last two terms of (B.1) match with the operator of hypergeometric differential equation (23) for $F_k^{(\nu)}(\vartheta_1)$

$$\left[-\left(\vartheta_1^2-1\right)\frac{\partial^2}{\partial\vartheta_1^2}-2(\nu+1)\vartheta_1\frac{\partial}{\partial\vartheta_1}\right]F_k^{(\nu)}(\vartheta_1) = \left(k^2+\nu(\nu+1)+\frac{1}{4}\right)F_k^{(\nu)}(\vartheta_1). \tag{B.3}$$

The orthogonality and completeness relations are, respectively,

$$\int_{-1}^{1} d\vartheta_0\left(1-\vartheta_0^2\right)^{\nu} P_n^{(\nu)}(\vartheta_0)P_{n'}^{(\nu)}(\vartheta_0)=\delta_{nn'}h_n^{(\nu)} \tag{B.4}$$

$$\sum_{n=0}^{\infty}\frac{P_n^{(\nu)}(\vartheta_0)P_n^{(\nu)}(\vartheta_0')}{h_n^{(\nu)}}=\frac{\delta(\vartheta_0-\vartheta_0')}{(1-\vartheta_0^2)^{\nu}} \tag{B.5}$$

and

$$\int_{1}^{\infty} d\vartheta_1\left(\vartheta_1^2-1\right)^{\nu}F_k^{(\nu)}(\vartheta_1)F_{k'}^{(\nu)}(\vartheta_1)=\frac{\delta(k-k')}{\left(A_k^{(\nu)}\right)^2} \tag{B.6}$$

$$\int_{0}^{\infty} dk\left(A_k^{(\nu)}\right)^2F_k^{(\nu)}(\vartheta_1)F_k^{(\nu)}(\vartheta_1')=\frac{\delta(\vartheta_1-\vartheta_1')}{(\vartheta_1^2-1)^{\nu}}, \tag{B.7}$$

where

$$\left(A_k^{(\nu)}\right)^2=\frac{|\Gamma(\nu+1/2+ik)|^2}{2^{2\nu}(\Gamma(\nu+1))^2|\Gamma(ik)|^2}. \tag{B.8}$$

Then the eigenfunctions and the eigenvalues of $\mathscr{H}$ are given by

$$\varphi_{nk}(\vartheta_0,\vartheta_1)=\frac{A_k^{(\nu)}}{\left(h_n^{(\nu)}\right)^{1/2}}P_n^{(\nu)}(\vartheta_0)F_k^{(\nu)}(\vartheta_1), \quad \varepsilon_{nk}=k^2+\left(n+\nu+\frac{1}{2}\right)^2 \tag{B.9}$$

with $-1\le\vartheta_0\le1$ and $1\le\vartheta_0\le\infty$. For the calculations of the moments of the heat conductance it is convenient to replace $F_k^{(\nu)}(\vartheta_1)$ by the following function in (B.3):

$$\widetilde{F}_k^{(\nu)}(\vartheta_1)\equiv\frac{1}{\omega(\vartheta_1)}\int_0^{\infty}dx\frac{\omega(x)F_k^{(\nu)}(x)}{x+\vartheta_1} = \frac{1}{\omega(\vartheta_1)}\left|\Gamma\left(\frac{1}{2}-\nu+ik\right)\right|^2F_k^{(-\nu)}(\vartheta_1), \tag{B.10}$$

so the new set of eigenfunctions $\varphi_{nk}(\vartheta_0,\vartheta_1)$ correspond to (22).

C.

Let us now show the equivalence between (27) and (29). First, we calculate the integrals shown in (27). We know from [25] that

$$I_{nl}^{(0)}=\frac{(-1)^n2^{l+2\nu+1}l!\Gamma(l+\nu+1)\Gamma(n+\nu+1)}{(l-n)!n!\Gamma(n+l+2\nu+2)}. \tag{C.1}$$

Using the identity $F(a,b;c;z)=\Gamma(1-a)\Gamma(c)P_{-a}^{(c-1,a+b-c)}(1-2z)/\Gamma(c-a)$ we can write

$$I_{nl}^{(1)}=\frac{\Gamma(1+n')\Gamma(-2n'-\mu-\nu)}{\Gamma(-n'-\mu-\nu)}\int_{-1}^{1}d\vartheta_0(1-\vartheta_0)^{\nu+l}\cdot(1+\vartheta_0)^{\nu}P_n^{(\nu)}(\vartheta_0)P_{n'}^{(-2n'-\mu-\nu-1,\nu)}(\vartheta_0), \tag{C.2}$$

where $n'=N-l-1$. This integral can be solved by means of (20) of Chapter 16.4 of [28]. We find

$$I_{nl}^{(1)}=\frac{2^{2\nu+l+1}\Gamma(-l+n)\Gamma(n+\nu+1)\Gamma(l+\nu+1)}{n!\Gamma(-l)\Gamma(2\nu+n+l+2)}\times{}_4F_3\left[-n',-n'-\mu,\nu+l+1,l+1;-2n'-\mu-\nu,2\nu+l+n+2,-n+l+1;1\right] \tag{C.3}$$

or after using $\Gamma(a-n)=(-1)^n\Gamma(-a)\Gamma(1+a)/\Gamma(n+1-a)$

$$I_{nl}^{(1)}=(-1)^n\frac{2^{2\nu+l+1}\Gamma(l+1)\Gamma(n+\nu+1)\Gamma(l+\nu+1)}{\Gamma(n-1)\Gamma(1+l-n)\Gamma(2\nu+n+l+2)}\times{}_4F_3\left[-n',-n'-\mu,\nu+l+1,l+1;-2n'-\mu-\nu,2\nu+l+n+2,-n+l+1;1\right]. \tag{C.4}$$

The remaining integrals can be calculated representing the hypergeometric functions as Meijer G functions by using the identity

$$F[a,b,c,z]=\frac{\Gamma(c)}{\Gamma(b)\Gamma(a)}G_{2\,2}^{1\,2}\left(\begin{matrix}1-a,1-b\\0,1-c\end{matrix}\;\middle|\;-z\right) \tag{C.5}$$

and integration identities of the Meijer G functions [29]. Performing the change of variables $-x=(1-\vartheta_1)/2$ and using the identity $F[a,b;c;z]=(1-z)^{c-a-b}F[c-a,c-b;c;z]$ [30], we get

$$\begin{aligned} J_{kl}^{(0)}&=\frac{1}{(-1)^{l+1}2^{2\nu+l}}\int_0^{\infty}dx\frac{x^{-\nu-l-1}}{(x+1)^{\nu}}\cdot F\left[-\nu+\frac{1}{2}+ik,-\nu+\frac{1}{2}-ik;-\nu+1;-x\right]=\frac{1}{(-1)^{l+1}2^{2\nu+l}}\cdot\int_0^{\infty}dx\frac{F\left[\frac{1}{2}+ik,\frac{1}{2}-ik;-\nu+1;-x\right]}{x^{\nu+l+1}}\\ &=\frac{\Gamma(-\nu+1)}{(-1)^{l+1}2^{2\nu+l}\left|\Gamma\left(\frac{1}{2}+ik\right)\right|^2}\int_0^{\infty}dx\cdot\frac{G_{2\,2}^{1\,2}\left(\begin{matrix}1/2+ik,1/2-ik\\0,\nu\end{matrix}\;\middle|\;x\right)}{x^{\nu+l+1}}=\frac{\Gamma(-\nu+1)}{(-1)^{l+1}2^{2\nu+l}|\Gamma(1/2+ik)|^2}\\ &\cdot\int_0^{\infty}dx\,G_{2\,2}^{1\,2}\left(\begin{matrix}-\frac{1}{2}+ik-\nu-l,-\frac{1}{2}-ik-\nu-l\\-\nu-l-1,-l-1\end{matrix}\;\middle|\;x\right)\\ &=\frac{\Gamma(-\nu+1)\Gamma(-\nu-l)\left|\Gamma\left(\frac{1}{2}+ik+\nu+l\right)\right|^2}{(-1)^{l+1}2^{2\nu+l}\left|\Gamma\left(\frac{1}{2}+ik\right)\right|^2\Gamma(l+1)}. \end{aligned} \tag{C.6}$$

Finally, similarly to (C.6), we get

$$\begin{aligned} J_{kl}^{(1)} &= \frac{1}{(-1)^{l+1}2^{2\nu+l}}\int_0^\infty dx \frac{x^{-\nu-l-1}}{(x+1)^\nu} F\left[-\nu+\frac{1}{2}+ik,-\nu+\frac{1}{2}-ik;-\nu+1;-x\right] \\ &\quad \times F\left[n'+1,n'+1+\mu;2n'+\mu+\nu+2;-x\right] \\ &= \frac{\Gamma(-\nu+1)}{(-1)^{l+1}2^{2\nu+l}\left|\Gamma\left(\frac{1}{2}+ik\right)\right|^2}\frac{\Gamma\left(2n'+\mu+\nu+2\right)}{\Gamma(n'+1)\Gamma(n'+\mu+1)} \\ &\quad \times \int_0^\infty dx G^{1\ 2}_{2\ 2}\left(\begin{matrix} -\frac{1}{2}+ik-\nu-l,-\frac{1}{2}-ik-\nu-l \\ -\nu-l-1,-l-1\end{matrix} \mid x\right) G^{1\ 2}_{2\ 2}\left(\begin{matrix} -n',-n'-\mu \\ 0,-2n'-\mu-\nu-1\end{matrix} \mid x\right) \\ &= \frac{\Gamma(-\nu+1)}{(-1)^{l+1}2^{2\nu+l}\left|\Gamma\left(\frac{1}{2}+ik\right)\right|^2}\frac{\Gamma\left(2n'+\mu+\nu+2\right)}{\Gamma(n'+1)\Gamma(n'+\mu+1)} \\ &\quad \times G^{3\ 3}_{4\ 4}\left(\begin{matrix} \nu+l+1,-n',-n'-\mu,l+1 \\ \frac{1}{2}-ik+\nu+l,\frac{1}{2}+ik+\nu+l,0,-2n'-\mu-\nu-1\end{matrix} \mid 1\right). \end{aligned} \tag{C.7}$$

Once the integrals have been obtained in terms of Meijer G functions, Theorem 1 follows from using the simple identity

$$\begin{aligned} &\left(A_k^{(-\nu)}\right)^2 \sum_{l=0}^{N-1}\left(I_{nl}^{(1)}J_{kl}^{(1)} - I_{nl}^{(0)}J_{kl}^{(0)}\right) \\ &= \frac{2P_n^{(\nu)}(1)\left|\Gamma(1/2-\nu+ik)\right|^2}{\left|\Gamma(ik)\right|^2 \varepsilon_{nk}} c_{nk}^{(\nu)}(N_1)\,c_{nk}^{(\nu)}(N_2) \end{aligned} \tag{C.8}$$

or equivalently

$$\begin{aligned} &\sum_{l=0}^{N-1}\left(I_{nl}^{(1)}J_{kl}^{(1)} - I_{nl}(0)\,J_{kl}(0)\right) \\ &= \frac{\Gamma(n+\nu+1)\left(\Gamma(-\nu+1)\right)^2}{2^{2\nu-1}\Gamma(n+1)\Gamma(\nu+1)}\frac{c_{nk}^{(\nu)}(N_1)\,c_{nk}^{(\nu)}(N_2)}{k^2+(n+\nu+1/2)^2}, \end{aligned} \tag{C.9}$$

which can be checked with the Meijer G function representation of algebraic computer systems such as Mathematica [31].

Conflicts of Interest

The authors declare that they have no conflicts of interest.

Acknowledgments

This work was supported by the Brazilian agencies CNPq and CAPES.

References

[1] C. W. J. Beenakker, "Random-matrix theory of quantum transport," *Reviews of Modern Physics*, vol. 69, no. 3, pp. 731–808, 1997.

[2] A. Altland and M. R. Zirnbauer, "Nonstandard symmetry classes in mesoscopic normal-superconducting hybrid structures," *Physical Review B: Condensed Matter and Materials Physics*, vol. 55, no. 2, pp. 1142–1161, 1997.

[3] M. Caselle, "A new classification scheme for random matrix theories," 1996, https://arxiv.org/abs/cond-mat/9610017.

[4] P. W. Brouwer and A. Altland, "Anderson localization from classical trajectories," *Physical Review B: Condensed Matter and Materials Physics*, vol. 78, no. 7, Article ID 075304, 2008.

[5] K. Efetov, *Supersymmetry in Disorder and Chaos*, Cambridge University Press, Cambridge, UK, 1997.

[6] K. Richter and M. Sieber, "Semiclassical theory of chaotic quantum transport," *Physical Review Letters*, vol. 89, article 206801, 2002.

[7] G. Berkolaiko and J. Kuipers, "Universality in chaotic quantum transport: the concordance between random-matrix and semiclassical theories," *Physical Review E: Statistical, Nonlinear, and Soft Matter Physics*, vol. 85, article 045201, 2012.

[8] S. R. Elliott and M. Franz, "Colloquium: Majorana fermions in nuclear, particle, and solid-state physics," *Reviews of Modern Physics*, vol. 87, no. 1, pp 137–163, 2015.

[9] J. P Dahlhaus, *Random-matrix theory and stroboscopic models of topological insulators and superconductors*, Lorentz Institute, Faculty of Science, Leiden University, Leiden, The Netherlands, 2012.

[10] C. W. Beenakker, "Random-matrix theory of Majorana fermions and topological superconductors," *Reviews of Modern Physics*, vol. 87, no. 3, pp 1037–1066, 2015.

[11] P. W. Brouwer, A. Furusaki, I. A. Gruzberg, and C. Mudry, "Localization and delocalization in dirty superconducting

wires," *Physical Review Letters*, vol. 85, no. 5, pp. 1064–1067, 2000.

[12] F. J. Dyson, "The dynamics of a disordered linear chain," *Physical Review A: Atomic, Molecular and Optical Physics*, vol. 92, no. 6, pp. 1331–1338, 1953.

[13] O. Motrunich, K. Damle, and D. A. Huse, "Griffiths effects and quantum critical points in dirty superconductors without spin-rotation invariance: one-dimensional examples," *Physical Review B: Condensed Matter and Materials Physics*, vol. 63, no. 22, article 224204, 2001.

[14] I. A. Gruzberg, N. Read, and S. Vishveshwara, "Localization in disordered superconducting wires with broken spin-rotation symmetry," *Physical Review B: Condensed Matter and Materials Physics*, vol. 71, no. 24, article 245124, 2005.

[15] C. Mudry, "Fractional abelian topological phases of matter for fermions in two-dimensional space," *Topological Aspects of Condensed Matter Physics*, vol. 103, article 265, 2017.

[16] A. F. Macedo-Junior and A. M. Macêdo, "Brownian-motion ensembles of random matrix theory: a classification scheme and an integral transform method," *Nuclear Physics*, vol. 752, no. 3, pp. 439–475, 2006.

[17] A. F. Macedo-Junior, *Transporte Em Nanoestruturas: Métodos De Movimento Browniano E Teoria De Circuitos*, Universidade Federal de Pernambuco, 2006.

[18] S. Datta, *Electronic Transport in Mesoscopic Systems*, Cambridge University Press, Cambridge, UK, 1997.

[19] J. P. Dahlhaus, B. Béri, and C. W. Beenakker, "Random-matrix theory of thermal conduction in superconducting quantum dots," *Physical Review B: Condensed Matter and Materials Physics*, vol. 82, no. 1, Article ID 014536, 2010.

[20] M. I. Sena-Junior, F. A. Almeida, and A. M. Macêdo, "Counting statistics and an anomalous metallic phase in a network of quantum dots," *Journal of Physics A*, vol. 47, Article ID 235101, 2014.

[21] S. Iida, H. A. Weidenmüller, and J. A. Zuk, "Statistical scattering theory, the supersymmetry method and universal conductance fluctuations," *Annals of Physics*, vol. 200, no. 2, pp. 219–270, 1990.

[22] A. Macêdo, "Quantum dot to disordered wire crossover: A complete solution in all length scales for systems with unitary symmetry," *Physical Review B: Condensed Matter and Materials Physics*, vol. 61, no. 7, pp. 4453–4456, 2000.

[23] G. C. Duarte-Filho, A. F. Macedo-Junior, and A. M. Macêdo, "Circuit theory and full counting statistics of charge transfer through mesoscopic systems: A random-matrix approach," *Physical Review B: Condensed Matter and Materials Physics*, vol. 76, no. 7, Article ID 075342, 2007.

[24] C. E. Souza and A. F. Macedo-Junior, "Quantum transport in chaotic ballistic cavity: a Brownian-motion approach," *Journal of Physics A: Mathematical and General*, vol. 48, no. 41, Article ID 415101, 2015.

[25] A. F. Macedo-Junior and A. M. Macêdo, "Brownian-motion ensembles: correlation functions of determinantal processes," *Journal of Physics A: Mathematical and General*, vol. 41, no. 1, Article ID 015004, 2008.

[26] A. M. Macêdo, "Nonanalytic scaling of conductance cumulants in dirty superconducting wires," *Physical Review B: Condensed Matter and Materials Physics*, vol. 65, no. 13, Article ID 132510, 2002.

[27] A. D. Mirlin, A. Muller-Groeling, and M. R. Zirnbauer, "Conductance fluctuations of disordered wires: Fourier analysis on supersymmetric spaces," *Annals of Physics*, vol. 236, no. 2, pp. 325–373, 1994.

[28] H. Bateman and A. Erdelyi, *Higher Transcendental Functions*, vol. 2, McGraw-Hill, New York, NY, USA, 1953.

[29] I. S. Gradshteyn and I. M. Ryzhik, *Table of Integrals, Series, and Products*, Academic Press, 2014.

[30] F. W. J. Olver, *NIST Handbook of Mathematical Functions Hardback and CD-ROM*, Cambridge University Press, Cambridge, UK, 2010.

[31] Wolfram Research, Inc., Mathematica, Version 10.2, Champaign, IL, 2017.

11

Macroscopic Irreversibility and Decay to Kinetic Equilibrium of the 1-Body PDF for Finite Hard-Sphere Systems

Massimo Tessarotto[1,2] **and Claudio Cremaschini**[3]

[1]*Department of Mathematics and Geosciences, University of Trieste, Via Valerio 12, 34127 Trieste, Italy*
[2]*Institute of Physics, Faculty of Philosophy and Science, Silesian University in Opava, Bezručovo nám. 13, 74601 Opava, Czech Republic*
[3]*Institute of Physics and Research Center for Theoretical Physics and Astrophysics, Faculty of Philosophy and Science, Silesian University in Opava, Bezručovo nám. 13, 74601 Opava, Czech Republic*

Correspondence should be addressed to Massimo Tessarotto; maxtextss@gmail.com

Academic Editor: Yannis Dimakopoulos

The conditions for the occurrence of the so-called macroscopic irreversibility property and the related phenomenon of decay to kinetic equilibrium which may characterize the 1-body probability density function (PDF) associated with hard-sphere systems are investigated. The problem is set in the framework of the axiomatic "ab initio" theory of classical statistical mechanics developed recently and the related establishment of an exact kinetic equation realized by the Master equation for the same kinetic PDF. As shown in the paper the task involves the introduction of a suitable functional of the 1-body PDF, identified here with the *Master kinetic information*. It is then proved that, provided the same PDF is prescribed in terms of suitably smooth, i.e., stochastic, solution of the Master kinetic equation, the two properties indicated above are indeed realized.

1. Introduction

The axiomatic theory of classical statistical mechanics (CSM) recently proposed in a series of papers (see [1–4]) and referred to as *ab initio theory of CSM* provides a self-consistent pathway to the kinetic theory of hard-sphere systems, as well as in principle also point particles subject to finite-range interactions [5]. Its theoretical basis and conditions of validity are indeed founded on a unique physical realization of the axioms which are set at the foundations of CSM [1–3], a fact which permits the treatment of phase-space and kinetic probability density functions (PDF) which are realized by either stochastic (i.e., ordinary) functions or distributions such as the $N-$body Dirac delta (or certainty function [6]). This feature is physically based being due to the prescription of the collision boundary conditions (CBC, [2]), i.e., the relationship occurring at collision events between incoming and outgoing multibody probability density functions PDF. The choice of the appropriate CBC indicated in [2], denoted as modified collision boundary condition (MCBC), is actually of crucial importance and departs from the customary realization/interpretation (of the same axioms) originally adopted in Boltzmann [7], Enskog [8, 9], and Grad [10] kinetic approaches (for a review of Grad's kinetic theory based on CSM see also Cercignani [11, 12]). The same choice implies, in fact, a number of theoretical and physically relevant consequences. In particular, it follows that the new theory

(i) unlike Enskog theory [3] applies also to *finite* $N-$*body hard-sphere systems* S_N, namely, systems formed by N like smooth hard-spheres of diameter σ and mass m, in which the parameters (N, σ, m) remain all constant and finite [3]. On the other hand, the same particles are assumed as usual: (A) subject to instantaneous (unary, binary, and multiple) elastic collisions which leave unchanged the particles angular momenta and (B) immersed in a stationary bounded domain Ω of the Euclidean space $\mathbb{R}^3$ with finite canonical measure;

(ii) has led to the discovery [3] of an exact kinetic equation holding globally in time [13] (i.e., for all $t \in I \equiv \mathbb{R}$) for these systems and denoted as Master kinetic equation (recalled in Appendix A). In other words the Master equation is nonasymptotic in character with respect to the parameters the (finite) parameters (N, σ, m). In addition the same equation holds under suitable maximal entropy conditions for the statistical treatment of the so-called Boltzmann-Sinai classical dynamical system (CDS), which implies that initial (binary or multibody) phase-space statistical correlations are assumed identically vanishing, while at the same time only suitable uniquely prescribed configuration-space correlations can arise. As such the equation generalizes and extends the validity of the Boltzmann and Enskog kinetic equations and notably applies to arbitrary 1–body PDFs which can be realized either in terms of stochastic functions or distributions;

(iii) is time-reversal invariant [4], namely, the Master kinetic equation is time-reversal ($\mathcal{TR}-$) symmetric. In other words, the same equation is invariant with respect to the $\mathcal{TR}-$transformation

$$\begin{aligned} \tau \equiv t - t_o &\longrightarrow \tau' \equiv t' - t_o = -t + t_o \equiv -\tau, \\ \mathbf{r}_1 &\longrightarrow \mathbf{r}_1' = \mathbf{r}_1, \\ \mathbf{v}_1 &\longrightarrow \mathbf{v}_1' = -\mathbf{v}_1. \end{aligned} \tag{1}$$

Thus, representing the absolute time as $t = \tau + t_o$, with t_o being a prescribed (arbitrary) initial time, it follows that the $\mathcal{TR}-$transformation leaves invariant the initial time t_o and the instantaneous position $\mathbf{r}_1 = \mathbf{r}_1(t \equiv t_o + \tau)$ of an arbitrary particle, while reversing the signature (i.e., versus) of its velocity $\mathbf{v}_1 = \mathbf{v}_1(t \equiv t_o + \tau)$. Accordingly, thanks to $\mathcal{TR}-$symmetry the two initial-value problems associated with the Master kinetic equation in the two cases are related in such a way that, respectively, the initial 1–body PDF at time t_o, $\rho_1^{(N)}(\mathbf{x}_1, t_o) \equiv \rho_{1(o)}^{(N)}(\mathbf{r}_1, \mathbf{v}_1)$, and the corresponding time-evolved PDF $\rho_1^{(N)}(\mathbf{x}_1, t)$ are carried into the $\mathcal{TR}-$transformed 1–body PDFs $\rho_{1(o)\mathcal{TR}}^{(N)}(\mathbf{r}_1', \mathbf{v}_1')$ and $\rho_{1\mathcal{TR}}^{(N)}(\mathbf{r}_1', \mathbf{v}_1', t')$, respectively, prescribed according to the law

$$\begin{gathered} \rho_{1(o)}^{(N)}(\mathbf{r}_1, \mathbf{v}_1) \\ \rho_1^{(N)}(\mathbf{x}_1, t) \equiv \rho_1^{(N)}(\mathbf{r}_1, \mathbf{v}_1, t_o + \tau) \\ \downarrow \\ \rho_{1(o)\mathcal{TR}}^{(N)}(\mathbf{r}_1', \mathbf{v}_1') \equiv \rho_{1(o)}^{(N)}(\mathbf{r}_1, -\mathbf{v}_1) \\ \rho_{1\mathcal{TR}}^{(N)}(\mathbf{r}_1', \mathbf{v}_1', t') \equiv \rho_1^{(N)}(\mathbf{r}_1, -\mathbf{v}_1, t_o - \tau); \end{gathered} \tag{2}$$

(iv) conserves the corresponding Boltzmann-Shannon (BS) statistical entropy [4]. This is identified with the phase-space moment

$$S\left(\rho_1^{(N)}(t)\right) - \int_{\Gamma_1} d\mathbf{x}_1 \rho_1^{(N)}(\mathbf{x}_1, t) \ln \frac{\rho_1^{(N)}(\mathbf{x}_1, t)}{A_1}, \tag{3}$$

with $\rho_1^{(N)}(t) \equiv \rho_1^{(N)}(\mathbf{x}_1, t)$ being an arbitrary stochastic PDF solution of the Master kinetic equation and A_1 an arbitrary positive constant such that the initial PDF

$$\rho_1^{(N)}(t_o) \equiv \rho_1^{(N)}(\mathbf{x}_1, t_o) = \rho_{10}^{(N)}(\mathbf{x}_1) \tag{4}$$

is such that the corresponding BS functional $S(\rho_1^{(N)}(t_o))$ is defined. As a consequence it follows that an arbitrary smooth solution of the Master kinetic equation satisfies the constant H-theorem

$$\frac{\partial}{\partial t} S\left(\rho_1^{(N)}(t)\right) = 0 \tag{5}$$

for all $t \in I \equiv \mathbb{R}$ (see again related discussion in [4]).

Based on the ab initio theory of CSM, in this paper the problem is posed of the existence of two phenomena which are expected to characterize the statistical description of finite $N-$body hard-sphere systems and therefore should lay at the very foundation of CSM and kinetic theory. These are related to the physical conditions for the possible occurrence of the so-called *property of macroscopic irreversibility* (PMI) and the consequent one represented by the *decay to kinetic equilibrium* (DKE) which characterize the 1–body (kinetic) PDF in these $N-$body systems, i.e., when 1–body-factorized initial conditions are considered for the $N-$body Liouville equation [3]. The conjecture is that—in some sense in analogy with the ubiquitous character of the ergodicity property which characterizes hard-sphere systems and hence the S_N-CDS [14, 15]—the occurrence of such phenomena should be independent of the number N of constituent particles of the system and therefore apply to actual physical systems for which the parameters (N, σ, m) are obviously all finite.

1.1. Motivations and Background. Both properties indicated above concern the statistical behavior of an ensemble S_N of like particles which are advanced in time by a suitable $N-$body classical dynamical system, identified here with the S_N-CDS. Specifically they arise in the context of the kinetic description of the same CDS, i.e., in terms of the corresponding 1–body (kinetic) probability density function (PDF) $\rho_1^{(N)}(t) \equiv \rho_1^{(N)}(\mathbf{x}_1, t)$. The latter is required to belong to the functional class $\{\rho_1^{(N)}(\mathbf{x}_1, t)\}$ of suitably smooth and strictly positive ordinary functions which are particular solutions of the relevant kinetic equation.

In fact, PMI should be realized by means of a suitable, but still possibly nonunique, functional which should be globally defined in the future (i.e., for all times $t \geq t_o$, being t_o the initial time) bounded and nonnegative, and therefore to be identified with the notion of information measure. Most importantly, however, the same functional, to be referred to here as *Master kinetic information* (MKI), should also exhibit a continuously differentiable and monotonic, i.e., in particular decreasing, time-dependence.

Regarding, instead, the second property of DKE this concerns the asymptotic behavior of the 1–body PDF $\rho_1^{(N)}(t) \equiv \rho_1^{(N)}(\mathbf{x}_1, t)$ which, accordingly, should be globally defined and decay for $t \longrightarrow +\infty$ to a stationary and spatially uniform Maxwellian PDF

$$\rho_{1M}^{(N)}(\mathbf{v}_1) = \frac{n_o}{\pi^{3/2}(2T_o/m)^{3/2}} \exp\left\{-\frac{m(\mathbf{v}_1 - \mathbf{V}_o)^2}{2T_o}\right\}, \quad (6)$$

where $\{n_o > 0, T_o > 0, \mathbf{V}_o\}$ are constant fluid fields.

Both PMI and DKE correspond to physical phenomena which are actually expected to arise in disparate classical N–body systems. The clue for their realization is represented by the ubiquitous occurrence of kinetic equilibria and consequently, in principle, also of the corresponding possible manifestation of macroscopic irreversibility and decay processes. Examples of the former ones are in principle easy to be found, ranging from neutral fluids [16] to collisional/collisionless and nonrelativistic/relativistic gases and plasmas [17–19].

However, the most notable example is perhaps provided by dilute hard-sphere systems ("gases") characterized by a large number of particles ($N \equiv 1/\varepsilon \gg 1$) and a small (i.e., infinitesimal) diameter $\sigma \sim O(\varepsilon^{1/2})$ of the same hard-spheres, for which the Boltzmann equation applies. Indeed the Boltzmann equation is actually specialized to the treatment of dilute hard-sphere systems in the Boltzmann-Grad limit discussed in the Lanford theorem [20–22] (for a detailed discussion of the topics in the context of the ab initio theory see also [4]). In such a case the 1–body PDF can be formally obtained by introducing the Boltzmann-Grad limit operator [4]

$$\mathscr{L}_{BG} \equiv \lim_{\substack{N \equiv 1/\varepsilon \longrightarrow \infty \\ \sigma \sim O(\varepsilon^{1/2})}}, \quad (7)$$

whereby the limit function $\rho_1(\mathbf{x}_1, t)$ is denoted

$$\rho_1(\mathbf{x}_1, t) = \mathscr{L}_{BG}\rho_1^{(N)}(\mathbf{x}_1, t), \quad (8)$$

and $\rho_1(\mathbf{x}_1, t)$ identifies a particular solution of the Boltzmann kinetic equation.

Historically, the property of irreversibility indicated above is known to be related to the Carnot's second Law of Classical Thermodynamics. More precisely, it is related to the first-principle-proof originally attempted by Ludwig Boltzmann in 1872 [7]. Actually it is generally agreed that both phenomena lie at the very heart of Boltzmann and Grad kinetic theories [7, 10] and the related original construction of the Boltzmann kinetic equation (1872). In particular, the goal set by Boltzmann himself in his 1872 paper was the proof of Carnot's Law providing at the same time also a possible identification of thermodynamic entropy. This was achieved in terms of what is nowadays known as Boltzmann-Shannon (BS) statistical entropy, which is identified with the phase-space moment

$$\begin{aligned} M_{X_E}(\rho_1(t)) &\equiv \int_{\Gamma_1} d\mathbf{x}_1 \rho_1(\mathbf{x}_1, t) X_E(\mathbf{x}_1, t) \\ &= -\int_{\Gamma_1} d\mathbf{x}_1 \rho_1(\mathbf{x}_1, t) \ln \frac{\rho_1(\mathbf{x}_1, t)}{A_1} \\ &\equiv S(\rho_1(t)). \end{aligned} \quad (9)$$

Here $X_E(\mathbf{x}_1, t) \equiv -\ln(\rho_1(\mathbf{x}_1, t)/A_1)$, $\rho_1(\mathbf{x}_1, t)$, and A_1 denote, respectively, the BS entropy density, an arbitrary particular solution of the Boltzmann equation for which the same phase-space integral exists and an arbitrary positive constant. In fact, according to the Boltzmann H-theorem [7] the same functional should satisfy the *entropic inequality*

$$\frac{\partial}{\partial t} S(\rho_1(t)) \geq 0, \quad (10)$$

while, furthermore, the *entropic equality condition*

$$\begin{aligned} &\frac{\partial}{\partial t} S(\rho_1(t)) = 0 \Longleftrightarrow \\ &\rho_1(\mathbf{x}_1, t) = \rho_{1M}^{(N)}(\mathbf{v}_1) \end{aligned} \quad (11)$$

should hold. The latter equation implies therefore that, provided $\rho_1(t)$ and $S(\rho_1(t))$ exist globally [23], then necessarily $\lim_{t \longrightarrow +\infty} \rho_1(\mathbf{x}_1, t) = \rho_{1M}(\mathbf{v}_1)$, with $\rho_{1M}(\mathbf{v}_1)$ denoting the stationary and spatially uniform Maxwellian PDF (6).

In this reference, however, the question arises of the precise characterization of the concept of irreversibility, i.e., whether it should be regarded as a purely macroscopic phenomenon ("macroscopic irreversibility"), i.e., affecting only the BS entropy $S(\rho_1(t))$ through the Boltzmann H-theorem indicated above, or microscopic in the sense that the same Boltzmann equation should be considered as irreversible ("microscopic irreversibility"). Thus, in principle, in the second case the further issue emerges of the possible physical origin of microscopic irreversibility in special reference to the Lanford's derivation of the Boltzmann equation and subsequent related comments discussed respectively by Uffink and Valente and Ardourel in [24, 25] (see also Drory [26]).

However, as shown in [4], the Boltzmann equation is actually $\mathscr{T}\mathscr{R}$–symmetric. Such a conclusion is of basic importance since it overcomes the so-called Loschmidt paradox, i.e., the objection raised by Loschmidt in 1876 [27] regarding the original Boltzmann formulation of his namesake kinetic equation and H-theorem [7]. In fact, Loschmidt claimed that the Boltzmann H-theorem inequality should change sign under time-reversal and thus violate the microscopic time-reversibility of the underlying hard-sphere classical dynamical system. In his long-pondered reply given in 1896 [28] Boltzmann himself introduced what was later referred to as the modified form of the Boltzmann H-theorem [29].

The key implication is therefore that, in contrast to Boltzmann's own statement and the traditional subsequent mainstream literature interpretation (see, for example, by Cercignani, Lebowitz in [30, 31] and more recently the review given by Gallavotti [32]), the Boltzmann H-theorem indicated—together with the modified form indicated above—cannot be interpreted as an intrinsic irreversibility property occurring at the microscopic level, namely, holding for the Boltzmann equation itself. On the contrary, consistent with the physical interpretation of the Loschmidt paradox provided in [4], this must be regarded only as *property of macroscopic irreversibility* (or PMI) of the 1–body PDF solution of the Boltzmann equation. In other words, the Boltzmann inequality (10) necessarily holds *independent of the orientation of the time axis* (arrow of time) and therefore cannot represent a true (i.e., microscopic) property which as such should uniquely determine the arrow of time.

Nevertheless, the possible realization of either PMI or DKE is more subtle. In fact they actually depend in a critical way on the prescription of the functional class $\{\rho^{(N)}(\mathbf{x}_1,t)\}$, so that their occurrence is actually nonmandatory. Indeed, both cannot occur—in principle also for Boltzmann and Grad kinetic theories—if the N–body probability density function $\rho^{(N)}(\mathbf{x},t)$ is identified with the deterministic N–body PDF [1], namely, the N–body phase-space Dirac delta. This is defined as $\delta(\mathbf{x}-\mathbf{x}(t)) \equiv \prod_{1=1,N}\delta(\mathbf{x}_i-\mathbf{x}_i(t))$, with $\mathbf{x} \equiv \{\mathbf{x}_1,\ldots,\mathbf{x}_N\}$ denoting the state of the N–body system and $\mathbf{x}(t) \equiv \{\mathbf{x}_1(t),\ldots,\mathbf{x}_N(t)\}$ is the image of an arbitrary initial state $\mathbf{x}(t_o) \equiv \mathbf{x}_o$ generated by the same N–body CDS. That such a PDF necessarily must realize an admissible particular solution of the N–body Liouville equation follows, in fact, as a straightforward consequence of the axioms of classical statistical mechanics [1].

Despite these premises, however, the case of a finite Boltzmann-Sinai CDS, which is characterized by a finite number of particles N and/or a finite-size of the hard-spheres and/or a dense or locally dense system, is more subtle and—as explained below—even unprecedented since it has actually remained unsolved to date. The reasons are as follows. First, Boltzmann and Grad kinetic theories are inapplicable to the finite Boltzmann-Sinai CDS. Second, the Boltzmann-Shannon entropy associated with an arbitrary particular solution $\rho^{(N)}(t) \equiv \rho^{(N)}(\mathbf{x}_1,t)$ of the Master kinetic equation, i.e., the functional $S(\rho_1^{(N)}(t)) \equiv M_{X_E}(\rho_1^{(N)}(t))$, in contrast to $S(\rho_1(t)) \equiv M_{X_E}(\rho_1(t))$, is exactly conserved in the sense that identically

$$\frac{\partial S\left(\rho_1^{(N)}(t)\right)}{\partial t} \equiv 0 \tag{12}$$

must hold. As a consequence the validity itself of Boltzmann H-theorem breaks down in the case of the Master kinetic equation. Third, an additional motivation is provided by the conjecture that both PMI and DKE might occur only if the Boltzmann-Grad limit is actually performed, i.e., only in validity of Boltzmann equation and H-theorem.

Hence the question which arises is whether in the case of a finite Boltzmann-Sinai CDS the phenomenon of DKE may still arise. Strong indications seem to be hinting at such a possibility. In this regard the example-case which refers to the statistical description of a Navier-Stokes fluid described by the incompressible Navier-Stokes equations (INSE) in terms of the Master kinetic equation is relevant and suggests that this may be indeed the case. In fact, thanks also to comparisons with the mean-field inverse kinetic approach to INSE [16], in such a case the decay of the fluid velocity field occurring in a bounded domain necessarily demands the existence of DKE. In other words, in the limit $t \longrightarrow +\infty$ the 1–body PDF must decay uniformly to the stationary and spatially uniform Maxwellian PDF (6).

However, besides the construction of the kinetic equation appropriate for such a case, a further unsolved issue lies in the determination of the functional class $\{\rho_1^{(N)}(\mathbf{x}_1,t)\}$ for which both PMI and DKE should/might be realized. In particular, the possible occurrence of both PMI and DKE should correspond to suitably smooth, but nonetheless still arbitrary, initial conditions $\{\rho_1^{(N)}(\mathbf{x}_1,t_o)\}$. These should warrant that in the limit $t \longrightarrow +\infty$, $\rho_1^{(N)}(\mathbf{x}_1,t)$ uniformly converges to the spatially homogeneous and stationary Maxwellian PDF $\rho_{1M}(\mathbf{v}_1)$ (6). Such a result, however, is highly nontrivial since it should rely on the establishment of a global existence theorem for the same 1–body PDF $\rho_1^{(N)}(\mathbf{x}_1,t)$—namely, holding in the whole time axis $I \equiv \mathbb{R}$, besides the same 1–body phase space Γ_1—for the involved kinetic equation which is associated with the S_N–CDS. In the context of the Boltzmann equation in particular, despite almost-endless efforts this task has actually not been accomplished yet, the obstacle being intrinsically related to the asymptotic nature of the Boltzmann equation [13]. In fact for the same equation it is not known in satisfactory generality whether smooth enough solutions of the same equation exist which satisfy the H–theorem inequality and decay asymptotically to kinetic equilibrium [23, 30].

1.2. Goals and Organization of the Paper. Based on these premises, the crucial new results that we intend to display in this paper concern the *proof-of-principle* of two phenomena which are expected to characterize the statistical description of finite N–body hard-sphere systems and therefore should lay at the very foundation of classical statistical mechanics and kinetic theory alike. These are related to the physical conditions for the possible occurrence of both PMI and the consequent one represented by the possible occurrence of DKE which should characterize the kinetic PDF in these systems. These phenomena are well known to occur in the case of dilute hard-sphere systems, i.e., in the Boltzmann-Grad limit. In particular, for an exhaustive treatment of the related issues which arise in the context of the ab initio theory we refer to discussions reported in [4]. Nevertheless, as indicated above, their existence in the case of finite hard-sphere systems is partly motivated by a previous investigation dealing with the kinetic description incompressible Navier-Stokes granular fluids [33].

Therefore, main goal of the paper is to show that these properties actually emerge as necessary implications of the ab initio theory of CSM. Incidentally, in doing so, the Master kinetic equation must be necessarily adopted. In fact, the finiteness requirement on the S_N–CDS rules out for further possible consideration either the Boltzmann or the Enskog kinetic equations, these equations being inapplicable to the treatment of systems of this type [3]. Specifically, in the following the case $N > 2$ is considered everywhere, which is by far the most physically relevant one. In this occurrence, in fact, nontrivial 2–body occupation coefficients arise (see related notations which are applicable for $N > 2$ recalled in Appendices A and B below). For completeness the case $N = 2$ is nevertheless briefly discussed in Appendix D.

For this purpose, first, in Section 2, the MKI functional is explicitly determined. We display in particular its construction method (see *No.#1- #4 MKI Prescriptions*). Based on the theory of the Master kinetic equation earlier developed [3] and suitable integral and differential identities (see Appendices A, B, and C), the properties of the MKI functional are investigated. These concern in particular the establishment of appropriate inequalities holding for the same functional (Theorem 1, Section 2.1), the signature of the time derivative of the same functional (Theorem 2, Section 2.2) and the property of DKE holding for a suitable class of 1–body PDFs (Theorem 3, Section 2.3). In the subsequent Sections 3 and 4, the issue of the consistency of the phenomena of PMI and DKE with microscopic dynamics is posed together with the physical interpretation and implications of the theory. The goal is to investigate the relationship of the DKE theory developed here with the microscopic reversibility principle and the Poincaré recurrence theorem. Finally in Section 5 the conclusions of the paper are drawn and possible applications/developments of the theory are pointed out.

2. Axiomatic Prescription of the MKI Functional

In view of the considerations given above in this section the problem is posed of the explicit realization of the MKI functional in terms of suitable axiomatic prescriptions. The same functional, denoted $I_M(\rho_1^{(N)}(t))$, should depend on the 1–body PDF $\rho_1^{(N)}(t)$, with $\rho_1^{(N)}(t) \equiv \rho_1^{(N)}(\mathbf{x}_1, t)$ being identified with a particular solution of the Master kinetic equation (see (A.4) in Appendix A holding for $N > 2$ and Appendix D for the case $N = 2$).

Unlike Boltzmann kinetic equation, the Master kinetic equation actually deals with the treatment of finite hard-sphere N–body systems, i.e., in which both the number of particles N and their diameter σ remain finite [3]. To achieve such a goal suitably prescribed physical collision boundary conditions (CBC) of the N–body PDF need to be adopted. More precisely, this concerns the prescription for arbitrary collision events of the relationship between incoming (−) and outgoing (+) PDFs, i.e., respectively, the left and right limits $\rho^{(\pm)(N)}(\mathbf{x}^{(\pm)}(t_i), t_i) = \lim_{t \longrightarrow t_i^{(\pm)}} \rho^{(N)}(\mathbf{x}(t), t)$, with $\mathbf{x}^{(\pm)}(t_i) = \lim_{t \longrightarrow t_i^{(\pm)}} \mathbf{x}(t)$ denoting the corresponding incoming (−) and outgoing (+) states. In particular, upon invoking due to causality the assumption of left-continuity, i.e., the requirement

$$\rho^{(-)(N)}\left(\mathbf{x}^{(-)}(t_i), t_i\right) \equiv \rho^{(N)}\left(\mathbf{x}^{(-)}(t_i), t_i\right), \tag{13}$$

the incoming PDF is required to coincide with the same N–body PDF evaluated in terms of the incoming state and time [1, 3]. Hence, as recalled in Appendix C (see also [2]) from (C.1) it follows that the so-called causal form of the modified collision boundary condition (MCBC [2])

$$\rho^{(+)(N)}\left(\mathbf{x}^{(+)}(t_i), t_i\right) = \rho^{(N)}\left(\mathbf{x}^{(+)}(t_i), t_i\right) \tag{14}$$

is mandatory. A further important requirement concerns precisely setting also the related *functional class of admissible solutions* $\{\rho_1^{(N)}(\mathbf{x}_1, t)\}$ in such a way that, besides $\rho_1^{(N)}(t)$, also the same functional $I_M(\rho_1^{(N)}(t))$ exists globally for arbitrary $t \in I \equiv \mathbb{R}$. For definiteness, we shall consider for this purpose the case of 1–body PDFs which satisfy the initial condition

$$\rho_1^{(N)}(t_o) \equiv \rho_1^{(N)}(\mathbf{x}_1, t_o) = \rho_{1(o)}^{(N)}(\mathbf{x}_1), \tag{15}$$

with $\rho_{1(o)}^{(N)}(\mathbf{x}_1)$ belonging to the functional class of *stochastic 1–body PDFs* $\{\rho_1^{(N)}(t_o)\}$. For a generic $t \in I$ belonging to the time axis $I \equiv \mathbb{R}$ this is the ensemble of 1–body PDFs $\rho_1^{(N)}(t)$ which are, respectively, (A) smoothly differentiable; (B) strictly positive; (C) summable, in the sense that the velocity—or phase-space—moments for the same PDF exist which correspond either to arbitrary monomial functions of $\mathbf{v}_1$ (or its components v_{1i}, for $i = 1, 2, 3$) or to the entropy density $\ln \rho_1^{(N)}(t)$, thus yielding the Boltzmann-Shannon (BS) entropy evaluated in terms of $\rho_1^{(N)}(t)$.

Concerning the choice of the setting $\{\rho_1^{(N)}(t_o)\}$ the following remarks are in order. As a first remark, the previous requirements (A), (B), and (C) for $\{\rho_1^{(N)}(t_o)\}$, together with validity of MCBC (14), actually should warrant that the corresponding solution of the Master kinetic equation $\rho_1^{(N)}(t)$ exists globally in the extended phase-space $(\mathbf{x}_1, t) \in \Gamma_1 \times I$ and that for all $t \in I$ the same PDF belongs to the class of stochastic PDFs $\{\rho_1^{(N)}(\mathbf{x}_1, t)\}$ indicated above and also fulfills identically the constant H-theorem (12). Indeed, one can show [13] that global existence of solutions for the Master kinetic equation follows in elementary way from the N–body Liouville equation. Indeed, an arbitrary 1–body PDF which is a particular solution of the Master kinetic equation realizes by construction also a particular factorized solution of the N–body Liouville equation, i.e., of the N–body PDF [3]. The same PDF evolves uniquely in time along arbitrary phase-space Lagrangian trajectories, its Lagrangian time evolution being determined at arbitrary collision times by MCBC (14) [13].

As a second remark, the validity of assumptions (A) and (B) for $\rho_{1(o)}^{(N)}(\mathbf{x}_1)$ and $\rho_1^{(N)}(\mathbf{x}_1, t)$ implies also suitable assumptions to apply all $t \in I$ for the local characteristic scale length L_ρ which characterize the same PDF $\rho_1^{(N)}(\mathbf{x}_1, t)$. More precisely, this is associated with the spatial variations of the 1–body PDF prescribed as

$$L_\rho(t) = \inf_{\mathbf{x}_1 \in \Gamma_1} \left\{ \left| \frac{\partial \ln \rho_1^{(N)}(\mathbf{x}_1, t)}{\partial \mathbf{r}_1} \right|^{-1} \right\}, \tag{16}$$

which necessarily assumed nonzero at all time $t \in I$. Hence $\partial \ln \rho_1^{(N)}(\mathbf{x}_1, t)/\partial \mathbf{r}_1$ is assumed to be bounded for all $(\mathbf{x}_1, t)$ spanning the extended 1–body phase space $\Gamma_{1(1)} \times I$.

Finally, as a third remark (see also the further related discussion in Theorem 2 below), the previous requirements are expected to warrant also the global existence of the MKI functional $I_M(\rho_1^{(N)}(t))$, so that $\{\rho_1^{(N)}(\mathbf{x}_1, t)\}$ effectively realizes the functional class of admissible solutions indicated above.

Given these premises let us pose now the problem of the identification of the functional $I_M(\rho_{1o}^{(N)}(\mathbf{x}_1))$, based on the introduction of 'ad hoc' physical requirements, to be referred to here as *MKI Prescriptions No.#1-#4*. The prescriptions are as follows:

(i) *MKI Prescription No.#1:* the first one is that the functional $I_M(\rho_1^{(N)}(t))$ should be determined in such a way that the existence of $I_M(\rho_{1o}^{(N)}(\mathbf{x}_1))$ at a suitable initial time $t_o \in I$ should warrant also that $I_M(\rho_1^{(N)}(t))$ must necessarily exist globally in the future, i.e., for all $t \geq t_o$. As a consequence the functional class $\{\rho_1^{(N)}(\mathbf{x}_1, t)\}$ must be suitably prescribed.

(ii) *MKI Prescription No.#2:* second, we shall require that $I_M(\rho_1^{(N)}(t))$ is real, nonnegative, and bounded in the sense that

$$0 \leq I_M\left(\rho_1^{(N)}(t)\right) \leq 1. \tag{17}$$

This implies that $I_M(\rho_1^{(N)}(t))$ can be interpreted as an information measure associated with the 1−body PDF $\rho_1^{(N)}(t) \equiv \rho_1^{(N)}(\mathbf{x}_1, t)$. For this reason the previous inequalities will be referred to as *information-measure inequalities.*

(iii) *MKI Prescription No.#3:* third, for consistency with the property of macroscopic irreversibility, $I_M(\rho_1^{(N)}(t))$ is prescribed in terms of a smoothly time-differentiable and monotonically time-decreasing functional in the sense that in the same time-subset the inequality

$$\frac{\partial}{\partial t} I_M\left(\rho_1^{(N)}(t)\right) \leq 0 \tag{18}$$

should identically apply $\forall t \geq t_o$, so that

$$0 \leq I_M\left(\rho_1^{(N)}(t)\right) \leq I_M\left(\rho_{1o}^{(N)}(\mathbf{x}_1)\right) \leq 1. \tag{19}$$

This implies that $I_M(\rho_1^{(N)}(t))$ is also globally defined for all $t \in I \equiv \mathbb{R}$ with $t \geq t_o$. In addition, if $(\partial/\partial t) I_M(\rho_1^{(N)}(t))|_{t=t_o} \neq 0$, without loss of generality its initial value $I_M(\rho_{1o}^{(N)}(\mathbf{x}_1))$, can always be set such that

$$I_M\left(\rho_{1o}^{(N)}(\mathbf{x}_1)\right) = 1. \tag{20}$$

(iv) *MKI Prescription No.#4:* fourth, in order to warrant the existence of DKE we shall require the functional $I_M(\rho_1^{(N)}(t))$ to be prescribed in such a way that at an arbitrary time $t \in I$, with $t \gtrsim t_o$, the vanishing of both $I_M(\rho_1^{(N)}(t))$ and its time derivative $(\partial/\partial t) I_M(\rho_1^{(N)}(t))$ should occur if and only if the 1−body PDF solution of the Master kinetic equation coincides with kinetic equilibrium. As a consequence, for the functional $I_M(\rho_1^{(N)}(t))$ the following propositions should be equivalent:

$$\begin{gathered} I_M\left(\rho_1^{(N)}(t)\right) = 0 \\ \frac{\partial}{\partial t} I_M\left(\rho_1^{(N)}(t)\right) = 0 \\ \Updownarrow \\ \rho_1^{(N)}(\mathbf{x}_1, t) \equiv \rho_{1M}^{(N)}(\mathbf{v}_1), \end{gathered} \tag{21}$$

with $\rho_{1M}^{(N)}(\mathbf{v}_1)$ being a kinetic equilibrium PDF of the form (6).

The immediate obvious implication of the previous prescriptions is that—provided a nontrivial realization of the MKI can be found in the functional class $\{\rho_{1o}^{(N)}(\mathbf{x}_1)\}$—the existence of both PMI and DKE for the Master kinetic equation is actually established. In the sequel the goal is to show, in particular, that the MKI functional can be identified by means of the prescription

$$I_M\left(\rho_1^{(N)}(t), \mathbf{b}\right) \equiv \frac{K_M\left(\rho_1^{(N)}(t), \mathbf{b}\right)}{K_{Mo}}, \tag{22}$$

where $K_M(\rho_1^{(N)}(t), \mathbf{b})$ and K_{Mo} denote, respectively, a suitable (and possibly nonunique) moment-dependent phase-space functional and an appropriate normalization constant to be chosen in such a way to satisfy all the MKI prescriptions indicated above. In particular, as shown below, an admissible choice for $K_M(\rho_1^{(N)}(t), b)$ and K_{Mo} is provided by

$$\begin{aligned} K_M\left(\rho_1^{(N)}(t), \mathbf{b}\right) = &-\int_{\Gamma_{1(1)}} d\mathbf{x}_1 \overline{\Theta}_1^{(\partial\Omega)}(\overline{\mathbf{r}}_1) M(\mathbf{v}_1, \mathbf{b}) \\ &\cdot \frac{\rho_1^{(N)}(\mathbf{x}_1, t)}{\widehat{\rho}_1^{(N)}(\mathbf{x}_1, t)} \frac{\partial^2 \widehat{\rho}_1^{(N)}(\mathbf{x}_1, t)}{\partial \mathbf{r}_1 \cdot \partial \mathbf{r}_1}, \\ K_{Mo} = &\sup\left\{1, K_M\left(\rho_{1o}^{(N)}(\mathbf{x}_1), \mathbf{b}\right)\right\}, \end{aligned} \tag{23}$$

while $M(\mathbf{v}_1, \mathbf{b})$ denotes the *directional kinetic energy* (along the unit vector $\mathbf{b}$) carried by particle 1, namely, the dynamical variable

$$M(\mathbf{v}_1, \mathbf{b}) \equiv (\mathbf{v}_1 \cdot \mathbf{b})^2, \tag{24}$$

with $\mathbf{b}$ denoting a still arbitrary constant unit vector. Hence,

$$M(\mathbf{v}_1, \mathbf{v}_2, \mathbf{b}) = \frac{1}{2}\left[M(\mathbf{v}_1, \mathbf{b}) + M(\mathbf{v}_2, \mathbf{b})\right] \tag{25}$$

identifies the corresponding *total directional kinetic energy* carried by particles 1 and 2. Here the remaining notation is standard. Thus, $\rho_1^{(N)}(t) \equiv \rho_1^{(N)}(\mathbf{x}_1, t)$, $\rho_{1o}^{(N)}(\mathbf{x}_1)$, and $\widehat{\rho}_1^{(N)}(t) \equiv \widehat{\rho}_1^{(N)}(\mathbf{x}_1, t)$ are, respectively, the 1−body PDF solution of the initial problem associated with the Master kinetic equation (see (A.4) in Appendix A), the initial PDF, and the renormalized 1−body PDF

$$\widehat{\rho}_1^{(N)}(\mathbf{x}_1, t) \equiv \frac{\rho_1^{(N)}(\mathbf{x}_1, t)}{k_1^{(N)}(\mathbf{r}_1, t)}, \tag{26}$$

while furthermore $k_1^{(N)}(\mathbf{r}_1, t)$ is the 1−body occupation coefficient recalled in Appendix B (see (B.1)). As a consequence in the previous equation it follows that $\rho_1^{(N)}(\mathbf{x}_1, t)/\widehat{\rho}_1^{(N)}(\mathbf{x}_1, t) \equiv k_1^{(N)}(\mathbf{r}_1, t)$. Furthermore, $\overline{\Theta}_1^{(\partial\Omega)}(\overline{\mathbf{r}}_1)$ is the boundary theta function given by (A.10) (see Appendix A). Finally, regarding the initial value $K_M(\rho_{1o}^{(N)}(\mathbf{x}_1), \mathbf{b})$ it follows that if, respectively, $K_M(\rho_{1o}^{(N)}(\mathbf{x}_1), \mathbf{b}) \geq 1$ or

$$0 \leq K_M\left(\rho_{1o}^{(N)}(\mathbf{x}_1), \mathbf{b}\right) < 1, \tag{27}$$

then correspondingly one obtains, consistent with (19), that the initial value of MKI functional $I_M(\rho_{1o}^{(N)},\mathbf{b})$ is

$$I_M\left(\rho_{1o}^{(N)},\mathbf{b}\right)=\begin{cases}1,\\ K_M\left(\rho_{1o}^{(N)}(\mathbf{x}_1),\mathbf{b}\right).\end{cases} \tag{28}$$

2.1. Proof of the Nonnegativity of the MKI Information Measure. The strategy adopted for the proof of the MKI Prescriptions No.#1 and No.#2 is to show initially the validity of the information-measure left inequality in (17), namely, that $I_M(\rho_1^{(N)}(t),\mathbf{b})$ cannot acquire negative values for arbitrary $t\geq t_o$. The result is established by the following theorem.

Theorem 1 (nonnegativity of $K_M(\rho_{1o}^{(N)}(\mathbf{x}_1),\mathbf{b})$, $K_M(\rho_1^{(N)}(t),\mathbf{b})$, and $I_M(\rho_1^{(N)}(t),\mathbf{b})$). *Let us assume that $\rho_1^{(N)}(\mathbf{x}_1,t)$ is an arbitrary stochastic and suitably smoothly differentiable, particular solution of the Master kinetic equation (A.4) prescribed so that the integral (23) expressed in terms of the initial PDF, namely, $K_M(\rho_{1o}^{(N)}(\mathbf{x}_1),\mathbf{b})$, is nonvanishing. Then, it follows necessarily that*

(i) *Proposition $P1_1$:*

$$K_M\left(\rho_{1o}^{(N)}(\mathbf{x}_1),\mathbf{b}\right)>0. \tag{29}$$

(ii) *Proposition $P1_2$: the corresponding time-evolved functional $K_M(\rho_1^{(N)}(t),\mathbf{b})$ for all $t\in I$ with $t>t_o$ is such that*

$$K_M\left(\rho_1^{(N)}(t),\mathbf{b}\right)\geq 0. \tag{30}$$

(iii) *Proposition $P1_3$: for all $t\in I$ with $t>t_o$ the functional $I_M(\rho_1^{(N)}(t),\mathbf{b})$ fulfills the inequality*

$$I_M\left(\rho_1^{(N)}(t),\mathbf{b}\right)\geq 0. \tag{31}$$

(iv) *Proposition $P1_4$: the following necessary and sufficient condition holds at a given time $t\in I$ with $t\geq t_o$:*

$$\begin{gathered}K_M\left(\rho_1^{(N)}(t),\mathbf{b}\right)=0\Longleftrightarrow\\ \rho_1^{(N)}(\mathbf{x}_1,t)\equiv\rho_{1M}^{(N)}(\mathbf{v}_1).\end{gathered} \tag{32}$$

Proof. One first notices that $K_M(\rho_1^{(N)}(t),\mathbf{b})$ can be equivalently written in the form

$$\begin{aligned}K_M\left(\rho_1^{(N)}(t),\mathbf{b}\right)\equiv&-\int_{\Gamma_{1(1)}}d\mathbf{x}_1\overline{\Theta}_1^{(\partial\Omega)}(\overline{\mathbf{r}}_1)M(\mathbf{v}_1,\mathbf{b})\\ &\cdot k_1^{(N)}(\mathbf{r}_1,t)\frac{\partial^2\widehat{\rho}_1^{(N)}(\mathbf{x}_1,t)}{\partial\mathbf{r}_1\cdot\partial\mathbf{r}_1},\end{aligned} \tag{33}$$

where in order that the same functional exists it is obvious that the renormalized 1−body PDF $\widehat{\rho}_1^{(N)}(\mathbf{x}_1,t)$ must be of class $C^{(2)}$. Integrating by parts and noting that the gradient term $\partial\overline{\Theta}_1^{(\partial\Omega)}(\overline{\mathbf{r}})/\partial\mathbf{r}_1$ gives a vanishing contribution to the phase-space integral, this yields equivalently

$$\begin{aligned}&K_M\left(\rho_1^{(N)}(t),\mathbf{b}\right)\\ &\quad\equiv\int_{\Gamma_{1(1)}}d\mathbf{x}_1\overline{\Theta}_1^{(\partial\Omega)}(\overline{\mathbf{r}}_1)M(\mathbf{v}_1,\mathbf{b})\frac{\partial k_1^{(N)}(\mathbf{r}_1,t)}{\partial\mathbf{r}_1}\\ &\quad\cdot\frac{\partial\widehat{\rho}_1^{(N)}(\mathbf{x}_1,t)}{\partial\mathbf{r}_1}.\end{aligned} \tag{34}$$

Therefore, upon invoking (B.9) reported in Appendix B, direct substitution delivers

$$\begin{aligned}K_M\left(\rho_1^{(N)}(t),\mathbf{b}\right)=&(N-1)\int_{\Gamma_{1(1)}}d\mathbf{x}_1\overline{\Theta}_1^{(\partial\Omega)}(\overline{\mathbf{r}}_1)\\ &\cdot M(\mathbf{v}_1,\mathbf{b})\frac{\partial\widehat{\rho}_1^{(N)}(\mathbf{x}_1,t)}{\partial\mathbf{r}_1}\cdot\int_{\Gamma_{1(2)}}d\mathbf{x}_2\mathbf{n}_{12}\\ &\times\delta\left(|\mathbf{r}_2-\mathbf{r}_1|-\sigma\right)\overline{\Theta}_2^{(\partial\Omega)}(\overline{\mathbf{r}}_2)\widehat{\rho}_1^{(N)}(\mathbf{x}_2,t)\\ &\cdot k_2^{(N)}(\mathbf{r}_1,\mathbf{r}_2,t).\end{aligned} \tag{35}$$

Next, invoking the identity $\mathbf{n}_{12}\delta(|\mathbf{r}_2-\mathbf{r}_1|-\sigma)=-(\partial/\partial\mathbf{r}_2)\overline{\Theta}(|\mathbf{r}_2-\mathbf{r}_1|-\sigma)$ and noting again that $(\partial/\partial\mathbf{r}_2)\overline{\Theta}_2^{(\partial\Omega)}(\overline{\mathbf{r}})$ gives vanishing contribution, one can perform a further integration by parts with respect to $\mathbf{r}_2$. This permits to cast the rhs of previous equation in the form

$$\begin{aligned}K_M\left(\rho_1^{(N)}(t),\mathbf{b}\right)\equiv&K_M^{(1)}\left(\rho_1^{(N)}(\mathbf{x}_1,t),\mathbf{b}\right)\\ &+\Delta K_M^{(1)}\left(\rho_1^{(N)}(\mathbf{x}_1,t),\mathbf{b}\right).\end{aligned} \tag{36}$$

Here the two terms on the rhs of (36) are defined as follows: (1) the first term $K_M^{(1)}(\rho_1^{(N)}(t),\mathbf{b})$ is symmetric and nonnegative, so that it can be expressed so to carry the total directional kinetic energy $M(\mathbf{v}_1,\mathbf{v}_2,\mathbf{b})$ of particles 1 and 2 (see (25)). Hence, it takes the form

$$\begin{aligned}&K_M^{(1)}\left(\rho_1^{(N)}(\mathbf{x}_1,t),\mathbf{b}\right)=(N-1)\int_{\Gamma_{1(1)}}d\mathbf{x}_1\\ &\quad\cdot\int_{\Gamma_{1(2)}}d\mathbf{x}_2\overline{\Theta}_1^{(\partial\Omega)}(\overline{\mathbf{r}}_1)\overline{\Theta}_2^{(\partial\Omega)}(\overline{\mathbf{r}}_2)\times\frac{\partial\widehat{\rho}_1^{(N)}(\mathbf{x}_1,t)}{\partial\mathbf{r}_1}\\ &\quad\cdot\frac{\partial\widehat{\rho}_1^{(N)}(\mathbf{x}_2,t)}{\partial\mathbf{r}_2}k_2^{(N)}(\mathbf{r}_1,\mathbf{r}_2,t)M(\mathbf{v}_1,\mathbf{v}_2,\mathbf{b})\\ &\quad\cdot\overline{\Theta}\left(|\mathbf{r}_2-\mathbf{r}_1|-\sigma\right).\end{aligned} \tag{37}$$

(2) The second term $\Delta K_M^{(1)}(\rho_1^{(N)}(\mathbf{x}_1,t),\mathbf{b})$ reads instead

$$\begin{aligned}&\Delta K_M^{(1)}\left(\rho_1^{(N)}(\mathbf{x}_1,t),\mathbf{b}\right)\equiv(N-1)\\ &\quad\cdot\int_{\Gamma_{1(1)}}d\mathbf{x}_1\overline{\Theta}_1^{(\partial\Omega)}(\overline{\mathbf{r}}_1)\frac{\partial\widehat{\rho}_1^{(N)}(\mathbf{x}_1,t)}{\partial\mathbf{r}_1}\cdot\int_{\Gamma_{1(2)}}d\mathbf{x}_2\end{aligned}$$

$$\times \overline{\Theta}_2^{(\partial\Omega)}(\overline{\mathbf{r}}_2)\, M(\mathbf{v}_1,\mathbf{b})\, \overline{\Theta}(|\mathbf{r}_2-\mathbf{r}_1|-\sigma)\, \widehat{\rho}_1^{(N)}(\mathbf{x}_2, t)\frac{\partial}{\partial \mathbf{r}_2} k_2^{(N)}(\mathbf{r}_1,\mathbf{r}_2,t), \tag{38}$$

where $(\partial/\partial\mathbf{r}_2)k_2^{(N)}(\mathbf{r}_1,\mathbf{r}_2,t)$ is given by the differential identity (B.10) reported in Appendix B. Thus, upon invoking the identity $\mathbf{n}_{23}\delta(|\mathbf{r}_3-\mathbf{r}_2|-\sigma) = -(\partial/\partial\mathbf{r}_3)\overline{\Theta}(|\mathbf{r}_3-\mathbf{r}_2|-\sigma)$, one notices that an integration by parts can be performed also with respect to $\mathbf{r}_3$. This means that a procedure analogous to the one used for the calculation of $\partial k_1^{(N)}(\mathbf{r}_1,t)/\partial\mathbf{r}_1$ can be invoked and iterated at all orders, i.e., up to the $(N-1)$–body occupation coefficient (see (B.11) in Appendix B). As a consequence the functional $K_M(\rho_1^{(N)}(t),\mathbf{b})$ can be represented in terms of a finite sum of the form $K_M(\rho_1^{(N)}(t),\mathbf{b}) \equiv \sum_{j=1,N-1} K_M^{(j)}(\rho_1^{(N)}(\mathbf{x}_1,t),\mathbf{b})$ in which each term of the sum $K_M^{(j)}(\rho_1^{(N)}(\mathbf{x}_1,t),\mathbf{b})$ is nonnegative and symmetric. This implies therefore that the same functional $K_M(\rho_1^{(N)}(t),\mathbf{b})$ can be cast in the form

$$\begin{aligned} K_M(\rho_1^{(N)}(t),\mathbf{b}) &= \int_{\Gamma_{1(1)}} d\mathbf{x}_1 \cdot \int_{\Gamma_{1(2)}} d\mathbf{x}_2 \overline{\Theta}_1^{(\partial\Omega)}(\overline{\mathbf{r}}_1)\,\overline{\Theta}_2^{(\partial\Omega)}(\overline{\mathbf{r}}_2)\, M(\mathbf{v}_1,\mathbf{v}_2,\mathbf{b}) \\ &\times F(\mathbf{r}_1,\mathbf{r}_2,t)\frac{\partial\widehat{\rho}_1^{(N)}(\mathbf{x}_1,t)}{\partial\mathbf{r}_1} \cdot \frac{\partial\widehat{\rho}_1^{(N)}(\mathbf{x}_2,t)}{\partial\mathbf{r}_2}\overline{\Theta}(|\mathbf{r}_2-\mathbf{r}_1|-\sigma), \end{aligned} \tag{39}$$

with $M(\mathbf{v}_1,\mathbf{v}_2,\mathbf{b}) \geq 0$ being the total directional kinetic energy (25) and $F(\mathbf{r}_1,\mathbf{r}_2,t)$ a suitable real scalar kernel which is symmetric in the variables $\mathbf{r}_1$ and $\mathbf{r}_2$. Hence $K_M(\rho_1^{(N)}(t),\mathbf{b})$ actually defines a nonnegative functional. This proves the validity of the inequality (31) (Proposition $P1_1$).

In a similar way also the remaining propositions can be established. In fact, invoking (28) it follows that the inequalities (30) and (31)—and hence also Propositions $P1_2$ and $P1_3$—manifestly hold too. Finally, regarding the proof of Proposition $P1_4$, one notices that $K_M(\rho_1^{(N)}(t),\mathbf{b}) \equiv 0$ if and only if identically $(\partial/\partial\mathbf{r}_1)\widehat{\rho}_1^{(N)}(\mathbf{x}_1,t) \equiv 0$. Since $\widehat{\rho}_1^{(N)}(\mathbf{x}_1,t)$ is by construction a solution of the Master kinetic equation it follows that this requires necessarily that $\widehat{\rho}_1^{(N)}(\mathbf{x}_1,t)$ must coincide with the local Maxwellian $\rho_{1M}^{(N)}(\mathbf{v}_1)$ (see (6)) and hence (32) must hold too under the same realization (Proposition $P1_4$). □

The conclusion is therefore that the definition of the MKI functional (22) given above in terms of $K_M(\rho_1^{(N)}(t),\mathbf{b})$ and K_{Mo} (see (23)) is indeed consistent with the physical prerequisites represented by the MKI Prescriptions No.#1 and No.#2.

2.2. Proof of PMI for the Master Kinetic Equation. The next step is to prove that the functional $I_M(\rho_{1o}^{(N)},\mathbf{b})$ defined above (see (28)) indeed exhibits a monotonic time-decreasing behavior which is consistent with the MKI Prescriptions No.#3 and No.#4, which are realized respectively by

(i) the time derivative inequality (18) and the conditions of existence of kinetic equilibrium (21);

(ii) the validity of the inequality $I_M(\rho_1^{(N)}(t)) \leq 1$.

In order to reach the proofs of these properties let us preliminarily determine the variation across a binary collision occurring between particles 1 and 2 of the total directional kinetic energy $M(\mathbf{v}_1,\mathbf{v}_2,\mathbf{b})$ (see (25)), namely, the phase-space scalar function $\Delta M(\mathbf{v}_1,\mathbf{v}_2,\mathbf{b}) \equiv M(\mathbf{v}_1^{(+)},\mathbf{v}_2^{(+)},\mathbf{b}) - M(\mathbf{v}_1\mathbf{v}_2,\mathbf{b})$. One obtains

$$\begin{aligned} \Delta M(\mathbf{v}_1,\mathbf{v}_2,\mathbf{b}) &= \mathbf{b}\cdot\mathbf{n}_{12}\left|\mathbf{n}_{12}\cdot\mathbf{v}_{12}^{(+)}\right|\mathbf{v}_{12}^{(+)}\cdot\mathbf{b} \\ &- (\mathbf{b}\cdot\mathbf{n}_{12})^2\left(\mathbf{n}_{12}\cdot\mathbf{v}_{12}^{(+)}\right)^2, \end{aligned} \tag{40}$$

the rhs being expressed in terms of the outgoing particle velocities $(\mathbf{v}_1^{(+)},\mathbf{v}_2^{(+)})$ only. Then, the following proposition holds.

Theorem 2 (property of macroscopic irreversibility (master equation PMI theorem)). *Let us assume that $\rho_1^{(N)}(\mathbf{x}_1,t) \equiv \rho_1^{(N)}(\mathbf{r}_1,\mathbf{v}_1,t)$ is an arbitrary stochastic particular solution of the Master kinetic equation (A.4) with initial condition $\rho_{1o}^{(N)}(\mathbf{x}_1)$ such that the integral $K_M(\rho_{1o}^{(N)}(\mathbf{x}_1),\mathbf{b})$ exists and is nonvanishing. Then it follows that*

(i) *Proposition $P2_1$: one finds that for all $t \geq t_o$*

$$\begin{aligned} \frac{\partial}{\partial t}K_M(\rho_1^{(N)}(t),\mathbf{b}) &= -(N-1)\sigma^2\int_{U_{1(1)}} d\mathbf{v}_1\int_{U_{1(2)}} d\mathbf{v}_2 \cdot \int_\Omega d\mathbf{r}_1\overline{\Theta}_1^{(\partial\Omega)}(\overline{\mathbf{r}}_1)\int^{(-)} d\Sigma_{21}\left|\mathbf{v}_{12}^{(+)}\cdot\mathbf{n}_{12}\right|(\mathbf{b}\cdot\mathbf{n}_{12})^2 \\ &\cdot\left(\mathbf{n}_{12}\cdot\mathbf{v}_{12}^{(+)}\right)^2\frac{\partial\widehat{\rho}_1^{(N)}(\mathbf{r}_1,\mathbf{v}_1^{(+)},t)}{\partial\mathbf{r}_1}\cdot\frac{\partial}{\partial\mathbf{r}_2} \\ &\cdot\widehat{\rho}_1^{(N)}(\mathbf{r}_2=\mathbf{r}_1+\sigma\mathbf{n}_{21},\mathbf{v}_2^{(+)}t) \\ &\cdot k_2^{(N)}(\mathbf{r}_1,\mathbf{r}_2=\mathbf{r}_1+\sigma\mathbf{n}_{21},t) \leq 0. \end{aligned} \tag{41}$$

(ii) *Proposition $P2_2$: the inequality*

$$\frac{\partial}{\partial t}I_M(\rho_1^{(N)}(t),\mathbf{b}) \leq 0 \tag{42}$$

holds globally (i.e., identically for all $t \geq t_o$) so that necessarily $I_M(\rho_1^{(N)}(t),\mathbf{b})$ is globally defined too, being also prescribed so that

$$I_M(\rho_1^{(N)}(t),\mathbf{b}) \leq 1. \tag{43}$$

(iii) *Proposition $P2_3$: one finds that a given time $t \in I$ with $t \geq t_o$*

$$\frac{\partial}{\partial t}K_M(\rho_1^{(N)}(t),\mathbf{b}) = 0 \Longleftrightarrow \rho_1^{(N)}(\mathbf{x}_1,t) \equiv \rho_{1M}^{(N)}(\mathbf{v}_1). \tag{44}$$

Proof. Consider first the proof of Proposition $P2_1$ which requires evaluation of the partial time derivative $(\partial/\partial t)K_M(\rho_1^{(N)}(t),\mathbf{b})$. Upon invoking the first form of the Master kinetic equation (see (A.1) in Appendix A), explicit differentiation of $K_M(\rho_1^{(N)}(t),\mathbf{b})$ delivers

$$\frac{\partial}{\partial t}K_M\left(\rho_1^{(N)}(t),\mathbf{b}\right) = -\int_{\Gamma_{1(1)}} d\mathbf{x}_1 M(\mathbf{v}_1,\mathbf{b})\,\overline{\Theta}_1^{(\partial\Omega)}(\overline{\mathbf{r}}_1) \cdot k_1^{(N)}(\mathbf{r}_1,t)\left(-\mathbf{v}_1\cdot\frac{\partial}{\partial\mathbf{r}_1}\right)\frac{\partial^2\widehat{\rho}_1^{(N)}(\mathbf{x}_1,t)}{\partial\mathbf{r}_1\cdot\partial\mathbf{r}_1} \cdot k_1^{(N)}(\mathbf{r}_1,t), \tag{45}$$

namely, upon integration by parts in the first integral on the rhs,

$$\frac{\partial}{\partial t}K_M\left(\rho_1^{(N)}(t),\mathbf{b}\right) = -\int_{\Gamma_{1(1)}} d\mathbf{x}_1 M(\mathbf{v}_1,\mathbf{b}) \cdot \overline{\Theta}_1^{(\partial\Omega)}\frac{\partial^2\widehat{\rho}_1^{(N)}(\mathbf{x}_1,t)}{\partial\mathbf{r}_1\cdot\partial\mathbf{r}_1}\left(\frac{\partial}{\partial t}+\mathbf{v}_1\cdot\frac{\partial}{\partial\mathbf{r}_1}\right)k_1^{(N)}(\mathbf{r}_1,t). \tag{46}$$

Hence, thanks to the differential identity (B.12) it follows

$$\frac{\partial}{\partial t}K_M\left(\rho_1^{(N)}(t),\mathbf{b}\right) = -\int_{\Gamma_{1(1)}} d\mathbf{x}_1 M(\mathbf{v}_1,\mathbf{b})\,\overline{\Theta}_1^{(\partial\Omega)}\frac{\partial^2\widehat{\rho}_1^{(N)}(\mathbf{x}_1,t)}{\partial\mathbf{r}_1\cdot\partial\mathbf{r}_1}\left(\frac{\partial}{\partial t}\right) = -(N-1)\int_{\Gamma_{1(1)}} d\mathbf{x}_1 M(\mathbf{v}_1,\mathbf{b})\,\overline{\Theta}_1^{(\partial\Omega)}(\overline{\mathbf{r}}_1)\frac{\partial^2\widehat{\rho}_1^{(N)}(\mathbf{x}_1,t)}{\partial\mathbf{r}_1\cdot\partial\mathbf{r}_1}\int_{\overline{\Gamma}_{1(2)}} d\mathbf{x}_2\mathbf{v}_{12}\cdot\mathbf{n}_{12}\times\delta\left(\left|\mathbf{r}_1-\mathbf{r}_2\right|-\sigma\right)k_2^{(N)}(\mathbf{r}_1,\mathbf{r}_2,t)\,\widehat{\rho}_1^{(N)}(\mathbf{x}_2,t). \tag{47}$$

Performing an integration by parts with respect to $\mathbf{r}_1$ and upon invoking the first differential identity (B.14) reported in Appendix B one obtains therefore

$$\frac{\partial}{\partial t}K_M\left(\rho_1^{(N)}(t),\mathbf{b}\right) = W_M\left(\rho_1^{(N)}(t),\mathbf{b}\right), \tag{48}$$

$$W_M\left(\rho_1^{(N)}(t),\mathbf{b}\right) \equiv \int_{\Gamma_{1(1)}} d\mathbf{x}_1 M(\mathbf{v}_1,\mathbf{b}) \cdot \overline{\Theta}_1^{(\partial\Omega)}(\overline{\mathbf{r}}_1)\frac{\partial\widehat{\rho}_1^{(N)}(\mathbf{x}_1,t)}{\partial\mathbf{r}_1}(N-1) \times\int_{\overline{\Gamma}_{1(2)}} d\mathbf{x}_2\mathbf{v}_{12}\cdot\mathbf{n}_{12}\frac{\partial}{\partial\mathbf{r}_1}\left[\delta\left(\left|\mathbf{r}_1-\mathbf{r}_2\right|-\sigma\right)\right] \cdot k_2^{(N)}(\mathbf{r}_1,\mathbf{r}_2,t)\,\widehat{\rho}_1^{(N)}(\mathbf{x}_2,t), \tag{49}$$

where $(\partial/\partial\mathbf{r}_1)[\delta(|\mathbf{r}_1-\mathbf{r}_2|-\sigma)] = -(\partial/\partial\mathbf{r}_2)[\delta(|\mathbf{r}_1-\mathbf{r}_2|-\sigma)]$. Hence performing a further integration by parts with respect to $\mathbf{r}_2$ and using the second differential identity on (B.14) (see Appendix B) the previous equation finally yields

$$W_M\left(\rho_1^{(N)}(t),\mathbf{b}\right) = (N-1)\int_{\Gamma_{1(1)}} d\mathbf{x}_1\overline{\Theta}_1^{(\partial\Omega)}(\overline{\mathbf{r}}_1) \cdot\int_{\overline{\Gamma}_{1(2)}} d\mathbf{x}_2\mathbf{v}_{12}\cdot\mathbf{n}_{12}M(\mathbf{v}_1,\mathbf{v}_2,\mathbf{b}) \times\delta\left(\left|\mathbf{r}_1-\mathbf{r}_2\right|-\sigma\right) \cdot k_2^{(N)}(\mathbf{r}_1,\mathbf{r}_2,t)\frac{\partial\widehat{\rho}_1^{(N)}(\mathbf{x}_1,t)}{\partial\mathbf{r}_1} \cdot\frac{\partial\widehat{\rho}_1^{(N)}(\mathbf{x}_2,t)}{\partial\mathbf{r}_2}, \tag{50}$$

where the symmetry property with respect to the exchange of states $(\mathbf{x}_1,\mathbf{x}_2)$ has been invoked. In the previous equation the integration on the Dirac delta can be performed at once letting

$$\int_{\Gamma_{1(1)}} d\mathbf{x}_1\overline{\Theta}_1^{(\partial\Omega)}\int_{\Gamma_{1(2)}} d\mathbf{x}_2\delta\left(\left|\mathbf{r}_1-\mathbf{r}_2\right|-\sigma\right) = \sigma^2\int_{U_{1(1)}} d\mathbf{v}_1 \cdot\int_{U_{1(2)}} d\mathbf{v}_2\int_\Omega d\mathbf{r}_1\overline{\Theta}_1^{(\partial\Omega)}(\overline{\mathbf{r}}_1) \cdot\left[\int^{(+)} d\mathbf{\Sigma}_{21}\left|\mathbf{v}_{12}\cdot\mathbf{n}_{12}\right| - \int^{(-)} d\mathbf{\Sigma}_{21}\left|\mathbf{v}_{12}\cdot\mathbf{n}_{12}\right|\right], \tag{51}$$

where the solid-angle integrations in the two integrals on the rhs are performed, respectively, on the outgoing (+) and incoming (−) particles. Furthermore, it is obvious that thanks to the causal form of MCBC (see (C.3) in Appendix C) the integral on outgoing particles $\int^{(+)} d\mathbf{\Sigma}_{21}$ can be transformed to a corresponding integration on incoming ones, namely, $\int^{(-)} d\mathbf{\Sigma}_{21}$. Thus, the contributions in the two phase-space integrals only differ because of the variation $\Delta M(\mathbf{v}_1,\mathbf{v}_2,\mathbf{b})$ of the total directional kinetic energy of particles 1 and 2. This implies that

$$W_M\left(\rho_1^{(N)}(t),\mathbf{b}\right) = (N-1)\sigma^2\int_{U_{1(1)}} d\mathbf{v}_1\int_{U_{1(2)}} d\mathbf{v}_2 \cdot\int_\Omega d\mathbf{r}_1\overline{\Theta}_1^{(\partial\Omega)}(\overline{\mathbf{r}}_1)\int^{(-)} d\mathbf{\Sigma}_{21}\times\left|\mathbf{v}_{12}\cdot\mathbf{n}_{12}\right| \cdot\Delta M(\mathbf{v}_1,\mathbf{v}_2,\mathbf{b})\frac{\partial\widehat{\rho}_1^{(N)}\left(\mathbf{r}_1,\mathbf{v}_1^{(+)},t\right)}{\partial\mathbf{r}_1}$$

$$\cdot \frac{\partial \hat{\rho}_1^{(N)}\left(\mathbf{r}_2=\mathbf{r}_1+\sigma\mathbf{n}_{21},\mathbf{v}_2^{(+)},t\right)}{\partial \mathbf{r}_2} \times k_2^{(N)}\left(\mathbf{r}_1,\mathbf{r}_2=\mathbf{r}_1+\sigma\mathbf{n}_{21},t\right), \tag{52}$$

where the solid-angle integration is performed on the incoming particles whereas $\Delta M(\mathbf{v}_1,\mathbf{v}_2,\mathbf{b})$ is evaluated in terms of the outgoing particles (+) and therefore must be identified with the second equation on the rhs of (40). Consider now the dependence in terms of the outgoing particle velocities $\mathbf{v}_1^{(+)}$ and $\mathbf{v}_2^{(+)}$ in the previous phase-space integral. The velocity dependence contained in the factors $|\mathbf{v}_{12}\cdot\mathbf{n}_{12}|$ and $\partial\hat{\rho}_1^{(N)}(\mathbf{r}_1,\mathbf{v}_1^{(+)},t)/\partial\mathbf{r}_1\cdot\partial\hat{\rho}_1^{(N)}(\mathbf{r}_2,\mathbf{v}_2^{(+)}t)/\partial\mathbf{r}_2$ is symmetric with respect to the variables $\mathbf{v}_1^{(+)}$ and $\mathbf{v}_2^{(+)}$. On the other hand, as a whole, the same integral should remain unaffected with respect to the exchange of the outgoing particle velocities $\mathbf{v}_1^{(+)}\Longleftrightarrow\mathbf{v}_2^{(+)}$. This means that the only term in $\Delta M(\mathbf{v}_1,\mathbf{v}_2,\mathbf{b})$ which gives a (possibly) nonvanishing contribution is $-(\mathbf{b}\cdot\mathbf{n}_{12})^2(\mathbf{n}_{12}\cdot\mathbf{v}_{12}^{(+)})^2$. As a consequence it is found that

$$\begin{aligned}\frac{\partial}{\partial t}K_M\left(\rho_1^{(N)}(t),\mathbf{b}\right) &\equiv W_M\left(\rho_1^{(N)}(t),\mathbf{b}\right) = -(N-1)\\ &\cdot\sigma^2\int_{U_{1(1)}}d\mathbf{v}_1\int_{U_{1(2)}}d\mathbf{v}_2\times\int_\Omega d\mathbf{r}_1\overline{\Theta}_1^{(\partial\Omega)}(\overline{\mathbf{r}}_1)\\ &\cdot\int^{(-)}d\mathbf{\Sigma}_{21}\frac{\partial\hat{\rho}_1^{(N)}\left(\mathbf{r}_1,\mathbf{v}_1^{(+)},t\right)}{\partial\mathbf{r}_1}\\ &\cdot\frac{\partial\hat{\rho}_1^{(N)}\left(\mathbf{r}_2=\mathbf{r}_1+\sigma\mathbf{n}_{21},\mathbf{v}_2^{(+)},t\right)}{\partial\mathbf{r}_2}\times\left|\mathbf{v}_{12}^{(+)}\cdot\mathbf{n}_{12}\right|\\ &\cdot(\mathbf{b}\cdot\mathbf{n}_{12})^2\left(\mathbf{n}_{12}\cdot\mathbf{v}_{12}^{(+)}\right)^2k_2^{(N)}\left(\mathbf{r}_1,\mathbf{r}_2=\mathbf{r}_1+\sigma\mathbf{n}_{21},t\right)\\ &\le 0,\end{aligned} \tag{53}$$

and hence $(\partial/\partial t)K_M(\rho_1^{(N)}(t),\mathbf{b})$ is necessarily negative or null, the second case occurring only if $\partial\hat{\rho}_1^{(N)}(\mathbf{r}_1,\mathbf{v}_1^{(+)},t)/\partial\mathbf{r}_1\equiv 0$ and consequently $\partial\hat{\rho}_1^{(N)}(\mathbf{r}_2=\mathbf{r}_1+\sigma\mathbf{n}_{21},\mathbf{v}_2^{(+)}t)/\partial\mathbf{r}_2\equiv 0$ too.

The proof of Proposition $P2_2$ follows in a similar way. In fact, first, one notices that thanks to the global validity of the 1–body PDF [13] the 1–body PDF $\rho_1^{(N)}(t)$ necessarily belongs to the functional class of stochastic PDFs $\{\rho_1^{(N)}(\mathbf{x}_1,t)\}$ prescribed so that *also the local characteristic scale length defined above* $L_\rho(t)$ (see (16)) *is larger than zero and finite.* As a consequence it follows that both the functional $K_M(\rho_1^{(N)}(t),\mathbf{b})$ and $I_M(\rho_1^{(N)}(t),\mathbf{b})$ (see (22)) are globally defined too. Consider in fact the representation of $K_M(\rho_1^{(N)}(t),\mathbf{b})$ achieved in Theorem 1 and given by (35). Next, let us notice that thanks to (16) the characteristic scale length

$$L_{\mu,\min}\equiv\inf\left\{\left|\frac{\partial\hat{\rho}_1^{(N)}(\mathbf{x}_1,t)}{\partial\mathbf{r}_1}\right|^{-1}\right\} \tag{54}$$

is necessarily strictly positive. Then, upon noting that $k_2^{(N)}(\mathbf{r}_1,\mathbf{r}_2,t)\le 1$ and $\int_{\Gamma_{1(2)}}d\mathbf{x}_2\delta(|\mathbf{r}_2-\mathbf{r}_1|-\sigma)\hat{\rho}_1^{(N)}(\mathbf{x}_2,t)\le\sup(\hat{n}_1^{(N)}(\mathbf{r}_2,t))<+\infty$, with $\hat{n}_1^{(N)}(\mathbf{r}_2,t)$ being the velocity moment $\hat{n}_1^{(N)}(\mathbf{r}_2,t)=\int_{U_{1(2)}}d\mathbf{v}_2\hat{\rho}_1^{(N)}(\mathbf{r}_2,\mathbf{v}_2,t)$, it follows that

$$\begin{aligned}K_M\left(\rho_1^{(N)}(t),\mathbf{b}\right)&\le\frac{(N-1)}{L_{\mu,\min}}\sup\left(\hat{n}_1^{(N)}(\mathbf{r}_2,t)\right)\\ &\cdot\int_{\Gamma_{1(1)}}d\mathbf{x}_1\overline{\Theta}_1^{(\partial\Omega)}(\overline{\mathbf{r}}_1)M(\mathbf{v}_1,\mathbf{b})\\ &\cdot\hat{\rho}_1^{(N)}(\mathbf{x}_1,t),\end{aligned} \tag{55}$$

where the integral on the rhs is necessarily bounded. This happens because $\rho_1^{(N)}(\mathbf{x}_1,t)$ belongs to the functional class $\{\rho_1^{(N)}(t)\}$ and therefore $\hat{n}_1^{(N)}(\mathbf{r}_2,t)$ is bounded, while, at the same time, the phase-space moments indicated above necessarily exist. Furthermore, since $(\partial/\partial t)I_M(\rho_1^{(N)}(t),\mathbf{b})\equiv(1/K_{Mo})(\partial/\partial t)K_M(\rho_1^{(N)}(t),\mathbf{b})$, the inequality (53) implies (42) and (43) too. Finally, since $\hat{\rho}_1^{(N)}(\mathbf{r}_1,\mathbf{v}_1,t)$ is a solution of the Master kinetic equation $(\partial/\partial\mathbf{r}_1)\hat{\rho}_1^{(N)}(\mathbf{x}_1,t)\equiv 0$ occurs if and only if $\hat{\rho}_1^{(N)}(\mathbf{x}_1,t)$ coincides with a Maxwellian kinetic equilibrium of the type (6). This result proves therefore also Proposition $P2_3$. □

The implication of Theorem 2 is therefore that provided the initial value $K_M(\rho_{1o}^{(N)}(\mathbf{x}_1),\mathbf{b})$ is nonvanishing then necessarily:

(i) the functional $K_M(\rho_1^{(N)}(t),\mathbf{b})$ is monotonically decreasing and thus $K_M(\rho_1^{(N)}(t),\mathbf{b})\le K_M(\rho_{1o}^{(N)}(\mathbf{x}_1),\mathbf{b})$;

(ii) similarly the MKI functional $I_M(\rho_1^{(N)}(t),\mathbf{b})$ is monotonically decreasing too, i.e., $I_M(\rho_1^{(N)}(t),\mathbf{b})\le I_M(\rho_{1o}^{(N)}(\mathbf{x}_1),\mathbf{b})$;

(iii) both $K_M(\rho_1^{(N)}(t),\mathbf{b})$ and $I_M(\rho_1^{(N)}(t),\mathbf{b})$ are nonnegative.

2.3. Proof of the DKE Property for the Master Kinetic Equation. Let us now show that in validity of Theorems 1 and 2 the time-evolved $\rho_1^{(N)}(\mathbf{x}_1,t)$ necessarily must decay asymptotically for $\tau\equiv t-t_o\longrightarrow+\infty$ to kinetic equilibrium, i.e., that the limit function $\lim_{\tau\longrightarrow+\infty}\rho_1^{(N)}(\mathbf{x}_1,t)\equiv\rho_{1\infty}^{(N)}(\mathbf{x}_1)$ exists and it necessarily coincides with a Maxwellian kinetic equilibrium of the type (6). In this regard the following proposition holds.

Theorem 3 (asymptotic behavior of $I_M(\rho_1^{(N)}(t),\mathbf{b})$ (master equation-DKE theorem)). *Let us assume that the initial condition* $\rho_{1o}^{(N)}(\mathbf{x}_1)\in\{\rho_{1o}^{(N)}(\mathbf{x}_1)\}$ *is such that the corresponding functional* $K_M(\rho_{1o}^{(N)}(\mathbf{x}_1),\mathbf{b})$ *is nonvanishing, i.e., in view of Theorem 1 necessarily* >0. *Then it follows that the corresponding time-evolved solution of the Master kinetic equation* $\rho_1^{(N)}(\mathbf{x}_1,t)$ *in the limit* $\tau\equiv t-t_o\longrightarrow+\infty$ *necessarily must decay to kinetic equilibrium, i.e.,*

$$\lim_{\tau\longrightarrow+\infty}\rho_1^{(N)}(\mathbf{x}_1,t)=\rho_{1M}^{(N)}(\mathbf{v}_1). \tag{56}$$

Proof. In order to reach the thesis it is sufficient to prove that necessarily

$$\lim_{\tau\longrightarrow+\infty}\frac{\partial}{\partial t}I_M\left(\rho_1^{(N)}(t),\mathbf{b}\right)=0. \quad (57)$$

In fact, let us assume "*ad absurdum*" that $(\partial/\partial t)I_M(\rho_1^{(N)}(t),\mathbf{b})\leq -k^2$ with $k^2>0$ being a real constant. Then Theorem 2 (Proposition $P2_2$) requires that

$$\lim_{\tau\longrightarrow+\infty}I_M\left(\rho_1^{(N)}(t),\mathbf{b}\right)\leq-\lim_{\tau\longrightarrow+\infty}(t-t_o)k^2=-\infty, \quad (58)$$

a result which contradicts Theorem 1. This proves the validity of (57). Furthermore, by construction $(\partial/\partial t)I_M \equiv (1/K_{Mo})(\partial/\partial t)K_M$ and furthermore $(\partial/\partial t)K_M$ is identified with the functional $W_M(\rho_1^{(N)}(t),\mathbf{b})\leq 0$ which is determined by (53). At this point one notices that, thanks to continuity of the functional $W_M(\rho_1^{(N)}(t),\mathbf{b})$, the identity

$$\lim_{\tau\longrightarrow+\infty}\frac{\partial}{\partial t}I_M\left(\rho_1^{(N)}(\mathbf{x}_1,t),\mathbf{b}\right)=W_M\left(\rho_{1\infty}^{(N)}(\mathbf{x}_1),\mathbf{b}\right) \quad (59)$$

holds, where, thanks to global existence of the 1−body PDF (see [13]), the limit function

$$\lim_{\tau\longrightarrow+\infty}\rho_1^{(N)}(\mathbf{x}_1,t)\equiv\rho_{1\infty}^{(N)}(\mathbf{x}_1) \quad (60)$$

necessarily exists. As a consequence (57) requires also the equation

$$W_M\left(\rho_{1\infty}^{(N)}(\mathbf{x}_1),\mathbf{b}\right)=0 \quad (61)$$

to hold. Upon invoking Proposition $P2_3$ of Theorem 2 this implies that necessarily $\rho_{1\infty}^{(N)}(\mathbf{x}_1)=\rho_{1M}^{(N)}(\mathbf{v}_1)$ so the thesis (56) is proved. Incidentally, thanks to Theorem 1, this requires also that

$$\lim_{\tau\longrightarrow+\infty}I_M\left(\rho_1^{(N)}(t),\mathbf{b}\right)=I_M\left(\rho_{1\infty}^{(N)}(\mathbf{x}_1),\mathbf{b}\right)=0. \quad (62)$$

2.4. Remarks. A few remarks are worth being pointed out regarding the results presented above.

(1) *Remark #1.* The choice of the MKI functional considered here (see (22)) is just one of the infinite particular admissible realizations which meet the complete set of MKI prescriptions indicated above. In particular, the choice of the velocity moment $M(\mathbf{v}_1,\mathbf{b})$ considered here (see (24)) remains in principle arbitrary, since $|\mathbf{v}_1\cdot\mathbf{b}|^2$ can be equivalently replaced, for example, by any factor of the form $|\mathbf{v}_1\cdot\mathbf{b}|^{2n}$, with $n\geq 1$. Furthermore it is obvious that $M(\mathbf{v}_1,\mathbf{b})$ can be replaced by any function of the form $M(\mathbf{v}_1,\mathbf{b})+\Delta M(\mathbf{v}_1,\mathbf{b})$, being $\Delta M(\mathbf{v}_1,\mathbf{b})$ prescribed in such a way that its contribution to $(\partial/\partial t)I_M$ vanishes identically so that the validity of the inequality (42) in Theorem 2 is preserved. This implies in turn that the prescription of the MKI functional $I_M(\rho_1^{(N)}(t))$ remains in principle nonunique.

(2) *Remark #2.* A possible issue is related to the requirement that the renormalized 1−body PDF, as the 1−body PDF itself, are strictly positive at all times and are nonvanishing. Here it is sufficient to state that an elementary consequence of the theory of the Master kinetic equation developed in [3] is that, provided the corresponding initial N−body PDF set at a prescribed initial time t_o is strictly positive in the whole N−body phase-space, both the corresponding renormalized 1−body PDF as the 1−body PDF remain necessarily strictly positive globally in time too and everywhere in the 1−body phase-space.

(3) *Remark #3.* It must be stressed that the signature of the time derivative $(\partial/\partial t)I_M(\rho_1^{(N)}(t),\mathbf{b})$ actually depends crucially on the adoption of the causal form of MCBC (i.e., see (C.1) or (C.3) in Appendix C) rather than the anticausal one (given instead by (C.2)). The first choice is mandatory in view of the causality principle. Indeed, it is immediate to prove that $(\partial/\partial t)I_M(\rho_1^{(N)}(t),\mathbf{b})$ changes signature if the anticausal MCBC (C.2) is invoked.

(4) *Remark #4.* Theorem 2 warrants that macroscopic irreversibility, namely, the inequality $(\partial/\partial t)I_M(\rho_1^{(N)}(t),\mathbf{b})\leq 0$, occurs specifically because of (a) the time-variation of the $\mathbf{b}$−directional total kinetic energy which occurs at arbitrary binary-collision events; (b) the occurrence of a velocity-space anisotropy in the 1−body PDF, i.e., the fact that the same PDF may not coincide with a local Maxwellian PDF.

(5) *Remark #5.* The existence of the limit function $\lim_{t\longrightarrow+\infty}\rho_1^{(N)}(\mathbf{x}_1,t)=\rho_{1\infty}^{(N)}(\mathbf{x}_1)$ follows uniquely as a consequence of the global existence theorem holding for the Master kinetic equation [13].

(6) *Remark #6.* Last but not least, the fact that the same limit function may coincide or not with the Maxwellian kinetic equilibrium (6) depends crucially on the functional setting prescribed for the same PDF $\rho_1^{(N)}(\mathbf{x}_1,t)$. More precisely, DKE can only occur provided $\rho_1^{(N)}(\mathbf{x}_1,t)$ is a suitably smooth stochastic PDF such that the MKI functional exists for the corresponding initial PDF at time t_o, i.e., $\rho_1^{(N)}(\mathbf{x}_1,t_o)=\rho_{1o}^{(N)}(\mathbf{x}_1)$.

Theorems 1–3 represent the main results reached in the paper of what may be referred to as the *PMI/DKE theory* for finite hard-sphere systems and which have concerned the axiomatic formulation in such a context of the notion of macroscopic irreversibility and the related one of decay to kinetic equilibrium.

3. Consistency of MPI/DKE Theory with Microscopic Dynamics

The crucial problem which arises in the context of the ab initio theory is in some sense analogous to that occurring in the Boltzmann and Grad kinetic theories. The question is in fact whether these phenomena are actually consistent with the fundamental symmetry properties of the underlying Boltzmann-Sinai CDS. The problem posed in the present section concerns, more precisely, the consistency with the time-reversible, energy-conserving, evolution of the underlying N−body Boltzmann-Sinai classical dynamical system S_N−CDS.

(1) *First issue: consistency with the microscopic reversibility principle*. This is related to the famous objection raised by Loschmidt to the Boltzmann equation and Boltzmann H-theorem: i.e., *whether* and possibly also *how* it may be possible to reconcile the validity of the reversibility principle for the S_N−CDS with the manifestation of a decay of the 1−body PDF to kinetic equilibrium, i.e., the uniform Maxwellian PDF of the form (6), as predicted by the above Master equation-DKE Theorem. That a satisfactory answer to this question is actually possible follows from considerations which are based on the axiomatic (ab initio) statistical description realized by the Master kinetic equation. In this regard it is worth recalling the discussion reported above concerning the role of MCBC regarding the functional $(\partial/\partial t)I_M(\rho_1^{(N)}(t))$. In particular, it is obvious that the signature depends on whether the causal (or anticausal) form of MCBC is invoked (see Appendix C). Such a choice is not arbitrary since, for consistency with the causality principle, it must depend on the microscopic arrow of time, i.e., the orientation of the time axis chosen for the reference frame. Based on these premises, consistency between the occurrence of macroscopic irreversibility associated with the DKE phenomenon and the principle of microscopic reversibility can immediately be established. Indeed, it is sufficient to notice that when a time-reversal or a velocity-reversal is performed on the S_N−CDS the form of the collision boundary conditions (i.e., in the present case the MCBC provided by (C.1) in Appendix C) must be changed, replacing them with the corresponding anticausal ones, i.e., (C.2). This implies that MKI functional decreases in both cases, i.e., after performing the time-reversal, so that no contradiction can possibly arise in this case between Theorem 3 and the microscopic reversibility principle.

(2) *Second issue: consistency with Poincaré recurrence theorem*. Similar considerations concern the consistency with the recurrence theorem due to Poincaré as well as the conservation of total (kinetic) energy for the S_N−CDS. In fact, first, as it follows from [3], by construction the Master collision operator admits the customary Boltzmann collisional invariants, including total kinetic energy of colliding particles. Hence, total energy conservation is again warranted for S_N−CDS. Second, regarding Poincaré recurrence theorem, it concerns the Lagrangian phase-space trajectories of the S_N−CDS, i.e., the fact that almost all of these trajectories return arbitrarily close—in a suitable sense to be prescribed in terms of a distance defined on the N−body phase-space—to their initial condition after a suitably large "recurrence time". Incidentally, its magnitude depends strongly both on the same initial condition and the notion of distance to be established on the same phase-space. Nevertheless, such a "recurrence effect" influences only the Lagrangian time evolution of the N−body PDF which occurs along the same Lagrangian N−body phase-space trajectories. Instead, the same recurrence effect has manifestly no influence on the time evolution of the Eulerian 1−body PDF which is advanced in time in terms of the Eulerian kinetic equation represented by the Master kinetic equation. Therefore the mutual consistency of DKE and Poincaré recurrence theorem remains obvious.

Hence, in the framework of the axiomatic ab initio theory based on the Master kinetic equation the full consistency is warranted with the microscopic dynamics of the underlying Boltzmann-Sinai CDS.

4. Physical Implications

Let us now investigate the physical interpretation and main implications emerging from the PMI/DKE theory developed here. The first issue is related to the physical mechanism at the basis of the PMI/DKE phenomenology.

It is well known that in the context of Boltzmann kinetic theory the property of macroscopic irreversibility and the occurrence of the DKE phenomenon are both ascribed to the Boltzmann H-theorem, both in its original formulation [7] and in its modified form introduced by Boltzmann himself while attempting to reply [28] to Loschmidt objection [27] (see also [30, 31] together with different views on the matter given in [24, 26]). As recalled above, this is expressed in terms of the production rate for the Boltzmann-Shannon entropy $(\partial/\partial t)S(\rho_1(t))$, with $S(\rho_1(t))$ being interpreted as a measure of the ignorance associated with a solution of the Boltzmann equation. In fact the customary interpretation is that they arise specifically because of the validity of the entropic inequality (10), i.e., the monotonic increase of $S(\rho_1(t))$, and the corresponding entropic equality (11) stating a necessary and sufficient condition for kinetic equilibrium. Such a theorem is actually intimately related to the equation itself. In fact both the theorem and the equation generally hold only for stochastic PDFs $\rho_1(t) = \rho_1(\mathbf{x}_1, t)$ which are suitably smooth and not for distributions [1]. According to Boltzmann's original interpretation, however, both the Boltzmann equation and Boltzmann H-theorem should only hold when the so-called Boltzmann-Grad limit is invoked, i.e., based on the limit operator $L_{BG} \equiv \lim_{\substack{N\longrightarrow+\infty \\ N\sigma^2\sim O(1)}}$ (see [4]).

In striking departure from such a picture, one has the following:

(i) The axiomatic ab initio theory based on the Master kinetic equation and the present PMI/DKE theory are applicable to an arbitrary finite Boltzmann-Sinai CDS. This means that they hold for hard-sphere systems having a finite number of particles and with finite diameter and mass, i.e., without the need of invoking validity of asymptotic conditions.

(ii) The main departure with respect to Boltzmann kinetic theory arises because, as earlier discovered (see in particular the related discussion reported in [4]), the Boltzmann-Shannon entropy associated with an arbitrary stochastic 1−body PDF $\rho_1^{(N)}(t) = \rho_1^{(N)}(\mathbf{x}_1, t)$ solution of the Master kinetic equation is identically conserved. Thus both PMI and DKE are essentially unrelated to the Boltzmann-Shannon entropy.

(iii) In the case of the Master kinetic equation the physical mechanism responsible for the occurrence of both PMI and DKE is unrelated to the Boltzmann-Shannon entropy. In fact, as shown here, it arises because of the properties of the MKI functional $I_M(\rho_1^{(N)}(t), \mathbf{b})$ when it is expressed in terms of an arbitrary stochastic PDF $\rho_1^{(N)}(t) = \rho_1^{(N)}(\mathbf{x}_1, t)$ solution of the Master kinetic equation. The only requirement is that the initial PDF $\rho_{1o}^{(N)}(\mathbf{x}_1)$ is prescribed so that the corresponding MKI functional $I_M(\rho_{1o}^{(N)}(\mathbf{x}_1), \mathbf{b})$ exists.

(iv) As shown here the MKI functional is a suitably weighted phase-space moment of $\rho_1^{(N)}(\mathbf{x}_1, t)$ which can be interpreted as an information measure for the same PDF, namely, belongs to the interval $[0, 1]$, and exhibits a monotonic-decreasing time-dependence, i.e., the property of macroscopic irreversibility.

(v) In addition both $I_M(\rho_1^{(N)}(t), \mathbf{b})$ and its time derivative $(\partial/\partial t) I_M(\rho_1^{(N)}(t), \mathbf{b})$ vanish identically if and only if the 1−body PDF coincides with a Maxwellian kinetic equilibrium of the type (6). This warrants in turn also the occurrence of the DKE phenomenon for $\rho_1^{(N)}(\mathbf{x}_1, t)$, i.e., that for $t - t_o \longrightarrow +\infty$ the same PDF must decay to a Maxwellian kinetic equilibrium of this type.

(vi) Finally, it is interesting to point out the peculiar behavior of the MKI functional $I_M(\rho_1^{(N)}(t), \mathbf{b})$ and its time derivative $(\partial/\partial t) I_M(\rho_1^{(N)}(t), \mathbf{b})$ when the Boltzmann-Grad limit is considered. In particular the 1− and 2−body occupation coefficients $k_1^{(N)}(\mathbf{r}_1, t)$ and $k_2^{(N)}(\mathbf{r}_1, \mathbf{r}_2, t)$ which appear in the Master kinetic equation (see Appendix B, (B.1) and (B.2)) become respectively

$$\begin{aligned} L_{BG} k_1^{(N)}(\mathbf{r}_1, t) &= 1 \\ L_{BG} k_2^{(N)}(\mathbf{r}_1, \mathbf{r}_2, t) &= 1. \end{aligned} \tag{63}$$

As a consequence the limit functionals $L_{BG} I_M(\rho_1^{(N)}(t), \mathbf{b})$ and $L_{BG}(\partial/\partial t) I_M(\rho_1^{(N)}(t), \mathbf{b})$, are necessarily identically vanishing. This means that the present theory applies properly when the exact Master kinetic equation is considered and not to its asymptotic approximation obtained in the Boltzmann-Grad limit, namely the Boltzmann kinetic equation (see [3, 13]).

An interesting issue, in the context of the PMI/DKE theory for the Master kinetic equation, is the role of MCBC in giving rise to the phenomena of macroscopic irreversibility and decay to kinetic equilibrium. Let us analyze for this purpose the two cases represented by unary and binary hard-sphere elastic collisions.

First, let us recall the customary treatment of collision boundary conditions for unary collision events (also referred to as the so-called mirror reflection CBC; see for example Cercignani [6, 11]). This refers to the occurrence at a collision time t_i of a single unary elastic collision for particle 1 at the boundary $\partial\Omega$. Let us denote by $\mathbf{n}_1$ the inward normal to the stationary rigid boundary $\partial\Omega$ at the point of contact with the same particle and respectively $\mathbf{x}_1^{(-)}(t_1) = (\mathbf{r}_1(t_1), \mathbf{v}_1^{(-)}(t_1))$ and $\mathbf{x}_1^{(+)}(t_1) = (\mathbf{r}_1(t_1), \mathbf{v}_1^{(+)}(t_1))$ the incoming and outgoing particle states while $\mathbf{v}_1^{(+)}$ is determined by the elastic collision law for unary collisions, namely,

$$\mathbf{v}_1^{(+)} = \mathbf{v}_1^{(-)} - 2\mathbf{n}_1 \mathbf{n}_1 \cdot \mathbf{v}_1^{(-)}. \tag{64}$$

Then, the PDF-conserving CBC for the 1−body PDF requires that the following identity holds:

$$\rho_1^{(N)}\left(\mathbf{x}_1^{(+)}(t_1), t_i\right) = \rho^{(N)}\left(\mathbf{x}_1^{(-)}(t_i), t_i\right), \tag{65}$$

with $\rho_1^{(N)}(\mathbf{x}_1^{(+)}(t_1), t_i) \equiv \rho_1^{(N)(+)}(\mathbf{x}_1^{(+)}(t_1), t_i)$ and $\rho^{(N)}(\mathbf{x}_1^{(-)}(t_i), t_i) \equiv \rho^{(N)(-)}(\mathbf{x}_1^{(-)}(t_i), t_i)$ denoting the outgoing and incoming 1−body PDF, respectively. This identifies the PDF-conserving CBC usually adopted in Boltzmann kinetic theory [7] (Grad [10]; see also [2, 3]). The obvious physical implication of (65) is that $\rho_1^{(N)}(\mathbf{x}_1^{(+)}(t_1), t_i)$ (and $\rho^{(N)}(\mathbf{x}_1^{(-)}(t_i), t_i)$) should be necessarily an even function of the velocity component $\mathbf{n}_1 \cdot \mathbf{v}_1^{(-)}$. Indeed as shown in [2, 3] the PDF-conserving CBC (65) should be replaced with a suitable CBC identified with the MCBC condition (see also Appendix C). When realized in terms of its causal form (predicting the outgoing PDF in terms of the incoming one) the MCBC for unary collisions is just

$$\rho_1^{(N)(+)}\left(\mathbf{x}_1^{(+)}(t_1), t_i\right) = \rho^{(N)(-)}\left(\mathbf{x}_1^{(+)}(t_i), t_i\right), \tag{66}$$

with $\rho^{(N)(-)}(\mathbf{x}_1^{(+)}(t_i), t_i)$ denoting the incoming 1−body PDF evaluated in terms of the outgoing state $\mathbf{x}_1^{(+)}(t_i)$. Assuming left-continuity (see related discussion in [2]), this can then be identified with $\rho^{(N)(-)}(\mathbf{x}_1^{(+)}(t_i), t_i) \equiv \rho^{(N)}(\mathbf{x}_1^{(+)}(t_i), t_i)$, thus yielding

$$\rho_1^{(N)(+)}\left(\mathbf{x}_1^{(+)}(t_1), t_i\right) = \rho^{(N)}\left(\mathbf{x}_1^{(+)}(t_i), t_i\right). \tag{67}$$

Equation (67) provides the physical prescription for the collision boundary condition, which is referred to as MCBC, holding for the 1−body PDF at arbitrary unary collision events. It is immediate to realize that the function $\rho^{(N)}(\mathbf{x}_1^{(+)}(t_i), t_i)$ needs not generally be even with respect to the velocity component $\mathbf{n}_1 \cdot \mathbf{v}_1^{(-)}$. In addition (67), just as

(65), also permits the existence of the customary collisional invariants which in the case of unary collisions are $X = 1, |\mathbf{n}_1 \cdot \mathbf{v}_1^{(-)}|, \mathbf{v}_1 \cdot [\underline{\underline{1}} - \mathbf{n}_1\mathbf{n}_1], v_1^2$. As a consequence, one can show that (67) warrants at the same time also the validity of the so-called no-slip boundary conditions for the fluid velocity field $\mathbf{V}(\mathbf{r}_1, t)$ carried by the 1−body PDF $\rho_1^{(N)}(\mathbf{x}_1, t)$.

The treatment of MCBC holding for the 2−body PDF in case of binary-collision events is analogous and is recalled for convenience in (C.2) of Appendix C.

Let us briefly analyze the qualitative physical implications of (67) and (C.2) as far as the DKE theory is concerned. First, we notice that unary collisions cannot produce in a proper sense a velocity-isotropization effect since, as shown by (67), in such a case MCBC gives rise only to a change in the velocity distribution occurring during a unary collision due to a single component of the particle velocity, namely, $\mathbf{n}_1 \cdot \mathbf{v}_1^{(-)}$. As a consequence, this explains why unary collisions do not affect the rate of change of the MKI functional (see Theorem 2). Second, (C.2) shows—on the contrary—that binary collisions actually do affect by means of MCBC a velocity-spreading for the 1− and 2−body PDF. In particular, since the spreading effect occurs in principle for all components of particle velocities affecting both particles 1 and 2, this explains why binary collisions are actually responsible for the irreversible time evolution of the MKI functional (see Theorems 2 and 3).

In turn, as implied by Theorem 3, DKE arises because of the phenomenon of macroscopic irreversibility (Theorem 2). The latter arises due specifically to the possible occurrence of a velocity-space anisotropy which characterizes the 1−body PDF when the same PDF differs locally from kinetic equilibrium. In turn, this requires also that the 1−body PDF belongs to the functional class of admissible stochastic PDFs $\{\rho_1^{(N)}(\mathbf{x}_1, t)\}$. In difference to Boltzmann kinetic theory, however, the key physical role is actually ascribed to the MKI functional $I_M(\rho_1^{(N)}(t))$ rather than the Boltzmann-Shannon entropy $S_1(\rho_1^{(N)}(t))$. In fact, recalled above, the same functional remains constant in time once the Master kinetic equation is adopted. Rather, as shown by Theorem 2, it is actually the Master kinetic information $I_M(\rho_1^{(N)}(t))$ which exhibits the characteristic signatures of macroscopic irreversibility.

The key differences arising between the two theories, i.e., the Boltzmann equation-DKE and the Master equation-DKE, are of course related to the different and peculiar intrinsic properties of the Boltzmann and Master kinetic equations. In particular, as discussed at length elsewhere (see [1, 3]), precisely because the Boltzmann equation is only an asymptotic approximation of the Master kinetic equation explains why a loss of information occurs in Boltzmann kinetic theory and consequently the related Boltzmann-Shannon entropy is not conserved.

The present investigation shows that in the context of the Master kinetic equation, the macroscopic irreversibility property, i.e., the monotonic time-decay behavior of the MKI functional, can be explained at a more fundamental level, i.e., based specifically on the time-variation of the $\mathbf{b}$−directional total kinetic energy which occurs at arbitrary binary-collision events.

The Master equation-DKE theorem (Theorem 3) given above provides a first-principle proof of the existence of the phenomenon of DKE occurring for the kinetic description of a finite number of extended hard-spheres, i.e., described by means of the Master kinetic equation. More precisely, the DKE phenomenon affects the 1−body PDFs belonging to the admissible functional class $\{\rho_1^{(N)}(\mathbf{x}_1, t)\}$ determined according to Theorem 1 and requiring also that the local characteristic scale length $L_\rho(t)$ associated with $\rho_1^{(N)}(\mathbf{x}_1, t)$ is nonzero at all times.

5. Conclusions

In this paper the problem of the property of microscopic irreversibility (PMI) and decay to kinetic equilibrium (DKE) of the 1−body PDF has been addressed. In doing so original ideas and methods are adopted of the new ab initio theory for hard-sphere systems recently developed in the context of classical statistical mechanics [1, 2].

These are not just small deviations from standard literature approaches. Such developments, in fact, have opened up a host of exciting new subjects of investigation and theoretical challenges in kinetic theory which arise thanks to, or in the context of, the ab initio approach to kinetic theory. Both are based in particular on the discovery of the Master kinetic equation first reported in [3], equation which has been adopted also in the present paper.

The ab initio theory and specifically the present paper represent the attempt to reach a new foundational basis and axiomatic physical description of the classical statistical mechanics for hard-sphere systems. The topic which has been pursued here—which represents also a challenging test of the ab initio theory itself—concerns the investigation of the physical origins of PMI and the related DKE phenomenon arising *in finite N−body hard-sphere systems.* These issues refer in particular to

(i) *the proof of the nonnegativity of Master kinetic information (Theorem 1, Section 2.1) together with the property of macroscopic irreversibility (PMI; Theorem 2, Section 2.2);*

(ii) *the establishment of Theorem 3 (Section 2.3) and the related proof of the property of decay to kinetic equilibrium (DKE);*

(iii) *the consistency of PMI and DKE with microscopic dynamics (Section 3);*

(iv) *the analysis of the main physical implications of DKE (Section 4).*

The theory presented here departs in several respects from previous literature and notably from Boltzmann kinetic theory. The main differences actually arise because of the nonasymptotic character of the new theory, i.e., the fact that it applies to arbitrary dense or rarefied systems for which the finite number and size of the constituent particles is accounted for [3]. In this paper basic consequences of the new theory have been investigated which concern the phenomenon of decay to global kinetic equilibrium.

The present results are believed to be crucial, besides in mathematical research, for the physical applications of the ab initio theory statistical theory, i.e., the Master kinetic equation. Indeed, regarding challenging future developments of the theory one should mention among others the following examples of possible (and mutually related) routes worth to be explored. One is related to the investigation of the possible effects due to arbitrarily prescribed, i.e., nonvanishing, initial (binary or multibody) *phase-space* statistical correlations. As recalled above, in fact, the Master equation is appropriate only when suitably prescribed configuration-space statistical correlations are taken into account. The second goal concerns the investigation of the time-asymptotic properties of the same kinetic equation, for which the present paper may represent a useful basis. The third goal refers to the possible extension of the theory to mixtures formed by hard-spheres of different masses and diameter which possibly undergo both elastic and anelastic collisions. Finally, the fourth one concerns the investigation of hydrodynamic regimes for which a key prerequisite is provided by the DKE theory established here.

Appendix

A. Realizations of the Master Kinetic Equation

For completeness we recall here the two equivalent forms of the Master kinetic equation [3]. In terms of the renormalized 1–body PDF $\widehat{\rho}_1^{(N)}(\mathbf{x}_1,t)$ (see (26)) the *first form* of the same equation reads

$$L_{1(1)}\widehat{\rho}_1^{(N)}(\mathbf{x}_1,t)=0, \tag{A.1}$$

with $L_{1(1)} = \partial/\partial t + \mathbf{v}_1 \cdot \partial/\partial \mathbf{r}_1$ denoting the 1–body free-streaming operator. Hence it follows

$$L_{1(1)}\rho_1^{(N)}(\mathbf{x}_1,t)=\widehat{\rho}_1^{(N)}(\mathbf{x}_1,t)\,L_{1(1)}k_1^{(N)}(\mathbf{r}_1,t), \tag{A.2}$$

where explicit evaluation of the rhs the last equation (see also (B.12) below) yields

$$\begin{aligned}\widehat{\rho}_1^{(N)}(\mathbf{x}_1,t)\,L_{1(1)}k_1^{(N)}(\mathbf{r}_1,t)&=(N-1)\,\sigma^2\int_{U_{1(2)}}d\mathbf{v}_2\\&\cdot\int d\Sigma_{21}\mathbf{v}_{21}\cdot\mathbf{n}_{21}\overline{\Theta}^*(\mathbf{r}_2)\,k_2^{(N)}(\mathbf{r}_1,\mathbf{r}_2,t)\,\widehat{\rho}_1^{(N)}(\mathbf{x}_1,t)\\&\cdot\widehat{\rho}_1^{(N)}(\mathbf{x}_2,t),\end{aligned} \tag{A.3}$$

with $\overline{\Theta}^*(\mathbf{r}_2) \equiv \overline{\Theta}_i^{(\partial\Omega)}(\overline{\mathbf{r}})$ and $k_2^{(N)}(\mathbf{r}_1,\mathbf{r}_2,t)$ being identified with the definitions given respectively by (A.6) and (B.1) in Appendix B. Then, consistent with [3] and upon invoking the causal form of MCBC (see (C.3) in Appendix C) the same equation can be written in the equivalent *second form* of the Master kinetic equation [3]. The corresponding initial-value problem, taking the form

$$\begin{aligned}L_{1(1)}\rho_1^{(N)}(\mathbf{x}_1,t)-\mathscr{C}_1\left(\rho_1^{(N)}\mid\rho_1^{(N)}\right)&=0,\\\rho_1^{(N)}(\mathbf{x}_1,t_o)&=\rho_{1o}^{(N)}(\mathbf{x}_1),\end{aligned} \tag{A.4}$$

can be shown to admit a unique global solution [13]. Here the notation is standard [3]. Thus

$$\begin{aligned}\mathscr{C}_1\left(\rho_1^{(N)}\mid\rho_1^{(N)}\right)&\equiv(N-1)\,\sigma^2\int_{U_{1(2)}}d\mathbf{v}_2\int^{(-)}d\Sigma_{21}\\&\cdot\left[\widehat{\rho}_1^{(N)}\left(\mathbf{r}_1,\mathbf{v}_1^{(+)},t\right)\widehat{\rho}_1^{(N)}\left(\mathbf{r}_2,\mathbf{v}_2^{(+)},t\right)\right.\\&\left.-\widehat{\rho}_1^{(N)}(\mathbf{r}_1,\mathbf{v}_1,t)\,\widehat{\rho}_1^{(N)}(\mathbf{r}_2,\mathbf{v}_2,t)\right]\times\left|\mathbf{v}_{21}\cdot\mathbf{n}_{21}\right|\\&\cdot k_2^{(N)}(\mathbf{r}_1,\mathbf{r}_2,t)\,\overline{\Theta}^*(\mathbf{r}_2)\end{aligned} \tag{A.5}$$

identifies the Master collision operator, while $\rho_{1o}^{(N)}(\mathbf{x}_1)$ is the initial 1–body PDF which belongs to the functional class $\{\rho_{1o}^{(N)}(\mathbf{x}_1)\}$ of stochastic, i.e., strictly positive, smooth ordinary functions, 1–body PDFs. Furthermore, the solid-angle integral on the rhs of (A.5) is now evaluated on the subset in which $\mathbf{v}_{12}\cdot\mathbf{n}_{12}<0$, while $\mathbf{r}_2$ identifies $\mathbf{r}_2=\mathbf{r}_1+\sigma\mathbf{n}_{21}$, while $k_1^{(N)}(\mathbf{r}_1,t)$ and $k_2^{(N)}(\mathbf{r}_1,\mathbf{r}_2,t)$ coincide, respectively, with the 1– and 2–body occupation coefficients [3] and $\overline{\Theta}^*\equiv\overline{\Theta}^*(\mathbf{r}_i)$ is prescribed by

$$\overline{\Theta}^*(\mathbf{r}_i)\equiv\overline{\Theta}_i^{(\partial\Omega)}(\overline{\mathbf{r}})\equiv\overline{\Theta}\left(\left|\mathbf{r}_i-\frac{\sigma}{2}\mathbf{n}_i\right|-\frac{\sigma}{2}\right), \tag{A.6}$$

with $\overline{\Theta}(x)$ being the strong Heaviside theta function

$$\overline{\Theta}(x)=\begin{cases}1 & y>0\\0 & y\leq 0.\end{cases} \tag{A.7}$$

Regarding the specific identification of the occupation coefficients (recalled in Appendix B) let us preliminarily recall the notion of S_N– *ensemble strong theta function* $\overline{\Theta}^{(N)}$. The latter is prescribed, according to [3], by requiring that

$$\overline{\Theta}^{(N)}(\overline{\mathbf{r}})=1 \tag{A.8}$$

for all configuration vectors $\overline{\mathbf{r}}\equiv\{\mathbf{r}_1,\ldots,\mathbf{r}_N\}$ belonging to the collisionless subset of $\Omega^{(N)}$. This is identified with the open subset of the N–body configuration domain $\Omega^{(N)}\equiv\prod_{i=1,N}\Omega$ in which each of the particles of S_N is not in mutual contact with any other particle of S_N or with the boundary $\partial\Omega$ of Ω. This means that, at any configuration $\overline{\mathbf{r}}$, $\overline{\Theta}^{(N)}(\overline{\mathbf{r}})$ can be prescribed as

$$\overline{\Theta}^{(N)}(\overline{\mathbf{r}})\equiv\prod_{i=1,N}\overline{\Theta}_i(\overline{\mathbf{r}})\,\overline{\Theta}_i^{(\partial\Omega)}(\overline{\mathbf{r}}). \tag{A.9}$$

Here $\overline{\Theta}_i^{(\partial\Omega)}(\overline{\mathbf{r}})$ identifies the i–th particle "boundary" theta function

$$\overline{\Theta}_i^{(\partial\Omega)}(\overline{\mathbf{r}})\equiv\overline{\Theta}_i^{(\partial\Omega)}(\overline{\mathbf{r}}_i)=\overline{\Theta}\left(\left|\mathbf{r}_i-\mathbf{r}_{Wi}\right|-\frac{\sigma}{2}\right), \tag{A.10}$$

with $\mathbf{r}_{Wi}=\mathbf{r}_i-\rho\mathbf{n}_i$ and $\rho\mathbf{n}_i$ being the inward vector normal to the boundary belonging to the center of the i–th particle having a distance $\rho/2$ from the same boundary. Furthermore

$\overline{\Theta}_i(\overline{\mathbf{r}})$ is the "binary-collision" theta function. A possible identification of $\overline{\Theta}_i(\overline{\mathbf{r}})$ which warrants validity of (A.8) is given by the expression

$$\overline{\Theta}_i(\overline{\mathbf{r}}) \equiv \prod_{\substack{j=1,N;\\ i<j}} \overline{\Theta}\left(\left|\mathbf{r}_i - \mathbf{r}_j\right| - \sigma\right), \tag{A.11}$$

Namely,

$$\begin{aligned} \overline{\Theta}_i(\overline{\mathbf{r}}) &\equiv \prod_{\substack{j=1,N;\\ i<j}} \overline{\Theta}_{ij}(\overline{\mathbf{r}}), \\ \overline{\Theta}_{ij}(\overline{\mathbf{r}}) &\equiv \overline{\Theta}\left(\left|\mathbf{r}_i - \mathbf{r}_j\right| - \sigma\right). \end{aligned} \tag{A.12}$$

However, an equivalent possible prescription of $\overline{\Theta}_i(\overline{\mathbf{r}})$ is also provided by the alternative realization obtained letting

$$\begin{aligned} \overline{\Theta}_i(\overline{\mathbf{r}}) &\equiv \prod_{\substack{j=1,N;\\ i<j}} \prod_{\substack{m,n=1,N\\ i<m<n}} \overline{\Theta}_{ij}^{mn}(\overline{\mathbf{r}}), \\ \overline{\Theta}_{ij}^{mn}(\overline{\mathbf{r}}) &\equiv \overline{\Theta}\left(\left|\mathbf{r}_i - \mathbf{r}_j\right| - \sigma\right) \\ &\quad \times \overline{\Theta}\left(\left|\mathbf{r}_i - \mathbf{r}_m\right| + \left|\mathbf{r}_i - \mathbf{r}_n\right| - 2\sigma\right) \\ &\quad \cdot \overline{\Theta}\left(\left|\mathbf{r}_m - \mathbf{r}_n\right| + \left|\mathbf{r}_i - \mathbf{r}_m\right| - 2\sigma\right). \end{aligned} \tag{A.13}$$

Indeed, in the subset of $\Omega^{(N)}$ in which for all $i = 1, N$ the rhs of (A.11) is identically equal to unity, the factor $\prod_{\substack{m,n=1,N\\ i<m<n}} \overline{\Theta}(|\mathbf{r}_i - \mathbf{r}_m| + |\mathbf{r}_i - \mathbf{r}_n| - 2\sigma)\overline{\Theta}(|\mathbf{r}_m - \mathbf{r}_n| + |\mathbf{r}_i - \mathbf{r}_m| - 2\sigma)$ is necessarily equal to unity too. Incidentally, we notice in fact that the latter factor carries the contributions due to triple collisions which are ruled out in the domain of validity of (A.8).

B. Integral and Differential Identities for the Occupation Coefficients

One notices that although the definitions (A.13) and (A.11) given in Appendix A for $\overline{\Theta}_i(\overline{\mathbf{r}})$ coincide in the collisionless subset of $\Omega^{(N)}$, only the first one is applicable in the complementary collision subset. Based on these premises in this appendix a number of integral and differential identities holding for the 1– and 2–body occupation coefficients are displayed.

First, recalling [3], one notices that the realizations of the 1– and s–body occupation coefficients $k_1^{(N)}(\mathbf{r}_i, t)$, $k_2^{(N)}(\mathbf{r}_1, \mathbf{r}_2, t), \ldots, k_s^{(N)}(\mathbf{r}_1, \mathbf{r}_2, \ldots \mathbf{r}_s, t)$ remain uniquely prescribed by the 1–body PDF, being given by

$$k_1^{(N)}(\mathbf{r}_1, t) \equiv F_1 \left\{ \prod_{j=2,N} \frac{\rho_1^{(N)}\left(\mathbf{x}_j, t\right)}{k_1^{(N)}\left(\mathbf{r}_j, t\right)} \right\}, \tag{B.1}$$

$$k_2^{(N)}(\mathbf{r}_1, \mathbf{r}_2, t) \equiv F_2 \left\{ \prod_{j=s+1,N} \frac{\rho_1^{(N)}\left(\mathbf{x}_j, t\right)}{k_1^{(N)}\left(\mathbf{r}_j, t\right)} \right\}, \tag{B.2}$$

$$\cdots \tag{B.3}$$

$$k_s^{(N)}(\mathbf{r}_1, \mathbf{r}_2, \ldots \mathbf{r}_s, t) \equiv F_s \left\{ \prod_{j=s+1,N} \frac{\rho_1^{(N)}\left(\mathbf{x}_j, t\right)}{k_1^{(N)}\left(\mathbf{r}_j, t\right)} \right\}, \tag{B.4}$$

where F_s denotes the integral operator

$$F_s \equiv \int_{\Gamma_N} d\overline{\mathbf{x}}\overline{\Theta}^{(N)}(\overline{\mathbf{r}}) \prod_{i=1,s} \delta\left(\mathbf{x}_i - \overline{\mathbf{x}}_i\right). \tag{B.5}$$

Therefore, since in the collisionless subset of $\Omega^{(N)}$ the prescriptions (A.11) and (A.13) are equivalent, in the same subset the 1– and 2–body occupation coefficients, written in terms of (A.11), become explicitly

$$\begin{aligned} k_1^{(N)}(\mathbf{r}_1, t) &= \int_{\Gamma_{1(2)}} d\mathbf{x}_2 \frac{\rho_1^{(N)}(\mathbf{x}_2, t)}{k_1^{(N)}(\mathbf{r}_2, t)} \Theta_2^{(\partial\Omega)}(\overline{\mathbf{r}}) \\ &\quad \cdot \overline{\Theta}\left(\left|\mathbf{r}_2 - \mathbf{r}_1\right| - \sigma\right) \times \int_{\Gamma_{1(3)}} d\mathbf{x}_3 \frac{\rho_1^{(N)}(\mathbf{x}_3, t)}{k_1^{(N)}(\mathbf{r}_3, t)} \Theta_3^{(\partial\Omega)}(\overline{\mathbf{r}}) \\ &\quad \cdot \prod_{j=1,2} \overline{\Theta}\left(\left|\mathbf{r}_3 - \mathbf{r}_j\right| - \sigma\right) \\ &\quad \cdots \int_{\Gamma_{1(N)}} d\mathbf{x}_N \frac{\rho_1^{(N)}(\mathbf{x}_N, t)}{k_1^{(N)}(\mathbf{r}_N, t)} \Theta_N^{(\partial\Omega)}(\overline{\mathbf{r}}) \\ &\quad \cdot \prod_{j=1,N-1} \overline{\Theta}\left(\left|\mathbf{r}_N - \mathbf{r}_j\right| - \sigma\right), \end{aligned} \tag{B.6}$$

and

$$\begin{aligned} k_2^{(N)}(\mathbf{r}_1, \mathbf{r}_2, t) &= \int_{\Gamma_{1(3)}} d\mathbf{x}_3 \frac{\rho_1^{(N)}(\mathbf{x}_3, t)}{k_1^{(N)}(\mathbf{r}_3, t)} \overline{\Theta}_3^{(\partial\Omega)}(\overline{\mathbf{r}}) \\ &\quad \cdot \prod_{j=1,2} \overline{\Theta}\left(\left|\mathbf{r}_3 - \mathbf{r}_j\right| - \sigma\right) \\ &\quad \times \int_{\Gamma_{1(4)}} d\mathbf{x}_4 \frac{\rho_1^{(N)}(\mathbf{x}_4, t)}{k_1^{(N)}(\mathbf{r}_4, t)} \Theta_4^{(\partial\Omega)}(\overline{\mathbf{r}}) \\ &\quad \cdot \prod_{j=1,3} \overline{\Theta}\left(\left|\mathbf{r}_4 - \mathbf{r}_j\right| - \sigma\right) \end{aligned} \tag{B.7}$$

$$\begin{aligned} &\cdots \int_{\Gamma_{1(N)}} d\mathbf{x}_N \frac{\rho_1^{(N)}(\mathbf{x}_N, t)}{k_1^{(N)}(\mathbf{r}_N, t)} \Theta_N^{(\partial\Omega)}(\overline{\mathbf{r}}) \\ &\quad \cdot \prod_{j=1,N-1} \overline{\Theta}\left(\left|\mathbf{r}_N - \mathbf{r}_j\right| - \sigma\right). \end{aligned} \tag{B.8}$$

Accordingly letting $\mathbf{n}_{jj} = \mathbf{r}_{uij}/|\mathbf{r}_{ij}|$ with $\mathbf{r}_{ij} = \mathbf{r}_i - \mathbf{r}_j$, one notices that in the collisionless subset of $\Omega^{(N)}$ the following differential identities hold for all $s = 1, N-1$:

$$\begin{aligned} \frac{\partial}{\partial \mathbf{r}_1} k_1^{(N)}(\mathbf{r}_1, t) &= (N-1) \int_{\Gamma_{1(2)}} d\mathbf{x}_2 \mathbf{n}_{12} \delta \\ &\quad \cdot \left(\left|\mathbf{r}_2 - \mathbf{r}_1\right| - \sigma\right) \times k_2^{(N)}(\mathbf{r}_1, \mathbf{r}_2, t)\, \overline{\Theta}_2^{(\partial\Omega)}(\overline{\mathbf{r}}) \end{aligned}$$

$$\cdot \widehat{\rho}_1^{(N)}(\mathbf{x}_2, t), \tag{B.9}$$

$$\begin{aligned}\frac{\partial}{\partial \mathbf{r}_1} k_2^{(N)}(\mathbf{r}_1, \mathbf{r}_2, t) &= (N-2) \\ &\cdot \int_{\Gamma_{1(3)}} d\mathbf{x}_3 \mathbf{n}_{13} \delta(|\mathbf{r}_3 - \mathbf{r}_1| - \sigma) \\ &\times \prod_{j=1,2; j\neq 1} \overline{\Theta}(|\mathbf{r}_3 - \mathbf{r}_j| - \sigma) \overline{\Theta}_3^{(\partial\Omega)}(\overline{\mathbf{r}}) \\ &\cdot k_3^{(N)}(\mathbf{r}_1, \mathbf{r}_2, \mathbf{r}_3, t) \widehat{\rho}_1^{(N)}(\mathbf{x}_3, t), \\ \frac{\partial}{\partial \mathbf{r}_2} k_2^{(N)}(\mathbf{r}_1, \mathbf{r}_2, t) &= (N-2) \int_{\Gamma_{1(3)}} d\mathbf{x}_3 \mathbf{n}_{23} \delta \\ &\cdot (|\mathbf{r}_3 - \mathbf{r}_2| - \sigma) \times \prod_{j=1,2; j\neq 2} \overline{\Theta}(|\mathbf{r}_3 - \mathbf{r}_j| - \sigma) \overline{\Theta}_3^{(\partial\Omega)} \\ &\cdot (\overline{\mathbf{r}}) k_3^{(N)}(\mathbf{r}_1, \mathbf{r}_2, \mathbf{r}_3, t) \widehat{\rho}_1^{(N)}(\mathbf{x}_3, t), \\ &\ldots \\ \frac{\partial}{\partial \mathbf{r}_1} k_s^{(N)}(\mathbf{r}_1, \mathbf{r}_2, \ldots, \mathbf{r}_s, t) &= (N-s) \int_{\Gamma_{1(s+1)}} d\mathbf{x}_{s+1} \mathbf{n}_{1s+1} \\ &\cdot \delta(|\mathbf{r}_{s+1} - \mathbf{r}_1| - \sigma) \times \prod_{j=1,s; j\neq 1} \overline{\Theta}(|\mathbf{r}_{s+1} - \mathbf{r}_j| - \sigma) \\ &\cdot \overline{\Theta}_3^{(\partial\Omega)}(\overline{\mathbf{r}}) k_{s+1}^{(N)}(\mathbf{r}_1, \mathbf{r}_2, \ldots, \mathbf{r}_{s+1}, t) \widehat{\rho}_1^{(N)}(\mathbf{x}_{s+1}, t) \\ \frac{\partial}{\partial \mathbf{r}_2} k_s^{(N)}(\mathbf{r}_1, \mathbf{r}_2, \ldots, \mathbf{r}_s, t) &= (N-s) \int_{\Gamma_{1(2)}} d\mathbf{x}_{s+1} \mathbf{n}_{2s+1} \delta \\ &\cdot (|\mathbf{r}_{s+1} - \mathbf{r}_2| - \sigma) \\ &\times \prod_{j=1,s; j\neq 2} \overline{\Theta}(|\mathbf{r}_{s+1} - \mathbf{r}_j| - \sigma) \overline{\Theta}_3^{(\partial\Omega)}(\overline{\mathbf{r}}) \\ &\cdot k_{s+1}^{(N)}(\mathbf{r}_1, \mathbf{r}_2, \ldots, \mathbf{r}_{s+1}, t) \widehat{\rho}_1^{(N)}(\mathbf{x}_{s+1}, t) \\ &\ldots \\ \frac{\partial}{\partial \mathbf{r}_s} k_s^{(N)}(\mathbf{r}_1, \mathbf{r}_2, \ldots, \mathbf{r}_s, t) &= (N-s) \\ &\cdot \int_{\Gamma_{1(s+1)}} d\mathbf{x}_{s+1} \mathbf{n}_{ss+1} \delta(|\mathbf{r}_{s+1} - \mathbf{r}_s| - \sigma) \\ &\times \prod_{j=1,s; j\neq s} \overline{\Theta}(|\mathbf{r}_{s+1} - \mathbf{r}_j| - \sigma) \overline{\Theta}_3^{(\partial\Omega)}(\overline{\mathbf{r}}) \\ &\cdot k_{s+1}^{(N)}(\mathbf{r}_1, \mathbf{r}_2, \ldots, \mathbf{r}_{s+1}, t) \widehat{\rho}_1^{(N)}(\mathbf{x}_{s+1}, t)\end{aligned}$$

(B.10) (B.11)

As a consequence the following identities (the first one needed to evaluate the rhs of (A.2) in Appendix A)

$$\begin{aligned}L_{1(1)} k_1^{(N)}(\mathbf{r}_1, t) &= (N-1) \int_{\Gamma_{1(2)}} d\mathbf{x}_2 \mathbf{v}_{21} \\ &\cdot \mathbf{n}_{21} \delta(|\mathbf{r}_2 - \mathbf{r}_1| - \sigma) \overline{\Theta}^*(\mathbf{r}_2) k_2^{(N)}(\mathbf{r}_1, \mathbf{r}_2, t) \\ &\cdot \widehat{\rho}_1^{(N)}(\mathbf{x}_2, t),\end{aligned} \tag{B.12}$$

$$\begin{aligned}\frac{\partial^2 k_1^{(N)}(\mathbf{r}_1, t)}{\partial \mathbf{r}_1 \cdot \partial \mathbf{r}_1} &= -(N-1) \int_{\Gamma_{1(2)}} d\mathbf{x}_2 k_2^{(N)}(\mathbf{r}_1, \mathbf{r}_2, t) \\ &\cdot \delta(|\mathbf{r}_2 - \mathbf{r}_1| - \sigma) \overline{\Theta}_2^{(\partial\Omega)}(\overline{\mathbf{r}}) \mathbf{n}_{21} \cdot \frac{\partial}{\partial \mathbf{r}_2} \widehat{\rho}_1^{(N)}(\mathbf{x}_2, t),\end{aligned} \tag{B.13}$$

hold too. However, the alternative realization of the factor $\overline{\Theta}_i(\overline{\mathbf{r}})$ given by (A.13) (see Appendix A) has the virtue of excluding explicitly multiple collisions. The consequence is that when such a definition is adopted the differential identities

$$\begin{aligned}\delta(|\mathbf{r}_2 - \mathbf{r}_1| - \sigma) \frac{\partial}{\partial \mathbf{r}_1} k_2^{(N)}(\mathbf{r}_1, \mathbf{r}_2, t) &= 0, \\ \delta(|\mathbf{r}_2 - \mathbf{r}_1| - \sigma) \frac{\partial}{\partial \mathbf{r}_2} k_2^{(N)}(\mathbf{r}_1, \mathbf{r}_2, t) &= 0,\end{aligned} \tag{B.14}$$

both hold identically. The latter equations, in fact, manifestly hold also in the collision subset where $\delta(|\mathbf{r}_2 - \mathbf{r}_1| - \sigma) \neq 0$.

C. Causal and Anticausal Forms of Collisional Boundary Conditions

For definiteness, let us denote, respectively, the outgoing and incoming N−body PDFs $\rho^{(-)(N)}(\mathbf{x}^{(-)}(t_i), t_i)$ and $\rho^{(+)(N)}(\mathbf{x}^{(+)}(t_i), t_i)$, with $\rho^{(\pm)(N)}(\mathbf{x}^{(\pm)}(t_i), t_i) = \lim_{t \longrightarrow t_i^{(\pm)}} \rho^{(N)}(\mathbf{x}(t), t)$, where $\mathbf{x}^{(-)}(t_i)$ and $\mathbf{x}^{(+)}(t_i)$, with $\mathbf{x}^{(\pm)}(t_i) = \lim_{t \longrightarrow t_i^{(\pm)}} \mathbf{x}(t)$, are the incoming and outgoing Lagrangian N−body states, their mutual relationship being again determined by the collision laws holding for the S_N−CDS. Here it is understood that

(i) the S_N−CDS is referred to a reference frame $O(\mathbf{r}, \tau \equiv t - t_o)$, having, respectively, spatial and time origins at the point O which belongs to the Euclidean space $\mathbb{R}^3$ and at time $t_o \in I$.

(ii) in addition, by assumption the time axis is oriented. Such an orientation is referred to as *microscopic arrow of time*.

For an arbitrary N−body PDF $\rho^{(N)}(\mathbf{x}, t)$ belonging to the extended functional setting and an arbitrary collision event occurring at time t_i two possible realizations of the MCBC can in principle be given, both yielding a relationship between the PDFs $\rho^{(+)(N)}$ and $\rho^{(-)(N)}$. In the context of the ab initio statistical approach based on the Master kinetic equation [1–3] these are provided by the two possible realizations of the so-called *modified CBC* (MCBC). When expressed in Lagrangian form they are realized, respectively, by either the *causal* and *anticausal MCBC*, namely,

$$\rho^{(+)(N)}(\mathbf{x}^{(+)}(t_i), t_i) = \rho^{(-)(N)}(\mathbf{x}^{(+)}(t_i), t_i), \tag{C.1}$$

or

$$\rho^{(-)(N)}(\mathbf{x}^{(-)}(t_i), t_i) = \rho^{(+)(N)}(\mathbf{x}^{(-)}(t_i), t_i). \tag{C.2}$$

The corresponding Eulerian forms of the MCBC can easily be determined (see [2]). The one corresponding to (C.1) is, for example, provided by the condition

$$\rho^{(+)(N)}\left(\mathbf{x}^{(+)},t\right)=\rho^{(-)(N)}\left(\mathbf{x}^{(+)},t\right), \tag{C.3}$$

where now $\mathbf{x}^{(+)}$ denotes again an arbitrary outgoing collision state.

Once the time axis is oriented, *i.e.*, the *microscopic arrow of time* is prescribed, the validity of the causality principle in the reference frame $(\mathbf{r},\tau\equiv t-t_o)$ manifestly requires invoking (C.1). Indeed, (C.1) predicts the future (*i.e.*, outgoing) PDF from the past (incoming) one. Therefore the choice (C.1) is the one which is manifestly consistent with the causality principle. On the other hand, if the arrow of time is changed, *i.e.*, the time-reversal transformation with respect to the initial time (or time-origin) t_o, *i.e.*, the map between the two reference frames

$$O\left(\mathbf{r},\tau\equiv t-t_o\right)\longrightarrow O\left(\mathbf{r},\tau'\right), \tag{C.4}$$

with $\tau'=-\tau$ is performed, it is obvious that for the transformed reference frame $O(\mathbf{r},\tau')$ the form of CBC consistent with causality principle becomes that given by (C.2). Analogous conclusions hold if a velocity-reversal is performed, implying the incoming states and corresponding PDF must be exchanged with corresponding outgoing ones and vice versa.

D. Treatment of Case $N=2$

For completeness let us briefly comment on the particular realization of MPI/DKE theory which is achieved in the special case $N=2$. For this purpose, one notices that—thanks to (B.2) recalled in Appendix B (see also [3])—in this case by construction $k_2^{(N)}(\mathbf{r}_1,\mathbf{r}_2,t)$ simply reduces to

$$k_2^{(N)}\left(\mathbf{r}_1,\mathbf{r}_2,t\right)\equiv 1. \tag{D.1}$$

Accordingly, once the same prescription is invoked, both the Master kinetic equation (A.4) and the corresponding Master collision operator (A.5) remain formally unchanged. In a similar way it is important to remark that the expression of the functional $(\partial/\partial t)K_M(\rho_1^{(N)}(t),\mathbf{b})\equiv W_M(\rho_1^{(N)}(t),\mathbf{b})$ given by (53) is still correct also in such a case, being now given by

$$\begin{aligned}
&\frac{\partial}{\partial t}K_M\left(\rho_1^{(N)}(t),\mathbf{b}\right)\equiv W_M\left(\rho_1^{(N)}(t),\mathbf{b}\right)\\
&\quad=-(N-1)\sigma^2\int_{U_{1(1)}}d\mathbf{v}_1\int_{U_{1(2)}}d\mathbf{v}_2\\
&\quad\times\int_{\Omega}d\mathbf{r}_1\int^{(-)}d\mathbf{\Sigma}_{21}\frac{\partial\widehat{\rho}_1^{(N)}\left(\mathbf{r}_1,\mathbf{v}_1^{(+)},t\right)}{\partial\mathbf{r}_1}\\
&\quad\cdot\frac{\partial\widehat{\rho}_1^{(N)}\left(\mathbf{r}_2=\mathbf{r}_1+\sigma\mathbf{n}_{21},\mathbf{v}_2^{(+)}t\right)}{\partial\mathbf{r}_2}\\
&\quad\times\left|\mathbf{v}_{12}^{(+)}\cdot\mathbf{n}_{12}\right|\left(\mathbf{b}\cdot\mathbf{n}_{12}\right)^2\left(\mathbf{n}_{12}\cdot\mathbf{v}_{12}^{(+)}\right)^2\leq 0.
\end{aligned} \tag{D.2}$$

It is then immediate to infer the validity of both the PMI theorem (Theorem 2) and the DKE property for the Master kinetic equation (Theorem 3). As a consequence one concludes that MPI/DKE theory holds also in the special case $N=2$. This conclusion is not unexpected. In fact, binary collisions, as indicated above, are responsible for the MPI/DKE phenomenology and in such a case can only occur between particles 1 and 2.

Conflicts of Interest

The authors declare that they have no conflicts of interest.

Acknowledgments

This work is dedicated to the dearest memory of Flavia, wife of Massimo Tessarotto, recently passed away. The investigation was developed in part within the research projects: (A) the Albert Einstein Center for Gravitation and Astrophysics, Czech Science Foundation no. 14-37086G; (B) the research projects of the Czech Science Foundation GAČR Grant no. 14-07753P; (C) the Grant no. 02494/2013/RRC "*Kinetický řístup k proudění tekutin*" (Kinetic approach to fluid flow) in the framework of the "Research and Development Support in Moravian-Silesian Region", Czech Republic. Initial framework and motivations of the investigation were based on the research projects developed by the Consortium for Magnetofluid Dynamics (University of Trieste, Italy) and the MIUR (Italian Ministry for Universities and Research) PRIN Research Program "*Problemi Matematici delle Teorie Cinetiche e Applicazioni*" (Mathematical Problems of Kinetic Theories and Applications), University of Trieste, Italy. One of the authors (Massimo Tessarotto) is grateful to the International Center for Theoretical Physics (Miramare, Trieste, Italy) for the hospitality during the preparation of the manuscript.

References

[1] M. Tessarotto, C. Cremaschini, and M. Tessarotto, "On the conditions of validity of the Boltzmann equation and Boltzmann H-theorem," *The European Physical Journal Plus*, vol. 128, no. 3, 2013.

[2] M. Tessarotto and C. Cremaschini, "First-principle proof of the modified collision boundary conditions for the hard-sphere system," *Physics Letters A*, vol. 378, no. 26-27, pp. 1760–1766, 2014.

[3] Tessarotto M. and C. Cremaschini, *The European Physical Journal Plus*, vol. 129, 2014.

[4] M. Tessarotto, C. Cremaschini, M. Mond, C. Asci, A. Soranzo, and G. Tironi, "On the Boltzmann-Grad limit for smooth hard-sphere systems," *Foundations of Physics. An International Journal Devoted to the Conceptual Bases and Fundamental Theories of Modern Physics*, vol. 48, no. 3, pp. 271–294, 2018.

[5] R. Clemente and M. Tessarotto, "On the derivation of the Boltzmann equation from the BBGKY hierarchy," *Transport Theory*

and Statistical Physics. An International Journal for Rapid Communication, vol. 8, no. 1, pp. 1–16, 1979.

[6] C. Cercignani, *Mathematical Methods in Kinetic Theory*, Plenum Press, New York, NY, USA, 1969.

[7] S. G. Brush, *Weitere Studien über das W ärmegleichgewicht unter Gasmolekülen*, Wiener Berichte, 66, 275–370, in WA I, paper 23, 1872.

[8] D. Enskog, "Kungl. Svensk Vetenskps Akademiens," English translation by S. G. Brush, vol. 4, 1921.

[9] S. Chapman and T. G. Cowling, *The Mathematical Theory of Nonuniform Gases*, Cambridge University Press, 1951.

[10] H. Grad, "Thermodynamics of gases," *Handbook der Physik*, vol. 205, 1958.

[11] C. Cercignani, *Theory And Application of The Boltzmann Equation*, Scottish Academic Press, Edinburgh and London, 1975.

[12] C. Cercignani, *The Boltzmann Equation and Its Applications*, vol. 67 of *Applied Mathematical Sciences*, Springer, New York, NY. USA, 1988.

[13] M. Tessarotto, C. Cremaschini, C. Asci, A. Soranzo, and G. Tironi, "Global validity of the Master kinetic equation for hard-sphere systems," *The European Physical Journal Plus*, vol. 130, no. 8, 2015.

[14] Y. G. Sinai, "Dynamical systems with elastic reflections," *Russian Mathematical Surveys*, vol. 25, p. 137, 1970.

[15] Y. G. Sinai, "Dynamical Systems II," in *Ergodic Theory with Applications to Dynamical Systems and Statistical Mechanics*, Springer-Verlag, Berlin, Germany, 1989.

[16] M. Tessarotto and M. Ellero, "Unique representation of an inverse-kinetic theory for incompressible Newtonian fluids," in *Proceedings of the AIP Conference*, vol. 762, p. 108, 2005.

[17] C. Cremaschini and M. Tessarotto, "Theory of spatially non-symmetric kinetic equilibria for collisionless plasmas," *Physics of Plasmas*, vol. 20, no. 1, 2013.

[18] C. Cremaschini, Z. Stuchlík, and M. Tessarotto, "Kinetic theory of quasi-stationary collisionless axisymmetric plasmas in the presence of strong rotation phenomena," *Physics of Plasmas*, vol. 20, no. 5, 2013.

[19] C. Cremaschini, M. Tessarotto, and Z. Stuchlík, "Kinetic equilibria of relativistic collisionless plasmas in the presence of non-stationary electromagnetic fields," *Physics of Plasmas*, vol. 21, no. 3, 2014.

[20] O. E. Lanford, "Time evolution of large classical systems," in *Proceedings of Dynamical Systems, Theory and Applications, 1974 Battelle Rencontre on Dynamical Systems*, J. Moser, Ed., Lecture Notes in Physics, Vol. 38, Springer,Berlin, p. 1, 1975.

[21] O. E. Lanford and Soc., "Société mathématique de France," *Astérisque*, vol. 40, p. 117, 1976.

[22] O. E. Lanford, "The hard sphere gas in the Boltzmann-Grad limit," *Physica A: Statistical Mechanics and its Applications*, vol. 106, no. 1-2, pp. 70–76, 1981.

[23] C. Villani, "Entropy production and convergence to equilibrium for the Boltzmann equation," in *Proceedings of the XIVth International Congress on Mathematical Physics*, pp. 130–144, Lisbon, Portugal, 2006.

[24] J. Uffink and G. Valente, "Lanford's theorem and the emergence of irreversibility," *Foundations of Physics. An International Journal Devoted to the Conceptual Bases and Fundamental Theories of Modern Physics*, vol. 45, no. 4, pp. 404–438, 2015.

[25] V. Ardourel, "Irreversibility in the derivation of the Boltzmann equation," *Foundations of Physics. An International Journal Devoted to the Conceptual Bases and Fundamental Theories of Modern Physics*, vol. 47, no. 4, pp. 471–489, 2017.

[26] A. Drory, "Is there a reversibility paradox? Recentering the debate on the thermodynamic time arrow," *Studies in History and Philosophy of Science. Part B. Studies in History and Philosophy of Modern Physics*, vol. 39, no. 4, pp. 889–913, 2008.

[27] J. Loschmidt, *Akademie der Wissenschaften zu Wien*, vol. 73, p. 128, 1876.

[28] L. Boltzmann, *Wissenschaftliche Abhandlungen*, 275, 1877, in Boltzmann Vol. II, pp. 112–148, 1909.

[29] P. Ehrenfest and T. Ehrenfest, *The Conceptual Foundations of The Statistical Approach in Mechanics*, Cornell University Press, New York , NY, USA, 1912.

[30] C. Cercignani, "H-theorem and trend to equilibrium in the kinetic theory of gases," *Polish Academy of Sciences. Institute of Fundamental Technological Research. Archives of Mechanics (Archiwum Mechaniki Stosowanej)*, vol. 34, no. 3, pp. 231–241 (1983), 1982.

[31] J. L. Lebowitz and Phys. Today, *Physics Today (Letters Section)*, p. 115, 1994.

[32] G. Gallavotti, *Nonequilibrium and irreversibility*, Theoretical and Mathematical Physics, Springer, 2014.

[33] M. Tessarotto, M. Mond, and C. Asci, "MOND as the weak field limit of an extended metric theory of gravity with torsion," *The European Physical Journal Plus*, vol. 213, p. 132, 2017.

12

Autonomous Jerk Oscillator with Cosine Hyperbolic Nonlinearity: Analysis, FPGA Implementation, and Synchronization

Karthikeyan Rajagopal,[1] Sifeu Takougang Kingni,[2] Gaetan Fautso Kuiate,[3] Victor Kamdoum Tamba,[4] and Viet-Thanh Pham[5]

[1] *Center for Nonlinear Dynamics, Defense University, Ethiopia*
[2] *Department of Mechanical, Petroleum and Gas Engineering, Faculty of Mines and Petroleum Industries, University of Maroua, P.O. Box 46, Maroua, Cameroon*
[3] *Department of Physics, Higher Teacher Training College, The University of Bamenda, P.O. Box 39 Bamenda, Cameroon*
[4] *Laboratory of Electronics and Signal Processing (LETS), Department of Physics, Faculty of Science, University of Dschang, P. O. Box 67, Dschang, Cameroon*
[5] *School of Electronics and Telecommunications, Hanoi University of Science and Technology, 01 Dai Co Viet, Hanoi, Vietnam*

Correspondence should be addressed to Gaetan Fautso Kuiate; fautso_kuiate@yahoo.com

Academic Editor: Dimitrios J. Frantzeskakis

A two-parameter autonomous jerk oscillator with a cosine hyperbolic nonlinearity is proposed in this paper. Firstly, the stability of equilibrium points of proposed autonomous jerk oscillator is investigated by analyzing the characteristic equation and the existence of Hopf bifurcation is verified using one of the two parameters as a bifurcation parameter. By tuning its two parameters, various dynamical behaviors are found in the proposed autonomous jerk oscillator including periodic attractor, one-scroll chaotic attractor, and coexistence between chaotic and periodic attractors. The proposed autonomous jerk oscillator has period-doubling route to chaos with the variation of one of its parameters and reverse period-doubling route to chaos with the variation of its other parameter. The proposed autonomous jerk oscillator is modelled on Field Programmable Gate Array (FPGA) and the FPGA chip statistics and phase portraits are derived. The chaotic and coexistence of attractors generated in the proposed autonomous jerk oscillator are confirmed by FPGA implementation of the proposed autonomous jerk oscillator. A good qualitative agreement is illustrated between the numerical and FPGA results. Finally synchronization of unidirectional coupled identical proposed autonomous jerk oscillators is achieved using adaptive sliding mode control method.

1. Introduction

There has been a significant increase in studies of chaotic oscillators over the past decades because of their complex behaviors and their promising applications [1–9]. There are increasing works on chaos in jerk oscillators [10–12]. It is worth noting that if the scalar $x(t)$ is denoted as a physical variable at time t, the third derivative $\dddot{x}(t)$ presents the jerk [13]. Different jerk oscillators with chaos were summarized by Sprott [14]. By using Josephson junctions, Yalcin constructed a general jerk circuit with multiscroll and hypercube attractors [15]. Multiscroll attractors in jerk circuits were presented by Ma et al. [16]. An elegant chaotic oscillators based on jerk equation was implemented in a circuit where the nonlinearity was provided by a single diode [17]. A simple chaotic jerk circuit was used in a sound encryption scheme [18]. Especially, a three-dimensional novel jerk chaotic oscillator with two hyperbolic sinusoidal nonlinearities was reported in [19].

Researchers have shown an increased interest in multistability [20–26]. Multistability is associated with the presence

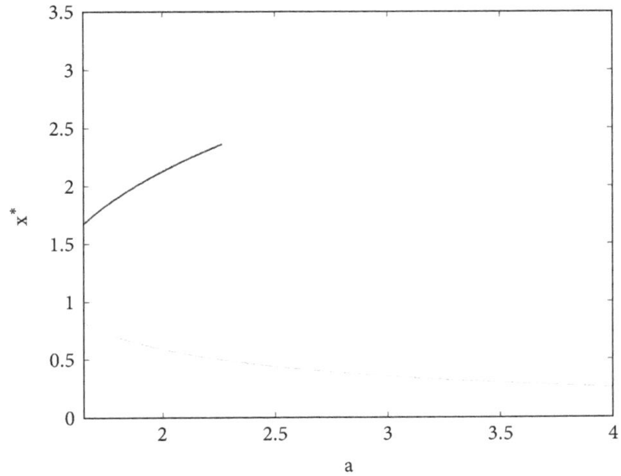

FIGURE 1: (Color online) Stability diagram of equilibrium points $E_{1,2}$ versus the parameter a for $b = 3$. Solid black lines indicate the stable branches and the dashed red lines the unstable branches.

of coexisting attractors with different initial conditions [20]. Jerk oscillators can exhibit multistability despite their simplicity [27–31]. Recently, Njitacke et al. discovered the coexistence of multiple attractors and crisis route to chaos in a chaotic jerk circuit [27] and a simple autonomous jerk oscillator with multiple attractors was introduced in [28]. Kengne et al. performed an analysis of a jerk circuit including a pair of semiconductor diodes connected in antiparallel and investigated its multiple attractors [29]. It is interesting that, by applying the diode bridge memristor into the original jerk circuit, authors proposed a novel hybrid diode based jerk circuit [30]. This work revealed that the hybrid diode based jerk circuit exhibits rich dynamical behaviors including multiple coexisting self-excited attractors. Derived from the autonomous jerk circuit, an autonomous memristor-based jerk circuit was constructed [31]. Interestingly, for the same values of system parameters, the coexistence of four different attractors was obtained in such a memristor-based jerk circuit.

Motived by published results related to jerk oscillators, some questions arose as to know, e.g., if a jerk oscillator with cosine hyperbolic nonlinearity can exhibit multistability or how such oscillator can synchronize. The aim of our work is to explore aspects of the unanswered questions. Our paper is structured as follows. Section 2 is devoted to the theoretical analysis of proposed autonomous jerk oscillator with a cosine hyperbolic term. FPGA implementation of proposed autonomous jerk oscillator is presented in Section 3. In Section 4, synchronization of unidirectional coupled jerk oscillators is studied by applying adaptive sliding mode control method. Finally, Section 5 concludes our work.

2. Theoretical Analysis of Proposed Autonomous Jerk Oscillator

In chapter three of "Elegant Chaos: Algebraically Simple flow" book published in 2010, Sprott proposed a list of sixteen autonomous chaotic jerk oscillators with different nonlinearities called memory oscillators [14]. The nonlinearities of memory oscillators include quadratic, cubic, quintic, absolute, maximum, sign, exponential, sine, and tangent hyperbolic functions. Inspired by [14], in this work we introduce an autonomous jerk oscillator with a cosine hyperbolic nonlinearity described by

$$\frac{dx}{dt} = z, \tag{1a}$$

$$\frac{dy}{dt} = ax - y - bz - \cosh(x), \tag{1b}$$

$$\frac{dz}{dt} = y, \tag{1c}$$

where x, y, z are state variables of the oscillator and a, b two positive parameters. System (1a), (1b), and (1c) can be converted to a jerk oscillator, as follows:

$$\frac{d^3x}{dt^3} = -\frac{d^2x}{dt^2} - b\frac{dx}{dt} + ax - \cosh(x). \tag{2}$$

System (1a), (1b), and (1c) is dissipative because $\nabla V = \partial(dx/dt)/\partial x + \partial(dy/dt)/\partial y + \partial(dz/dt)/\partial z = -1$. The equilibrium points of system (1a), (1b), and (1c) are obtained by solving $dx/dt = 0$, $dy/dt = 0$, $dz/dt = 0$, which gives

$$y^* = z^* = 0, \tag{3a}$$

$$ax^* - \cosh(x^*) = 0. \tag{3b}$$

Equation (3b) cannot be solved analytically. We thus use the Newton-Raphson method to find the value of x^*. Depending on the value of parameter a, (3b) presents no roots for $0 \le a < 1.509$ and one root (x^*) for $a \ge 1.509$. Therefore, for $0 \le a < 1.509$, system (1a), (1b), and (1c) has no equilibrium points while, for $a \ge 1.509$, it has two equilibrium points $E_1 = (x_1^*, 0, 0)$ and $E_2 = (x_2^*, 0, 0)$ with $x_1^* > x_2^*$ (see Figure 1). The characteristic equation evaluated at the equilibrium point $E = (x^*, 0, 0)$ is

$$\lambda^3 + \lambda^2 + b\lambda - a + \sinh(x^*) = 0. \tag{4}$$

According to the Routh-Hurwitz criterion, the real parts of all the roots λ of (4) are negative if and only if

$$-a + \sinh(x^*) > 0 \tag{4b}$$

$$a + b - \sinh(x^*) > 0 \tag{4c}$$

The stability analysis of equilibrium point E_1 as function of the parameter a of system (1a), (1b), and (1c) is depicted in Figure 1.

From Figure 1, one can notice that equilibrium point E_2 is always unstable for $1.65 \le a \le 4$, whereas the equilibrium point E_1 is stable for $1.65 \le a \le 2.269$ and unstable for $a > 2.269$. Since the equilibrium points E_1 is stable for $1.65 \le a \le 2.269$ but it changes stability properties at $a \approx 2.269$, system (1a), (1b), and (1c) can exhibit a Hopf bifurcation from equilibrium point E_1 when the parameter a varies.

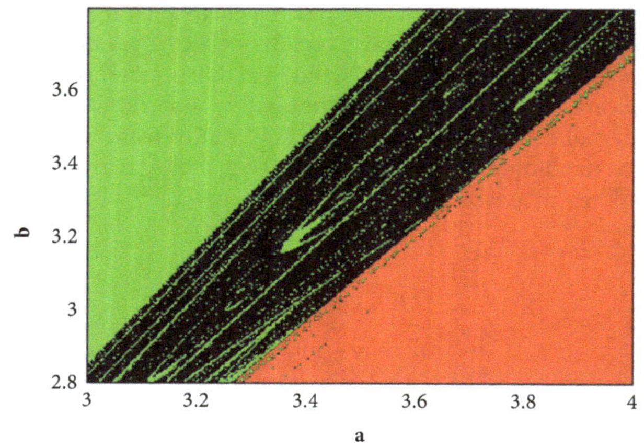

FIGURE 2: (Color online) Two-parameter (a, b) bifurcation diagram depicting the dynamical behaviors of system (1a), (1b), and (1c). Unbounded orbits are in red; periodic oscillations are in green and chaos is in black.

Theorem 1. *System (1a), (1b), and (1c) undergoes a Hopf bifurcation at the equilibrium point E_1 if $b > 0$ and the parameter a passes through the critical value $a_H = -b + \sinh(x_1^*)$.*

Proof. Substituting $\lambda = i\omega(\omega > 0)$ into (4) and separating real and imaginary parts, we obtain

$$\omega = \omega_0 = \sqrt{b}, \tag{5a}$$

$$a = a_H = -b + \sinh\left(x_1^*\right). \tag{5b}$$

Differentiating both sides of (4) with respect to a, we can obtain

$$3\lambda^2 \frac{d\lambda}{da} + 2\lambda \frac{d\lambda}{da} + b\frac{d\lambda}{da} - 1 = 0 \tag{6a}$$

and

$$\frac{d\lambda}{da} = \frac{1}{3\lambda^2 + 2\lambda + b} \tag{6b}$$

then

$$\operatorname{Re}\left(\left.\frac{d\lambda}{da}\right|_{a=a_H,\lambda=i\omega_0}\right) = \frac{-1}{2(b+1)} \neq 0. \tag{7}$$

All the conditions for Hopf bifurcation to occur are met; therefore system (1a), (1b), and (1c) undergoes a Hopf bifurcation at E_1 when $a_H = -b + \sinh(x_1^*)$ and periodic solutions will exist in a neighbourhood of the point a_H (provided that $b > 0$ holds). If $b = 3$, the critical value is $a = a_H \approx 2.269$. For $a = 2.2 < a_H$, the trajectories of jerk system (1a), (1b), and (1c) converge to the equilibrium point E_1 whereas, for $a = 2.5 > a_H$, system (1a), (1b), and (1c) exhibits a limit cycle (not shown).

We plot in Figure 2 a two-parameter (a, b) bifurcation diagram depicting the dynamical behaviors of system (1a), (1b), and (1c).

Figure 2 indicated that system (1a), (1b), and (1c) can display unbounded orbits and periodic and chaotic behaviors. In Figure 3, we firstly fix $b = 3$ and plot the bifurcation diagram with respect to a and the related largest Lyapunov exponent.

The bifurcation diagram of the output $x(t)$ in Figure 2(a) shows that the trajectories of system (1a), (1b), and (1c) converge to the equilibrium point E_1 up to $a \approx 2.269$ where a Hopf bifurcation occurs followed by a limit cycle motion (i.e., period-1 oscillations) and period-doubling to chaos interspersed with periodic windows. The largest Lyapunov exponent shown in Figure 3(b) confirms the chaotic behavior found in Figure 3(a). The chaotic behavior is illustrated in Figure 4 for a specific value of parameter a.

The trajectories of chaotic attractor are swirling around one of the two equilibrium points in Figure 4. This is a signature of one-scroll chaotic attractor.

Secondly, for $a = 3.3$, sample results showing the bifurcation diagram versus b and the related plot of Largest Lyapunov exponent are provided in Figure 5.

By varying parameter b from 2.87 to 3.8, the bifurcation diagram of the output $x(t)$ in Figure 5(a) displays chaotic behavior interspersed with periodic windows. For $b > 3.18$, period-12-oscillations, period-6-oscillations, and period-3-oscillations are found, respectively, up to $b < 3.2012$ followed by reverse period-doubling to chaos interspersed with periodic windows. By varying parameter b from 3.8 to 2.87 [see red dots of Figure 5(a)], the output $x(t)$ displays the same dynamical behaviors as shown by black dots of Figure 5(a) in the ranges $2.87 \le b \le 3.182$ and $3.2012 \le b \le 3.8$, while, in the range $3.182 < b < 3.2012$, the output $x(t)$ presents chaotic behavior. By comparing black dots of Figure 5(a) and red dots of Figure 5(a), one can notice that system (1a), (1b), and (1c) displays coexistence of period-6-oscillations and chaotic attractors in the range $3.182 < b < 3.188$ and coexistence of period-3-oscillations and chaotic attractors in the range $3.188 \le b < 3.2012$. The largest Lyapunov exponent shown in Figure 5(b) confirms the chaotic behavior found in Figure 5(a). Figure 6 depicts the phase portraits of coexisting attractors found in Figure 5(a) for specific values of parameter b.

For $b = 3.185$, system (1a), (1b), and (1c) can exhibit either one-scroll chaotic attractor or period-6-oscillations depending on initial conditions as shown in Figure 6(a), while, in Figure 6(b) for $b = 3.189$, system (1a), (1b), and (1c) can exhibit either one-scroll chaotic attractor or period-3-oscillations depending on initial conditions. Figure 7 presents the basin of attraction of system (1a), (1b), and (1c) in the plane $y = 0$ for $a = 3.3$ and $b = 3.189$.

In Figure 7, red, green, and black regions contain initial conditions that lead to unbounded orbits, periodic and chaotic attractors, respectively. One can see from the basin of attraction that the possibility of occurrence of unbounded orbits is greater than the ones of periodic and chaotic attractors. □

3. FPGA Implementation of Proposed Autonomous Jerk Oscillator

There have been many literatures about implementation of chaotic oscillators using FPGA like FPGA based multiscroll attractor discussed in [32–39], digital chaotic oscillators and

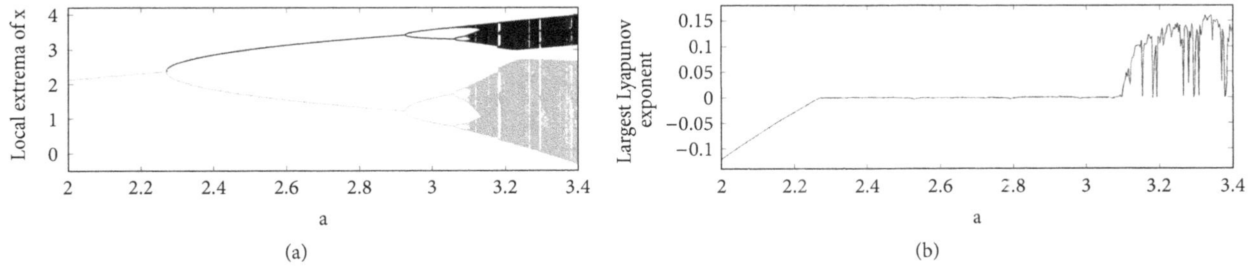

FIGURE 3: The bifurcation diagrams depicting the local maxima (black dots) and local minima (gray dots) of $x(t)$ (a) and the largest Lyapunov exponents (b) versus the parameter a for $b = 3$.

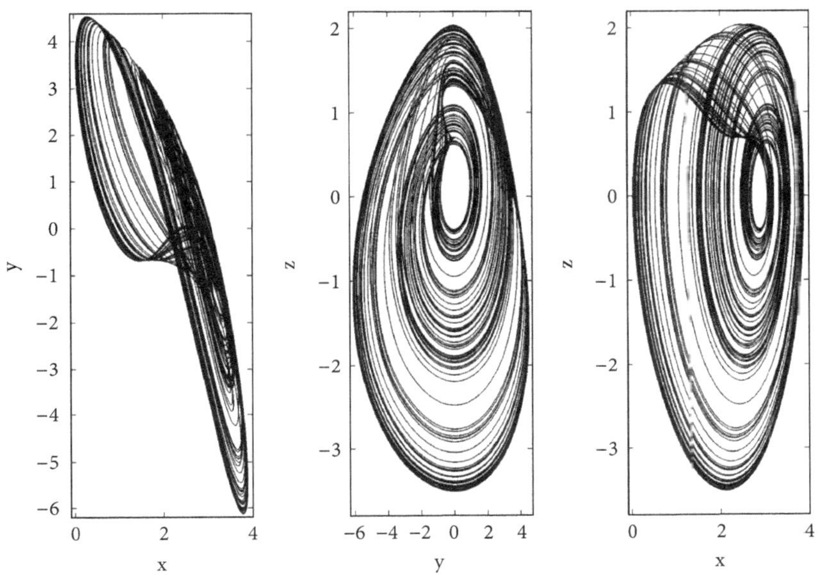

FIGURE 4: 2D phase portrait of system (1a), (1b), and (1c) for $a = 3.3$ and $b = 3$. The initial conditions are (x(0), y(0), z(0)) = (1, 0, 0.5).

its FPGA model [**36, 40**], FPGA based four-wing chaotic attractor [**37**], and image cryptography with chaotic ciphers and its FPGA implementations and FPGA implemented memristor-based chaotic attractors [**41**]. Hyperchaotic oscillators with complex mathematical models are also implemented using FPGA and show that the FPGA performs better in power efficiency [**42, 43**]. More complex synchronization algorithms are also implemented in FPGA such as synchronization of Ikeda oscillators [**44**], genetically optimized PID and sliding mode control based synchronizations [**45, 46**], and adaptive sliding mode synchronization [**47**]. These FPGA realizations are mainly based on hardware-software cosimulations [**45–47**] and with experimental analysis [**44**]. FPGA are also used in implementing various chaos-based encryption schemes like chaos-based pseudorandom number generator [**48**]. In this section we propose to implement the proposed autonomous jerk oscillator using FPGA. The major task in this implementation is to decide the type of numerical method to solve the dynamical system. We choose the forward Euler method [**40**] to solve the proposed autonomous jerk oscillator and the set of discretized system equations is given as follows:

$$x_{k+1} = x_k + h\left[z_{k-1}\right], \tag{8a}$$

$$y_{k+1} = y_k + h\left[ax_{k-1} - y_{k-1} - bz_{k-1} - \cosh\left(x_{k-1}\right)\right], \tag{8b}$$

$$z_{k+1} = z_k + h\left[y_{k-1}\right], \tag{8c}$$

where a, b are the system parameters and h is the step size for the discrete numerical solution. To implement the hyperbolic cosine function, we use the Taylor series expansion with the first three polynomials as

$$\cosh(x) = 1 + \frac{x^2}{2!} + \frac{x^4}{4!} + \frac{x^6}{6!}. \tag{9}$$

To achieve the accuracy we select the step size $h = 0.001$ and as per the IEEE754 standards, we use the calculation of 32-bit finite integer instead of floating-point operation. By using the arithmetic operations multiplying, adding or subtracting

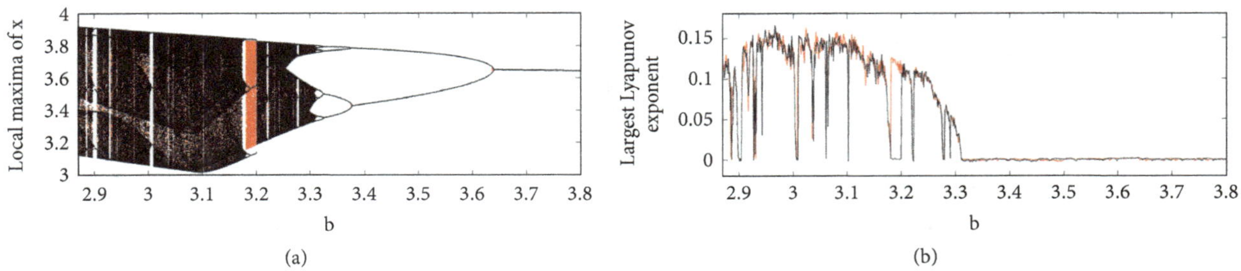

FIGURE 5: (Color online) The bifurcation diagrams depicting the local maxima of $x(t)$ (a) and the largest Lyapunov exponents (b) versus the parameter b for $a = 3.3$. In the graph (a), two sets of data corresponding, respectively, to increasing (black dot) and decreasing (red dot) values of b are superimposed. Largest Lyapunov exponents are obtained by scanning parameter b upwards (black line) and downwards (red line).

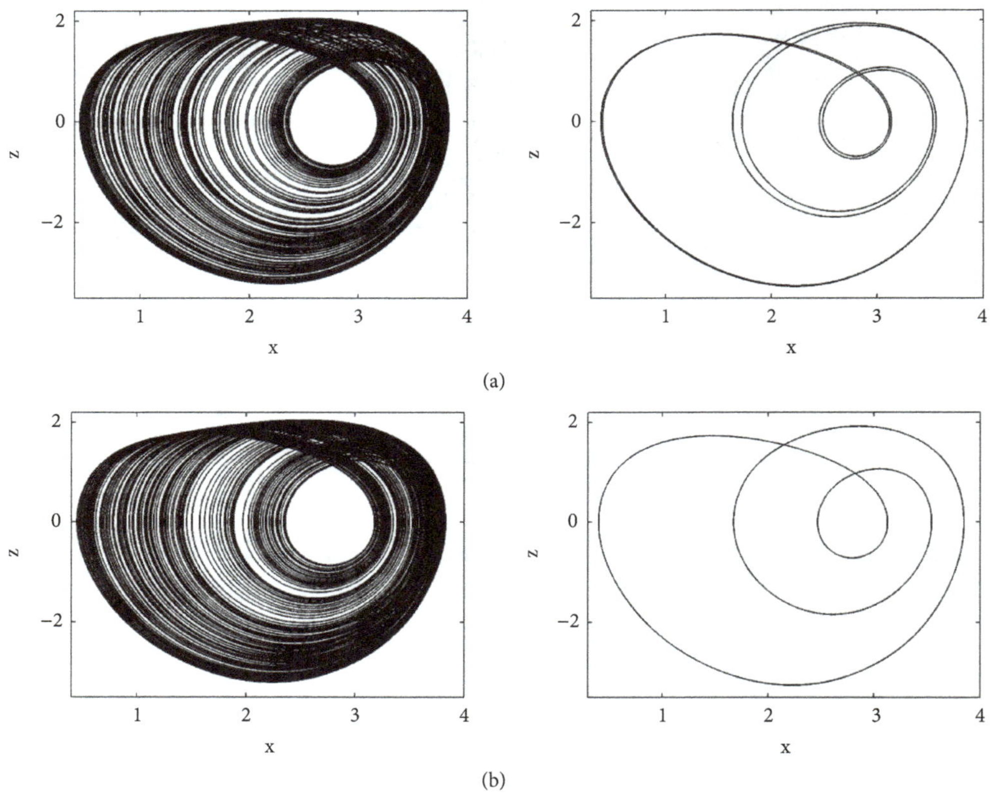

FIGURE 6: Phase portraits in the plane (x, z) for specific values of parameter b: (a) $b = 3.185$ and (b) $b = 3.189$. The initial conditions are (x(0), y(0), z(0)) = (1.0, 0, 0.5) in left column and (x(0), y(0), z(0)) = (3.0, 0, 0.5) in right column. The other parameter value is $a = 3.3$.

the system (8a), (8b), and (8c) is implemented and the initial conditions for the next iteration can be obtained. By repeating this process of iteration we can derive the discrete chaotic system. The absolute value operation only needs to set the first bit of the 32-bit integer to 0. The discrete state equations are implemented using hardware and software cosimulations and the needed basic arithmetic operators are implemented using the Xilinx system generator toolkit. We used the Kintex 7 chipset (xc7k160tfbg)) for the cosimulations and Matlab Simulink is used to plot the phase plots. The latency of the arithmetic blocks is kept at 3 and the maximum clock frequency of the FPGA used is 437 MHz. The register-transfer level (RTL) schematic of hyperbolic cosine function given by (9) is not shown while the RTL schematic of proposed autonomous jerk oscillator is shown in Figure 8.

Table 1 shows the resources utilized by the proposed autonomous jerk oscillator for implementing in Kintex-7.

Figure 9 shows the 2D phase portraits of the proposed autonomous jerk oscillator.

Figure 10 shows the coexisting attractors exhibited by the FPGA implemented proposed autonomous jerk oscillator for two different initial conditions.

One-scroll chaotic attractor and coexistence of attractors are clearly seen from Figures 9 and 10, respectively. By

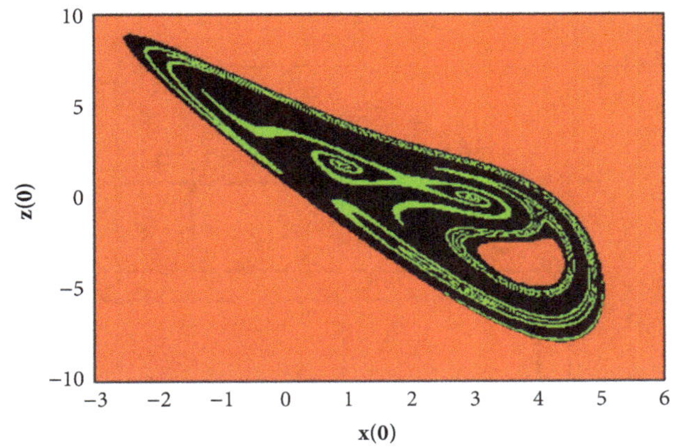

FIGURE 7: (Color online) Cross section of the basin of attraction of system (1a), (1b), and (1c) in the x-z-plane at $y = 0$ for $a = 3.3$ and $b = 3.189$. Initial conditions in the black region lead to unbounded orbits, those in the green region lead to the periodic attractor and those in the red region lead to the strange attractor.

TABLE 1: Resources consumed by the FPGA implemented of proposed autonomous jerk oscillator.

Resource	Utilization	Available	Utilization %
LUT	880	101400	0.87
FF	522	202800	0.26
DSP	20	600	3.33
IO	97	285	34.04
BUFG	1	32	3.13

comparing Figures 9 and 10 with Figures 9 and 10, it can be noticed that there is a good qualitative agreement between the numerical simulations and FPGA results.

4. Synchronization of Unidirectional Coupled Proposed Autonomous Jerk Oscillators Using Adaptive Sliding Mode Control Method

Here we present synchronization results of coupled chaotic proposed jerk oscillators in order to promote chaos-based synchronization designs of this type of proposed jerk oscillator. Synchronization of chaotic systems has applications in secure communication and cryptography. The highly sensitive nature of chaotic oscillators to initial conditions makes it difficult to synchronize the oscillators with uncertainties and disturbance. Some well-known ways to synchronize chaotic oscillators are using active control method **[49, 50]** adaptive control method **[51, 52]**, extended back stepping control method **[53, 54]**, sliding mode control method **[55, 56]**, adaptive sliding mode control method **[56–59]**, etc. Sliding mode control method is applied to provide robustness in the face of internal and external disturbances. In section, we use adaptive sliding mode control method for synchronization of unidirectional coupled identical proposed autonomous jerk oscillators using adaptive sliding mode control method.

4.1. Problem Statement. Let the generalized master system be defined as

$$\dot{x} = f(x) + F(x)c \tag{10}$$

and the slave system as

$$\dot{y} = g(y) + G(y)d + u(t) \tag{11}$$

where $f(x), g(y)$ are $n \times 1$ row vector and $F(x), G(y)$ are $m \times n$ matrix elements of the master and slave systems, respectively, c, d are the unknown parameters of the systems, and $u(t)$ is the controller to synchronize the systems. The control objective is to synchronize the slave system with initial condition $y(0)$ with master system of initial condition $x(0)$ such that the synchronization errors (12) approaches zero.

$$\lim_{t\to\infty} e_i = y_i - x_i. \tag{12}$$

The sliding mode controller design for synchronizing the two systems involves selection of sliding surface for the desired dynamics and designing the reaching law such that any point on the phase space is brought to the sliding surface in the presence of uncertainties.

Let us define the proportional integral sliding surface **[60]** as

$$s_i = e_i + K\int e_i(\tau)\,d\tau \tag{13}$$

where K is the proportional constant vector $[k_1, k_2, k_3]$.

The first derivative of the sliding surface is derived as

$$\dot{s}_i = \dot{e}_i + Ke_i. \tag{14}$$

For the existence of the sliding mode, it is necessary and sufficient that the sliding surface and its first derivative should be equal to zero. The error dynamics can be derived as

$$\dot{e} = g(x_s) + G(x_s)d + u(t) - f(x_m) - F(x_m)c. \tag{15}$$

In order to avoid the chattering phenomenon caused by discontinuous control signals, the adaptive controller used to synchronize the master and slave system is chosen as

$$u(t) = -g(x_s) - G(x_s)\hat{d} + f(x_m) - F(x_m)\hat{c} - ke - \eta\tanh(s) - \rho s \tag{16}$$

where k, η, ρ are positive gain values, $\hat{c}, \hat{d}$ are parameter estimates of master and slave systems, and s is the sliding surface.

Using (16) in (15), the error dynamics simplifies to

$$\dot{e} = G(x_s)\left[d - \hat{d}\right] - F(x_m)[c - \hat{c}] - \eta\tanh(s) - \rho s. \tag{17}$$

The stability of the proposed controller can be analyzed using the Lyapunov candidate function:

$$V = \frac{1}{2}s^2 + \frac{1}{2}\left(d - \hat{d}\right)^2 + \frac{1}{2}(c - \hat{c})^2. \tag{18}$$

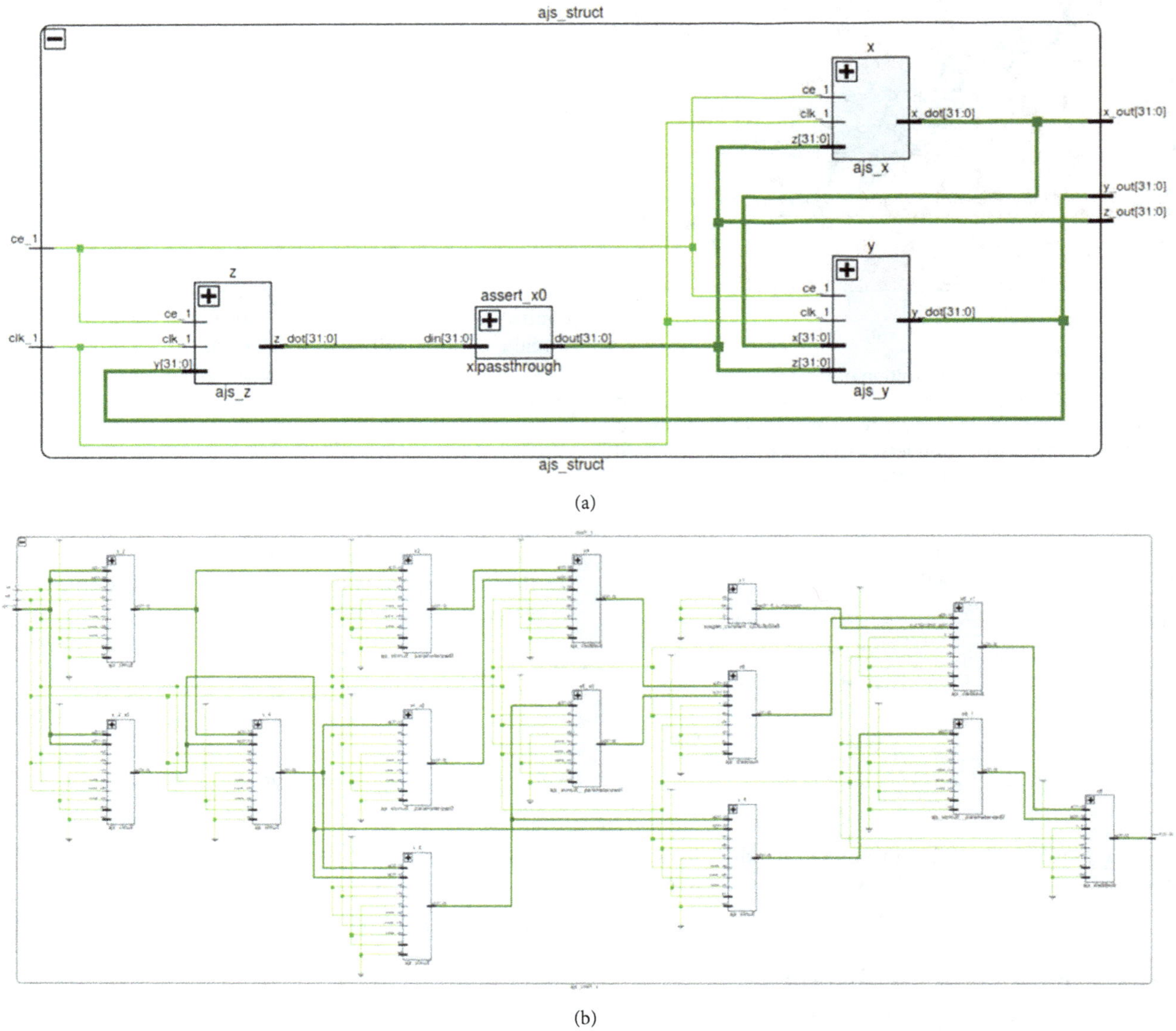

FIGURE 8: (Color online) RTL schematic of proposed autonomous jerk oscillator implemented in Kintex 7.

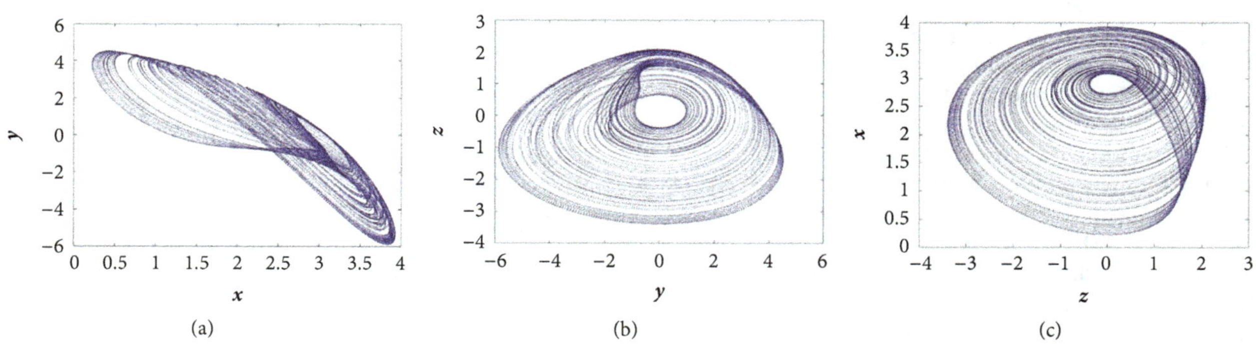

FIGURE 9: (Color online) Phase portraits of the FPGA implemented of proposed autonomous jerk oscillator with hardware-software co-simulation for parameter values $a = 3.3$, $b = 3$ and initial conditions $[1, 0, 0.5]$ with $h = 0.001$.

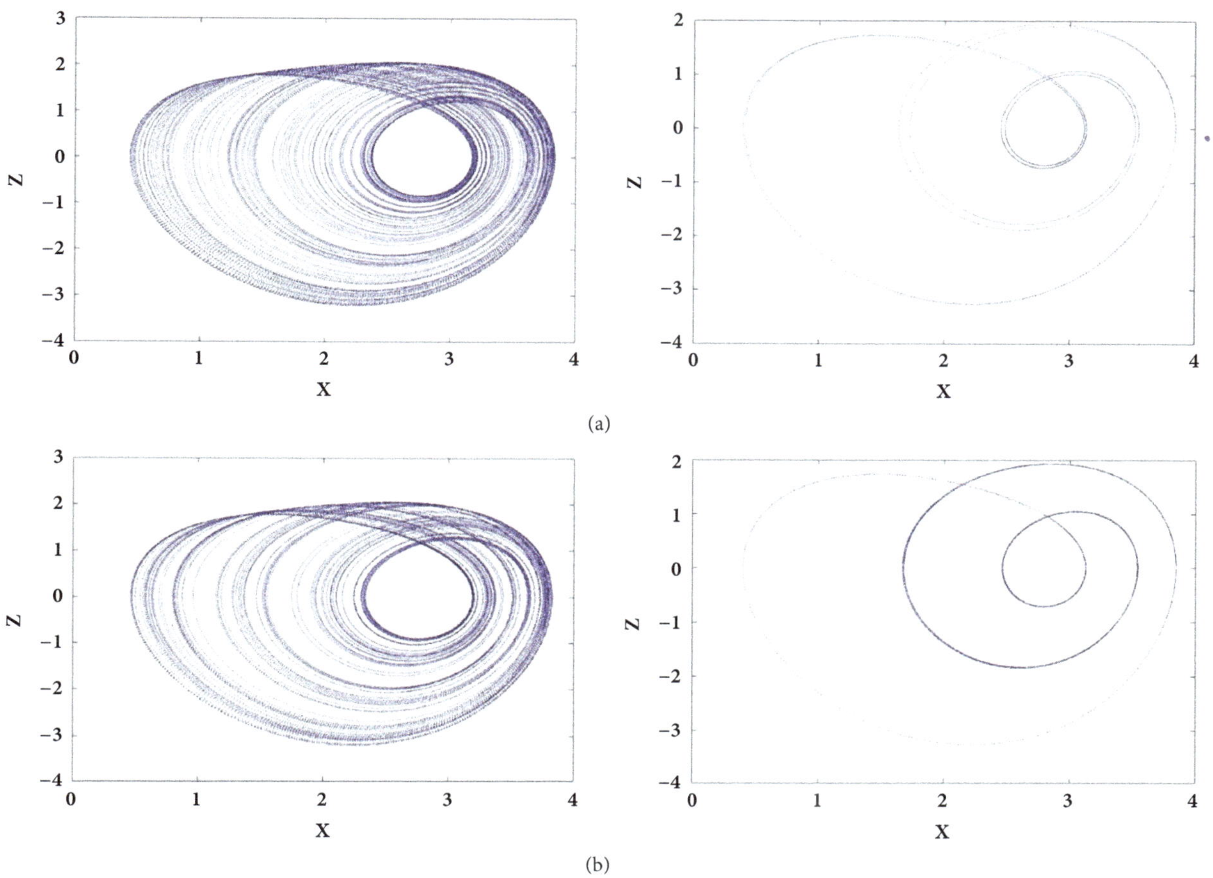

FIGURE 10: (Color online) Phase portraits in the plane (x, z) of the FPGA implemented proposed autonomous jerk oscillator for specific values of parameter b: (a) $b = 3.185$ and (b) $b = 3.189$ for the initial conditions (x(0), y(0), z(0)) = (1.0, 0, 0.5) in left column and (x(0), y(0), z(0)) = (3.0, 0, 0.5) in right column with the value of the parameter $a = 3.3$. The step size is taken as $h = 0.001$.

The dynamics of the Lyapunov candidate function can be derived as follows:

$$\dot{V} = s.\dot{s} + \left(d - \widehat{d}\right)\left(-\dot{\widehat{d}}\right) + \left(c - \widehat{c}\right)\left(-\dot{\widehat{c}}\right) \tag{19}$$

Using (17), (16), and (14) in (19),

$$\dot{V} \leq ks\left[G\left(x_s\right)\left(d - \widehat{d}\right) - F\left(x_m\right)\left(c - \widehat{c}\right) - \eta \tanh\left(s\right) - \rho s\right] - \left(d - \widehat{d}\right)\left(\dot{\widehat{d}}\right) - \left(c - \widehat{c}\right)\left(\dot{\widehat{c}}\right). \tag{20}$$

Let us define the parameter estimate laws as

$$\dot{\widehat{c}} = K_c s.G\left(x_s\right), \tag{21a}$$

$$\dot{\widehat{d}} = K_d s.F\left(x_m\right), \tag{21b}$$

where K_c, K_d are positive constants.

Using (21a) and (21b) in (20), we can solve the Lyapunov function dynamics as follows:

$$\dot{V} \leq -\eta\left|s_w\right| - \rho s^2, \tag{22}$$

where η and ρ are all positive, and $\dot{V}$ is negative definite.

4.2. Numerical Verifications. For numerical validation of the proposed synchronization method, we use the proposed autonomous jerk oscillators as master and slave and apply the adaptive sliding mode control to achieve the synchronization. Let us define the master system as

$$\dot{x}_m = z_m, \tag{23a}$$

$$\dot{y}_m = ax_m - y_m - bz_m - \cosh\left(x_m\right), \tag{23b}$$

$$\dot{z}_m = y_m, \tag{23c}$$

where a, b are the system parameters. The slave system with the adaptive sliding mode controllers (u_i) are defined as

$$\dot{x}_s = z_s + u_x, \tag{24a}$$

$$\dot{y}_s = ax_s - y_s - bz_s - \cosh\left(x_s\right) + u_y, \tag{24b}$$

$$\dot{z}_s = y_s + u_z. \tag{24c}$$

The parameters of the slave system are assumed to be unknown with parameter estimates $\widehat{a}, \widehat{b}$. Using the master

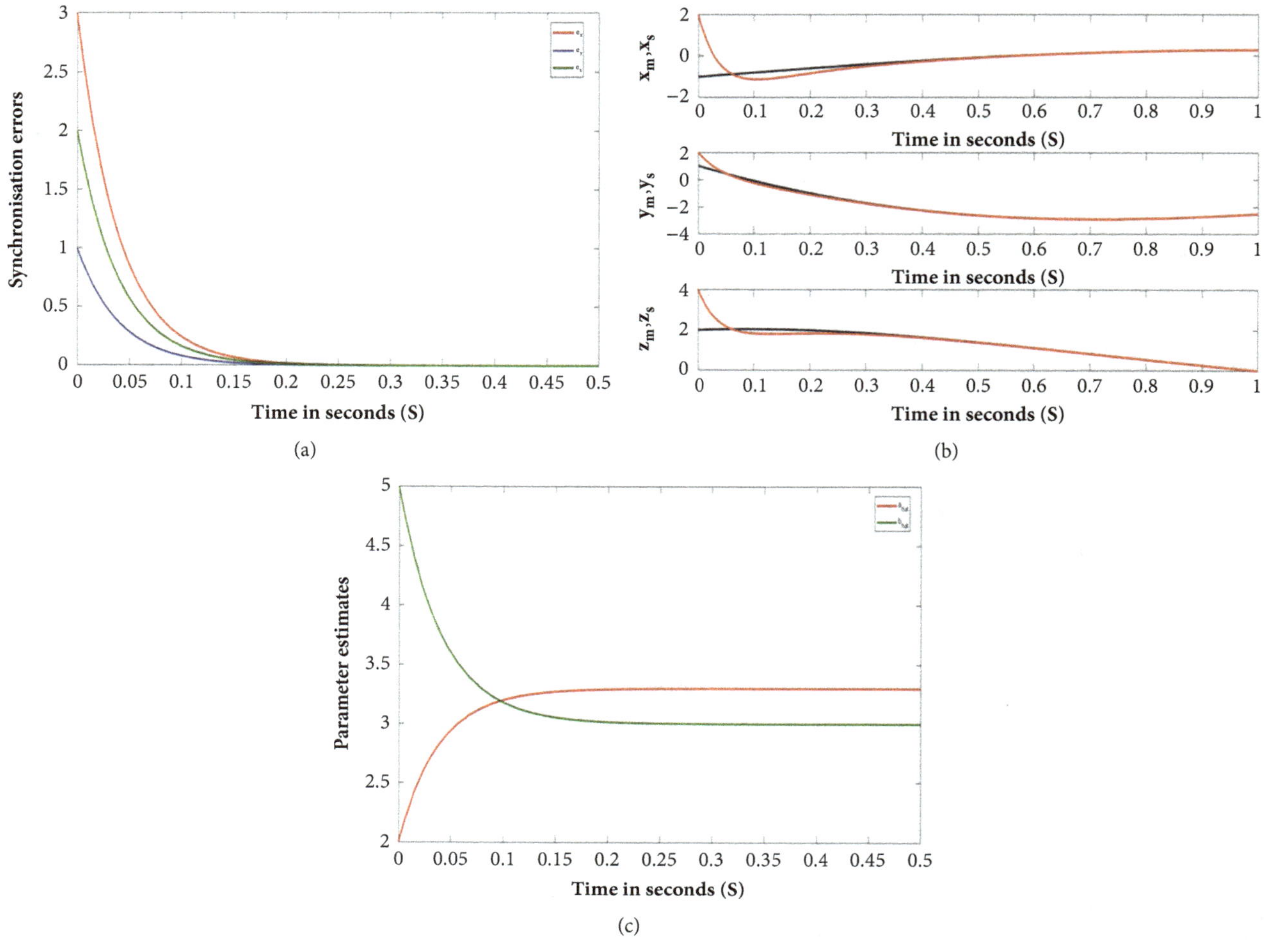

FIGURE 11: Time history of synchronization error variables (a), synchronized states (b), and estimated parameters (c).

system (23a), (23b), and (23c) and slave system (24a), (24b), and (24c) in the error dynamics can be derived as follows:

$$\dot{e}_x = e_z + u_x, \tag{25a}$$

$$\dot{e}_y = \hat{a}x_s - \hat{b}z_s - e_y - \cosh(x_s) - ax_m + bz_m + \cosh(x_m) + u_y, \tag{25b}$$

$$\dot{e}_z = e_y + u_z. \tag{25c}$$

Let the adaptive sliding mode controllers be chosen as follows:

$$u_x = -e_z - \gamma_x \tanh(s_x) - \rho_x s_x - k_x e_x, \tag{26a}$$

$$u_y = -\hat{a}e_x + \hat{b}e_z + e_y + \cosh(x_s) - \cosh(x_m) - \gamma_y \tanh(s_y) - \rho_y s_y - k_y e_y, \tag{26b}$$

$$u_z = -e_y - \gamma_z \tanh(s_z) - \rho_z s_z - k_z e_z, \tag{26c}$$

where $\rho_i > 0$, $\gamma_i > 0$ are the sliding surface gains and k_i is the controller gain for $i = x, y, z$

Using (17) and (18) with (13), the parameter update laws can be defined as follows:

$$\dot{\hat{a}} = -s_y x_m, \tag{27a}$$

$$\dot{\hat{b}} = s_y z_m, \tag{27b}$$

where $\dot{\hat{a}}, \dot{\hat{b}}$ are the dynamics of the parameter estimates $\hat{a}, \hat{b}$. For numerical simulations, we take the initial conditions of the master system (25a), (25b), and (25c) as $[-1, 1, 2]$, slave system as $[2, 2, 4]$, parameter estimates as $\hat{a}(0) = 2, \hat{b}(0) = 5$, and the proportional constants $K = [1, 1, 1]$. Figure 11 shows the dynamics of the synchronization errors and synchronized states of the master and slave systems and estimated parameters.

When t>0.3, the synchronization error variables converge to zero with exponentially asymptotical speed [see Figure 11(a)] and thereby guaranteeing the synchronization between the master system (23a), (23b), and (23c) and slave system (24a), (24b), and (24c) [see Figure 11(b)]. The unknown parameters of slave system (24a), (24b), and (24c) are simultaneously successfully estimated to their values [see Figure 11(c)].

5. Conclusion

In this paper, an autonomous jerk oscillator with a cosine hyperbolic nonlinearity and two parameters was proposed and studied. It was demonstrated that the Hopf bifurcation occurs near the equilibrium point as one of jerk oscillator parameter crosses the critical value. The proposed autonomous jerk oscillator exhibits periodic attractors, one-scroll chaotic attractors,, and coexistence between chaotic and periodic attractors. Period-doubling route to chaos and reverse period-doubling route to chaos was found in proposed autonomous jerk oscillator with the variation of each of its two parameters. Further, the proposed autonomous jerk oscillator was implemented using Field Programmable Gate Array in order to show that the proposed autonomous jerk oscillator is hardware realizable. Finally, synchronization of unidirectional coupled identical proposed autonomous jerk oscillators was achieved using adaptive sliding mode control method.

Conflicts of Interest

The authors declare that there are no conflicts of interest regarding the publication of this paper.

References

[1] E. N. Lorenz, "Deterministic nonperiodic flow," *Journal of the Atmospheric Sciences*, vol. 20, no. 2, pp. 130–141, 1963.

[2] A. Tamaševičius, A. Čenys, G. Mykolaitis, A. Namajunas, and E. Lindberg, "Hyperchaotic oscillator with gyrators," *IEEE Electronics Letters*, vol. 33, no. 7, pp. 542–544, 1997.

[3] M. Lakshmanan and D. V. Senthilkumar, *Dynamics of Nonlinear Time-Delay Systems*, Springer Series in Synergetics, Springer, Berlin, Germany, 2010.

[4] A. Buscarino, L. Fortuna, M. Frasca, and L. V. Gambuzza, "A chaotic circuit based on Hewlett-Packard memristor," *Chaos: An Interdisciplinary Journal of Nonlinear Science*, vol. 22, no. 2, Article ID 023136, 2012.

[5] C.-L. Li, H.-M. Li, W. Li et al., "Dynamics, implementation and stability of a chaotic system with coexistence of hyperbolic and non-hyperbolic equilibria," *AEÜ - International Journal of Electronics and Communications*, vol. 84, pp. 199–205, 2018.

[6] D. Lathrop, *Physics Today*, vol. 68, no. 4, pp. 54-55, 2015.

[7] B. Muthuswamy and S. Banerjee, "Designing Hardware for FPGAs," in *A Route to Chaos Using FPGAs*, vol. 16 of *Emergence, Complexity and Computation*, pp. 29–53, Springer International Publishing, Cham, 2015.

[8] V.-T. Pham, S. Jafari, C. Volos, T. Gotthans, X. Wang, and D. V. Hoang, "A chaotic system with rounded square equilibrium and with no-equilibrium," *Optik - International Journal for Light and Electron Optics*, vol. 130, pp. 365–371, 2017.

[9] F. Nazarimehr, K. Rajagopal, J. Kengne, S. Jafari, and V.-T. Pham, "A new four-dimensional system containing chaotic or hyperchaotic attractors with no equilibrium, a line of equilibria and unstable equilibria," *Chaos, Solitons & Fractals*, vol. 111, pp. 108–118, 2018.

[10] J. C. Sprott, "Some simple jerk functions," *American Journal of Physics*, vol. 65, no. 6, pp. 537–543, 1997.

[11] R. Eichhorn, S. J. Linz, and P. Hänggi, "Simple polynomial classes of chaotic jerky dynamics," *Chaos, Solitons & Fractals*, vol. 13, no. 1, pp. 1–15, 2002.

[12] J.-M. Malasoma, "What is the simplest dissipative chaotic jerk equation which is parity invariant?" *Physics Letters A*, vol. 264, no. 5, pp. 383–389, 2000.

[13] S. H. Schot, "Jerk: the time rate of change of acceleration," *American Journal of Physics*, vol. 46, no. 11, pp. 1090–1094, 1978.

[14] J. C. Sprott, *Elegant Chaos: Algebraically Simple Chaotic Flows*, World Scientific, 2010.

[15] M. E. Yalçin, "Multi-scroll and hypercube attractors from a general jerk circuit using Josephson junctions," *Chaos, Solitons & Fractals*, vol. 34, no 5, pp. 1659–1666, 2007.

[16] J. Ma, X. Y. Wu, R. T. Chu, and L. Zhang, "Selection of multi-scroll attractors in Jerk circuits and their verification using Pspice," *Nonlinear Dynamics*, vol. 76, no. 4, pp. 1951–1962, 2014.

[17] J. C. Sprott, "A new chaotic jerk circuit," *IEEE Transactions on Circuits and Systems II: Express Briefs*, vol. 58, no. 4, pp. 240–243, 2011.

[18] C. Volos, A. Akgul, V.-T. Pham, I. Stouboulos, and I. Kyprianidis, "A simple chaotic circuit with a hyperbolic sine function and its use in a sound encryption scheme," *Nonlinear Dynamics*, vol. 89, no. 2, pp. 1047–1061, 2017.

[19] S. Vaidyanathan, C. Volos, V.-T. Pham, K. Madhavan, and B. A. Idowu, "Adaptive backstepping control, synchronization and circuit simulation of a 3-D novel jerk chaotic system with two hyperbolic sinusoidal nonlinearities," *Archives of Control Sciences*, vol. 24, no. 3, pp. 375–403, 2014.

[20] A. N. Pisarchik and U. Feudel, "Control of multistability," *Physics Reports*, vol. 540, no. 4, pp. 167–218, 2014.

[21] X. Luo, M. Small, M.-F. Danca, and G. Chen, "On a dynamical system with multiple chaotic attractors," *International Journal of Bifurcation and Chaos*, vol. 17, no. 9, pp. 3235–3251, 2007.

[22] Z. T. Njitacke, J. kengne, H. B. Fotsin, A. NguomkamNegou, and D. Tchiotsop, *Coexistence of multiple attractors and crisis route to chaos in a novel memristive diode bidge-based Jerk circuit, Chaos, Solitons & Fractals*, vol. 99, 2017.

[23] B.-C. Bao, Q. Xu, H. Bao, and M. Chen, "Extreme multistability in a memristive circuit," *IEEE Electronics Letters*, vol. 52, no. 12, pp. 1008–1010, 2016.

[24] B. Bao, T. Jiang, Q. Xu, M. Chen, H. Wu, and Y. Hu, "Coexisting infinitely many attractors in active band-pass filter-based memristive circuit," *Nonlinear Dynamics*, vol. 86, no. 3, pp. 1711–1723, 2016.

[25] A. Massoudi, M. G. Mahjani, and M. Jafarian, "Multiple attractors in Koper-Gaspard model of electrochemical periodic and chaotic oscillations," *Journal of Electroanalytical Chemistry*, vol. 647, no. 1, pp. 74–86, 2010.

[26] V.-T. Pham, C. Volos, S. Jafari, and T. Kapitaniak, "Coexistence of hidden chaotic attractors in a novel no-equilibrium system," *Nonlinear Dynamics*, vol. 87, no. 3, pp. 2001–2010, 2017.

[27] Z. T. Njitacke, J. Kengne, N. A. Nguomkam, T. M. Fouodji, and H. B. Fotsin, "Coexistence of multiple attractorsand crisis route to chaos in a novel chaotic jerk circuit," *International Journal of Bifurcation and Chaos*, vol. 25, pp. 1550052–1550060, 2015.

[28] J. Kengne, Z. T. Njitacke, and H. B. Fotsin, "Dynamical analysis of a simple autonomous jerk system with multiple attractors," *Nonlinear Dynamics*, vol. 83, no. 1-2, pp. 751–765, 2016.

[29] J. Kengne, A. Nguomkam Negou, and Z. T. Njitacke, "Antimonotonicity, chaos and multiple attractors in a novel autonomous jerk circuit," *International Journal of Bifurcation and Chaos*, vol. 27, no. 7, 1750100, 20 pages, 2017.

[30] Z. T. Njitacke, J. kengne, and L. Kamdjeu Kengne, "Antimonotonicity, chaos and multiple coexisting attractors in a simple hybrid diode-based jerk circuit," *Chaos, Solitons & Fractals*, vol. 105, pp. 77–91, 2017.

[31] J. Kengne, A. N. Negou, and D. Tchiotsop, "Antimonotonicity, chaos and multiple attractors in a novel autonomous memristor-based jerk circuit," *Nonlinear Dynamics*, vol. 88, no. 4, pp. 2589–2608, 2017.

[32] E. Tlelo-Cuautle, A. D. Pano-Azucena, J. J. Rangel-Magdaleno, V. H. Carbajal-Gomez, and G. Rodriguez-Gomez, "Generating a 50-scroll chaotic attractor at 66 MHz by using FPGAs," *Nonlinear Dynamics*, vol. 85, no. 4, pp. 2143–2157, 2016.

[33] K. Rajagopal, A. Akgul, S. Jafari, A. Karthikeyan, and I. Koyuncu, "Chaotic chameleon: Dynamic analyses, circuit implementation, FPGA design and fractional-order form with basic analyses," *Chaos, Solitons & Fractals*, vol. 103, pp. 476–487, 2017.

[34] E. Dong, Z. Liang, S. Du, and Z. Chen, "Topological horseshoe analysis on a four-wing chaotic attractor and its FPGA implement," *Nonlinear Dynamics*, vol. 83, no. 1-2, pp. 623–630, 2016.

[35] S. M. Ismail, L. A. Said, A. A. Rezk et al., "Generalized fractional logistic map encryption system based on FPGA," *AEÜ - International Journal of Electronics and Communications*, vol. 80, pp. 114–126, 2017.

[36] E. Tlelo-Cuautle, J. J. Rangel-Magdaleno, A. D. Pano-Azucena, P. J. Obeso-Rodelo, and J. C. Nunez-Perez, "FPGA realization of multi-scroll chaotic oscillators," *Communications in Nonlinear Science and Numerical Simulation*, vol. 27, no. 1–3, pp. 66–80, 2015.

[37] Y.-M. Xu, L.-D. Wang, and S.-K. Duan, "A memristor-based chaotic system and its field programmable gate array implementation," *Acta Physica Sinica*, vol. 65, no. 12, Article ID 120503, 2016.

[38] E. Tlelo-Cuautle, L. G. de la Fraga, V.-T. Pham, C. Volos, S. Jafari, and A. D. J. Quintas-Valles, "Dynamics, FPGA realization and application of a chaotic system with an infinite number of equilibrium points," *Nonlinear Dynamics*, vol. 89, no. 2, pp. 1129–1139, 2017.

[39] Z. Wei, A. Akgul, U. E. Kocamaz, I. Moroz, and W. Zhang, *Control, electronic circuit application and fractional-order analysis of hidden chaotic attractors in the self-exciting homopolar disc dynamo, Chaos, Solitons & Fractals*, vol. 111, 2018.

[40] D. Valli, B. Muthuswamy, S. Banerjee et al., "Synchronization in coupled Ikeda delay systems," *The European Physical Journal Special Topics*, vol. 223, no. 8, pp. 1465–1479, 2014.

[41] Karthikeyan Rajagopal, Anitha Karthikeyan, and Prakash Duraisamy, "Hyperchaotic Chameleon: Fractional Order FPGA Implementation," *Complexity*, vol. 2017, Article ID 8979408, 16 pages, 2017.

[42] Karthikeyan Rajagopal, Laarem Guessas, Sundarapandian Vaidyanathan, Anitha Karthikeyan, and Ashokkumar Srinivasan, "Dynamical Analysis and FPGA Implementation of a Novel Hyperchaotic System and Its Synchronization Using Adaptive Sliding Mode Control and Genetically Optimized PID Control," *Mathematical Problems in Engineering*, vol. 2017, Article ID 7307452, 14 pages, 2017.

[43] Karthikeyan Rajagopal, Laarem Guessas, Anitha Karthikeyan, Ashokkumar Srinivasan, and Girma Adam, "Fractional Order Memristor No Equilibrium Chaotic System with Its Adaptive Sliding Mode Synchronization and Genetically Optimized Fractional Order PID Synchronization," *Complexity*, vol. 2017, Article ID 1892618, 19 pages, 2017.

[44] K. Rajagopal, A. Karthikeyan, and A. K. Srinivasan, "FPGA implementation of novel fractional-order chaotic systems with two equilibriums and no equilibrium and its adaptive sliding mode synchronization," *Nonlinear Dynamics*, vol. 87, no. 4, pp. 2281–2304, 2017.

[45] J. M. Bahi, X. Fang, C. Guyeux, and L. Larger, "FPGA design for pseudorandom number generator based on chaotic iteration used in information hiding application," *Applied Mathematics & Information Sciences*, vol. 7, no. 6, pp. 2175–2188, 2013.

[46] C. Chen, H. Ma, H. Chen, Y. Meng, and Q. Ding, "FPGA implementation of a UPT chaotic signal generator for image encryption," *Pacific Science Review A: Natural Science and Engineering*, vol. 17, no. 3, pp. 97–102, 2015.

[47] M. L. Barakat, A. G. Radwan, and K. N. Salama, "Hardware realization of chaos based block cipher for image encryption," in *Proceedings of the 2011 23rd International Conference on Microelectronics, ICM 2011*, Tunisia, December 2011.

[48] A. M. Atteya and A. H. Madian, "A hybrid Chaos-AES encryption algorithm and its impelmention based on FPGA," in *Proceedings of the 2014 12th IEEE International New Circuits and Systems Conference, NEWCAS 2014*, pp. 217–220, Canada, June 2014.

[49] S. T. Kingni, V.-T. Pham, S. Jafari, and P. Woafo, "A chaotic system with an infinite number of equilibrium points located on a line and on a hyperbola and its fractional-order form," *Chaos, Solitons & Fractals*, vol. 99, pp. 209–218, 2017.

[50] B. A. Idowu, U. E. Vincent, and A. N. Njah, "Synchronization of chaos in non-identical parametrically excited systems," *Chaos, Solitons & Fractals*, vol. 39, no. 5, pp. 2322–2331, 2009.

[51] V. Sundarapandian and R. Karthikeyan, "Anti-synchronization of Lü and Pan chaotic systems by adaptive nonlinear control," *European Journal of Scientific Research*, vol. 64, no. 1, pp. 94–106, 2011.

[52] V. Sundarapandian and R. Karthikeyan, "Adaptive anti-synchronization of Uncertain Tigan and Li Systems," *Journal of Engineering and Applied Sciences*, vol. 7, no. 1, pp. 45–52, 2012.

[53] S. S. Majidabad and H. T. Shandiz, "Discrete-time Terminal Sliding Mode Control of Chaotic Lorenz System," *Journal of Control and Systems Engineering*, vol. 1, no. 1, pp. 1–8, 2013.

[54] S. Vaidyanathan, "A new 3-D jerk chaotic system with two cubic nonlinearities and its adaptive backstepping control," *Archives of Control Sciences. Polish Academy of Sciences. Committee of Automatic Control and Robotics*, vol. 27(63), no. 3, pp. 409–439, 2017.

[55] O. S. Onma, O. I. Olusola, and A. N. Njah, "Control and Synchronization of Chaotic and Hyperchaotic Lorenz Systems via Extended Backstepping Techniques," *Journal of Nonlinear Dynamics*, vol. 2014, Article ID 861727, 15 pages, 2014.

[56] B. Wang, Y. Li, and D. L. Zhu, "Simplified sliding mode of a novel class of four-dimensional fractional-order chaos," *International Journal of Control and Automation*, vol. 8, no. 8, pp. 425–438, 2015.

[57] C. Yin, S. Dadras, S.-M. Zhong, and Y. Chen, "Control of a novel class of fractional-order chaotic systems via adaptive sliding mode control approach," *Applied Mathematical Modelling*, vol. 37, no. 4, pp. 2469–2483, 2013.

[58] H. Liu and J. Yang, "Sliding-mode synchronization control for uncertain fractional-order chaotic systems with time delay," *Entropy*, vol. 17, no. 6, pp. 4202–4214, 2015.

13

Scaling Laws of Droplet Coalescence: Theory and Numerical Simulation

M. Irshad Khodabocus,[1] Mathieu Sellier,[2] and Volker Nock[3]

[1]*Department of Mathematics, Faculty of Science, University of Mauritius, Réduit 80837, Mauritius*
[2]*Department of Mechanical Engineering, University of Canterbury, Private Bag 4800, Christchurch 8140, New Zealand*
[3]*Department of Electrical and Computer Engineering, University of Canterbury, Private Bag 4800, Christchurch 8140, New Zealand*

Correspondence should be addressed to M. Irshad Khodabocus; ikhodabo@gmail.com

Academic Editor: Luigi C. Berselli

When two Newtonian liquid droplets are brought into contact on a solid substrate, a highly curved meniscus neck is established between the two which transforms the bihemispherically shaped fluid domain to a hemispherically shaped domain. The rate at which such topological transformation, called coalescence phenomenon, evolves results from a competition between the inertial force which resists the transformation, the interfacial force which promotes the rate, and the viscous force which arrests it. Depending on the behaviour of these forces, different scaling laws describing the neck growth can be observed, predicted theoretically, and proved numerically. The twofold objective of the present contribution is to propose a simple theoretical framework which leads to an Ordinary Differential Equation, the solution of which predicts the different scaling laws in various limits, and to validate these theoretical predictions numerically by modelling the phenomenon in the commercial Finite Element software COMSOL Multiphysics.

1. Introduction

When two Newtonian liquid droplets are brought into contact on a solid substrate, a highly curved meniscus bridge is established and thereafter driven by surface tension so as to transform the bulk flow region from bihemispherically shaped at early-time regimes to more hemispherically shaped at late-time regimes. Such topological transformation, involving the merging of two or more sessile droplets into a single one on a solid surface, is called *droplet coalescence phenomenon*.

Physically speaking, the coalescence process consists of three regimes, called the *viscous*, *visco-inertia*, and *inertial coalescence regimes* [1]. In each regime, a range of surprisingly complex behaviours can be observed, so to speak topological changes of the interface [2]. Thus, the subject of droplets coalescence is a fascinating, multifaceted subject of inquiry with many interesting phenomena to learn and discover. Indeed, the subject has intrigued many physicists [3–6] and mathematicians [7, 8].

In actual fact, authors of [9] appear to be the first scientists who have addressed the subject. The authors study the formation of vortex rings by droplets falling into liquids and some allied phenomena. For a miscible droplet falling from not too great a height into its bulk fluid, it is observed that the droplet descends through the bulk fluid as a ring, whereas for the case of an immiscible droplet, [9] observed the droplet to descend through the bulk fluid as a spherical droplet. Thereafter, there have been many attempts on experimental, theoretical, and numerical sides to elucidate the underlying physics governing the dynamics of coalescence.

Authors of [10] have studied the partial coalescence of a water droplet, with radius $R \in [0.02, 1.3]$ mm, with the free-surface of a bulk fluid composed of water-glycerol mixtures experimentally. For cases $R \longrightarrow 0.02, 1.3$ mm, the authors drew attention that the coalescence process is arrested by gravitational and viscous effects and, depending on these two effects, three coalescence regimes were observed, which [10] termed *gravity*, *inertio-capillary*, and *viscous* regimes. Furthermore, the authors argued that the coalescence time is strongly dependent upon the Bond number $\mathrm{Bo} = 4(\rho_1 - \rho_2)gR^2/\gamma$ and the Ohnesorge number $\mathrm{Oh} = \mu_1/(2(\rho_1 + \rho_2)\gamma R)^{1/2}$, where ρ_1, μ_1, and γ denote the density, dynamic viscosity, and surface tension of the droplet, respectively; ρ_1 and μ_1 denote the properties of its vapour phase surrounding, and g denotes the gravitational acceleration.

Reference [7] addressed the more general case of the coalescence of droplets surrounded by a viscous outer fluid, both analytically and numerically, using asymptotic methods and the integral equation given by [11]. Describing the initial coalescence regime by the Stokes equations, they showed that the three-dimensional solution is asymptotically equivalent to the two-dimensional one reported in the works of [12–16]. When the viscosity of the outer fluid is disregarded, the radius and width of the highly curved meniscus bridge were found to obey the following laws: $r_m \sim (t\gamma/\pi\mu)\ln[t\gamma/(\mu R)]$ and $\Delta \propto r_m^3$, respectively, where t is the time, and properties R, γ, and μ are the radius, surface tension, and viscosity constant of the droplet; when the viscosity of the outer fluid is taken into account, a toroidal bubble of radius $\Delta \propto r_m^{3/2}$ is formed by the outer fluid, and $r_m \sim (t\gamma/4\pi\mu)\ln[t\gamma/(\mu R)]$. Employing the Euler equations in the inertial regime, they showed that $r_m \propto (\gamma R/\rho)^{1/4}t^{1/2}$, where ρ denotes the density of the coalescing droplets. (In the present study, the property $b(\boldsymbol{\pi};t)$ will stand for r_m; $\boldsymbol{\pi}$ designates the set of property constants upon which it depends.)

The paper of [7] is one classic paper which treats the coalescence of droplets both theoretically and numerically. The method used by [7] is termed *conventional* by [8]. The work of [8] is another classic paper on the subject. They addressed the subject using the conventional method and a different method called *interface formation/disappearance method* and comparing their corresponding results with experimental findings. The authors argued that their proposed model shows better agreement with experimental data than the conventional one.

In the literature, [4] has given a combined theoretical and experimental account of the coalescence of water droplets on a solid substrate surrounded by an atmosphere of nitrogen saturated with water. For small contact angle $\theta < 20^\circ$ and large contact angle $\theta \in (70^\circ, 90^\circ)$, the bidroplet undergoes very fast coalescence. Moreover, the relaxation time characterising the coalescence flow is found to be $t_c \propto R$ (or, according to [4], $t_c = \pi\mu/(3K\gamma)$), and the bulk capillary relaxation time, $t_b = R\mu/\gamma \approx t_c \times 10^{-7}$, where R, γ, and μ are the radius, surface tension, and viscosity constant of the bidroplet system.

From a hydrodynamical viewpoint, [3] found that the time evolution of the characteristic scale $b(\boldsymbol{\pi};t)$ of that small liquid bridge—where the meniscus of the bidroplet system is highly curved—obeys a time-dependent scaling/power law: $b(\boldsymbol{\pi};t) \propto t^k$, where $k > 0$. For instance, in the *inertial coalescence regime* it was found that $k \approx 1/2$, whereas in the *viscous coalescence regime* $k \approx 1$. The confirmation of these laws as regards the growth of the property $b(\boldsymbol{\pi};t)$ in those coalescence regimes was also reported by earlier scientists. Reference [2] addressed the early-time evolution of the highly curved meniscus bridge experimentally. They found that $b(\boldsymbol{\pi};t) \propto K^{1/4}t^{1/2}$, where K is the inverse of the curvature of the meniscus bridge; by comparison, $K = \gamma R/\rho$ [7].

On increasing the viscosity of the fluid, [17] has shown the linear dependence of $b(\boldsymbol{\pi};t)$ on time. Reference [18] has studied the influence of the geometrical properties of low viscosity liquid droplets experimentally. For contact angles $\theta < 90^\circ$, they deduced $b(\boldsymbol{\pi};t) \propto t^{2/3}$, while for $\theta > 90^\circ$ their analysis yielded to $b(\boldsymbol{\pi};t) \propto t^{1/2}$. Most interestingly, they demonstrated the 1/2 exponent scaling law in the limit that $\theta \longrightarrow 90^\circ$, thereby unifying the coalescence of liquid droplets and freely suspended droplets in the inertial regime. The 2/3 exponent scaling law was also argued by [19] who attributed it to the self-similar description of the shape of inviscid pinchoff of a liquid droplet in a still air.

On the other hand, [20] has considered that the influence of surface tension gradient on droplet coalescence was addressed on a combined experimental and numerical basis. The situation considered by these authors was the coalescence of a droplet deposited on the surface of a miscible liquid reservoir. Their results reveal three distinct coalescence regimes based on the reservoir-to-droplet surface tension ratio $\lambda_\gamma > 0$. For instance, the situation $\lambda_\gamma < 0.42$ results in the ejection of a small daughter droplet from the top of the top of the coalescing droplet. Contrarily, it was only when $\lambda_\gamma \in [0.42, 0.93)$ that total coalescence could be observed.

Reference [21] has addressed the early-time coalescence of viscous droplets on a flat, wettable substrate. The authors have shown in terms of their respective radii R_0, initial heights $h_0 \ll R_0$, surface tensions $\gamma > 0$, and dynamics viscosities $\mu > 0$ that $b(\boldsymbol{\pi};t) \approx (\gamma h_0^3/\mu R_0^2)^{1/2}\sqrt{t}$, which led them to conclude that the evolution of $b(\boldsymbol{\pi};t)$ is highly sensible to the geometrical properties of the liquid droplets. Reference [22] considered the coalescence of air bubbles and salt water droplets systems in silicone oils experimentally. The proportionality constant and the α-exponent of the scaling law $b(\boldsymbol{\pi};t) \propto t^\alpha$ are analysed; the authors argued that the evolution of $b(\boldsymbol{\pi};t)$ is independent of the viscosity of its outer fluid, changing only the proportionality constant by a factor roughly equal to 1.5 and leaving the α-exponent unaffected.

The above references reveal extensive studies of the characters of the property $b(\boldsymbol{\pi};t) = \kappa \cdot t^\alpha$ with respect to some properly/carefully chosen pair of scalar fields (κ, α), paying little attention to the characters of the coalescence flow fields in three-dimensional space and on cutting planes and leaving the simultaneous establishment of these power law regimes from one and only one generic equation untouched. Moreover, should an attempt be made to study a class of physical situations ranging from droplets spreading to self-propulsion and coalescence, the several modelling frameworks proposed by those scientists are inconvenient to employ without reconstructing either the mathematical framework or the numerical framework, though the outcomes of their works are no doubt striking. Thus, on the one hand, we must admit that a b-evolution equation which permits the deduction of those power laws has never been derived and, on the other hand, the models proposed by those authors make it difficult, if not impossible, to display the global characters of the coalescence flow fields as well as to adapt from one physical situation to another.

In this part, we develop a sound framework which upon variations of an interdependent leading parameter permits to study either the coalescence of two liquid droplets or the spreading of a single droplet on a three-dimensional solid substrate. Taking advantage of the proposed geometrical model, we then establish through theoretical analysis that b-evolution equation and derive from it the power law relationships $b(\boldsymbol{\pi}; t) \propto t^{\alpha}$ ($\alpha = 1/2, 2/3, 1$).

To demonstrate that the proposed model might be adopted with advantage, the $1/2, 2/3, 1$-exponent scaling laws are proved numerically, and the characters of the coalescence flow fields are illustrated in three-dimensional space and on cutting planes.

2. Theory

2.1. Mathematical Formulations. To formulate the problem mathematically, we reason in the following manner. Suppose two liquid droplets $\Omega_1, \Omega_2 \subset \mathbb{R}^3$ coalesce on a solid substrate $\mathcal{S} \subset \mathbb{R}^3$, at some time $t > 0$ they will occupy a *master* domain $\bigcup_{i=1,2} \Omega_i$ satisfying the geometrical condition

$$\operatorname{vol}\left(\bigcup_{i=1,2} \Omega_i\right) \leq \operatorname{vol}(\Omega_1) + \operatorname{vol}(\Omega_2) \quad \text{for all } t \in (0, T). \tag{1}$$

Following [8], equality follows from (1) only when the size of the *interfacial circular sector* centered at the point $\mathbf{x}_s \in \mathcal{S}$ (see Figure 1) is zero, and strict inequality whenever it is nonzero, as is easily understood. In the present work, the domain $\bigcup_{i=1,2} \Omega_i$, which can vary in time, will be termed *coalescence domain*, though the term *bidroplet system* is also attributed to; the set $(0, T)$ (with $0 < T < \infty$) denotes the period of time over which the study is carried out. Mathematically, the coalescence domain is bounded by two disjoint sets called the *interface* $\partial(\bigcup_{i=1,2} \Omega_i)$ and the *footprint* $\bigcup_{i=1,2} \Omega_i \cap \mathcal{S}$ of the coalescence domain. In Figure 1, we illustrate the coalescence of two viscous liquid droplets $\Omega_1, \Omega_2 \subset \mathbb{R}^3$ conditioned by *zero-flux* of momentum, mass and energy across the (free) surface $\partial(\bigcup_{i=1,2} \Omega_i)$ and the (contact) surface $\bigcup_{i=1,2} \Omega_i \cap \mathcal{S}$; the null flux vectors $\boldsymbol{\varphi}_i = \mathbf{0}$ (with $i = 1, 2, s$) stand for those boundary conditions. Figure 2 exposes to view the projection of the coalescence domain in the (y, z)-plane (top) and in the (x, z)-plane (bottom). Points $\mathbf{x}_0, \mathbf{x}_s \in \mathcal{S}$ designate the rear and centroid of the bidroplet system; and, points $\mathbf{x}_1, \mathbf{x}_2 \in \mathcal{S}$ the centroids of its first and second droplets, both taken at the instant of time $t = 0$. Quantities R_1, R_2 (with $R_2 \leq R_1$) are the initial radii of the coalescing droplets, and $b_0 \equiv b(\boldsymbol{\pi}; t = 0)$ the characteristic size at their corresponding points of contact. In particular, it is assumed in Figure 2 that the coalescence domain is initially *bihemispherical*, L units in length in the direction of $\mathbf{e}_y$, with respective radii R_1, R_2 (with $R_2 \leq R_1$, by hypothesis), and satisfying the condition $R_1 + R_2 < L/2$. (We agree that $R_1 + R_2 = L/2$ necessarily implies that $b_0 = 0$ and, $R_1 + R_2 > L/2$ implies that the coalescing droplets have not yet been brought into contact; see Figure 2.)

The dynamic state of the coalescence domain is characterised completely by the following macroscopic vector and scalar fields quantities: the *velocity* $\mathbf{u}(\mathbf{x}, t) = (u_x, u_y, u_z)^T$ and the *pressure* $p(\mathbf{x}, t)$. We draw attention that, the vector field $\mathbf{u}(\mathbf{x}, t)$ describes the coalescence flow within the coalescence domain $\bigcup_{i=1,2} \Omega_i$.

The *governing equations* describing these property variables in $\bigcup_{i=1,2} \Omega_i$ are the following generic momentum and mass transport equations of a divergence-free flow field:

$$\begin{aligned} \partial_t(\rho\mathbf{u}) + \operatorname{div}(\rho\mathbf{u} \otimes \mathbf{u}) + \nabla p = \rho\mathbf{f}_\gamma + \operatorname{div}\mathbb{S}(\mathbf{u}) \quad &\text{in } \bigcup_{i=1,2} \Omega_i \times (0, T); \\ \operatorname{div}(\rho\mathbf{u}) = 0 \quad &\text{in } \bigcup_{i=1,2} \Omega_i \times (0, T). \end{aligned} \tag{2}$$

In the above system of balance laws, the scalar field $p(\mathbf{x}, t)$ denotes the pressures field and the vector field $\mathbf{f}_\gamma(\mathbf{x}, t)$ denotes the density of volume force due to the surface tension. The tensor $\mathbb{S}(\mathbf{u})$ denotes the *incompressible Newtonian viscous stress tensor*, defined as

$$\mathbb{S}(\mathbf{u}) = \mu\left\{\operatorname{grad}\mathbf{u} + (\operatorname{grad}\mathbf{u})^T\right\} \quad \text{in } \bigcup_{i=1,2} \Omega_i \times (0, T). \tag{3}$$

In (2), the property constants $\rho, \mu > 0$ are, respectively, the *density* and *dynamic viscosity* of the mixed fluid occupying the domain $\bigcup_{i=1,2} \Omega_i$. The operator $\otimes$ designates the *tensor product operator*; for given vector field variables $\mathbf{u}, \mathbf{v} \in \mathbb{R}^3$, their tensor product yields to a 3×3 matrix with the following entries: $\mathbf{u} \otimes \mathbf{v} = u_i v_j$ (with $i, j = x, y, z$).

To model the vector field $\mathbf{f}_\gamma(\mathbf{x}, t)$, we employ the interfacial scalar field $\gamma(\mathbf{x}, t)$ into the following alternative expression

$$\begin{aligned} \mathbf{f}_\gamma(\mathbf{x}, t) = \operatorname{div}(\gamma(\nabla_s \cdot \mathbf{n})\mathbf{n} - \nabla_s\gamma) \quad &\text{on } \partial\left(\bigcup_{i=1,2} \Omega_i\right) \times (0, T); \\ \nabla_s(\cdot) = (\mathbb{I} - \mathbf{n} \otimes \mathbf{n})\nabla(\cdot). & \end{aligned} \tag{4}$$

The detailed derivation of (4) is found in [23]. The vector $\mathbf{n}(\mathbf{x}, t) = (n_x, n_y, n_z)^T$ is the unit outer vector normal to $\bigcup_{i=1,2} \Omega_i$ and $\bigcup_{i=1,2} \Omega_i \cap \mathcal{S}$ ($\neq \bigcup_{i=1,2} \Omega_i$, since, $\Omega_i \cap \mathcal{S} \neq \Omega_i$ for $i = 1, 2$), respectively. The operator $\nabla_s(\cdot)$ is the *surface gradient operator* on $\partial(\bigcup_{i=1,2} \Omega_i)$, and the matrix $\mathbb{I}$ stands for the *identity matrix*.

2.2. Boundary and Initial Conditions. At the instant of time $t = 0$ the distributions of the velocity field $\mathbf{u}(\mathbf{x}, t)$, the pressure field $p(\mathbf{x}, t)$ and the chemical concentration $c(\mathbf{x}, t)$ should be supplemented in $\bigcup_{i=1,2} \Omega_i$, and thereafter conditioned on boundaries $\partial(\bigcup_{i=1,2} \Omega_i)$ and $\bigcup_{i=1,2} \Omega_i \cap \mathcal{S}$ for all $t \in (0, T)$. These are discussed in the following two sections.

The dynamic boundary conditions imposed on the interface $\partial(\bigcup_{i=1,2} \Omega_i)$ for the velocity field $\mathbf{u}(\mathbf{x}, t)$ are the Neumann-type conditions

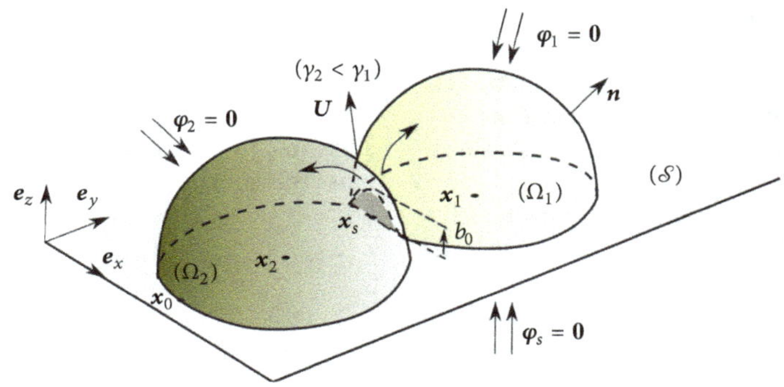

FIGURE 1: A freehanded illustration of the coalescence of two liquid droplets.

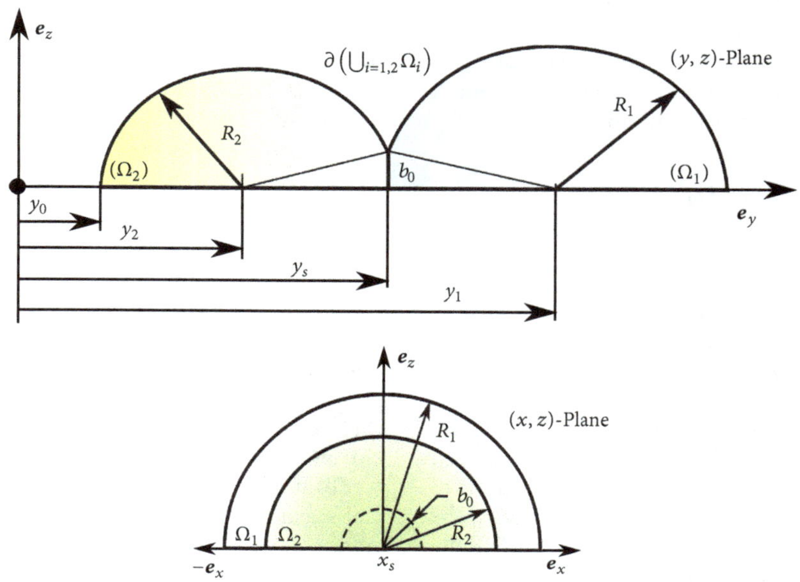

FIGURE 2: The projection of the coalescence domain in the (y, z)-plane (top) and in the (x, z)-plane (bottom).

$$(\mathbb{S}(\mathbf{u}) - p\mathbb{I})\,\mathbf{n} = (-p_{\text{ext}}\mathbb{I} + \gamma(\nabla_s\cdot\mathbf{n})\,\mathbb{I})\,\mathbf{n} \quad \text{on } \partial\left(\bigcup_{i=1,2}\Omega_i\right)\times(0,T), \tag{5}$$
$$(\mathbb{S}(\mathbf{u}) - p\mathbb{I})\,\mathbf{t} = |\nabla_s\gamma| \quad \text{on } \partial\left(\bigcup_{i=1,2}\Omega_i\right)\times(0,T).$$

In the above conditions, the property $p_{\text{ext}} > 0$ designates the *external pressure* applied on the interface $\partial(\Omega_1 \cup \Omega_2)$; $\gamma(\nabla_s \cdot \mathbf{n})$ is the force per unit area due to local curvature of that interface. The term on the right-hand side of the second equation designates the tangential stress associated with gradients in the surface tension coefficient. The proper boundary conditions to be considered on the contact surface $\bigcup_{i=1,2}\Omega_i \cap \mathcal{S}$ are now discussed. At some time $t > 0$ thereafter, the footprint of the coalescence domain $\bigcup_{i=1,2}\Omega_i$ is expected to deform from a *bidisc* shaped domain to a *monodisc* shaped domain. Therefore, a wetted wall boundary condition is considered, the *solid-fluid boundary conditions*. Thus, if the scalar quantity $\beta > 0$ designates the *Navier-slip* coefficient [24], considering the *Navier-slip-with-friction boundary condition* yields

$$(\mathbb{S}(\mathbf{u})\,\mathbf{n} + \beta\mathbf{u})\cdot\mathbf{t} = 0 \quad \text{on } \bigcup_{i=1,2}\Omega_i \cap \mathcal{S}\times(0,T),$$
$$\mathbf{u}\cdot\mathbf{n} = 0 \quad \text{on } \bigcup_{i=1,2}\Omega_i \cap \mathcal{S}\times(0,T). \tag{6}$$

In (5), (6), the vector $\mathbf{t}(\mathbf{x}, t) = (t_x, t_y, t_z)^T$ designates the tangent vectors to boundaries $\partial(\bigcup_{i=1,2}\Omega_i)$ and $\bigcup_{i=1,2}\Omega_i \cap \mathcal{S}$. The following initial condition is imposed on the vector field $\mathbf{u}(\mathbf{x}, t)$ at $t = 0$:

$$\mathbf{u}(\mathbf{x}, t) = \mathbf{u}_0(\mathbf{x}) \quad \text{in } \bigcup_{i=1,2}\Omega_i \times \{0\}, \tag{7}$$

The momentum and mass transport equations along with the foregoing set of boundary and initial conditions are solved using the COMSOL Moving Mesh Interface, discussed in what follows.

2.3. Moving Coalescence Domain Explained. At this stage, we have beforehand the mathematical framework and its respective domain of definition. Bearing in mind that the bulk motion of the coalescence domain results from topological changes of its interface from one instant of time to another, to adapt the mathematical descriptions with the movement of the coalescence domain $\bigcup_{i=1,2}\Omega_i$ we employ the *moving mesh method*, which we shall now throw some light upon. For more details, see the works of [23, 25, 26].

Let the interface $\partial(\bigcup_{i=1,2}\Omega_i)$ of the coalescence domain designates a *material point-set* consisting of material points $\mathbf{X} = (X, Y, Z)^T$ at the instant of time $t = 0$. Under this condition, let this coalescence domain at some time $t \geq 0$ be described as a *spatial point-set* consisting of points $\mathbf{x} = (x, y, z)^T$. On the other hand, the coalescence domain being subjected to smooth deformation from one instant to another, there follows that

$$\mathbf{x} = \mathbf{x}(\mathbf{X}, t), \quad \text{on } \partial\left(\bigcup_{i=1,2}\Omega_i\rfloor_{t=0}\right) \times (0, T), \tag{8}$$

where $\mathbf{X} = \mathbf{x}(\mathbf{X}, 0)$. Having these *material* and *spatial* vector fields at hand, to relate the topological changes of the interface to its bounded bulk fluid, it suffices to equate the associated velocities on the said interface. Thus, if the vector field $\mathbf{U}(\mathbf{x}, t) = (U_x, U_y, U_z)^T$ describes the movement of the interface, there results in

$$\mathbf{u} \cdot \mathbf{n} = \mathbf{U} \cdot \mathbf{n}, \quad \text{on } \partial\left(\bigcup_{i=1,2}\Omega_i\right) \times (0, T); \; \mathbf{x} = \mathbf{x}(\mathbf{X}, t). \tag{9}$$

Note that, we write $\partial(\bigcup_{i=1,2}\Omega_i\rfloor_{t=0})$ instead of $\partial(\bigcup_{i=1,2}\Omega_i)$ to arrest attention that the interface is undeformed initially.

3. Numerical Implementation

In this section, the construction of the physical domain $\bigcup_{i=1,2}\Omega_i$, described in the (x, y, z)-space by Figure 1 and in the (y, z)-plane by Figure 2, and the numerical implementation of those equations and conditions, namely, (1) to (9), describing its dynamic state is outlined.

The bidroplet system—that is, the coalescence domain—is built as a three-dimensional bihemispherical computational domain sitting on a (x, y)-plane solid substrate. The solid substrate is defined as a rectangular (x, y)-plane sized $L_i > 0$ units in length in the direction of $\mathbf{e}_i$ (with $i = x, y$). To realise these geometrical constructions, the radii R_1, R_2, and position vectors $\mathbf{x}_k = (x_k, y_k, z_k)^T \in \bigcup_{j=1,2}\Omega_j \cap \mathcal{S}$ (with $k = 1, 2, s$) must be supplied, and are tabulated in Table 1. They are based on the following line of reasoning.

Initially, the bidroplet system is assumed symmetrical with respect to the (y, z)-plane. On the solid substrate, the position vector $\mathbf{x}_s = (0, y_s, 0)^T$ denotes its centroid of the coalescence domain $\bigcup_{i=1,2}\Omega_i$; the position vectors $\mathbf{x}_1 = (0, y_1, 0)^T$, $\mathbf{x}_2 = (0, y_2, 0)^T$ denote the centroids of its first and second droplets, respectively. These descriptions are clearly represented in Figure 2. To stress the dependence of the property $b(\boldsymbol{\pi}; t)$ on those geometrical quantities, we consider the following expressions:

Table 1: Initial and boundary conditions imposed on the coalescence domains $\bigcup_{i=1,2}\Omega_i$. Geometrical properties used to generate the b_0-dependent coalescence domains $\bigcup_{i=1,2}\Omega_i$.

ρ	μ	γ	p_{ext}	$\mathbf{u}_0(\mathbf{x})$
[g·cm^{-3}]	[Pa·s]	[N·m^{-1}]	[Pa]	[m·s^{-1}]
1.3	1.4	6×10^{-2}	1.0	$(0, 0, 0)^T$
$L_x = 2L_y$	$R_1 = R_2$	$\mathbf{x}_1$	$\mathbf{x}_2$	$\mathbf{x}_s$
[mm]	[mm]	[mm]	[mm]	[mm]
10	1.0	$(6, 5, 0)^T$	$(4, 5, 0)^T$	$(5, 5, 0)^T$

$$b(\boldsymbol{\pi}; t) = b_0 = \left(R_i^2 - |y_i - y_s|^2\right)^{1/2} \quad \text{at } t = 0, \tag{10}$$

where $i = 1, 2$. Furthermore, to stress that the property $b_0 > 0$ is a relatively small quantity initially, we impose the condition $b_0 \ll \min(R_1, R_2)$; in regard to the freehand sketches of Figure 2, this means that $b_0 \ll R_2 \leq R_1$. We note that $b(\boldsymbol{\pi}; t) \geq b_0$ for all $t \geq 0$; thus the coalescence phenomenon is a *relative* process, since it is studied through the evolution of the relative property $b(\boldsymbol{\pi}; t) - b_0$. Consequently, the expansion rate at that region where the curvature is very large is very dependent upon the magnitude of $b_0 > 0$. Clearly, the smaller be the size of the property b_0, the higher be the curvature at the common point of contact of two droplets, and, consequently, the faster be the coalescence rate at early-time.

Topologically, the case for which $b_0 = 0$ prescribes a *one-point interface*, while for the other cases for which $b_0 \neq 0$, all prescribe *infinitely many-points interface*. Since those interfaces disappear for all $t \in (0, T)$, the first case might well be termed the *one-point model* and the other three cases *infinitely many-points model.*

Granted the above descriptions, the following set of equations is implemented: the COMSOL *Laminar Two-Phase Flow Moving Mesh* equations set, having appropriate interfaces of their initial and boundary conditions. This model, as its name implies, is used to describe the spatiotemporal descriptions of the variable field function $\mathbf{u}(\mathbf{x}, t)$. To adapt the COMSOL $\mathbf{u}$-equation model with the physical situation under consideration, the COMSOL default model in question is expanded so as to suit the $\mathbf{u}$-set of initial and boundary conditions given in earlier (theoretical) sections. The values attributed to those conditions are reported in Table 1.

A series of CFD experiments are ran on this ground and their results are presented in what follows.

4. Results and Discussion

The results to which the proposed model led on a combined theoretical and numerical basis are discussed here.

4.1. Preliminaries. The highly curved meniscus is a fundamental entity. On the grounds of earlier works [3, 18, 19, 21, 27], it can be characterised as follows. By hypothesis, let the d-dimensional sequence $\boldsymbol{\pi} = (\pi_1, \ldots, \pi_d)$ denotes the property-set of $d \geq 1$ positive property constants, upon which rests the character of the coalescence dynamics. Further, let $\psi_\alpha(\boldsymbol{\pi})$ be a $\boldsymbol{\pi}$-dependent property, where the subscript α stands for the associated α-exponent law. Then, following earlier works one can postulate the following expression:

$$b(\boldsymbol{\pi},t) = \psi_\alpha(\boldsymbol{\pi}) \cdot t^\alpha, \quad \text{for } t \in (0,\mathrm{T}), \text{ where } \alpha > 0. \quad (11)$$

Given L, T the fundamental units of length and time, the dimensional unit of the property $\psi_\alpha(\boldsymbol{\pi})$ is deduced as follows: We observe that $[\partial_t b] = [\mathrm{L} \cdot \mathrm{T}^{-1}]$, therefore, $[\psi_\alpha] = [\mathrm{L} \cdot \mathrm{T}^{-\alpha}]$, where $\alpha > 0$. Moreover, if by properties b_0, Ψ_{ref}, $\tau_{\text{ref}} > 0$ are meant the characteristic scaling of the scalar fields $b(\boldsymbol{\pi};t)$, $\psi(\boldsymbol{\pi})$, t, the dimensionless form of (11) reads: $b(\boldsymbol{\pi};t) = t^\alpha$, where $\mathrm{Ord}(b_0) \sim \Psi_{\text{ref}} \cdot \tau_{\text{ref}}^\alpha$ for some $\alpha > 0$, meaning that the law is independent of the working Newtonian fluids. Consequently, sufficient is that our CFD framework is capable of proving coalescence laws. These statements motivate us to study the coalescence of the bihemispherically shaped droplet system.

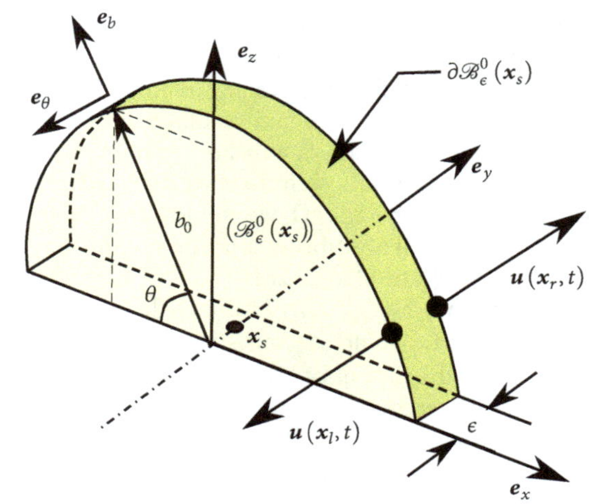

FIGURE 3: A freehanded sketch of the elementary domain $\mathscr{B}^0_\epsilon(\mathbf{x}_s)$ in $\mathbb{R}^3$.

5. The Laws of $b(\boldsymbol{\pi};t)$ Predicted Theoretically

Granted the proposed three-dimensional coalescence domain (see Figure 1), and the accompanying two-dimensional geometrical model (see Figure 2) wherein are reflected its geometrical properties in the (x,z)-plane and in the (y,z)-plane, one can predict the time evolution of the property $b(\boldsymbol{\pi};t)$ in the power law regimes. In the present section, the theoretical prediction of the property $b(\boldsymbol{\pi};t)$ in those regimes is undertaken. Let it be given $\epsilon \ll 1$, $b_0 > 0$, and $\mathbf{x}_s = (x_s, y_s, z_s)^T \in \bigcup_{i=1,2}\Omega_i \cap \mathscr{S}$, and consider the elementary domain defined by

$$\mathscr{B}^0_\epsilon(\mathbf{x}_s) = \{(x,y,z) : |y - y_s| \le \epsilon,\ (x - x_s) + (z - z_s) \le b_0\}, \quad (12)$$

with boundary $\partial\mathscr{B}^0_\epsilon(\mathbf{x}_s)$, at the instant of time $t = 0$ (see Figure 3 for geometrical insight). Note that, the elementary domain $\mathscr{B}^0_\epsilon(\mathbf{x}_s) \subset \bigcup_{i=1,2}\Omega_i$ may be conceived of as a 180 degree sector of ϵ-order thick aligned with the (x,z)-plane, and having its footprint centered at the point $\mathbf{x}_s$. The region $\mathscr{B}^0_\epsilon(\mathbf{x}_s)$ is characterised geometrically by the following length and width at time $t = 0$: $\mathrm{leng}(\partial\mathscr{B}^0_\epsilon) = \pi b_0$, $\mathrm{wid}(\partial\mathscr{B}^0_\epsilon) = \epsilon$. Following [7], the coalescence flow is driven by the time evolution of $\partial\mathscr{B}^0_\epsilon(\mathbf{x}_s)$. One can therefore postulate that

$$\mathrm{Ord}\left(\left|\mathbf{e}_y \cdot \mathbf{u}(\mathbf{x}_l,t)\right|\right) \sim \mathrm{Ord}\left(\left|\mathbf{e}_y \cdot \mathbf{u}(\mathbf{x}_r,t)\right|\right) \quad \text{at } \mathbf{x}_l, \mathbf{x}_r \in \partial\mathscr{B}^0_\epsilon;\ t \in (0,T). \quad (13)$$

The placement of the velocity vector fields $\mathbf{u}(\mathbf{x}_r,t)$, $\mathbf{u}(\mathbf{x}_l,t)$ is illustrated in Figure 3. The coalescence dynamics being dominated radially, consequently, the radial component of the equation of motion describing the evolution of the region $\mathscr{B}^0_\epsilon(\mathbf{x}_s)$—from the standpoint of Eulerian description—might at some time $t > 0$ be written in the form

$$\kappa_1\partial_t^2 b + \frac{\kappa_2}{b^2}\partial_t b + \kappa_3\partial_b\left(\frac{\gamma}{b}\right) = 0 \quad \text{for } t \in (0,\mathrm{T}). \quad (14)$$

The establishment of (14) is based on these chains of reasoning. In earlier sections, the standard basis $\{\mathbf{e}_i : i = x,y,z\}$ was used. Here, use of the cylindrical basis $\{\mathbf{e}_\alpha : \alpha = b,\theta,z\}$ will be made to prove (14). Thus, every point $\mathbf{x} = (x,y,z)^T \in \mathscr{B}^0_\epsilon(\mathbf{x}_s)$ on the basis $\{\mathbf{e}_i : i = x,y,z\}$ can be represented by $\mathbf{x} = (b,\theta,y)^T \in \mathscr{B}^0_\epsilon(\mathbf{x}_s)$ on the basis $\{\mathbf{e}_\alpha : \alpha = b,\theta,z\}$. For $t \in (0,T)$, it is postulated that the rate of change of linear momentum equates the differences of capillary and viscous forces across boundary $\partial\mathscr{B}^0_\epsilon(\mathbf{x}_0)$, *i.e.*,

$$\text{Rate of Change of Momentum} = \sum_{\alpha=\text{cap,visc}} \mathbf{f}_\alpha. \quad (15)$$

With this view in mind, one then reasons as follows: The curve traced by a (x,z)-plane $\Pi_{xz}(\mathbf{x}_s)$ cutting the boundary $\partial\mathscr{B}^0_\epsilon(\mathbf{x}_0)$ is described by an equation of the form

$$z(x) = z_s + \left(b^2 - |x - x_s|^2\right)^{1/2} \quad \text{on } \partial\mathscr{B}^0_\epsilon \cap \Pi_{xz}. \quad (16)$$

Consequently, the curvature of that curve yields

$$\mathrm{curv}\left(\partial\mathscr{B}^0_\epsilon(\mathbf{x}_0)\right) = \frac{\left|\partial_x^2 z(x)\right|}{\left(1 + \left|\partial_x z(x)\right|^2\right)^{3/2}} = \frac{1}{b}. \quad (17)$$

On the other hand, an ϵ-volume of $\mathscr{B}^0_\epsilon(\mathbf{x}_0)$ and an ϵ-surface of $\partial\mathscr{B}^0_\epsilon(\mathbf{x}_0)$ gives:

$$\mathrm{vol}_\epsilon\left(\mathscr{B}^0_\epsilon(\mathbf{x}_0)\right) = \epsilon \cdot b \cdot \delta b \cdot \delta\theta \text{ and}$$
$$\mathrm{surf}_\epsilon\left(\partial\mathscr{B}^0_\epsilon(\mathbf{x}_0)\right) = \epsilon \cdot b \cdot \delta\theta, \quad (18)$$

respectively. Note in passing that the plane $\Pi_{xz}(\mathbf{x}_s)$ and $\Pi_{xz}(\mathbf{x}_s)$ are clearly illustrated in Figure 4. Bearing in mind that the property b is a $(\boldsymbol{\pi};t)$-dependent scalar field, the rate of change of the linear momentum of $\mathrm{vol}_\epsilon(\mathscr{B}^0_\epsilon)$ along $\mathbf{e}_b$, $\boldsymbol{\varpi}(\mathscr{B}^0_\epsilon)$ say, then results in

$$\boldsymbol{\varpi}\left(\mathscr{B}^0_\epsilon\right) = \rho\partial_t^2 b \cdot \mathrm{vol}_\epsilon\left(\mathscr{B}^0_\epsilon\right)\mathbf{e}_b. \quad (19)$$

The capillary force difference sustained across the interface $\partial\mathscr{B}^0_\epsilon(\mathbf{x}_0)$ is given by

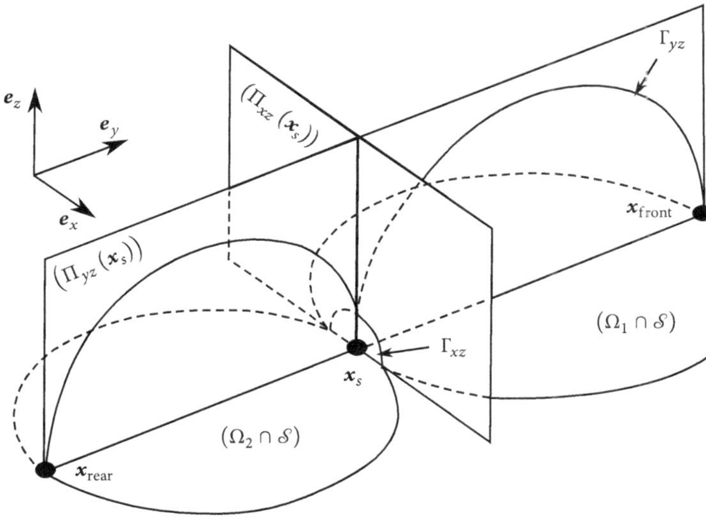

FIGURE 4: Illustration of the planes $\Pi_{\alpha z}(\mathbf{x}_s)$, the corresponding boundary curves $\Gamma_{\alpha z}$ (with $\alpha = x, y$), and the location of the points $\mathbf{x}_{\text{rear}}, \mathbf{x}_s, \mathbf{x}_{\text{front}} \in \bigcup_{i=1,2}(\Omega_i \cap \mathcal{S})$.

$$\begin{aligned}\delta\mathbf{f}_{\text{cap}}\left(\partial\mathscr{B}_\epsilon^0\right) &= -\delta p \cdot \text{surf}_\epsilon\left(\partial\mathscr{B}_\epsilon^0\right)\mathbf{e}_b \\ &= -\partial_b\left(2\gamma\cdot\text{curv}\left(\partial\mathscr{B}_\epsilon^0\right)\right)\cdot\text{vol}_\epsilon\left(\mathscr{B}_\epsilon^0\right)\mathbf{e}_b \\ &= -\partial_b\left(\frac{2\gamma}{b}\right)\cdot\text{vol}_\epsilon\left(\mathscr{B}_\epsilon^0\right)\mathbf{e}_b.\end{aligned} \tag{20}$$

That due to viscous force is given by

$$\begin{aligned}\delta\mathbf{f}_{\text{visc}}\left(\partial\mathscr{B}_\epsilon^0\right) &= \sum_{i=b,\theta,y}\left\{-\mathbb{S}_{bi}(\mathbf{u})\cdot\text{surf}_\epsilon\left(\partial\mathscr{B}_\epsilon^0\right)\big|_b\right. \\ &\quad\left.+\,\mathbb{S}_{bi}(\mathbf{u})\cdot\text{surf}_\epsilon\left(\partial\mathscr{B}_\epsilon^0\right)\big|_{b+\delta b}\right\}\mathbf{e}_b. \\ &= \sum_{i=b,\theta,y}\partial_b\left(b\mathbb{S}_{bi}(\mathbf{u})\right)\cdot\text{vol}_\epsilon\left(\mathscr{B}_\epsilon^0\right)\mathbf{e}_b\end{aligned} \tag{21}$$

In (21), the tensor $\mathbb{S}(\mathbf{u}) = (\mathbb{S}_{ij}(\mathbf{u}))_{i,j=b,\theta,y}$, with $\mathbf{u}$-dependent entries $\mathbb{S}_{ij}(\mathbf{u})$, where i is the row index and j the column index, stands for the viscous stress tensor. Substituting (19) through (21) into (15) yields

$$\rho\partial_t^2 b + \partial_b\left(\frac{2\gamma}{b}\right) - \sum_{i=b,\theta,y}\partial_b\left(b\mathbb{S}_{bi}(\mathbf{u})\right) = 0. \tag{22}$$

For points $\mathbf{x} = (b,\theta,y)^T \in \partial\mathscr{B}_\epsilon^0(\mathbf{x}_0)$, it is asserted that $\mathbf{u} = (u_b, u_\theta, u_y)^T \sim \partial_t b(1,\epsilon,\epsilon)^T$ on the basis of postulate (13), yielding

$$\begin{aligned}&\sum_{i=b,\theta,y}\partial_b\left(b\mathbb{S}_{bi}(\mathbf{u})\right) \\ &= \mu\left(\frac{1}{b}\partial_b\left(\partial_b u_b\right) + \frac{1}{b^2}\partial_\theta^2 u_b + \partial_y^2 u_b - \frac{u_b}{b^2} - \frac{2}{b^2}\partial_\theta u_\theta\right) \\ &\sim -\frac{\mu}{b^2}\partial_t b,\end{aligned} \tag{23}$$

$$\text{since } \mathbf{u} = \left(u_r, u_\theta, u_y\right)^T \sim \partial_t b\left(1,\epsilon,\epsilon\right)^T.$$

Approximating the last term of (22) by (23) and then setting the triple $(\kappa_1,\kappa_2,\kappa_3) = (\rho,\mu,2)$, (14) follows. It is to be noted here that $(\kappa_i)_{i=1}^3$ is a $\boldsymbol{\pi}$-dependent sequence because of the dependence of κ_1 on ρ, and κ_2 on μ, as is easily understood. One is now well prepared to discuss the so-called α-exponent laws, α reals.

Granted $b(\boldsymbol{\pi};t) = \psi(\boldsymbol{\pi})\cdot t^\alpha$, after its substitution into (14), one arrives at the following algebraic expression:

$$\begin{aligned}&\alpha(\alpha-1)\kappa_1\left|\psi(\boldsymbol{\pi})\right|t^{\alpha-2} + \frac{\alpha\kappa_2}{\left|\psi(\boldsymbol{\pi})\right|}t^{-\alpha-1} - \frac{\kappa_3\gamma}{\left|\psi(\boldsymbol{\pi})\right|^2}t^{-2\alpha} \\ &= 0.\end{aligned} \tag{24}$$

For that matter of analytical inquiries, set

$$\begin{aligned}\Phi_1(\boldsymbol{\pi};t) &= \alpha(\alpha-1)\kappa_1\left|\psi(\boldsymbol{\pi})\right|t^{\alpha-2}, \\ \Phi_2(\boldsymbol{\pi};t) &= \frac{\alpha\kappa_2}{\left|\psi(\boldsymbol{\pi})\right|}t^{-\alpha-1}, \\ \Phi_3(\boldsymbol{\pi};t) &= -\frac{\kappa_3\gamma}{\left|\psi(\boldsymbol{\pi})\right|^2}t^{-2\alpha}\end{aligned} \tag{25}$$

$$\text{for given } \boldsymbol{\pi}, \text{ and } \left(\kappa_i\right)_{i=1}^3,\ t\in(0,T),$$

respectively. Mathematically, it is proposed to predict the power law behaviour of the property $b(\boldsymbol{\pi};t)$ based upon the magnitudes of $(\Phi_i)_{i=1}^3$. Since, (24) is a three-term equation, the predictions are argued on three possible pairwise terms.

5.1. Case 1. Suppose that $\text{Ord}(\Phi_1) \sim \text{Ord}(\Phi_2)$, and $\text{Ord}(\Phi_3) \ll 1$. The latter may then be neglected. Thus, for every fixed set of property constants, the following conditions must be fulfilled, namely, $\alpha - 2 = -\alpha - 1$. The law describing this situation therefore writes

$$b(\boldsymbol{\pi};t) = \psi(\boldsymbol{\pi})\cdot t^{1/2} \quad \text{for } t\in(0,T). \tag{26}$$

5.2. Case 2. Let it be assumed that $\mathrm{Ord}(\Phi_1) \sim \mathrm{Ord}(\Phi_3)$, and $\mathrm{Ord}(\Phi_2) \ll 1$, respectively. Then, analogous to the above statements, one obtains $\alpha - 2 = -2\alpha$. Thus, $b(\boldsymbol{\pi}; t)$ is in this case of the form

$$b(\boldsymbol{\pi}; t) = \psi(\boldsymbol{\pi}) \cdot t^{2/3} \quad \text{for } t \in (0, T). \tag{27}$$

5.3. Case 3. Conceive the orders of magnitudes of $(\Phi_i)_{i=1}^3$ with respect to themselves to be as follows: $\mathrm{Ord}(\Phi_2) \sim \mathrm{Ord}(\Phi_3)$, and $\mathrm{Ord}(\Phi_1) \ll 1$. In this case, one deduces $-\alpha - 1 = -2\alpha$. Consequently, the scaling law describing the time evolution of the property $b(\boldsymbol{\pi}; t)$ writes

$$b(\boldsymbol{\pi}; t) = \psi(\boldsymbol{\pi}) \cdot t \quad \text{for } t \in (0, T). \tag{28}$$

It is to me pointed out that by $\mathrm{Ord}(\Phi_i)$ is meant the order of magnitude of the property Φ_i (with $i = 1, 2, 3$) with respect to time.

For liquid droplets whose viscosities are not high enough, and having contact angles $\theta < 90^\circ$, practical experiments have shown that $b(\boldsymbol{\pi}; t) \propto t^{2/3}$ [18, 19]. Contrarily, for liquids whose viscosities are high enough and having contact angles $\theta \geq 90^\circ$, many authors have established the law $b(\boldsymbol{\pi}; t) \propto t^{1/2}$ [18, 21, 27]. On the other hand, in purely viscous coalescence regime, the law $b(\boldsymbol{\pi}; t) \propto t$ has also been deduced experimentally [3].

In the next section are proved the power law regimes numerically, thereby giving agreement to experimental data.

6. The Laws of $b(\boldsymbol{\pi};t)$ Proved Numerically

Having established the power laws $b(\boldsymbol{\pi}; t) \propto t^{1/2}, t^{2/3}, t$ theoretically, the chief aim of the present section is to validate them numerically. Before doing so, the groundwork necessary for such endeavours is first laid.

Given the triple $\boldsymbol{\pi}$-dependent proportionality constants $(\psi_\alpha(\boldsymbol{\pi}))_{\alpha \in I_\alpha}$, where $I_\alpha = \{1/2, 2/3, 1\}$. Write $b(\boldsymbol{\pi}; t) = \psi_\alpha(\boldsymbol{\pi}) \cdot t^\alpha$ to stress that the prefactor referred to is the one associated with the α-exponent law and, let quantities b_{theo}, b_{num} stand for its theoretical and numerical predictions, respectively. Thus, to express an ϵ-order discrepancy between the theoretical predictions and the simulation results, one can write $|b_{\text{theo}} - b_{\text{num}}| \sim \mathrm{Ord}(\epsilon)$ and to negate the latter one can write $|b_{\text{theo}} - b_{\text{num}}| \neq \mathrm{Ord}(\epsilon)$. The question arises: Which of those α-exponent laws, $\alpha \in I_\alpha$, should be computed first? The answer to this question lies in the following paragraph.

Suppose two droplets Ω_1, Ω_2 having the same size coalesce and time evolves, the bidroplet system $\bigcup_{i=1,2} \Omega_i$ then will experience a *viscous-inertial coalescence regime* which is preceded by a *viscous coalescence regime* and followed by an *inertial coalescence regime*, as is easily understood. This shows the existence of two overlapping regimes. Moreover, if the set $(0, T_\alpha)$ designates the interval of time over which the law $b(\boldsymbol{\pi}; t) = \psi_\alpha(\boldsymbol{\pi}) \cdot t^\alpha$ holds for some $T_\alpha > 0$, $\alpha \in I_\alpha$, then it is clear that the time intervals corresponding to those three coalescence regimes interrelate themselves as follows:

$$(0, T_1) \subset (0, T_{2/3}) \subset (0, T_{1/2}) \subseteq (0, T). \tag{29}$$

These open sets are termed the *coalescence times* of the corresponding α-regimes, $\alpha \in I_\alpha$. For some $\delta T_\alpha > 0$, $\alpha \in I_\alpha$, it may be said that those two overlapping regimes occur over the time intervals

$$\begin{gathered}(T_1 - \delta T_1, T_1 + \delta T_{2/3}) \text{ and} \\ (T_{2/3} - \delta T_1, T_{2/3} + \delta T_{1/2}).\end{gathered} \tag{30}$$

In view of the above discussion, it follows, in chronological order, that the establishment of 2/3-exponent law must precede that of the 1-exponent law and follow that of the 1/2-exponent law. Indeed, if one is to express for every $\alpha \in I_\alpha$ the law $b(\boldsymbol{\pi}; t) = \psi_\alpha(\boldsymbol{\pi}) \cdot t^\alpha$ by making use of the open set $(0, T_\alpha)$, the results would resolve into three algebraic expressions of the forms:

$$\begin{aligned} b(\boldsymbol{\pi}; t) &\approx b_0 + \psi_1(\boldsymbol{\pi}) \cdot t \quad \text{for } t \in (0, T_1), \\ b(\boldsymbol{\pi}; t) &\approx b_0 + \psi_1(\boldsymbol{\pi}) \cdot T_1 + \psi_{2/3}(\boldsymbol{\pi}) \cdot t^{2/3} \\ &\qquad \text{for } t \in (T_1, T_{2/3}), \\ b(\boldsymbol{\pi}; t) &\approx b_0 + \psi_1(\boldsymbol{\pi}) \cdot T_1 + \psi_{2/3}(\boldsymbol{\pi}) \cdot T_{2/3}^{2/3} + \psi_{1/2}(\boldsymbol{\pi}) \\ &\quad \cdot t^{1/2} \quad \text{for } t \in (T_{2/3}, T_{1/2}). \end{aligned} \tag{31}$$

Carrying out the discussion a step further, one can generalise the form of the scalar field $b(\boldsymbol{\pi}; t)$ with respect to $(\boldsymbol{\pi}; t)$. Supposed granted $\boldsymbol{\pi} = (\pi_i)_{i=1}^d$, where $d > 1$, and $\{\psi_\alpha(\boldsymbol{\pi})\}_{\alpha \in I_\alpha}$ a carefully chosen set of $\boldsymbol{\pi}$-dependent proportionality constants, with the further supposition that $I_\alpha \in \mathbb{Q}$, *i.e.*, $I_\alpha - \{1/2, 2/3, 1\} \neq \emptyset$. Then, it is no error to state that the time evolution of $b(\boldsymbol{\pi}; t)$ in the most general case is a series function of the form:

$$b(\boldsymbol{\pi}; t) \approx b_0 + \sum_{0<\alpha\leq 1} \psi_\alpha(\boldsymbol{\pi}) \cdot t^\alpha \quad \text{for } t \in (0, T). \tag{32}$$

Leaving (32) out of consideration, thought it appears to demonstrate an important law, hence, might be a fairer, if not the fairest, approximation in practice, the following sections are devoted to the establishment of the laws $b(\boldsymbol{\pi}; t) = \psi_\alpha(\boldsymbol{\pi}) \cdot t^\alpha$, where $\alpha \in I_\alpha$.

6.1. The Viscous Regime Law $b(\boldsymbol{\pi}; t) = \psi_1(\boldsymbol{\pi}) \cdot t$. In this section, the numerical proof of the viscous coalescence power law $b(\boldsymbol{\pi}; t) = \psi_1(\boldsymbol{\pi}) \cdot t$ is proved. How the numerical study has been undertaken may be understood through the following lines.

Numerical experiments is first run. The geometrical properties used to construct the computational domain $\bigcup_{i=1,2} \Omega_i$ are those given in Table 1. The physical properties of the working fluid occupying the domain $\bigcup_{i=1,2} \Omega_i$ as well as the initial and boundary conditions imposed on that domain are those tabulated in Table 1. The said power law is then established in the following manner:

$$\begin{aligned} b_i &= b(\boldsymbol{\pi}; t_i), \\ t_i &= t_0 + i \cdot \delta t, \\ &\quad \text{where } \delta t > 0,\ i \geq 0, \end{aligned} \tag{33}$$

so that the pair of scalar fields $(t_i, b_i)_{i\geq 0}$ generates a sequence of points in the (t, b)-plane. Obviously, the dimensional units are such that $[b] = [\text{mm}]$ and $[t] = [\text{s}]$ (and $[b] = [L \cdot T^{-\alpha}]$, $\alpha \in I_\alpha$). For every $i \in [0, 12]$, $\delta t = 5$ [ms], the numerical value of (t_i, b_i) is extracted. This gives the necessary data required to describe the time evolution of the geometrical property $b(\boldsymbol{\pi}; t) = \psi_1(\boldsymbol{\pi}) \cdot t$. By means of (33), the pairs of scalar fields $(t_i, b_i)_{i\geq 0}$ are collected in the viscous regime. When plotted in the (t, b)-plane, one obtains the graphical result exposed in Figure 5.

In Figure 5, circular markers ∘ are used to represent the numerical data, and the solid line the corresponding fitted power law. This power law function is shaped approximately as follows:

$$b(\boldsymbol{\pi}; t) = \psi_1(\boldsymbol{\pi}) \cdot t \approx 0.0087t - 4.9 \times 10^{-6} \quad \text{for } t \in (0, T). \tag{34}$$

From Figure 5, one sees that $|b_{\text{theo}} - b_{\text{num}}| \sim \text{Ord}(\epsilon)$ only for $t \in (0, T_1)$. The critical instant of time at which $|b_{\text{theo}} - b_{\text{num}}| \neq \text{Ord}(\epsilon)$ is observed at $t \approx 0.05$ [s]; thus, approximately $T_1 \approx 0.05$ [s].

6.2. The Viscous-Inertial Regime Law $b(\boldsymbol{\pi}; t) = \psi_{2/3}(\boldsymbol{\pi}) \cdot t^{2/3}$. The *viscous-inertial* coalescence law $b(\boldsymbol{\pi}; t) = \psi_{2/3}(\boldsymbol{\pi}) \cdot t^{2/3}$ is demonstrated numerically in the present section. The procedure is similar to that discussed in the above section, differing only in the upper bound of the time T. That is, to demonstrate numerically the law $b(\boldsymbol{\pi}; t) = \psi_{2/3}(\boldsymbol{\pi}) \cdot t^{2/3}$, the property values exposed in Table 1 are used, and set $T \approx 0.07$ [s] (*i.e.*, $T > T_1$). Furthermore, since knowing the pair of scalar fields $(t_i, b_i)_{i\geq 0}$ yields the desired law in the (t, b)-plane numerically, to settle the 2/3-exponent law one then proceed in likely the same way as in the earlier section (see Section 6.1).

In Figure 6 are graphed the graphical results. Circular markers ∘ are used to represent the numerical data, and the solid line the equation corresponding fitted power law. That fitted power law curve is shaped as

$$b(\boldsymbol{\pi}; t) = \psi_{2/3}(\boldsymbol{\pi}) \cdot t^{2/3} \approx 0.0045t^{2/3} \quad \text{for } t \in (0, T). \tag{35}$$

It is understood that $|b_{\text{theo}} - b_{\text{num}}| \sim \text{Ord}(\epsilon)$ only for $t \in (0, T_{2/3})$. The critical instant of time at which $|b_{\text{theo}} - b_{\text{num}}| \neq \text{Ord}(\epsilon)$ occurs at $t \approx 0.06$ [s]; thus, approximately $T_{2/3} \approx 0.06$ [s].

6.3. The Inertial Regime Law $b(\boldsymbol{\pi}; t) = \psi_{1/2}(\boldsymbol{\pi}) \cdot t^{1/2}$. The power law $b(\boldsymbol{\pi}; t) = \psi_{1/2}(\boldsymbol{\pi}) \cdot t^{1/2}$ is here validated numerically. To do so numerical experiments were first run. The geometrical properties, initial and boundary conditions imposed, and the physical properties of the working fluid are those tabulated in Table 1. The said power law is then established in exactly the same way as in the foregoing two sections (see Sections 6.1 and 6.2).

In Figure 7 are exposed to view their graphical results. Circular markers ∘ are used to represent the numerical data, and the solid line the corresponding fitted power law. Through numerical experiments, it is believed that the following scalar function—representative of that solid line curve—gives good agreement:

$$b(\boldsymbol{\pi}; t) = \psi_{1/2}(\boldsymbol{\pi}) \cdot t^{1/2} \approx 0.0026t^{0.49} \quad \text{for } t \in (0, T). \tag{36}$$

In regard to (36), it follows that $\psi_{1/2}(\boldsymbol{\pi}) \approx 0.0026$. The power-the law $b(\boldsymbol{\pi}; t) = \psi_{1/2}(\boldsymbol{\pi}) \cdot t^{1/2}$ is thus proved numerically, demonstrated experimentally by [18, 21, 27]. It may be remarked that $|b_{\text{theo}} - b_{\text{num}}| \sim \text{Ord}(\epsilon)$ only for $t \in (0, T_{1/2})$. The critical instant of time at which $|b_{\text{theo}} - b_{\text{num}}| \neq \text{Ord}(\epsilon)$ occurs at $t \approx 0.17$ [s]; thus, approximately $T_{1/2} \approx 0.17$ [s]. The values of the scalar fields $T_i > 0$ ($i = 1/2, 2/3, 1$) are observed to be roughly equal to these values:

$$\begin{aligned} T_1 &\approx 0.05 \text{ [s]}, \\ T_{2/3} &\approx 0.06 \text{ [s]}, \text{ and} \\ T_{1/2} &\approx 0.17 \text{ [s]}. \end{aligned} \tag{37}$$

Moreover, from (36), (34), (35), one also observes that

$$\text{Ord}(\psi_\alpha(\boldsymbol{\pi})) \sim 10^{-3} \, [\text{mm}\cdot\text{s}^{-\alpha}] \quad \text{for all } \alpha \in I_\alpha. \tag{38}$$

In what follows the $\boldsymbol{\pi}$-dependent sequence $(\psi_\alpha(\boldsymbol{\pi}))_{\alpha\in I_\alpha}$ are analysed.

6.4. Of the Characters of $(\psi_\alpha(\boldsymbol{\pi}))_{\alpha\in I_\alpha}$. When the literature was surveyed, one found no more than two elements of the sequence $(\psi_\alpha(\boldsymbol{\pi}))_{\alpha\in I_\alpha}$ have gained attention, yet their establishment in terms of the property constants involved in the problems they ultimately lie have been correctly settled. With this view in mind, it is proposed to discuss in the present section the characters of $(\psi_\alpha(\boldsymbol{\pi}))_{\alpha\in I_\alpha}$ through analysis.

The analysis commences by relating the leading property associated with the interfacial coalescence hydrodynamics to that associated with the bulk coalescence hydrodynamics, that is, it is proposed to relate $b(\boldsymbol{\pi}; t)$ to $\mathbf{u}(\mathbf{x}, t)$ on Γ_{xz}. It is clear that $b(\boldsymbol{\pi}; t) = \|\mathbf{x} - \mathbf{x}_s\|$, where $\mathbf{x}, \mathbf{x}_s \in \Pi_{xz}(\mathbf{x}_s)$. Therefore, taking its derivative, one obtains

$$\mathbf{u} \cdot (\mathbf{x} - \mathbf{x}_s) = \alpha |\psi_\alpha(\boldsymbol{\pi})|^2 t^{2\alpha-1} \quad \text{on } \Gamma_{xz} \times (0, T_\alpha), \text{ where } \alpha \in I_\alpha. \tag{39}$$

On the other hand, the leading property constants being $\rho, \mu, \gamma_0, b_0, R_1$, consequently, $\boldsymbol{\pi} = (\rho, \mu, \gamma_0, b_0, R_1)^T$. Given the characteristic scales $(T_\alpha, X_\alpha, U_\alpha, \Psi_\alpha)_{\alpha\in I_\alpha}$, one rescales the properties appearing in (39) as follows: $t \longmapsto T_\alpha t$, $\mathbf{x} \longmapsto \mathbf{x}_s + X_\alpha(\mathbf{x} - \mathbf{x}_s)$, $\mathbf{u} \longmapsto U_\alpha \mathbf{u}$, and $\psi_\alpha \longmapsto \Psi_\alpha \psi_\alpha$, yielding

$$\text{Ord}(|\Psi_\alpha(\boldsymbol{\pi})|) \sim \sqrt{\frac{U_\alpha X_\alpha}{\alpha T_\alpha^{2\alpha-1}}}, \quad \text{where } \alpha \in I_\alpha. \tag{40}$$

It now remains to express the right-hand side of (40) in terms of the property vector $\boldsymbol{\pi} = (\rho, \mu, \gamma_0, b_0, R_1)^T$. To do this, one reasons according to the α-regimes, where $\alpha \in I_\alpha$.

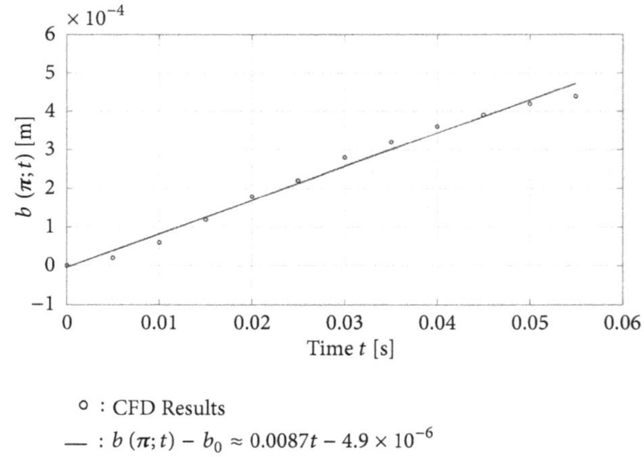

FIGURE 5: Time evolution of the neck heights $b(\pi; t)$ in the viscous regime. The *circular markers* ∘ indicate the numerical experiment results and the solid line represents the fitted curve.

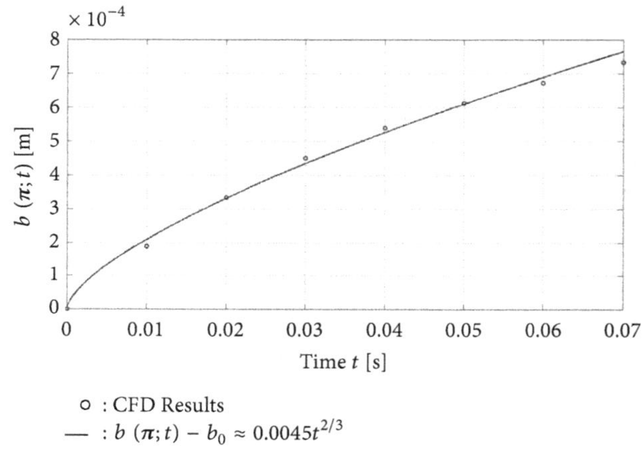

FIGURE 6: Time evolution of the neck heights $b(\pi; t)$ in the inertial-viscous regime. The *circular markers* ∘ indicate the numerical experiment results and the solid line represents the fitted curve.

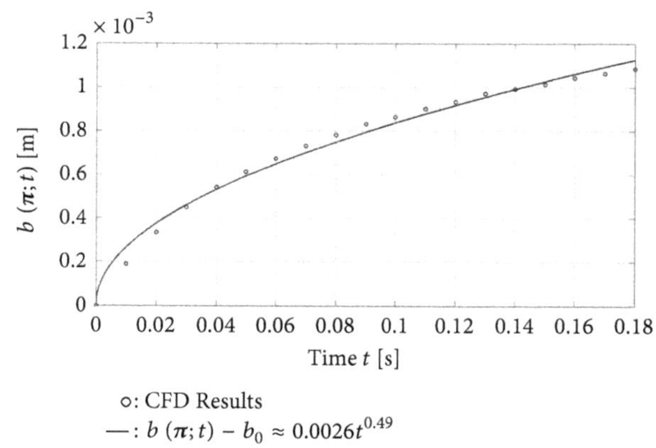

FIGURE 7: Time evolution of the neck heights $b(\pi; t)$ in the inertial regime. The *circular markers* ∘ indicate the numerical experiment results and the solid line represents the fitted curves.

Analysing the set of properties $(T_\alpha, X_\alpha, U_\alpha)_{\alpha\in I_\alpha}$ in terms of $\boldsymbol{\pi}$, the following orders of magnitude follow:

$$\begin{aligned}\text{Ord}\left(T_1, T_{2/3}, T_{1/2}\right) &\sim \left(\frac{\mu b_0}{\gamma_0}, \frac{\mu b_0}{\gamma_0}, \sqrt{\frac{\rho R_1^3}{\gamma_0}}\right),\\ \text{Ord}\left(X_1, X_{2/3}, X_{1/2}\right) &\sim \left(b_0, b_0, R_1\right),\\ \text{Ord}\left(U_1, U_{2/3}, U_{1/2}\right) &\sim \left(\frac{\gamma_0}{\mu}, \frac{\gamma_0}{\mu}, \sqrt{\frac{\gamma_0}{\rho R_1}}\right).\end{aligned} \tag{41}$$

These being so, substituting (41) into (40), one finds

$$\begin{aligned}&\text{Ord}\left(\left|\Psi_1(\boldsymbol{\pi})\right|, \left|\Psi_{2/3}(\boldsymbol{\pi})\right|, \left|\Psi_{1/2}(\boldsymbol{\pi})\right|\right)\\ &\sim \left(\frac{\gamma_0}{\mu}, \sqrt{\frac{3}{2}}\left(\frac{\gamma_0^2 b_0}{\mu^2}\right)^{1/3}, \sqrt{2}\left(\frac{\gamma_0 R_1}{\rho}\right)^{1/4}\right).\end{aligned} \tag{42}$$

Moreover, inserting the property constants of Table 1 into (42) and, thereafter, dividing each element of $(\psi_\alpha(\boldsymbol{\pi})\rfloor_{\text{theo}})_{\alpha\in I_\alpha}$ by the corresponding element of $(\psi_\alpha(\boldsymbol{\pi})\rfloor_{\text{num}})_{\alpha\in I_\alpha}$, the orders of magnitude of the ratio of $\Psi_\alpha(\boldsymbol{\pi})\rfloor_{\text{theo}}$ to $\Psi_\alpha(\boldsymbol{\pi})\rfloor_{\text{num}}$ yield

$$\text{Ord}\left(\frac{\Psi_\alpha(\boldsymbol{\pi})\rfloor_{\text{theo}}}{\Psi_\alpha(\boldsymbol{\pi})\rfloor_{\text{num}}}\right) \approx 4.5 \sim \text{Ord}(1) \quad \text{for all } \alpha \in I_\alpha. \tag{43}$$

In (43), it is understood that quantities $\Psi_\alpha(\boldsymbol{\pi})\rfloor_{\text{theo}}$, $\Psi_\alpha(\boldsymbol{\pi})\rfloor_{\text{num}}$ stand for the theoretical and numerical estimation, respectively, of the scalar field $\Psi_\alpha(\boldsymbol{\pi})$. Expressing (34), (35) and (36) in terms of (42) gives

$$\begin{aligned}b(\boldsymbol{\pi}; t) &\sim \frac{\gamma_0}{\mu} t \quad \text{for } t \in \left(0, T_1\right),\\ b(\boldsymbol{\pi}; t) &\sim \sqrt{\frac{3}{2}}\left(\frac{\gamma_0^2 b_0}{\mu^2}\right)^{1/3} t^{2/3} \quad \text{for } t \in \left(0, T_{2/3}\right),\\ b(\boldsymbol{\pi}; t) &\sim \sqrt{2}\left(\frac{\gamma_0 R_1}{\rho}\right)^{1/4} t^{1/2} \quad \text{for } t \in \left(0, T_{1/2}\right).\end{aligned} \tag{44}$$

Equation (44) at once justifies the reasonableness of the foregoing chains of reasoning and, this completes the discussion of the present section concerning the $(\boldsymbol{\pi}; t)$-dependent evolution of the field $b(\boldsymbol{\pi}; t)$. In the next section, it is proposed to enter upon the discussion of the characters of the curve $\mathbf{u}(\mathbf{x}, t)$ along a carefully chosen one-dimensional boundary.

6.5. Of the Characters of $\mathbf{u}(\mathbf{x}, t)$. If the velocity norm $\|\mathbf{u}\|$ of the interacting droplets systems is plotted on the curved boundary Γ_{yz}, the curve traced by the intersection of the interface $\partial(\Omega_1 \cup \Omega_2)$ with the plane $\Pi_{yz}(\mathbf{x}_s)$, one obtains Figure 8. The placement of arrows on the $\mathbf{u}$ curves of Figure 8 serve merely to indicate the evolution of time evolution of these curves. In Figure 8, the graphs of the velocity norm $\|\mathbf{u}\|$ correspond to the sequence of times $(t_k)_{k=0}^5 = (10k)$ [ms]. The nature of these curves are now explained.

Inspecting Figure 8, one sees that the highly curved meniscus induces an ϵ-order singularity into the coalescence flow in the early-time regime, peaking to magnitudes of the order of $\|\mathbf{u}\| \approx 20$ mm·s^{-1} inside that highly curved meniscus zone (indicated by dashed, vertical lines). It is perhaps because of this striking effect that some authors have characterised such motion as *singular motion* [7, 22].

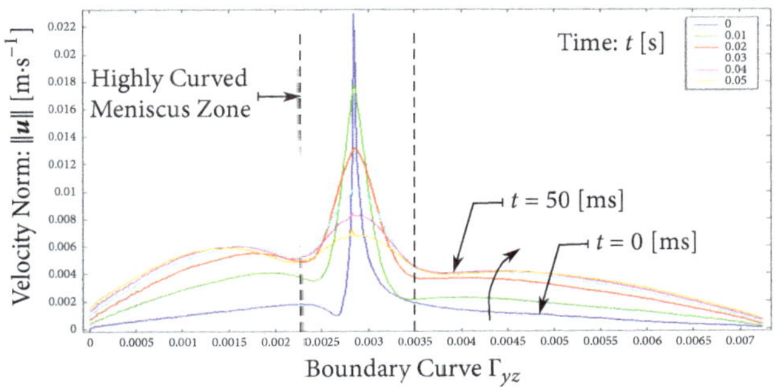

FIGURE 8: Time evolution of the velocity norm $\|\mathbf{u}\|$ of two interacting droplets with radii R_1, R_2 (with $R_2 = R_1$) along the boundary curve Γ_{yz} for the sequence of times $(t_k)_{k=0}^5 = (10k)_{k=0}^5$ [ms].

It is interesting to see the spatiotemporal distributions of the field variable $\mathbf{u}(\mathbf{x}, t)$. This is the purpose of the next section.

7. The Coalescence Flow Fields Illustrated

Having discussed the growth of the time-dependent function $b(\boldsymbol{\pi}; t)$ in the inertial, inertial-viscous and viscous regimes, the evolutions of the vector field function $\mathbf{u}(\mathbf{x}, t)$ with respect to $(\mathbf{x}, t) \in \bigcup_{i=1,2} \Omega_i \times (0, T)$ are now reported. In few words, the present endeavours is an attempt to lay additional stress on the qualitative aspects of the proposed model.

When the coalescence domain $\bigcup_{i=1,2} \Omega_i$ undergoes coalescence phenomenon, as soon as its pair of droplets are brought into contact, a mass movement is started, thereby causing the corresponding droplets to fuse. It is not necessary to go into details; sufficient is to demonstrate that the proposed model is capable of describing the underlying motion phenomena and the latter are exposed in the 3D space and on the 2D planes $\Pi_{\alpha z}(\mathbf{x}_s)$ (with $\alpha = x, y$).

In Figures 9 and 10, the time evolutions of the velocity norm $\|\mathbf{u}\|$ of the coalescence domain $\bigcup_{i=1,2} \Omega_i$ at times t = 5, 50, 100, 190 [ms] are exposed in the 3D space. The corresponding two-dimensional plots of the scalar field $\|\mathbf{u}\|$ are reported, respectively, on the 2D planes $\Pi_{\alpha z}(\mathbf{x}_s)$ (with $\alpha = x, y$) given in Figures 11–14, where solid lines stand for the iso-$\|\mathbf{u}\|$ contours.

8. Closing Remarks

In this part, a three-dimensional modelling methodology has been proposed to study droplets coalescence phenomenon. On the whole, the statements arrived at conclusively are:

When, upon variations of geometrical parameters, the footprint $\bigcup_{i=1,2}(\Omega_i \cap \mathcal{S})$ and the free-surface $\partial(\Omega_1 \cup \Omega_2)$ of the bidroplet system $\bigcup_{i=1,2} \Omega_i$ reduce to the footprint $\Omega_2 \cap \mathcal{S}$ and

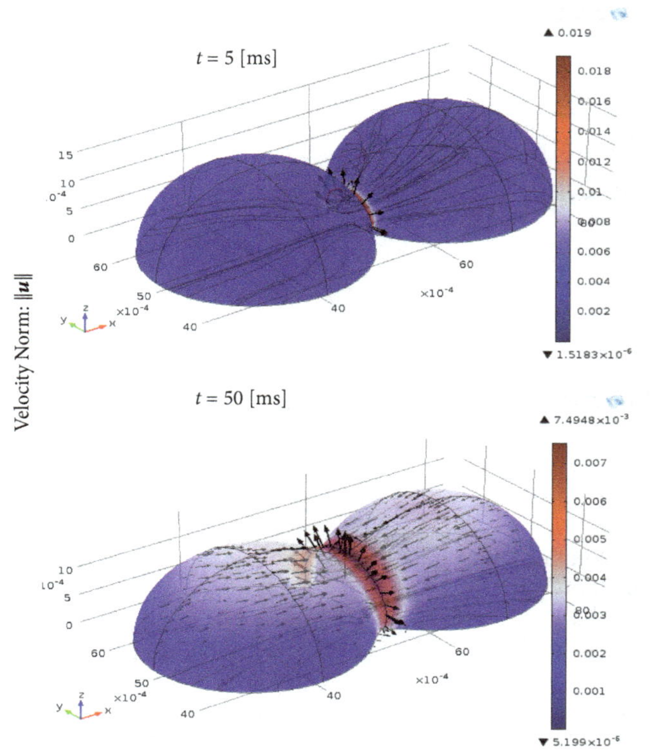

FIGURE 9: Three-dimensional plots of the velocity norm $\|\mathbf{u}\|$ of two interacting droplets of the same radius $R_1 = R_2$ at times $t = 5, 50$ [ms].

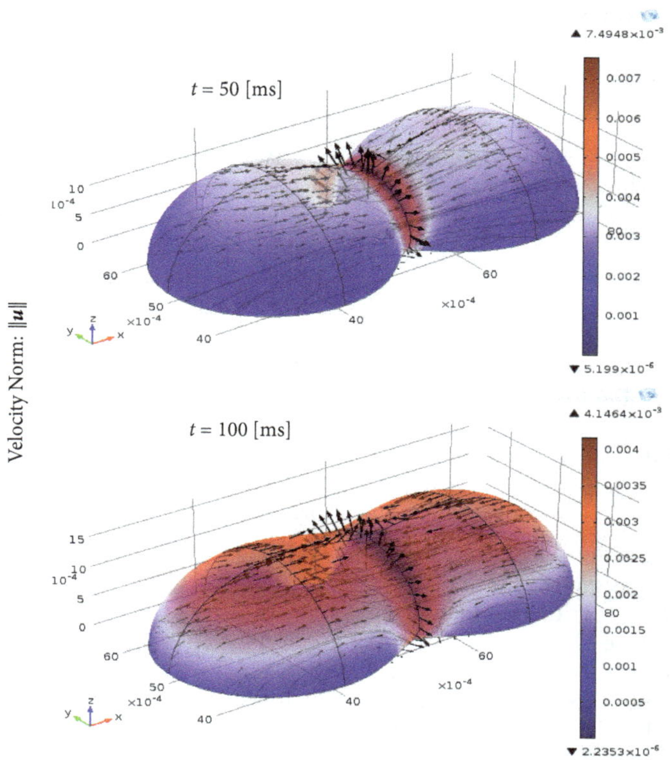

FIGURE 10: Three-dimensional plots of the velocity norm $\|\mathbf{u}\|$ of two interacting droplets of the same radius $R_1 = R_2$ at times $t = 100, 190$ [ms].

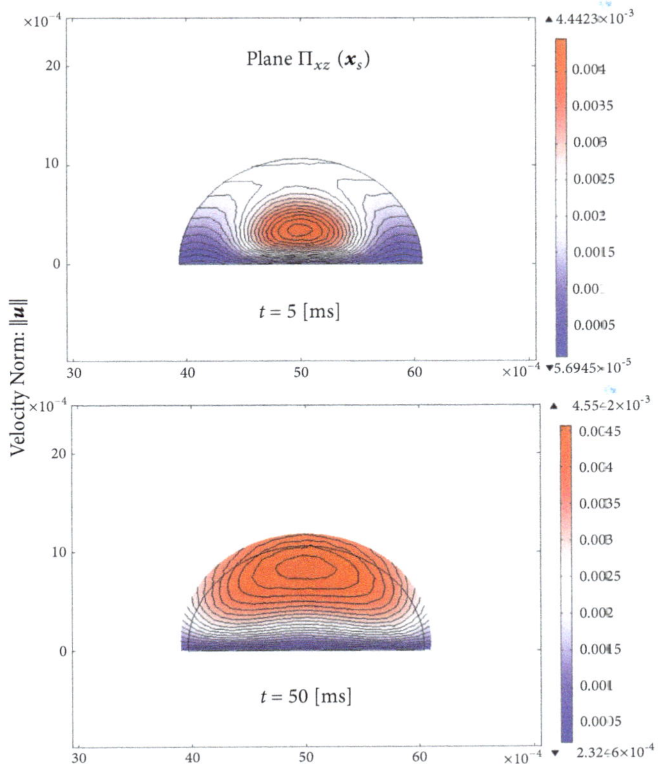

FIGURE 11: Two-dimensional plots of the velocity norm $\|\mathbf{u}\|$ of two interacting droplets of the same radius $R_1 = R_2$ on the plane $\Pi_{xz}(\mathbf{x}_s)$ at times $t = 5, 50$ [ms]. The solid lines stand for the iso-$\|\mathbf{u}\|$ contours.

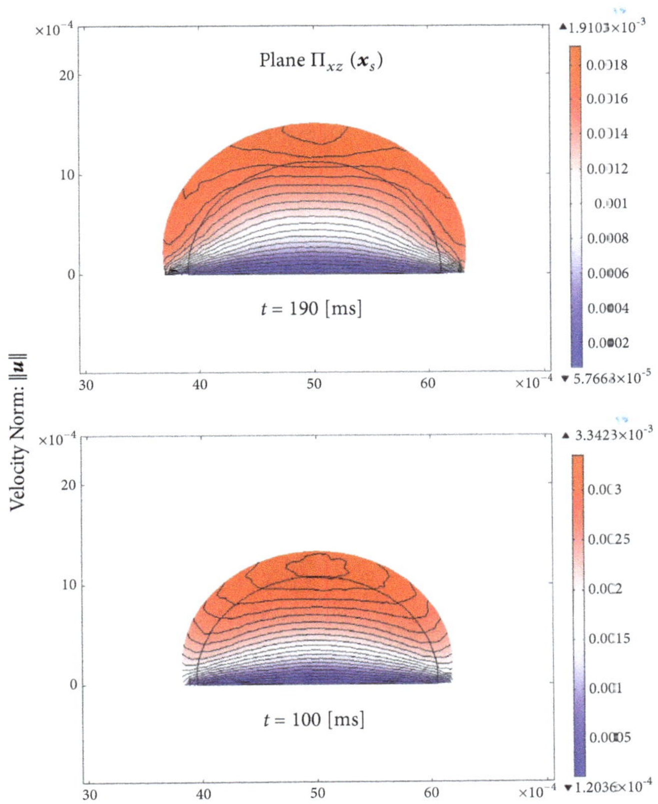

FIGURE 12: Two-dimensional plots of the velocity norm $\|\mathbf{u}\|$ of two interacting droplets of the same radius $R_1 = R_2$ on the plane $\Pi_{xz}(\mathbf{x}_s)$ at times $t = 100, 190$ [ms]. The solid lines stand for the iso-$\|\mathbf{u}\|$ contours.

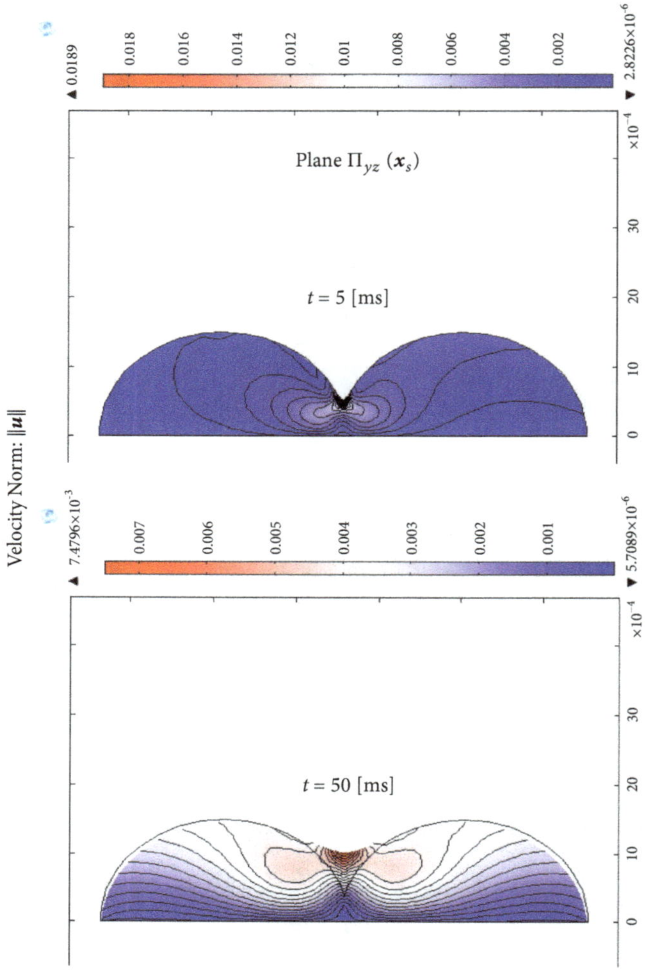

FIGURE 13: Two-dimensional plots of the velocity norm $\|\mathbf{u}\|$ of two interacting droplets of the same radius $R_1 = R_2$ on the plane $\Pi_{xz}(\mathbf{x}_s)$ at times $t = 5, 50$ [ms]. The solid lines stand for the iso-$\|\mathbf{u}\|$ contours.

the free-surface $\partial\Omega_2$, respectively, of a monodroplet system Ω_2, the corresponding boundary conditions are by default adapted into the COMSOL interfaces. As a result, with one numerical model, its property constants can be so arranged so as to compute many physical situations, a task which one would have undertaken had the proposed model been unversatile.

Taking advantage of the geometrical features of the proposed model, it is possible to deduce through theoretical analysis the scaling laws $b(\boldsymbol{\pi}; t) = \psi_\alpha(\boldsymbol{\pi}) \cdot t^\alpha$ (with $\alpha = 1/2$, $2/3$, 1), laws already reported in the (experimental) literature.

Carried out numerical proofs of these laws, $\alpha \approx 1$ is observed in the visco-dominated coalescence flow, $\alpha \approx 2/3$ in the inertia-visco-dominated coalescence flow, and $\alpha \approx 1/2$ in the inertia-dominated coalescence flow.

Through scaling analysis, it is found that the dependence of the sequence of prefactors $(\psi_\alpha(\boldsymbol{\pi}))_{\alpha \in I_\alpha}$ on $\boldsymbol{\pi} = (\rho, \mu, \gamma_0, b_0, R_1)^T$ can be approximated as follows: $\psi_1(\boldsymbol{\pi}) \approx \gamma_0/\mu$, $\psi_{2/3}(\boldsymbol{\pi}) \approx \sqrt{3/2}(\gamma_0^2 b_0/\mu^2)^{1/3}$ and $\psi_{1/2}(\boldsymbol{\pi}) \approx \sqrt{2}(\gamma_0 R_1/\rho)^{1/4}$.

The flow field variables illustrated by $2D$, $3D$ plots clearly demonstrate that the expected features associated with droplets coalescence phenomenon are well captured by the model.

On the whole, yet the approach in the present analysis has been to present the leading features of the proposed model by considering the coalescence that results under an idealized situation, it is no way contradictory to declare that it can be used to model more complex physical situations. For example, numerical experiments on droplets coalescence phenomenon can be carried out under appropriate CFD conditions to display useful information concerning mixing process driven by Marangoni flow.

Disclosure

The current address of M. I. Khodabocus and M. Sellier is Mechanical Engineering Department, University of

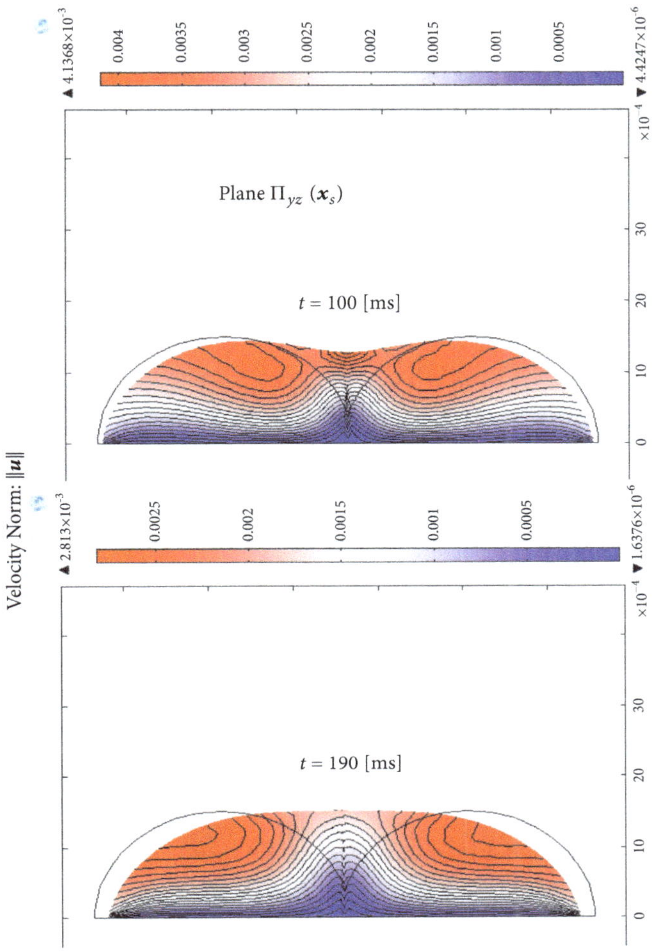

FIGURE 14: Two-dimensional plots of the velocity norm $\|\mathbf{u}\|$ of two interacting droplets of the same radius $R_1 = R_2$ on the plane $\Pi_{xz}(\mathbf{x}_s)$ at times $t = 100, 190$ [ms]. The solid lines stand for the iso-$\|\mathbf{u}\|$ contours.

Canterbury, Private Bag 4800, Christchurch 8140, New Zealand. The current address of V. Nock is Electrical and Computer Engineering Department, University of Canterbury, Private Bag 4800, Christchurch 8140, New Zealand.

Conflicts of Interest

The authors declare that they have no conflicts of interest.

References

[1] J. D. Paulsen, J. C. Burton, S. R. Nagel, S. Appathurai, M. T. Harris, and O. A. Basaran, "The inexorable resistance of inertia determines the initial regime of drop coalescence," *Proceedings of the National Acadamy of Sciences of the United States of America*, vol. 109, no. 18, pp. 6857–6861, 2012.

[2] M. M. Wu, T. Cubaud, and C.-M. Ho, "Scaling law in liquid drop coalescence driven by surface tension," *Physics of Fluids*, vol. 16, no. 7, pp. L51–L54, 2004.

[3] D. G. Aarts, H. N. Lekkerkerker, H. Guo, G. H. Wegdam, and D. Bonn, "Hydrodynamics of droplet coalescence," *Physical Review Letters*, vol. 95, no. 16, 2005.

[4] C. Andrieu, D. A. Beysens, V. S. Nikolayev, and Y. Pomeau, "Coalescence of sessile drops," *Journal of Fluid Mechanics*, vol. 453, pp. 427–438, 2002.

[5] M. Sellier, V. Nock, and C. Verdier, "Self-propelling, coalescing droplets," *International Journal of Multiphase Flow*, vol. 37, no. 5, pp. 462–468, 2011.

[6] M. Sellier, V. Nock, C. Gaubert, and C. Verdier, "Droplet actuation induced by coalescence: experimental evidences and phenomenological modeling," *The European Physical Journal Special Topics*, vol. 219, no. 1, pp. 131–141, 2013.

[7] J. Eggers, J. R. Lister, and H. A. Stone, "Coalescence of liquid drops," *Journal of Fluid Mechanics*, vol. 401, pp. 293–310, 1999.

[8] J. E. Sprittles and Y. D. Shikhmurzaev, "Coalescence of liquid drops: Different models versus experiment," *Physics of Fluids*, vol. 24, no. 12, 2012.

[9] J. J. Thomson and H. F. Newall, "On the formation of vortex rings by drops falling into liquids, and some allied phenomena," *Proceedings of The Royal Society of London*, vol. 39, no. 239-241, pp. 417–436, 1885.

[10] X. Chen, S. Mandre, and J. J. Feng, "Partial coalescence between a drop and a liquid-liquid interface," *Physics of Fluids*, vol. 18, no. 5, p. 051705, 2006.

[11] J. M. Rallison and A. Acrivos, "A numerical study of the deformation and burst of a viscous drop in an extensional flow," *Journal of Fluid Mechanics*, vol. 89, no. 1, pp. 191–200, 1978.

[12] R. W. Hopper, "Plane Stokes flow driven by capillarity on a free surface," *Journal of Fluid Mechanics*, vol. 213, pp. 349–375, 1990.

[13] R. W. Hopper, "Stokes flow of a cylinder and half-space driven by capillarity," *Journal of Fluid Mechanics*, vol. 243, pp. 171–181, 1992.

[14] R. W. Hopper, "Coalescence of two viscous cylinders by capillarity: Part I, theory," *Journal of the American Ceramic Society*, vol. 76, no. 12, pp. 2947–2952, 1993.

[15] R. W. Hopper, "Coalescence of two viscous cylinders by capillarity: Part II, shape evolution," *Journal of the American Ceramic Society*, vol. 76, no. 12, pp. 2953–2960, 1993.

[16] S. Richardson, "Two-dimensional slow viscous flows with time-dependent free boundaries driven by surface tension," *European Journal of Applied Mathematics*, vol. 3, no. 3, pp. 193–207, 1992.

[17] W. Yao, H. J. Maris, P. Pennington, and G. M. Seidel, "Coalescence of viscous liquid drops," *Physical Review E: Statistical, Nonlinear, and Soft Matter Physics*, vol. 71, no. 1, 2005.

[18] A. Eddi, K. G. Winkels, and J. H. Snoeijer, "Influence of droplet geometry on the coalescence of low viscosity drops," *Physical Review Letters*, vol. 111, no. 14, 2013.

[19] R. F. Day, E. J. Hinch, and J. R. Lister, "Self-similar capillary pinchoff of an inviscid fluid," *Physical Review Letters*, vol. 80, no. 4, pp. 704–707, 1998.

[20] F. Blanchette, L. Messio, and J. W. M. Bush, "The influence of surface tension gradients on drop coalescence," *Physics of Fluids*, vol. 21, no. 7, 2009.

[21] W. D. Ristenpart, P. M. McCalla, R. V. Roy, and H. A. Stone, "Coalescence of spreading droplets on a wettable substrate," *Physical Review Letters*, vol. 97, no. 6, 2006.

[22] J. D. Paulsen, R. Carmigniani, A. Kannan, J. C. Burton, and S. R. Nagel, "Coalescence of bubbles and drops in an outer fluid," *Nature Communications*, vol. 5, pp. 1–7, 2014.

[23] B. Lafaurie, C. Nardone, R. Scardovelli, S. Zaleski, and G. Zanetti, "Modelling merging and fragmentation in multiphase flows with SURFER," *Journal of Computational Physics*, vol. 113, no. 1, pp. 134–147, 1994.

[24] C. L. Navier, "Mémoire sur les Lois Mouvement des Fluides," *Mémoires de l'Académie Royale des Sciences de l'Institut de France*, vol. 6, pp. 389–440, 1823.

[25] J.-F. Gerbeau and T. Lelievre, "Generalized Navier boundary condition and geometric conservation law for surface tension," *Computer Methods Applied Mechanics and Engineering*, vol. 198, no. 5-8, pp. 644–656, 2009.

[26] R. Scardovelli and S. Zaleski, "Direct numerical simulation of freesurface and interfacial flow," *Annual Review of Fluid Mechanics*, vol. 31, no. 1, pp. 567–603, 1999.

[27] L. Duchemin, J. Eggers, and C. Josserand, "Inviscid coalescence of drops," *Journal of Fluid Mechanics*, no. 487, pp. 167–178, 2003.

14

Hom-Yang-Baxter Equations and Frobenius Monoidal Hom-Algebras

Yuanyuan Chen and Liangyun Zhang

College of Science, Nanjing Agricultural University, Nanjing, Jiangsu 210095, China

Correspondence should be addressed to Liangyun Zhang; zlyun@njau.edu.cn

Academic Editor: Andrei D. Mironov

It is shown that quasi-Frobenius Hom-Lie algebras are connected with a class of solutions of the classical Hom-Yang-Baxter equations. Moreover, a similar relation is discussed on Frobenius (symmetric) monoidal Hom-algebras and solutions of quantum Hom-Yang-Baxter equations. Monoidal Hom-Hopf algebras with Frobenius structures are studied at last.

1. Introduction

Hom-type algebras appeared first in physical contexts, in connection with twisted, discretized, or deformed derivatives and corresponding generalizations, discretizations, and deformations of vector fields and differential calculus. The notion of Hom-Lie algebras was introduced by Hartwig, Larsson, and Silvestrov in [1–3] as part of a study of deformations of Witt algebras and Virasoro algebras. Because of close relation to discrete and deformed vector fields and differential calculus, Hom-Lie algebras are widely studied recently; see [4–9].

Hom-associative algebras generalize the concept of associative algebras. They were introduced by Makhlouf and Silvestrov in [10]. Hom-associative algebras and their related structures have recently become rather popular, due to the prospect of having a general framework in which one can produce many types of natural deformations of algebras. Among them, there are such structures as Hom-coassociative coalgebras, Hom-Hopf algebras, Hom-alternative algebras, Hom-Jordan algebras, Hom-Poisson algebras, Hom-Leibniz algebras, infinitesimal Hom-bialgebras, Hom-power associative algebras, and quasi-triangular Hom-bialgebras (see [11–16]). Furthermore, some categories of Hom-modules on Hom-Hopf algebras are studied, such as the category of Hom-Hopf modules and the category of Yetter-Drinfel'd Hom-modules in [17].

The concept of Frobenius algebras is very important because of the connections to such diverse areas as group representations, homology of a compact oriented manifold, topological quantum field theories, quantum cohomology, Gorenstein rings in commutative algebras, Hopf algebras, coding theory, Lie quasi-Frobenius algebras, and the classical (quantum) Yang-Baxter equation (see [18–21]). In addition, there is a "quantum version" of the classical result that any finite dimensional Hopf algebra is Frobenius. The main properties of Frobenius algebras were developed by Nakayama in [22]. Nakayama automorphism is a distinguished k-algebra automorphism of a Frobenius algebra A which measures how far A is from being a symmetric algebra, where k is a fixed field. The automorphism is the identity if and only if A is symmetric. A connection between solutions of the classical Yang-Baxter equation and quasi-Frobenius Lie algebras was studied by Drinfel'd [23]. The authors in [19] gave the relation between Frobenius algebras and solutions of the quantum Yang-Baxter equation.

Naturally, we would like to do some research in Frobenius Hom-algebras. Some preliminary results about monoidal Hom-algebras with Frobenius property have been studied, such as when finite dimensional monoidal Hom-Hopf algebras are Frobenius associated with integral spaces [16], the structures of separable and Frobenius monoidal Hom-algebras related to the Hom-Frobenius-separability equation, as well as the Nakayama automorphism of Frobenius monoidal Hom-algebras [24], and so on.

The main purpose of this paper is to do further research in Frobenius Hom-algebras connected with solutions of the quantum Hom-Yang-Baxter equation. This paper is organized as follows. In Section 3, we give some preliminary results about surgenerate bilinear forms and the element $Q \in (M\otimes M, \mu\otimes\mu)$, where (M,μ) is an object in the Hom-category $\widetilde{\mathscr{H}}(\mathscr{M}_k)$. In addition, a connection between solutions of the classical Hom-Yang-Baxter equation (CHYBE) and quasi-Frobenius Hom-Lie algebras is also discussed. Specifically, for Hom-Lie algebra (L,α), such an element $Q \in L\otimes L$ satisfies CHYBE if and only if there exists a surgenerate bilinear form on (L,α) such that it is quasi-Frobenius. In Section 4, a similar relationship between Frobenius (symmetric) monoidal Hom-algebras and the quantum Hom-Yang-Baxter equation (QHYBE) is discussed. For any Frobenius monoidal Hom-algebra, there exists a class of solutions of the QHYBE. And a converse result when monoidal Hom-algebras are Frobenius is also presented. It is known that a finite dimensional monoidal Hom-Hopf algebra (H,α) is Frobenius if and only if the space of integrals in H is not zero. Last, the properties of monoidal Hom-Hopf algebras with Frobenius structures are studied in Section 5.

2. Preliminaries

Throughout the paper, all vector spaces, tensor products, and homomorphisms are over k. The symbol τ denotes the transposition map.

Let $\widetilde{\mathscr{H}}(\mathscr{M}_k)$ be the Hom-category associated with the category $\mathscr{M}_k$ of k-modules (see [15] cited by [16]). Throughout, we use the conception of [15] for convenience.

Definition 1. A *monoidal Hom-algebra* ([15, Proposition 2.1]) is a vector space A together with an element $1_A \in A$ and linear maps

$$m : A\otimes A \longrightarrow A; a\otimes b \longmapsto ab, \quad \alpha \in \mathrm{Aut}_k(A) \tag{1}$$

such that

$$\alpha(a)(bc) = (ab)\alpha(c), \tag{2}$$

$$\alpha(ab) = \alpha(a)\alpha(b), \tag{3}$$

$$a1_A = 1_A a = \alpha(a), \tag{4}$$

$$\alpha(1_A) = 1_A, \tag{5}$$

for all $a,b,c \in A$.

The definition of monoidal Hom-algebras is the same to the unital monoidal Hom-associative algebras defined in [16, Definition 2.1]. A monoidal Hom-algebra is indeed a unital and associative algebra in the Hom-category $\widetilde{\mathscr{H}}(\mathscr{M}_k)$. Let (A,α) be a monoidal Hom-algebra in the rest of this section.

Right (A,α)-Hom-modules have been introduced in [16, Definition 2.5]. In the following, we give the definition of a left (A,α)-Hom-module similarly.

Definition 2. A *left (A,α)-Hom-module* consists of (M,μ) in $\widetilde{\mathscr{H}}(\mathscr{M}_k)$ together with a morphism $\psi : A\otimes M \longrightarrow M, \psi(a\otimes m) = a\cdot m$ such that

$$\begin{aligned}\alpha(a)\cdot(b\cdot m) &= (ab)\cdot\mu(m),\\ \mu(a\cdot m) &= \alpha(a)\cdot\mu(m),\\ 1_A\cdot m &= \mu(m),\end{aligned} \tag{6}$$

for all $a,b \in A$ and $m \in M$.

A *morphism of left (A,α)-Hom-modules* is a morphism of left A-modules in the Hom-category $\widetilde{\mathscr{H}}(\mathscr{M}_k)$. $\widetilde{\mathscr{H}}({}_A\mathscr{M})$ will be denoted the category of left (A,α)-Hom-modules and left (A,α)-linear morphisms between them.

If (M,μ) is both a left (A,α)-Hom-module and a right (A,α)-Hom-module such that the following compatibility condition

$$\alpha(a)\cdot(m\cdot b) = (a\cdot m)\cdot\alpha(b) \tag{7}$$

holds, then (M,μ) is called an *(A,α)-Hom-bimodule*.

For any object $(M,\mu) \in \widetilde{\mathscr{H}}(\mathscr{M}_k)$, let $r : (M\otimes M, \mu\otimes\mu) \longrightarrow (M\otimes M, \mu\otimes\mu)$ be an morphism in $\widetilde{\mathscr{H}}(\mathscr{M}_k)$. Consider the maps

$$\begin{aligned} r^{12}, r^{13}, r^{23} &: (M\otimes M\otimes M, \mu\otimes\mu\otimes\mu)\\ &\longrightarrow (M\otimes M\otimes M, \mu\otimes\mu\otimes\mu)\end{aligned} \tag{8}$$

given by $r^{12} = r\otimes\mu$, $r^{13} = (\mathrm{id}\otimes\tau)(r\otimes\mu)(\mathrm{id}\otimes\tau)$, $r^{23} = r\otimes\mu$. We say that r is a *solution of the quantum Hom-Yang-Baxter equation* (or the QHYBE for convenience) if

$$r^{12}r^{13}r^{23} = r^{23}r^{13}r^{12}. \tag{9}$$

For elements $R = \sum_{i=1}^n a_i\otimes b_i \in A\otimes A$, where (A,α) is a monoidal Hom-algebra with unit 1_A. Then R acts on $A\otimes A$ via Hom-multiplication and is a *solution of the quantum Hom-Yang-Baxter equation* (or the QHYBE for convenience) if

$$\alpha^{\otimes 3}\left(R^{12}\right)\left(R^{13}R^{23}\right) = \left(R^{23}R^{13}\right)\alpha^{\otimes 3}\left(R^{12}\right). \tag{10}$$

where $R^{12} = \sum_{i=1}^n a_i\otimes b_i\otimes 1_A$, $R^{13} = \sum_{i=1}^n a_i\otimes 1_A\otimes b_i$, $R^{23} = \sum_{i=1}^n 1_A\otimes a_i\otimes b_i$. In what follows, we often omit the summation symbols for convenience.

3. Hom-Lie Algebras and Classical Hom-Yang-Baxter Equation

In this section, we give some preliminary results about surgenerate bilinear forms and a specific element $Q \in (M\otimes M, \mu\otimes\mu)$, where (M,μ) is an object in $\widetilde{\mathscr{H}}(\mathscr{M}_k)$. In addition, an equivalent relation between solutions of the classical Hom-Yang-Baxter equation (CHYBE) and quasi-Frobenius Hom-Lie algebras is also discussed.

Definition 3. A *Hom-Lie algebra* in [9] is a triple $(L,[-,-],\alpha)$ consisting of vector space L, bilinear map $[-,-] : L^{\otimes 2} \longrightarrow L$

(called the "bracket"), and a linear endomorphism $\alpha : L \longrightarrow L$ satisfying

$$\begin{aligned} &\alpha[x,y] = [\alpha(x),\alpha(y)], \quad (multiplicativity) \\ &[x,y]+[y,x]=0, \quad (anti-symmetric) \\ &\circlearrowleft [[x,y],\alpha(z)] = 0, \\ &(Hom-Jacobi\ identity) \end{aligned} \tag{11}$$

for any $x, y, z \in L$.

Now let $(L, [-,-], \alpha)$ be a Hom-Lie algebra. Element $R = \sum_{i=1}^{n} a_i \otimes b_i \in L \otimes L$ is called a solution of *the classical Hom-Yang-Baxter equation (or the CHYBE for convenience)* if

$$\sum_{i=1}^{n} a_i \otimes b_i = -\sum_{i=1}^{n} b_i \otimes a_i, \tag{12}$$

and

$$[R^{12}, R^{13}] + [R^{12}, R^{23}] + [R^{13}, R^{23}] = 0. \tag{13}$$

Here

$$\begin{aligned} [R^{12}, R^{13}] &= [a_i, a_j] \otimes \alpha(b_i) \otimes \alpha(b_j), \\ [R^{12}, R^{23}] &= \alpha(a_i) \otimes [b_i, a_k] \otimes \alpha(b_k), \\ [R^{13}, R^{23}] &= \alpha(a_j) \otimes \alpha(a_k) \otimes [b_j, b_k], \end{aligned} \tag{14}$$

where $R^{12} = R \otimes 1 = a_i \otimes b_i \otimes 1$, $R^{13} = (\tau \otimes id)(1 \otimes R) = a_j \otimes 1 \otimes b_j$ and $R^{23} = 1 \otimes R = 1 \otimes a_k \otimes b_k$.

Let $(M, \mu) \in \widetilde{\mathscr{H}}(\mathscr{M}_k)$ be finite dimensional. A bilinear form $B : (M \otimes M, \mu \otimes \mu) \longrightarrow (k, \mathrm{id})$ in $\widetilde{\mathscr{H}}(\mathscr{M}_k)$ is called *surgenerate* if there exists an isomorphism $\phi : (M, \mu) \longrightarrow (M^*, (\mu^*)^{-1})$ in $\widetilde{\mathscr{H}}(\mathscr{M}_k)$ given by

$$\phi(x)(y) = B(y, x). \tag{15}$$

Here B is a morphism in $\widetilde{\mathscr{H}}(\mathscr{M}_k)$ which means that $B(\mu(x), \mu(y)) = B(x, y)$, for all $x, y \in M$. And this makes sure that ϕ is also a morphism in $\widetilde{\mathscr{H}}(\mathscr{M}_k)$.

A Hom-Lie algebra $(L, [-,-], \alpha)$ is said to be *quasi-Frobenius* if L is finitely generated projective k-module having a skew-symmetric surgenerate bilinear form $B : L \otimes L \longrightarrow k$ satisfying the following equation:

$$\begin{aligned} &B([x,y],\alpha(z)) + B([y,z],\alpha(x)) + B([z,x],\alpha(y)) \\ &= 0, \end{aligned} \tag{16}$$

for all $x, y, z \in L$.

Lemma 4. *Let $(M, \mu) \in \widetilde{\mathscr{H}}(\mathscr{M}_k)$ be finite dimensional. Define a homomorphism $\rho_M : (M \otimes M, \mu \otimes \mu) \longrightarrow (Hom(M^*, M), \iota)$ by $\rho_M(a \otimes b)(f) = f(a)b$, for all $a, b \in M$, $f \in M^*$, where $\iota : Hom(M^*, M) \longrightarrow Hom(M^*, M)$ is the twisting given by $\iota(\psi) = \mu\psi\mu^*$. Then ρ_M is an isomorphism in $\widetilde{\mathscr{H}}(\mathscr{M}_k)$.*

Proof. First ρ_m is well defined in $\widetilde{\mathscr{H}}(\mathscr{M}_k)$, since $\rho_M(\mu(a) \otimes \mu(b))(f) = f\mu(a)\mu(b) = \mu(\rho_M(a \otimes b)\mu^*(f))$. Define a morphism $\lambda_M : Hom(M^*, M) \longrightarrow M \otimes M$ by $\psi \longmapsto e_i \otimes \psi(f^i)$, where (e_i, f^i) is a dual basis for M. We can easily check that λ_M is the inverse of ρ_M. Indeed, for any $a, b \in M$, $f \in M^*$,

$$\begin{aligned} \lambda_M \rho_M(a \otimes b) &= e_i \otimes \rho_M(a \otimes b)(f^i) = e_i \otimes f^i(a)(b) \\ &= a \otimes b, \end{aligned} \tag{17}$$

and

$$\begin{aligned} \rho_M \lambda_M(\psi)(f) &= \rho_M(e_i \otimes \psi(f^i))(f) \\ &= f(e_i)\psi(f^i) = \psi(f). \end{aligned} \tag{18}$$

□

The endomorphism algebra $\mathrm{End}_k(M, \mu)$ can be considered as a monoidal Hom-algebra, where the Hom-multiplication is the composite of morphisms, the unit is the identity homomorphism, and the twisting map $\iota : \mathrm{End}_k(M, \mu) \longrightarrow \mathrm{End}_k(M, \mu)$ is given by $\iota(\boldsymbol{\varphi}) = \mu\boldsymbol{\varphi}\mu^{-1}$, for $\boldsymbol{\varphi} \in \mathrm{End}_k(M, \mu)$.

Proposition 5. *Let $(M, \mu) \in \widetilde{\mathscr{H}}(\mathscr{M}_k)$ be finite dimensional with a surgenerate bilinear form $B : (M \otimes M, \mu \otimes \mu) \longrightarrow (k, \mathrm{id})$ in $\widetilde{\mathscr{H}}(\mathscr{M}_k)$. Further let $e_i \in M$, $f^i \in M^*$, $i = 1, 2, 3, \ldots, n$, be a dual basis of M, and $e^i = \phi^{-1}(f^i)$, $i = 1, 2, 3, \ldots, n$. We set $Q = e_i \otimes e^i \in M \otimes M$, and define a homomorphism in $\widetilde{\mathscr{H}}(\mathscr{M}_k)$*

$$\chi_M : (M \otimes M, \mu \otimes \mu) \longrightarrow (\mathrm{End}_k(M, \mu), \iota) \tag{19}$$

by

$$\begin{aligned} \chi_M(a \otimes b)(x) &= \langle \phi(b), \mu^{-1}(x) \rangle \mu(a) \\ &= B(x, \mu(b))\mu(a). \end{aligned} \tag{20}$$

Further we define the mapping $Tr : \mathrm{End}_k(M, \mu) \longrightarrow k$ as follows:

$$\begin{aligned} Tr(\boldsymbol{\varphi}) &= \langle f^i, \boldsymbol{\varphi}(e_i) \rangle = \langle \phi(e^i), \boldsymbol{\varphi}(e_i) \rangle \\ &= B(\boldsymbol{\varphi}(e_i), e^i), \end{aligned} \tag{21}$$

for any $\boldsymbol{\varphi} \in \mathrm{End}_k(M, \mu)$.

Then

(1) *$x = B(x, e^i)e_i = B(e_i, x)e^i$ for all $x \in M$.*

(2) *χ_M is an isomorphism in $\widetilde{\mathscr{H}}(\mathscr{M}_k)$.*

(3) *$Q = \rho_M^{-1}(\phi^{-1}) = \chi_M^{-1}(\mathrm{id}_M)$.*

(4) *The element Q does not depend on the choice of the dual bases e_i, e^i, $i = 1, 2, \ldots, n$, of M.*

(5) *The mapping Tr does not depend on the choice of the dual bases e_i, e^i, $i = 1, 2, \ldots, n$, of M.*

Proof. (1) Since $f^i(x) = \phi(e^i)(x) = B(x, e^i)$, $x = f^i(x)e_i = B(x, e^i)e_i$, for all $x \in M$. Given $x \in M$, set $y = B(e_i, x)e^i$. Then

$$\begin{aligned} B(z,y) &= B\left(z, B(e_i,x)e^i\right) = B(e_i,x)B\left(z,e^i\right) \\ &= B\left(B\left(z,e^i\right)e_i,x\right) = B(z,x), \end{aligned} \tag{22}$$

for all $z \in M$. Since B is surgenerate, $y = x$.

(2) Define a homomorphism $\varsigma : (M\otimes M^*, \mu\otimes(\mu^*)^{-1}) \longrightarrow (\mathrm{End}_k(M,\mu), \iota)$ in $\widetilde{\mathscr{H}}(\mathscr{M}_k)$ by

$$\varsigma(a\otimes f)(x) = f\left(\mu^{-1}\right)(x)\mu(a), a, x \in M, f \in M^*. \tag{23}$$

It is not difficult to show that ς is an isomorphism with an inverse $\varsigma^{-1} : f \longmapsto f(e_i)\otimes f^i$. Since $\chi_M = \varsigma(\mathrm{id}\otimes\phi)$, χ_M is an isomorphism in $\widetilde{\mathscr{H}}(\mathscr{M}_k)$.

(3) For all $k = 1, 2, \ldots, n$, we have

$$\begin{aligned} \rho_M(Q)\left(f^k\right) &= f^k(e_i)e^i = B\left(e_i,e^k\right)e^i = B\left(e_i,e^k\right)e^i \\ &= e^k \end{aligned} \tag{24}$$

by (1). On the other hand, $\phi^{-1}(f^k) = e^k$, for all $k = 1, 2, \ldots, n$, so $\rho_M(Q) = \phi^{-1}$, and hence $Q = \rho_M^{-1}(\phi^{-1})$. Further, $\chi_M(Q)(x) = B(\mu^{-1}(x), e^i)\mu(e_i) = x$ for all $x \in M$ by (1). Therefore $\chi_M(Q) = \mathrm{id}_M$, which implies $Q = \chi_M^{-1}(\mathrm{id}_M)$.

(4) follows directly from (3).

(5) For $\varphi \in \mathrm{End}(M,\mu)$, define the mapping $\Lambda : (M\otimes M, \mu\otimes\mu) \longrightarrow (k, \mathrm{id}_k)$ by $\Lambda(a\otimes b) = B(\varphi(a), b)$ in $\widetilde{\mathscr{H}}(\mathscr{M}_k)$. Obviously, $Tr(\varphi) = \Lambda(Q) = B(\varphi(e_i), e^i)$. It follows from (4) that $Tr(\varphi)$ does not depend on the choice of the dual bases $e_i, e^i, i = 1, 2, \ldots, n$, of M. □

Lemma 6. *Let $(M,\mu) \in \widetilde{\mathscr{H}}(\mathscr{M}_k)$ be finite dimensional with a surgenerate bilinear form $B : (M\otimes M, \mu\otimes\mu) \longrightarrow (k, \mathrm{id})$ in $\widetilde{\mathscr{H}}(\mathscr{M}_k)$. Further let $e_i \in M, f^i \in M^*, i = 1, 2, 3, \ldots, n$ be a dual basis of M. Choose $e^1, e^2, \ldots, e^n \in M$ such that $B(e_i, e^j) = \delta_{ij}$ for all $i, j = 1, 2, 3, \ldots, n$. We set $Q = e_i\otimes e^i \in M\otimes M$, and write $e^i = \sum_{j=1}^n l^{ji}e_j$, where $l^{ji} \in k$. Next we set $n_{ij} = \phi(e_j)(e_i)$, $N = (n_{ij})_{i,j=1}^n \in M_n(k)$, and $L = (l^{ji})_{i,j=1}^n$.*

Then

(1) *$LN = E = NL$, where E is the identity of $M_n(k)$;*

(2) *$Q = \sum_{i=1}^n l^{ji}e_i\otimes e_j$.*

Proof. Since $\delta_{ji} = B(e_j, e^i) = B(e_j, \sum_{k=1}^n l^{ki}e_k) = \sum_{k=1}^n B(e_j, e_k)l^{ki} = \sum_{k=1}^n b_{jk}r^{ki}$, the first assertion is true.

The second statement is obvious. □

Theorem 7. *Let $(L, [-,-], \alpha)$ be a Hom-Lie algebra with a skew-symmetric surgenerate bilinear form $B : L\otimes L \longrightarrow k$. Suppose that L is finite dimensional and $\{e_i, e^i\}, i = 1, 2, \ldots, n$ is a dual basis of L. We set $Q = e_i\otimes e^i \in L\otimes L$. Then*

(1) *$Q = \rho_L^{-1}(M)(\phi^{-1}) = \chi_L^{-1}(\mathrm{id}_L)$.*

(2) *The element Q does not depend on the choice of dual basis of L.*

(3) *The element Q satisfies CHYBE if and only if B satisfies (16).*

Proof. The first two statements follow from (3) and (4) of Proposition 5.

Next, if Q satisfies the CHYBE, then by (12) we have $e^j\otimes e_j = \tau(e_j\otimes e^j) = -e_j\otimes e^j$. Choose $w^i \in L$ such that $B(e^j, w^i) = \delta_{ij}$. Then from (4) of Proposition 5,

$$e^i\otimes w^i = e_i\otimes e^i = -e^i\otimes e_i, \tag{25}$$

so $w^i = -e_i$. Therefore we have

$$B\left(e_i,e^j\right) = B\left(e^j,w^i\right) = B\left(e^j,-e_i\right) = -B\left(e^j,e_i\right), \tag{26}$$

which proves that B is skew-symmetric.

Last, we only need to show that Q satisfies the equation $[Q^{12}, Q^{13}] + [Q^{12}, Q^{23}] + [Q^{13}, Q^{23}] = 0$ if and only if B satisfies (16). Since $e_i\otimes e^i = -e^i\otimes e_i$, we have

$$\begin{aligned} \left[Q^{12},Q^{13}\right] &= \left[e_i,e_j\right]\otimes\alpha\left(e^i\right)\otimes\alpha\left(e^j\right) \\ &= B\left(\left[e_i,e_j\right],e^k\right)e_k\otimes\alpha\left(e^i\right)\otimes\alpha\left(e^j\right) \\ &= B\left(\alpha^{-1}\left(\left[e_i,e_j\right]\right),e^k\right)\alpha\left(e_k\right)\alpha\left(e^i\right) \\ &\quad\otimes\alpha\left(e^j\right) \\ \left[Q^{12},Q^{23}\right] &= \alpha\left(e_k\right)\otimes\left[e^k,e_j\right]\otimes\alpha\left(e^j\right) \\ &= \alpha\left(e_k\right)\otimes B\left(e_i,\left[e^k,e_j\right]\right)e^i\otimes\alpha\left(e^j\right) \\ &= B\left(e_i,\left[e^k,e_j\right]\right)\alpha\left(e_k\right)\otimes e^i\otimes\alpha\left(e^j\right) \\ &= B\left(\left[e_j,e^k\right],e_i\right)\alpha\left(e_k\right)\otimes e^i\otimes\alpha\left(e^j\right) \\ &= B\left(\alpha^{-1}\left(\left[e_j,e^k\right]\right),e_i\right)\alpha\left(e_k\right)\otimes\alpha\left(e^i\right) \\ &\quad\otimes\alpha\left(e^j\right) \end{aligned} \tag{27}$$

and

$$\begin{aligned} \left[Q^{13},Q^{23}\right] &= -\alpha\left(e_k\right)\otimes\alpha\left(e^i\right)\otimes\left[e^k,e_i\right] \\ &= -\alpha\left(e_k\right)\otimes\alpha\left(e^i\right)\otimes B\left(e_j,\left[e^k,e_i\right]\right)e^j \\ &= -B\left(e_j,\alpha^{-1}\left(\left[e^k,e_i\right]\right)\right)\alpha\left(e_k\right)\otimes\alpha\left(e^i\right) \\ &\quad\otimes\alpha\left(e^j\right) \\ &= B\left(\alpha^{-1}\left(\left[e^k,e_i\right]\right),e_j\right)\alpha\left(e_k\right)\otimes\alpha\left(e^i\right) \\ &\quad\otimes\alpha\left(e^j\right). \end{aligned} \tag{28}$$

Then

$$\begin{aligned}&[Q^{12},Q^{13}]+[Q^{12},Q^{23}]+[Q^{13},Q^{23}]\\&=\left(B\left(\alpha^{-1}\left([e_i,e_j]\right),e^k\right)+B\left(\alpha^{-1}\left([e_j,e^k]\right),e_i\right)\right.\\&\left.+B\left(\alpha^{-1}\left([e^k,e_i]\right),e_j\right)\right)\alpha(e_k)\otimes\alpha(e^i)\otimes\alpha(e^j).\end{aligned} \tag{29}$$

So

$$[Q^{12},Q^{13}]+[Q^{12},Q^{23}]+[Q^{13},Q^{23}]=0 \tag{30}$$

if and only if

$$\begin{aligned}&B\left(\alpha^{-1}\left([e_i,e_j]\right),e^k\right)+B\left(\alpha^{-1}\left([e_j,e^k]\right),e_i\right)\\&+B\left(\alpha^{-1}\left([e^k,e_i]\right),e_j\right)=0\end{aligned} \tag{31}$$

which is equivalent to

$$\begin{aligned}&B\left([e_i,e_j],\alpha(e^k)\right)+B\left([e_j,e^k],\alpha(e_i)\right)\\&+B\left([e^k,e_i],\alpha(e_j)\right)=0.\end{aligned} \tag{32}$$

□

Note that if $(L,[-,-],\alpha)$ is a quasi-Frobenius Hom-Lie algebra, then Lemma 6 implies that $r^{ij}e_i\otimes e_j=e^j\otimes e_j=-Q$ is skew-symmetric.

4. Frobenius Monoidal Hom-Algebras and Quantum Hom-Yang-Baxter Equations

A relationship between Frobenius (symmetric) monoidal Hom-algebras and QHYBE is mainly discussed in this section.

For a finite dimensional monoidal Hom-algebra (A,α), we can check that $(A^*,(\alpha^*)^{-1})$ is a right (A,α)-Hom-module via the action $\leftharpoonup$ given by

$$\theta\leftharpoonup a=\theta\alpha^{-2}(\alpha(a)-)=\theta_1\alpha^{-1}(a)(\alpha^*)^{-2}(\theta_2), \tag{33}$$

where $\Delta(\theta)=\theta_1\otimes\theta_2$ and Δ is the dual Hom-coalgebra structure on $(A^*,(\alpha^*)^{-1})$, for all $a\in A,\theta\in A^*$.

Definition 8 (see [16, Definition 5.1]). A finite dimensional monoidal Hom-algebra (A,α) is called *Frobenius* if

$$\phi:(A,\alpha)\longrightarrow\left(A^*,(\alpha^*)^{-1}\right) \tag{34}$$

is an isomorphism as right (A,α)-Hom-modules.

Recall from [16, Proposition 5.2] that a finite dimensional monoidal Hom-algebra being Frobenius has several equivalent characterization as follows.

Lemma 9. *For a finite dimensional monoidal Hom-algebra (A,α), the following assertions are equivalent.*

(1) *(A,α) is Frobenius.*

(2) *$(A,\alpha)\cong(A^*,(\alpha^*)^{-1})$ as left (A,α)-Hom-modules, which is called Frobenius isomorphism.*

(3) *There exists a Frobenius structure of (A,α). Equivalently, there exist elements $\varphi\in A^*$ (called Frobenius homomorphism), and $a_i\otimes b_i\in A\otimes A$ such that $\varphi\alpha=\varphi,\alpha(a_i)\otimes\alpha(b_i)=a_i\otimes b_i$ and $\varphi(xa_i)b_i=a_i\varphi(b_ix)=x$, for any $x\in A$.*

(4) *There exists a Hom-associative, nondegenerate bilinear form for (A,α). That is, there exists a bilinear map $B:A\times A\longrightarrow k$ in $\widetilde{\mathscr{H}}(\mathscr{M}_k)$, such that $B(\alpha(a),bc)=B(ab,\alpha(c))$, and if $B(x,a)=0$ or $B(a,x)=0$ for any $x\in A$, then $a=0$.*

Let (A,α) be a finite dimensional monoidal Hom-algebra. And further we assume that (A,α) is Frobenius with Frobenius isomorphism $\phi:(A,\alpha)\longrightarrow(A^*,(\alpha^*)^{-1})$, Frobenius structure $\varphi\in A^*$ and $a_i\otimes b_i\in A\otimes A$, and the corresponding bilinear form $B:A\otimes A\longrightarrow k$. By the proof of the above lemma, we know that

$$B(x,y)=\varphi(xy)=\phi(y)(x), \tag{35}$$

for all $x,y\in A$. Then there exists a dual basis $e_1,e_2,\dots,e_n\in A$ and $f^1,f^2,\dots,f^n\in A^*$ such that $x=f^i(x)e_i=\sum_{i=1}^n\langle f^i,x\rangle e_i$. Setting $e^i=\phi^{-1}(f^i),i=1,2,\dots,n$, we note that $f^i(x)=\phi(e^i)(x)=\varphi(xe^i)$ for all $x\in A,i=1,2,\dots,n$. So $x=\varphi(xe^i)e_i$ for all $x\in A$. Thus we can also refer to $e_1,e_2,\dots,e_n$ and $e^1,e^2,\dots,e^n$ as dual basis of A.

Note that there is an automorphism $\sigma:(A,\alpha)\longrightarrow(A,\alpha)$ called *the Nakayama automorphism of* (A,α) defined in [16, Section 5]. It is easy to check that σ is a Hom-algebra automorphism satisfying $B(x,y)=B(y,\sigma(x))$, that is, $\varphi(xy)=\varphi(y\sigma(x))$, for all $x,y\in A$.

If a Frobenius monoidal Hom-algebra (A,α) has an augmentation ε, then the element N in A is called *a left norm of* (A,α) if $\varphi(aN)=\varepsilon(a)$, for any $a\in A$.

Given a monoidal Hom-algebra (A,α), we denote

$$\begin{aligned}Q(A)&=\{Q\in A\otimes A\mid(1\otimes a)Q=Q(a\otimes1),\forall a\in A\}\\&=\{Q=a_i\otimes b_i\in A\otimes A\mid\alpha(a_i)\otimes ab_i=a_ia\\&\otimes\alpha(b_i),\forall a\in A\}.\end{aligned} \tag{36}$$

We consider $(A\otimes A,\alpha\otimes\alpha)$ as an (A,α)-Hom-bimodule under the actions given by

$$a\cdot(x\otimes y)=\alpha^{-1}(a)x\otimes\alpha(y) \tag{37}$$

and

$$(x\otimes y)\cdot b=\alpha(x)\otimes y\alpha^{-1}(b), \tag{38}$$

for all $x,y,a,b\in A$.

Lemma 10. *Let (A,α) be a monoidal Hom-algebra and $Q=a_i\otimes b_i\in Q(A)$. Then*

(1) $Q^{12}Q^{13}=Q^{23}Q^{12}=Q^{13}Q^{23}$;

(2) $\alpha^{\otimes3}(Q^{12})(Q^{23}Q^{12})=(Q^{23}Q^{12})\alpha^{\otimes3}(Q^{23})$.

Proof. (1)

$$\begin{aligned} Q^{12}Q^{13} &= a_ia_j \otimes \alpha(b_i) \otimes \alpha(b_j) \\ &= \alpha(a_i) \otimes a_jb_i \otimes \alpha(b_j) = Q^{23}Q^{12} \\ &= \alpha(a_i) \otimes \alpha(a_j) \otimes b_ib_j = Q^{13}Q^{23}. \end{aligned} \quad (39)$$

(2) From (1), we have

$$\begin{aligned} \alpha^{\otimes 3}(Q^{12})(Q^{23}Q^{12}) &= \alpha^{\otimes 3}(Q^{12})(Q^{13}Q^{23}) \\ &= (Q^{12}Q^{13})\alpha^{\otimes 3}(Q^{23}) \\ &= (Q^{23}Q^{12})\alpha^{\otimes 3}(Q^{23}). \end{aligned} \quad (40)$$

□

Theorem 11. *Let (A,α) be a Frobenius monoidal Hom-algebra with Nakayama automorphism σ and let $e_i, e^i, i = 1,2,\ldots,n$, be dual bases of A. We set $Q = e_i \otimes e^i \in A \otimes A$. Then*

(1) $Q = \rho_A^{-1}(\phi^{-1}) = \chi_A^{-1}(\mathrm{id}_A)$.

(2) *The element Q does not depend on the choice of the dual bases of e_i, e^i of A.*

(3) $(a \otimes b)Q = Q(b \otimes \sigma(a))$, *for all $a, b \in A$.*

(4) $\alpha(x)((e_iz)e^i) = ((e_iz)e^i)\sigma(x)$.

(5) $Q^{12}Q^{13} = Q^{23}Q^{12} = Q^{13}Q^{23}$.

(6) *The element Q satisfies the braid relation*

$$\alpha^{\otimes 3}(Q^{12})(Q^{23}Q^{12}) = (Q^{23}Q^{12})\alpha^{\otimes 3}(Q^{23}). \quad (41)$$

Proof. The first two statements follow directly from (3) and (4) of Proposition 5.

(3) According to Proposition 5 (1), for any $a, b \in A$, we have

$$\begin{aligned} (a \otimes b)Q &= ae_i \otimes be^i = ae_i \otimes B(e_j, be^i)e^j \\ &= ae_i \otimes \varphi(e_j(be^i))e^j \\ &= a\varphi((\alpha^{-1}(e_j)b)\alpha(e^i))e_i \otimes e^j \\ &= a\varphi((\alpha^{-2}(e_j)\alpha^{-1}(b))e^i)e_i \otimes e^j \\ &= a(\alpha^{-2}(e_j)\alpha^{-1}(b)) \otimes e^j \\ &= \alpha^{-1}(a\alpha^{-1}(e_j))b \otimes e^j \\ &= \varphi(\alpha^{-1}(a\alpha^{-1}(e_j))e^i)e_ib \otimes e^j \\ &= e_ib \otimes \varphi(a\alpha^{-1}(\alpha^{-1}(e_j)e^i))e^j \\ &= e_ib \otimes \varphi(\alpha^{-1}(\alpha^{-1}(e_j)e^i)\sigma(a))e_j \\ &= e_ib \otimes \varphi(\alpha^{-1}(e_j)\alpha^{-1}(e_i\sigma(a)))e^j \\ &= e_ib \otimes \varphi(e_j(e_i\sigma(a)))e^j = e_ib \otimes e^i\sigma(a) \\ &= Q(b \otimes \sigma(a)), \end{aligned} \quad (42)$$

which proves (3).

(4) For any given $z \in A$, define a map $\lambda_z : A \otimes A \longrightarrow A$ given by $\lambda_z(a \otimes b) = (az)b$. λ_z is a homomorphism of (A,α)-Hom-bimodules, where the actions on $A \otimes A$ are given by (37) and (38). Then (3) implies that

$$\begin{aligned} \alpha^2(x)((e_i\alpha^{-1}(z))e^i) &= (\alpha(x)(e_i\alpha^{-1}(z)))\alpha(e^i) \\ &= ((xe_i)z)\alpha(e^i) = \lambda_z(xe_i \otimes \alpha(e^i)) \\ &= \lambda_z((x \otimes 1)(e_i \otimes e^i)) \\ &= \lambda_z((e_i \otimes e^i)(1 \otimes \sigma(x))) = \lambda_z(\alpha(e_i) \otimes e^i\sigma(x)) \\ &= (\alpha(e_i)z)(e^i\sigma(x)) = ((e_i\alpha^{-1}(z))e^i)\sigma(\alpha(x)), \end{aligned} \quad (43)$$

so $\alpha(x)((e_iz)e^i) = ((e_iz)e^i)\sigma(x)$, for any $x, z \in A$.

It is obvious that $Q \in Q(A)$ by setting $a = 1$ in the equality of (3). The statements (5) and (6) follow from Lemma 10. □

Theorem 12. *Let (A,α) be a finite dimensional monoidal Hom-algebra with two bases $e_1, e_2, \ldots, e_n$ and $e^1, e^2, \ldots, e^n$, and $Q = e_i \otimes e^i \in A \otimes A$. Then the following conditions are equivalent:*

(1) $Q^{12}Q^{13} = Q^{23}Q^{12}$;

(2) $Q^{12}Q^{13} = Q^{13}Q^{23}$;

(3) $Q^{23}Q^{12} = Q^{13}Q^{23}$;

(4) $(1 \otimes a)Q = Q(a \otimes 1)$;

(5) *A is a Frobenius algebra with Frobenius homomorphism $\varphi : A \longrightarrow k$ such that $\varphi(e_ie^j) = \delta_{ij}$, for all $i, j = 1,2,\ldots,n$.*

Proof. From Lemma 10, it is enough to show that one of the conditions (1), (2), and (3) implies (4). We may as well show that (3) implies (4), and the others can be proved analogously. From (3), it follows that

$$\begin{aligned} Q^{23}Q^{12} &= \alpha(e_i) \otimes e_je^i \otimes \alpha(e_j) \\ &= \alpha(e_i) \otimes \alpha(e_j) \otimes e^ie^j = Q^{13}Q^{23} \end{aligned} \quad (44)$$

and so

$$e_je^i \otimes \alpha(e^j) = \alpha(e_j) \otimes e^ie^j, \quad (45)$$

for all $i = 1,2,\ldots,n$, which proves (4).

Due to Theorem 11, it is enough to prove that (4) implies (5). Define a bilinear map $B : (A \otimes A, \alpha \otimes \alpha) \longrightarrow (k, \mathrm{id})$ in $\mathscr{H}(\mathscr{M}_k)$ as follows: $B(e_i, e^j) = B(\alpha(e_i), \alpha(e^j)) = \delta_{ij}$, for all $i, j = 1,2,\ldots,n$. Since $e_1, e_2, \ldots, e_n$ and $e^1, e^2, \ldots, e^n$ are two bases of A, it follows that B is a well-defined surgenerate bilinear form. And we claim that B is Hom-associative. Indeed, it is enough to show that $B(e_jx, \alpha(e^i)) = B(\alpha(e_j), xe^i)$, for all $i, j = 1,2,\ldots,n$. Since for any $a \in A$, $a = B(e_i, a)e^i = B(a, e^i)e_i$, then equality (45) implies that

$$\alpha(e_i) \otimes xe^i = e_jx \otimes \alpha(e^j) \quad (46)$$

and hence

$$\begin{aligned}\alpha(e_i)\otimes B\left(e_j,\alpha^{-1}\left(xe^i\right)\right)\alpha\left(e^j\right)\\ = B\left(\alpha^{-1}\left(e_jx\right),e^i\right)\alpha(e_i)\otimes\alpha\left(e^j\right),\end{aligned} \tag{47}$$

which implies that

$$B\left(e_j,\alpha^{-1}\left(xe^i\right)\right) = B\left(\alpha^{-1}\left(e_jx\right),e^i\right). \tag{48}$$

Thus $B(\alpha(e_j), xe^i) = B(e_jx, \alpha(e^i))$, for all $i, j = 1, 2, \ldots, n$ and $x \in A$, by the α-invariance of B. Setting $\varphi(x) = B(1, \alpha^{-1}(x))$, we conclude that φ is a Frobenius homomorphism and

$$\varphi\left(e_ie^j\right) = B\left(1,\alpha^{-1}\left(e_ie^j\right)\right) = B\left(e_i,e^j\right) = \delta_{ij}, \tag{49}$$

for all $i, j = 1, 2, \ldots, n$. □

A Frobenius monoidal Hom-algebra with Frobenius homomorphism $\varphi : A \longrightarrow k$ is called *symmetric* if $\varphi(xy) = \varphi(yx)$, for all $x, y \in A$. In this case, the Nakayama automorphism σ is the identity map of A.

Lemma 13. *Let (A,α) be a monoidal Hom-algebra with two bases $e_1, e_2, \ldots, e_n$ and $e^1, e^2, \ldots, e^n$, and $Q = e_i \otimes e^i \in A \otimes A$. Suppose that $(a\otimes b)Q = Q(b\otimes a)$. Then*

(1) $Q^{12}Q^{13} = Q^{23}Q^{12} = Q^{13}Q^{23}$ *and* $Q^{12}Q^{23} = Q^{23}Q^{13} = Q^{13}Q^{12}$;

(2) $\alpha^{\otimes 3}(Q^{12})(Q^{13}Q^{23}) = (Q^{23}Q^{13})\alpha^{\otimes 3}(Q^{12})$.

Proof. The first assertion of (1) is proved in Theorem 12, and the second one can be showed similarly.

(2) From (1), we have

$$\begin{aligned}\alpha^{\otimes 3}\left(Q^{12}\right)\left(Q^{13}Q^{23}\right) &= \left(Q^{12}Q^{13}\right)\alpha^{\otimes 3}\left(Q^{23}\right)\\ &= \left(Q^{23}Q^{12}\right)\alpha^{\otimes 3}\left(Q^{23}\right)\\ &= \alpha^{\otimes 3}\left(Q^{23}\right)\left(Q^{12}Q^{23}\right)\\ &= \alpha^{\otimes 3}\left(Q^{23}\right)\left(Q^{13}Q^{12}\right)\\ &= \left(Q^{23}Q^{13}\right)\alpha^{\otimes 3}\left(Q^{12}\right).\end{aligned} \tag{50}$$

□

Theorem 14. *Let (A,α) be a symmetric monoidal Hom-algebra with Frobenius homomorphism $\varphi : A \longrightarrow k$, and $\{e_i\}, \{e^j\}, i, j = 1, 2, \ldots, n$, be dual bases of A. We set $Q = e_i \otimes e^i \in A\otimes A$. Then*

(1) $(a\otimes b)Q = Q(b\otimes a)$, *for any* $a, b \in A$;

(2) $Q^{12}Q^{13} = Q^{23}Q^{12} = Q^{13}Q^{23}$ *and* $Q^{12}Q^{23} = Q^{23}Q^{13} = Q^{13}Q^{12}$;

(3) *the element* Q *is a solution of QHYBE, and* $\tau(Q) = Q$.

Proof. Since (A,α) is symmetric, the corresponding Nakayama automorphism is the identity. It follows from Theorem 11 (3) that $(a\otimes b)Q = Q(b\otimes a)$, for all $a, b \in A$. The last two statements are true from Lemma 13. □

Next we consider the converse of the above theorem.

Theorem 15. *Let (A,α) be a finite dimensional monoidal Hom-algebra with two bases $e_1, e_2, \ldots, e_n$ and $e^1, e^2, \ldots, e^n$. Further let $Q = e_i \otimes e^i \in A\otimes A$. Then the following conditions are equivalent:*

(1) $Q^{12}Q^{13} = Q^{23}Q^{12}$ *and* $Q^{12}Q^{23} = Q^{23}Q^{13}$;

(2) $Q^{12}Q^{13} = Q^{13}Q^{23}$ *and* $Q^{12}Q^{23} = Q^{13}Q^{12}$;

(3) $Q^{23}Q^{12} = Q^{13}Q^{23}$ *and* $Q^{23}Q^{13} = Q^{13}Q^{12}$;

(4) $(a\otimes b)Q = Q(b\otimes a)$, *for all* $a \in A$;

(5) *A is a symmetric monoidal Hom-algebra with Frobenius homomorphism $\varphi : A \longrightarrow k$ such that $\varphi(e^ie^j) = \delta_{ij}$, for all $i, j = 1, 2, \ldots, n$.*

Proof. We have proved that (4) implies the first three equalities in Lemma 13. And each of the first three conditions implies the fourth one in the same way as it was done in Theorem 12. Taking into account Theorem 14, we have only to prove that (4) implies (5). Set $B : (A\otimes A, \alpha\otimes\alpha) \longrightarrow (k, \mathrm{id})$ in $\widetilde{\mathscr{H}}(\mathscr{M}_k)$ as follows: $B(e_i, e^j) = B(\alpha(e_i), \alpha(e^j)) = \delta_{ij}$, for all $i, j = 1, 2, \ldots, n$. We have seen that B is a Hom-associative surgenerate bilinear form by Theorem 12, and $\varphi(x) = B(1, \alpha^{-1}(x))$ is a Frobenius homomorphism. To show $\varphi(xy) = \varphi(yx)$, we only need to show $B(1, xy) = B(1, yx)$, for $x, y \in A$, which is equivalent to prove that $B(x, y) = B(y, x)$, for $x, y \in A$. From (4) we see that $ae_i \otimes \alpha(e^i) = \alpha(e_j)\otimes e^ja$, for all $a \in A$, so

$$B\left(ae_i,e^j\right)e_j\otimes\alpha\left(e^i\right) = \alpha\left(e_j\right)\otimes B\left(e_i,e^ja\right)e^i \tag{51}$$

by Proposition 5 (1). Hence

$$\begin{aligned}B\left(\alpha^{-1}\left(ae_i\right),e^j\right)\alpha\left(e_j\right)\otimes\alpha\left(e^i\right)\\ = \alpha\left(e_j\right)\otimes B\left(e_i,\alpha^{-1}\left(e^ja\right)\right)\alpha\left(e^i\right),\end{aligned} \tag{52}$$

which implies

$$B\left(ae_i,\alpha\left(e^j\right)\right) = B\left(\alpha\left(e_i\right),e^ja\right), \tag{53}$$

for all $i, j = 1, 2, \ldots, n, a \in A$. Thus $B(ax, \alpha(y)) = B(\alpha(x), ya)$, for all $x, y \in A$. Taking $x = 1_A$, we get

$$B(\alpha(a),\alpha(y)) = B(1,ya) = B(\alpha(y),\alpha(a)), \tag{54}$$

so $B(a, y) = B(y, a)$, for all $a, y \in A$. □

5. Frobenius Monoidal Hom-Hopf Algebras

It is known that a finite dimensional monoidal Hom-Hopf algebra (H,α) is Frobenius if and only if the space of integrals in H is not zero. In this section, we do further research on monoidal Hom-Hopf algebras with Frobenius structures.

Definition 16 (see [16, Definition 2.2]). A *monoidal Hom-coalgebra* is an object (C,γ) in the category $\widetilde{\mathscr{H}}(\mathscr{M}_k)$ together with linear maps $\Delta : C \longrightarrow C\otimes C, \Delta(c) = c_1\otimes c_2$ and $\varepsilon : C \longrightarrow k$ such that

$$\gamma^{-1}(c_1)\otimes\Delta(c_2) = \Delta(c_1)\otimes\gamma^{-1}(c_2), \quad c_1\varepsilon(c_2) = \gamma^{-1}(c) = \varepsilon(c_1)c_2, \tag{55}$$

$$\Delta(\gamma(c)) = \gamma(c_1)\otimes\gamma(c_2), \quad \varepsilon(\gamma(c)) = \varepsilon(c), \tag{56}$$

for all $c \in C$.

Definition 17. A *monoidal Hom-bialgebra* [16, Definition 2.3] $H = (H, \alpha, m, \eta, \Delta, \varepsilon)$ is a bialgebra in the category $\widetilde{\mathscr{H}}(\mathscr{M}_k)$. This means that (H, α, m, η) is a monoidal Hom-algebra and $(H, \alpha, \Delta, \varepsilon)$ is a monoidal Hom-coalgebra such that Δ and ε are Hom-algebra maps; that is, for any $h, g \in H$,

$$\begin{aligned}\Delta(hg) &= \Delta(h)\Delta(g),\\ \Delta(1_H) &= 1_H\otimes 1_H,\\ \varepsilon(hg) &= \varepsilon(h)\varepsilon(g),\\ \varepsilon(1_H) &= 1_k.\end{aligned} \tag{57}$$

Definition 18 (see [16, Definition 2.5]). A monoidal Hom-bialgebra (H, α) is called a *monoidal Hom-Hopf algebra* if there exists a morphism (called antipode) $S : H \longrightarrow H$ in $\widetilde{\mathscr{H}}(\mathscr{M}_k)$ (i.e., $S\circ\alpha = \alpha\circ S$), such that, for any $h \in H$,

$$S(h_1)h_2 = \varepsilon(h)1_H = h_1S(h_2). \tag{58}$$

In fact, monoidal Hom-coalgebras, monoidal Hom-bialgebras, and monoidal Hom-Hopf algebras are coalgebras, bialgebras, and Hopf algebras in the category $\widetilde{\mathscr{H}}(\mathscr{M}_k)$, respectively. Further, the antipodes of monoidal Hom-Hopf algebras have similar properties (see [16]) of those of Hopf algebras. The self-duality of finite dimensional monoidal Hom-Hopf algebras also holds from [16], and the dual structure of monoidal Hom-Hopf algebra $(H, \alpha, m, \eta, \Delta, \varepsilon, S)$ is given as follows: for all $g, h \in H, \phi, \varphi \in H^*$,

$$\begin{aligned}\langle\phi\varphi, h\rangle &= \langle\phi, h_1\rangle\langle\varphi, h_2\rangle,\\ 1_{H^*} &= \varepsilon;\\ \langle\Delta(\phi), g\otimes h\rangle &= \langle\phi, gh\rangle,\\ \varepsilon_{H^*} &= \eta;\\ (\alpha^*)^{-1}(\phi) &= \phi\circ\alpha^{-1},\\ S^*(\phi) &= \phi\circ S^{-1}.\end{aligned} \tag{59}$$

Definition 19 (see [16, Definition 3.1]). A *left integral* in (H, α) is an α-invariant element $t \in H$ (i.e., $\alpha(t) = t$) such that

$$ht = \varepsilon(h)t, \tag{60}$$

for all $h \in H$.

Lemma 20. *Let (H, α) be a monoidal Hom-Hopf algebra with antipode S. Further let φ be an α^*-invariant element in H^*. Then for all $h \in H$,*

$$h_1\varphi(h_2) = \varphi(h)1_H \tag{61}$$

if and only if φ is a left integral in H^*.

Proof. Let $\psi \in H^*$. Since for any $h \in H$,

$$\begin{aligned}(\varepsilon_{H^*}(\psi)\varphi)(h) &= \varphi(h)\psi(1_H) = \psi(\varphi(h)1_H),\\ (\psi\varphi)(h) &= \psi(h_1)\varphi(h_2) = \psi(h_1\varphi(h_2)),\end{aligned} \tag{62}$$

we have $h_1\varphi(h_2) = \varphi(h)1_H$ if and only if $\psi\varphi = \varepsilon_{H^*}(\psi)\varphi$, which means that φ is a left integral in H^*. □

Lemma 21. *Let (H, α) be a monoidal Hom-bialgebra and Frobenius with Frobenius homomorphism $\varphi : (H, \alpha) \longrightarrow k$ such that $h_1\varphi(h_2) = \varphi(h)1_H$ and with left norm N, for all $h \in H$. Then (H, α) is a monoidal Hom-Hopf algebra.*

Proof. We define $S : (H, \alpha) \longrightarrow (H, \alpha)$ in $\widetilde{\mathscr{H}}(\mathscr{M}_k)$ by

$$S(h) = \alpha(N_1)\varphi(\alpha^{-1}(h)N_2), \tag{63}$$

for all $h \in H$. Then by the properties of φ, we see that

$$\begin{aligned}h_1S(h_2) &= h_1\alpha(N_1)\varphi(\alpha^{-1}(h_2)N_2)\\ &= h_1\alpha(N_1)\varphi(h_2\alpha(N_2)) = \varphi(h\alpha(N))1_H\\ &= \varphi(\alpha^{-1}(h)N)1_H = \varepsilon(\alpha^{-1}(h))1_H\\ &= \eta\varepsilon(h),\end{aligned} \tag{64}$$

for all $h \in H$.

It is not difficult to check that the endomorphism $\mathsf{End}_k(H, \alpha)$ in $\widetilde{\mathscr{H}}(\mathscr{M}_k)$ forms a monoidal Hom-algebra with Hom-multiplication convolution $*$, twisting map id and unit $\eta\varepsilon$. Indeed, for all $h \in H, f \in \mathsf{End}_k(H, \alpha)$

$$(f * \eta\varepsilon)(h) = f(h_1)\eta\varepsilon(h_2) = f(h). \tag{65}$$

And the Hom-associativity is true by the Hom-coassociativity of (H, α). The map

$$- * S : \mathsf{End}_k(H, \alpha) \longrightarrow \mathsf{End}_k(H, \alpha); f \longmapsto f * S \tag{66}$$

is epimorphism, since $f = f * \eta\varepsilon = f * (\mathsf{id} * S) = (f * \mathsf{id}) * S$. By Lemma 4 in [25], $- * S$ is an isomorphism with inverse map $- * \mathsf{id}$. So $S * \mathsf{id} = (\eta\varepsilon * S) * \mathsf{id} = \eta\varepsilon$. Thus S is an antipode of (H, α). □

Theorem 22. *Let (H, α) be a monoidal Hom-Hopf algebra with antipode S and also Frobenius with Frobenius homomorphism $\varphi : (H, \alpha) \longrightarrow k$ such that*

$$h_1\varphi(h_2) = \varphi(h)1_H, \tag{67}$$

for all $h \in H$ and with the Nakayama automorphism σ. Further let $\{e_i\}, \{e^i\}, i = 1, 2, \ldots, n$, be dual bases of H. Then

(1) $N = \varepsilon(e_i)e^i = S(e_i)\varepsilon(e^i)$ *is a left norm and a left integral in* H*. Further* $\Delta(N) = S\alpha^{-1}(e_i)\otimes\alpha^{-1}(e^i)$*, and* $\varepsilon(N)1_H = e^ie_i = S^2(e_i)e^i = \varepsilon(e_i)\varepsilon(e^i)1_H$*;*

(2) φ *generates the Hom-submodule of left integral in* H^* *and*

$$Tr\left(S^2\right) = \varepsilon(N)\,\varepsilon_{H^*}(\varphi). \tag{68}$$

Proof. (1) Setting $N = \varepsilon(e_i)e^i$, we have $\varepsilon(N) = \varepsilon(e_i)\varepsilon(e^i)$ and

$$\begin{aligned}\varphi(N) &= \varepsilon(e_i)\,\varphi\left(e^i\right) = \varepsilon\left(\varphi\left(e^i\right)e_i\right) = \varepsilon\left(\varphi\left(1_He^i\right)e_i\right)\\ &= \varepsilon\left(1_H\right) = 1\end{aligned} \tag{69}$$

by the α-invariance of φ and Proposition 5 (1). Further, for all $x\in H$,

$$\begin{aligned}xN &= \varepsilon(e_i)\,xe^i = \varepsilon(e_i)\,\varphi\left(e_j\left(xe^i\right)\right)e^j\\ &= \varepsilon\left(\varphi\left(e_j\left(xe^i\right)\right)e_i\right)e^j\\ &= \varepsilon\left(\varphi\left(\left(\alpha^{-1}\left(e_j\right)x\right)\alpha\left(e^i\right)\right)e_i\right)e^j\\ &= \varepsilon\left(\varphi\left(\alpha^{-1}\left(\alpha^{-1}\left(e_j\right)x\right)e^i\right)e_i\right)e^j\\ &= \varepsilon\left(\alpha^{-1}\left(\alpha^{-1}\left(e_j\right)x\right)\right)e^j = \varepsilon\left(\alpha^{-1}\left(e_j\right)x\right)e^j\\ &= \varepsilon\left(e_j\right)\varepsilon(x)\,e^j = \varepsilon(x)\,N.\end{aligned} \tag{70}$$

Since $\{\alpha(e_i)\}, \{\alpha^{-1}(e^i)\}, i = 1,2,\ldots,n$, can also be considered as dual bases of H, hence we see that

$$\begin{aligned}\alpha^{-1}(N) &= \varepsilon(e_i)\,\alpha^{-1}\left(e^i\right) = \varepsilon\alpha(e_i)\,\alpha^{-1}\left(e^i\right) = \varepsilon(e_i)\,e^i\\ &= N,\end{aligned} \tag{71}$$

which implies that N is α-invariant. Thus N is a left integral in H by Definition 19.

Next, $\varphi(xN) = \varphi(\varepsilon(x)N) = \varepsilon(x)\varphi(N) = \varepsilon(x)$, so N is a left norm.

From Lemma 21, we know that the antipode $S : (H,\alpha)\longrightarrow(H,\alpha)$ is given by

$$S(x) = \alpha(N_1)\,\varphi\left(\alpha^{-1}(x)\,N_2\right). \tag{72}$$

And S is bijective when H if finite dimensional by Theorem 5.3 of [16]. So $\{S\alpha^{-1}(e_i)\}$ forms a basis of H. Therefore $\Delta(N) = S\alpha^{-1}(e_i)\otimes a_i$ for some $a_i\in H$. Then we rewrite the expression of S as

$$S(x) = S(e_i)\,\varphi\left(\alpha^{-1}(x)\,a_i\right), \tag{73}$$

for all $x\in H$. Applying S^{-1} we get

$$x = e_i\varphi\left(\alpha^{-1}(x)\,a_i\right), \tag{74}$$

for all $x\in H$. On the other hand $x = e_i\varphi(xe^i)$. So $\varphi(x\alpha(a_i)) = \varphi(\alpha^{-1}(x)a_i) = \varphi(xe^i)$. It follows that $\alpha(a_i) = e^i$, so $a_i = \alpha^{-1}(e^i)$, and hence

$$\Delta(N) = S\alpha^{-1}(e_i)\otimes\alpha^{-1}\left(e^i\right). \tag{75}$$

By the counitary property of (H,α) we have $N = \alpha S\alpha^{-1}(e_i)\boldsymbol{\varepsilon}\alpha^{-1}(e^i) = S(e_i)\boldsymbol{\varepsilon}(e^i)$. And $N_1S(N_2) = S\alpha^{-1}(e_i)S\alpha^{-1}(e^i) = \boldsymbol{\varepsilon}(N)1_H$, so $\alpha^{-1}(e^ie_i) = \boldsymbol{\varepsilon}(N)S^{-1}(1_H) = \boldsymbol{\varepsilon}(N)1_H$ by applying S^{-1}. Therefore $e^ie_i = \boldsymbol{\varepsilon}(N)1_H$. Similarly, $S^2\alpha^{-1}(e_i)\alpha^{-1}(e^i) = \boldsymbol{\varepsilon}(N)1_H$, which implies that $S^2(e_i)e^i = \boldsymbol{\varepsilon}(N)1_H$. According to the above, so we obtain

$$e^ie_i = S^2(e_i)\,e^i = \boldsymbol{\varepsilon}(N)\,1_H = \boldsymbol{\varepsilon}(e_i)\,\boldsymbol{\varepsilon}\left(e^i\right)1_H. \tag{76}$$

(2) From Proposition 3.4 of [16], we know that the integral in finite dimensional monoidal Hom-Hopf algebras does not necessarily exist. But if it exists, it is unique up to a scalar multiplication. And φ is a left integral in H^* by Lemma 21, so φ generates the Hom-submodule of left integral in H^*. Since $\boldsymbol{\varepsilon}(N)1_H = S^2(e_i)e^i$ by (1), we see that

$$\begin{aligned}\boldsymbol{\varepsilon}(N)\,\varphi(1_H) &= \varphi(\boldsymbol{\varepsilon}(N)\,1_H) = \varphi\left(S^2(e_i)\,e^i\right)\\ &= Tr\left(S^2\right)\end{aligned} \tag{77}$$

by applying φ and the definition of Tr in Proposition 5. So $Tr(S^2) = \boldsymbol{\varepsilon}(N)\varphi(1_H) = \boldsymbol{\varepsilon}(N)\boldsymbol{\varepsilon}_{H^*}(\varphi)$. □

Acknowledgments

This work is supported by the Fundamental Research Funds for the Central Universities (no. KJQN201550), the National Natural Science Foundation of China (no. 11401311, no. 11571173), and the Science Foundation of Jiangsu Province (no. BK20140676, no. BK20141358).

References

[1] D. Larsson and S. D. Silvestrov, "Quasi-hom-Lie algebras, central extensions and 2-cocycle-like identities," *Journal of Algebra*, vol. 288, no. 2, pp. 321–344, 2005.

[2] D. Larsson and S. D. Silvestrov, "Quasi-deformations of sl2(F) using twisted derivations," *Communications in Algebra*, vol. 35, no. 12, pp. 4303–4318, 2007.

[3] J. T. Hartwig, D. Larsson, and S. D. Silvestrov, "Deformations of Lie algebras using σ-derivations," *Journal of Algebra*, vol. 295, no. 2, pp. 314–361, 2006.

[4] J. Arnlind, A. Makhlouf, and S. Silvestrov, "Ternary Hom–Nambu–Lie algebras induced by Hom–Lie algebras," *Journal of Mathematical Physics*, vol. 51, no. 4, Article ID 043515, 2010.

[5] D. Yau, "Enveloping algebras of Hom-Lie algebras," *Journal of Generalized Lie Theory and Applications*, vol. 2, no. 2, pp. 95–108, 2008.

[6] D. Yau, "The HOM-Yang-Baxter equation, HOM-Lie algebras, and quasi-triangular bialgebras," *Journal of Physics A: Mathematical and Theoretical*, vol. 42, no. 16, Article ID 165202, 2009.

[7] F. Ammar and A. Makhlouf, "Hom-Lie superalgebras and Hom-Lie admissible superalgebras," *Journal of Algebra*, vol. 324, no. 7, pp. 1513–1528, 2010.

[8] Q. Q. Jin and X. C. Li, "Hom-LIE algebra structures on semi-simple LIE algebras," *Journal of Algebra*, vol. 319, no. 4, pp. 1398–1408, 2008.

[9] Y. Chen, Y. Wang, and L. Zhang, "The construction of Hom-Lie bialgebras," *Journal of Lie Theory*, vol. 20, no. 4, pp. 767–783, 2010.

[10] A. Makhlouf and S. D. Silvestrov, "Hom-algebra structures," *Journal of Generalized Lie Theory and Applications*, vol. 2, no. 2, pp. 51–64, 2008.

[11] A. Gohr, "On hom-algebras with surjective twisting," *Journal of Algebra*, vol. 324, no. 7, pp. 1483–1491, 2010.

[12] A. Makhlouf, "Hom-alternative algebras and Hom-Jordan algebras," *International Electronic Journal of Algebra*, vol. 8, pp. 177–190, 2010.

[13] D. Yau, "Hom-algebras and homology," *Journal of Lie Theory*, vol. 19, no. 2, pp. 409–421, 2009.

[14] D. Yau, "Hom-quantum groups: I. Quasi-triangular Hom-bialgebras," *Journal of Physics A: Mathematical and Theoretical*, vol. 45, no. 6, Article ID 065203, 2012.

[15] S. Caenepeel and I. Goyvaerts, "Monoidal Hom-HOPf algebras," *Communications in Algebra*, vol. 39, no. 6, pp. 2216–2240, 2011.

[16] Y. Chen, Z. Wang, and L. Zhang, "Integrals for monoidal Hom-Hopf algebras and their applications," *Journal of Mathematical Physics*, vol. 54, no. 7, Article ID 073515, 2013.

[17] Y. Y. Chen and L. Y. Zhang, "The category of Yetter-Drinfel'd Hom-modules and the quantum Hom-Yang-Baxter equation," *Journal of Mathematical Physics*, vol. 55, no. 3, Article ID 031702, 2014.

[18] M. Atiyah, "An introduction to topological quantum field theories," *Turkish Journal of Mathematics*, vol. 21, no. 1, pp. 1–7, 1997.

[19] K. I. Beidar, Y. Fong, and A. Stolin, "On Frobenius algebras and the quantum Yang-Baxter equation," *Transactions of the American Mathematical Society*, vol. 349, no. 9, pp. 3823–3836, 1997.

[20] K. I. Beidar, Y. Fong, and A. A. Stolin, "On antipodes and integrals in Hopf algebras over rings and the quantum Yang-Baxter equation," *Journal of Algebra*, vol. 194, no. 1, pp. 36–52, 1997.

[21] Y. Chen and L. Zhang, "The structure and construction of bi-Frobenius Hom-algebras," *Communications in Algebra*, vol. 45, no. 5, pp. 2142–2162, 2017.

[22] T. Nakayama, "On Frobeniusean algebras," *Annals of Mathematics: Second Series*, vol. 40, pp. 611–633, 1939.

[23] V. G. Drinfel'd, "Hamiltonian structures on lie groups, lie bialgebras and the geometrical meaning of the classical yang-baxter equation," *Soviet Mathematics Doklady*, vol. 27, no. 2, pp. 68–71, 1983.

[24] Y. Chen and X. Zhou, "Separable and Frobenius monoidal Hom-algebras," *Colloquium Mathematicum*, vol. 137, no. 2, pp. 229–251, 2014.

[25] B. Pareigis, "When Hopf algebras are Frobenius algebras," *Journal of Algebra*, vol. 18, pp. 588–596, 1971.

An Approximate Analytical Solution of the Nonlinear Schrödinger Equation with Harmonic Oscillator Using Homotopy Perturbation Method and Laplace-Adomian Decomposition Method

Emad K. Jaradat,[1] **Omar Alomari**,[2] **Mohammad Abudayah**,[2] **and Ala'a M. Al-Faqih**[1]

[1] *Department of physics, Mutah University, Jordan*
[2] *German Jordanian University, Amman, Jordan*

Correspondence should be addressed to Omar Alomari; omar.alomari@gju.edu.jo

Academic Editor: Antonio Scarfone

The Laplace-Adomian Decomposition Method (LADM) and Homotopy Perturbation Method (HPM) are both utilized in this research in order to obtain an approximate analytical solution to the nonlinear Schrödinger equation with harmonic oscillator. Accordingly, nonlinear Schrödinger equation in both one and two dimensions is provided to illustrate the effects of harmonic oscillator on the behavior of the wave function. The available literature does not provide an exact solution to the problem presented in this paper. Nevertheless, approximate analytical solutions are provided in this paper using LADM and HPM methods, in addition to comparing and analyzing both solutions.

1. Introduction

The Schrödinger equation is often encountered in many branches of science and engineering, including quantum mechanics, nonlinear optics, plasma physics, hydrodynamics, and superconductivity. It is a mathematical partial differential equation used to describe the motion and behavior change of the physical system over time. In classical mechanics, it plays the role of Newton's law and conservation of energy. In quantum mechanics, we describe systems using wave function. The Schrödinger equation has two "forms"; one is the time-dependent wave equation that describes how the wave function of a particle will evolve in time. The other is the time independent wave equation in which the time dependence has been "removed"; it describes what the allowed energies are of the particle [1, 2].

In recent years, a considerable amount of research focused on finding analytical solution to the Schrödinger equations using various methods, among which are Adomian Decomposition Method [3–8], Elzaki decomposition method [9], Variation Iteration method [10], Nikiforod–Uvarov (NV) method [11], and Homotopy Perturbation Method [3, 4, 12–16]. Additionally, Borhanifar [17] solved the nonlinear Schrödinger and coupled Schrödinger equations with a differential transformation method. Shidfar and Molabahrami [18, 19] investigated the d-dimensional Schrödinger equation with a power-law nonlinearity, Zhenga et al. [20] solved the time-dependent Schrödinger equation using homotopy analysis method (HAM) and the Adomian decomposition technique (ADM), and Amador et al. [21] solved nonlinear Schrödinger equations with variable coefficients using Riccati equations and similarity transformations. Finally, Khan and Wu [22] applied Homotopy perturbation transform method (HPTM) to solve nonlinear equations; HPTM uses the Homotopy Perturbation Method together with the Laplace transformation to solve the nonlinear equations. Also, Hosseini et al. [23–27] investigated various forms of the nonlinear Schrödinger equation (NLSE).

This paper is organized in several sections. The HPM method is briefly explained in "Homotopy Perturbation Method". Then the LADM model is described in the "Laplace-Adomian Decomposition Method". Then in the

"One-Dimensional Nonlinear Schrödinger Equation with Harmonic Oscillator", the solution to the One-Dimensional Nonlinear Schrödinger Equation with Harmonic Oscillator equation in its nonlinear version is provided with a numerical example. Similarly, the solution of the Two-Dimensional Nonlinear Schrödinger Equation with Harmonic Oscillator is presented in the "Two-Dimensional Nonlinear Schrödinger Equation with Harmonic Oscillator". Finally, in the "Conclusion", we summarize our findings and present our final remarks. Since the exact solution to this problem is not available, we compare our numerical results with the results obtained using Mathematica function NDsolve.

2. Laplace-Adomian Decomposition Method

The Adomian Decomposition Method (ADM) is a method to solve differential equations by expressing the analytic solution in terms of a series. The method separates the linear and nonlinear parts of a differential equation. The nonlinear part can be expressed in terms of what is called Adomian Polynomials [28–30]. The initial condition and the terms that contain the independent variables will be used as the initial approximation. Then by means of a recurrence relation, it is possible to find the terms of the series that give the approximate solution of the differential equation.

The Laplace transform is an integral transform that is powerful and useful technique to solve differential equations, which transforms the original differential equation into an algebraic equation.

Below are the definitions of Laplace transform and inverse Laplace transform.

Definition 1. Given a function $f(t)$ defined for all $t \geq 0$, the Laplace transform of f is the function F defined by

$$F(s) = \mathscr{L}\{f(t)\} = \int_0^{\infty} f(t)\, e^{-st} dt, \tag{1}$$

and the inverse Laplace transform is defined as follows.

Definition 2. Given a continuous function $f(t)$, if $F(s) = \mathscr{L}\{f(t)\}$, then $f(t)$ is called the inverse Laplace transform of $F(s)$ denoted $f(t) = \mathscr{L}^{-1}\{F(s)\}$.

The Laplace-Adomian Decomposition Method (LADM) was first introduced by Suheil A. Khuri [31] and has been effectively used to find solutions to general nonlinear equations. The added value of this method utilizes the two methods (Laplace Transform and ADM) to obtain the solution for nonlinear equations. Consider the following equation:

$$\begin{gathered} L\Psi - iR\Psi - iN\Psi = 0, \\ \Psi(x,0) = f(x), \end{gathered} \tag{2}$$

where $L = \partial/\partial t$ and $R = \partial^2/\partial x^2$, L and R are Linear operators, and N is a nonlinear operator.

Laplace-Adomian Decomposition Method consists of applying Laplace transform to both sides of (2) and yields

$$\mathscr{L}\{L\Psi\} - \mathscr{L}\{iR\Psi\} - \mathscr{L}\{iN\Psi\} = 0. \tag{3}$$

From Laplace transform of first derivative and substituting the initial condition, we get

$$\mathscr{L}\{\Psi(x,t)\} = \frac{f(x)}{s} + \frac{i}{s}\mathscr{L}\{R\Psi + N\Psi\}. \tag{4}$$

Next step is replacing the wave function by an infinite series of terms to be determined later as per the Adomian Decomposition Method (ADM):

$$\Psi(x,t) = \sum_{n=0}^{\infty} \Psi_n(x,t), \tag{5}$$

and the nonlinear terms are replaced by the series:

$$N\Psi = \sum_{n=0}^{\infty} A_n(\Psi_0, \Psi_1, \ldots, \Psi_n), \tag{6}$$

where $A_n(\Psi_0, \Psi_1, \ldots)$'s are the Adomian Polynomials, defined by

$$A_n = \frac{1}{n!}\left[\frac{d^n}{d\lambda^n} N\left(\sum_{n=0}^{\infty} \lambda^i \Psi_i\right)\right]\Bigg|_{\lambda=0}, \quad n = 0, 1, 2\ldots \tag{7}$$

Substituting (5) and (6) into (4) and taking inverse Laplace transform, we get

$$\begin{aligned} &\sum_{n=0}^{\infty} \Psi_n(x,t) \\ &\quad = f(x) \\ &\quad + \mathscr{L}^{-1}\left\{\frac{i}{s}\left(\mathscr{L}\left\{R\sum_{n=0}^{\infty}\Psi_n(x,t) + \sum_{n=0}^{\infty} A_n\right\}\right)\right\}. \end{aligned} \tag{8}$$

From (8), one can obtain

$$\begin{gathered} \Psi_0(x,t) = \Psi(x,0) = f(x), \\ \Psi_1(x,t) = i\mathscr{L}^{-1}\left\{\frac{1}{s}\mathscr{L}\{R\Psi_0(x,t) + A_0\}\right\}, \\ \Psi_2(x,t) = i\mathscr{L}^{-1}\left\{\frac{1}{s}\mathscr{L}\{R\Psi_1(x,t) + A_1\}\right\}, \\ \vdots \\ \Psi_n(x,t) = i\mathscr{L}^{-1}\left\{\frac{1}{s}\mathscr{L}\{R\Psi_{n-1}(x,t) + A_{n-1}\}\right\}. \end{gathered} \tag{9}$$

3. Homotopy Perturbation Method

The Homotopy Perturbation Method (HPM) is a special case of the homotopy analysis method (HAM) [32]. HPM was presented by He [33] in 1999, and is considered as a strong and efficient technique in finding an exact or an approximate analytical solutions to nonlinear equations.

To demonstrate the idea of Homotopy Perturbation Method, we consider the general form nonlinear differential equation with initial conditions of the form [3]:

$$L(\Psi) + N(\Psi) - f(r) = 0, \quad r \in \Omega, \tag{10}$$

$$B\left(\Psi, \frac{\partial \Psi}{\partial n}\right) = 0, \quad r \in \Gamma, \tag{11}$$

where L and N are linear and nonlinear operators respectively, $f(r)$ is an analytic function, Γ is the boundary of the domain Ω, and $\partial\Psi/\partial n$ denotes the differentiation of Ψ with respect to n.

To apply the Homotopy concept to (10), we construct a suitable Homotopy,

$$v(r,p):\Omega\times[0,1]\longrightarrow\mathbb{R}, \tag{12}$$

that satisfies

$$H(v,p)=(1-p)\left[L(v)-L(\Psi_0)\right]+p\left[L(v)+N(v)-f(r)\right]=0,\quad r\in\Omega, \tag{13}$$

where $p\in[0,1]$ is a parameter which increases from 0 to 1, $\mathbb{R}$ represents all real numbers, and Ψ_0 is an initial approximate solution of (10), which satisfies the boundary conditions (11). Clearly, from (13) we have

$$H(v,0)=L(v)-L(\Psi_0)=0, \tag{14}$$

$$H(v,1)=L(v)+N(v)-f(r)=0. \tag{15}$$

Now, when the value of p changes from 0 to 1, $v(r,p)$ changes from Ψ_0 to $\Psi(r)$. According to the concept of topology, this is called deformation and $L(v)-N(\Psi_0)$ and $L(v)-f(r)$ are called homotopy. If we consider p as a small parameter, then applying the original perturbation technique method, we can assume that the solution of (13) can be defined as a power series in p:

$$v=v_0+v_1p+v_2p^2+\cdots \tag{16}$$

Then the solution Ψ to (10) is obtained as p approaches 1:

$$\Psi=\lim_{p\longrightarrow1}v=v_0+v_1+v_2+\cdots \tag{17}$$

The series in (17) is convergent for most cases, the rate of convergence however is dependent on the nonlinear operator $N(v)$ [33].

In the following two sections we apply the above two methods to solve the One- and Two-Dimensional Nonlinear Schrödinger Equation with Harmonic Oscillator.

4. One-Dimensional Nonlinear Schrödinger Equation with Harmonic Oscillator

The nonlinear Schrödinger equation with harmonic oscillator described by Ψ with identical initial condition can be expressed as [20]

$$\frac{\partial\Psi}{\partial t}-\frac{i}{2m}\frac{\partial^2\Psi}{\partial x^2}+\frac{i}{2}kx^2\Psi+i|\Psi|^2\Psi=0, \tag{18}$$

$$\Psi(x,0)=e^{ix}, \tag{19}$$

where Ψ is the wave function, m is the mass of the particle, i is the imaginary unit to describe motion, and k spring constant.

4.1. LADM. In this section we solve the One-Dimensional Nonlinear Schrödinger Equation with Harmonic Oscillator (18) using LADM method by first applying Laplace transform to both sides of the equation (18) as follows:

$$\mathscr{L}\left\{\frac{\partial\Psi}{\partial t}\right\}-i\mathscr{L}\left\{\frac{1}{2m}\frac{\partial^2\Psi}{\partial x^2}\right\}+i\mathscr{L}\left\{\frac{1}{2}kx^2\Psi\right\}+i\mathscr{L}\left\{|\Psi^2\Psi\right\}=0. \tag{20}$$

From the properties of Laplace transform of the first derivative and substituting the initial conditions (19), (20) becomes

$$\mathscr{L}\{\Psi(x,t)\}=\frac{e^{ix}}{s}+\frac{1}{s}\mathscr{L}\left\{\frac{i}{2m}\frac{\partial^2\Psi}{\partial x^2}\right\}-\frac{1}{s}\frac{ik}{2}\mathscr{L}\left\{x^2\Psi\right\}-\frac{1}{s}\mathscr{L}\left\{i|\Psi|^2\Psi\right\}. \tag{21}$$

Now substituting (5) and (6) into (21), we get

$$\mathscr{L}\left\{\sum_{n=0}^{\infty}\Psi_n(x,t)\right\}=\frac{e^{ix}}{s}+\frac{1}{s}\mathscr{L}\left\{\frac{i}{2m}\frac{\partial^2}{\partial x^2}\left(\sum_{n=0}^{\infty}\Psi_n(x,t)\right)\right\}-\frac{ik}{2s}\mathscr{L}\left\{x^2\sum_{n=0}^{\infty}\Psi_n(x,t)\right\}-\frac{i}{s}\mathscr{L}\left\{\sum_{n=0}^{\infty}A_n\right\}. \tag{22}$$

By applying inverse Laplace transform to (22) and taking into consideration

$$\mathscr{L}^{-1}\left\{\mathscr{L}\{\Psi_0(x,t)\}\right\}=\mathscr{L}^{-1}\left\{\frac{e^{ix}}{s}\right\}, \tag{23}$$

we have

$$\Psi_n(x,t)=\mathscr{L}^{-1}\left\{\frac{i}{2ms}\mathscr{L}\left\{\frac{\partial^2\Psi_{n-1}(x,t)}{\partial x^2}\right\}\right\}-\mathscr{L}^{-1}\left\{\frac{ik}{2s}\mathscr{L}\left\{x^2\Psi_{n-1}(x,t)\right\}\right\}-\mathscr{L}^{-1}\left\{\frac{i}{s}\mathscr{L}\{A_{n-1}\}\right\}, \tag{24}$$

where A_n are the so called Adomian Polynomials of $(\Psi_0,\Psi_1,\Psi_2,\ldots,\Psi_n)$ to replace $N\Psi=|\Psi|^2\Psi=\Psi^2\overline{\Psi}$ and $\overline{\Psi}$ is the conjugate of Ψ.

The Adomian Polynomials can be calculated using (7):

$$\begin{aligned}A_0&=\Psi_0^2\overline{\Psi}_0,\\A_1&=2\Psi_0\Psi_1\overline{\Psi}_0+\Psi_0^2\overline{\Psi}_1,\\A_2&=2\Psi_0\Psi_2\overline{\Psi}_0+\Psi_1^2\overline{\Psi}_0+2\Psi_0\Psi_1\overline{\Psi}_1+\Psi_0^2\overline{\Psi}_2.\end{aligned} \tag{25}$$

Now, using (24), we get

$$\Psi_0(x,t) = e^{ix},$$
$$\Psi_1(x,t) = -\frac{ite^{ix}}{2m} - \frac{ikx^2te^{ix}}{2} - ite^{ix},$$
$$\Psi_2(x,t) = \frac{t^2}{2!}\left(\frac{-e^{ix}}{4m^2} + \frac{ke^{ix}}{2m} + \frac{ikxe^{ix}}{m} - \frac{e^{ix}}{m} - \frac{kx^2e^{ix}}{2m} - \frac{k^2x^4e^{ix}}{4} - kx^2e^{ix} - e^{ix}\right),$$
$$\Psi_3(x,t) = \frac{t^3}{3!}\left(\frac{3ie^{ix}}{4m^2} + \frac{3ikx^2e^{ix}}{8m^2} + \frac{ie^{ix}}{8m^3} + \frac{3kxe^{ix}}{2m^2} - \frac{7ike^{ix}}{4m^2} - \frac{3ike^{ix}}{2m} + \frac{3ikx^2e^{ix}}{2m} + \frac{3ie^{ix}}{2m} + \frac{ik^3x^6e^{ix}}{8} + \frac{i3k^2x^4e^{ix}}{4} + ie^{ix} + \frac{3kxe^{ix}}{m} + \frac{3ik^2x^4e^{ix}}{8m} - \frac{7ik^2x^2e^{ix}}{4m} + \frac{3k^2x^3e^{ix}}{2m} + \frac{3ikx^2e^{ix}}{2}\right). \tag{26}$$

Therefore, the solution $\Psi(x,t)$ is given by

$$\Psi(x,t) = \Psi_0 + \Psi_1 + \Psi_2 + \cdots \Psi_n = e^{ix}\left[1 + \left(-\frac{it}{2m} - \frac{ikx^2t}{2} - it\right) + \frac{1}{2!}\left(\frac{-it}{2m} - \frac{ikx^2t}{2} - it\right)^2 + \frac{t^2}{2!}\left(\frac{k}{2m} + \frac{ikx}{m}\right) + \frac{1}{3!}\left(\frac{-it}{2m} - \frac{ikx^2t}{2} - it\right)^3 + \frac{t^3}{3!}\left(\frac{3kx}{2m^2} - \frac{7ik}{4m^2} - \frac{3ik}{2m} + \frac{3kx}{m} - \frac{7ik^2x^2}{4m} + \frac{3k^2x^3}{2m}\right) + \cdots\right] \tag{27}$$

4.2. HPM. In this section we apply the Homotopy Perturbation Method to obtain a solution to (18). Consider the following homotopy:

$$H(v,p) = (1-p)\left(\frac{\partial v}{\partial t} - \frac{\partial \Psi_0}{\partial t}\right) + p\left(\frac{\partial v}{\partial t} - \frac{i\partial^2 v}{2m\partial x^2} + \frac{i}{2}kx^2v + iv^2\overline{v}\right) = 0, \tag{28}$$

where $\Psi_0 = \Psi(x,0)$, $\partial\Psi_0/\partial t = \partial\Psi(x,0)/\partial t$, and $\overline{v}(x,t)$ is the complex conjugate of $v(x,t)$. Suppose that the series solution v of (28) and its conjugate $\overline{v}$ have the following forms:

$$v(x,t) = v_0(x,t) + v_1(x,t)\,p + v_2(x,t)\,p^2 + \cdots, \tag{29}$$

and

$$\overline{v}(x,t) = \overline{v}_0(x,t) + \overline{v}_1(x,t)\,p + \overline{v}_2(x,t)\,p^2 + \cdots, \tag{30}$$

and hence, the solution to (18) is

$$\Psi(x,t) = v_0(x,t) + v_1(x,t) + v_2(x,t) + \cdots. \tag{31}$$

Substituting (29) and (30) into (28) and equating the coefficients of p powers, we have

$$p^0: \frac{\partial v_0}{\partial t} - \frac{\partial \Psi_0}{\partial t} = 0,$$
$$\Psi_0(x,0) = e^{ix},$$
$$p^1: \frac{\partial v_1}{\partial t} = \frac{i\partial^2 v_0}{2m\partial x^2} - \frac{i}{2}kx^2v_0 - i\left|v_0\right|^2 v_0,$$
$$v_1(x,0) = 0,$$
$$p^2: \frac{\partial v_2}{\partial t} = \frac{i\partial^2 v_1}{2m\partial x^2} - \frac{i}{2}kx^2v_1 - i\left[v_0^2\overline{v_1} + 2v_0\overline{v_0}v_1\right],$$
$$v_2(x,0) = 0,$$
$$\vdots$$
$$p^j: \frac{\partial v_j}{\partial t} = \frac{i\partial^2 v_{j-1}}{2m\partial x^2} - \frac{i}{2}kx^2v_{j-1} - i\left(\sum_{i=0}^{i-1}\sum_{k=0}^{j-i-1}\left|v_i\right|\left|v_k\right|\left|v_{j-k-i-1}\right|\right),$$
$$v_j(x,0) = 0. \tag{32}$$

We finally obtain the general solution of (18) given the recurrence relation for $j = 1, 2, 3, \ldots$

$$v_j = \int_0^t\left(\frac{i\partial^2 v_{j-1}}{2m\partial x^2} - \frac{i}{2}kx^2v_{j-1} - i\left(\sum_{i=0}^{i-1}\sum_{k=0}^{j-i-1}\left|v_i\right|\left|v_k\right|\left|v_{j-k-i-1}\right|\right)\right)d\tau. \tag{33}$$

The first few terms of the solution are calculated as follows:

$$v_0(x,t) = e^{ix},$$
$$v_1(x,t) = -\frac{ite^{ix}}{2m} - \frac{ikx^2te^{ix}}{2} - ite^{ix},$$
$$v_2(x,t) = \frac{t^2}{2!}\left(\frac{-e^{ix}}{4m^2} + \frac{ke^{ix}}{2m} + \frac{ikxe^{ix}}{m} - \frac{e^{ix}}{m} - \frac{kx^2e^{ix}}{2m} - \frac{k^2x^4e^{ix}}{4} - kx^2e^{ix} - e^{ix}\right),$$

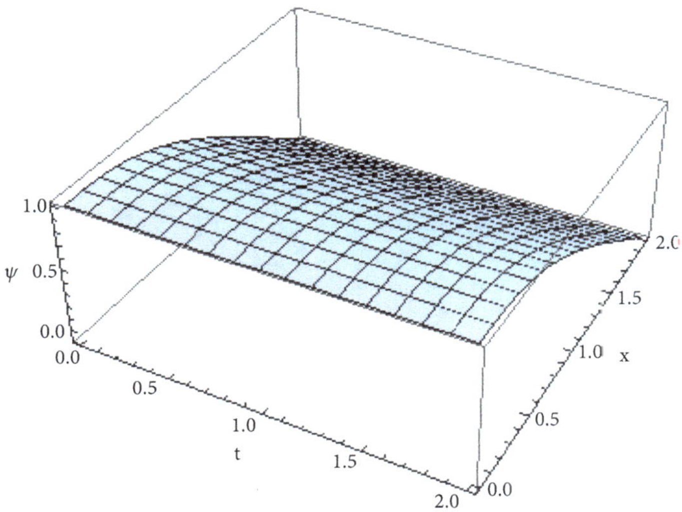

FIGURE 1: Real solution of the one-dimensional wave function obtained by LADM and HPM.

$$v_3(x,t) = \frac{t^3}{3!}\left(\frac{3ie^{ix}}{4m^2} + \frac{3ikx^2e^{ix}}{8m^2} + \frac{ie^{ix}}{8m^3} + \frac{3kxe^{ix}}{2m^2} - \frac{7ike^{ix}}{4m^2} - \frac{3ike^{ix}}{2m} + \frac{3ikx^2e^{ix}}{2m} + \frac{3ie^{ix}}{2m} + \frac{ik^3x^6e^{ix}}{8} + \frac{i3k^2x^4e^{ix}}{4} + ie^{ix} + \frac{3kxe^{ix}}{m} + \frac{3ik^2x^4e^{ix}}{8m} - \frac{7ik^2x^2e^{ix}}{4m} + \frac{3k^2x^3e^{ix}}{2m} + \frac{3ikx^2e^{ix}}{2}\right), \tag{34}$$

and hence, the solution to (18) is

$$\Psi(x,t) = v_0(x,t) + v_1(x,t) + v_2(x,t) + v_3(x,t) + \cdots = e^{ix}\left[1 + \left(-\frac{it}{2m} - \frac{ikx^2t}{2} - it\right) + \frac{1}{2!}\left(\frac{-it}{2m} - \frac{ikx^2t}{2} - it\right)^2 + \frac{t^2}{2!}\left(\frac{k}{2m} + \frac{ikx}{m}\right) + \frac{1}{3!}\left(\frac{-it}{2m} - \frac{ikx^2t}{2} - it\right)^3 + \frac{t^3}{3!}\left(\frac{3kx}{2m^2} - \frac{7ik}{4m^2} - \frac{3ik}{2m} + \frac{3kx}{m} - \frac{7ik^2x^2}{4m} + \frac{3k^2x^3}{2m}\right) + \cdots\right]. \tag{35}$$

The graphs of the one-dimensional wave function are shown in Figures 1, 2, 3, and 4. Figure 1 is the graph of the third order approximation for the solution of the real part of the wave function obtained by LADM and HPM. Figure 2 is the graph of the wave function obtained using Mathematica function NDsolve. Figures 3 and 4 show the imaginary part of the solution. We use $k = m = 1$ in the calculations.

5. Two-Dimensional Nonlinear Schrödinger Equation with Harmonic Oscillator

In this section we look at a particle movement in two dimensions, the nonlinear Schrödinger equation with harmonic oscillator when a particle moves in two dimensions with the initial condition can be written as [20]

$$\frac{\partial \Psi}{\partial t} - \frac{i}{2m}\left(\frac{\partial^2 \Psi}{\partial x^2} + \frac{\partial^2 \Psi}{\partial y^2}\right) + \frac{ik}{2}\left(x^2 + y^2\right)\Psi + i|\Psi|^2\Psi = 0, \tag{36}$$

$$\Psi(x,y,0) = e^{i(x+y)}. \tag{37}$$

5.1. LADM. Similar to the one-dimensional, after applying Laplace transform to (36), substituting the initial condition, and making $\mathscr{L}\{\Psi(x,y,t)\}$ the subject, we have

$$\mathscr{L}\{\Psi(x,y,t)\} = \frac{e^{i(x+y)}}{s} + \frac{1}{s}\mathscr{L}\left\{\frac{i}{2m}\left(\frac{\partial^2 \Psi}{\partial x^2} + \frac{\partial^2 \Psi}{\partial y^2}\right)\right\} - \frac{1}{s}\mathscr{L}\left\{\frac{ik}{2}\left(x^2 + y^2\right)\Psi\right\} - \frac{1}{s}\mathscr{L}\left\{i|\Psi|^2\Psi\right\}. \tag{38}$$

Replacing the wave function $\Psi(x, y)$ and $|\Psi|^2\Psi$ in (38) by the infinite series below,respectively,

$$\Psi(x,y,t) = \sum_{n=0}^{\infty}\Psi_n(x,y,t), \tag{39}$$

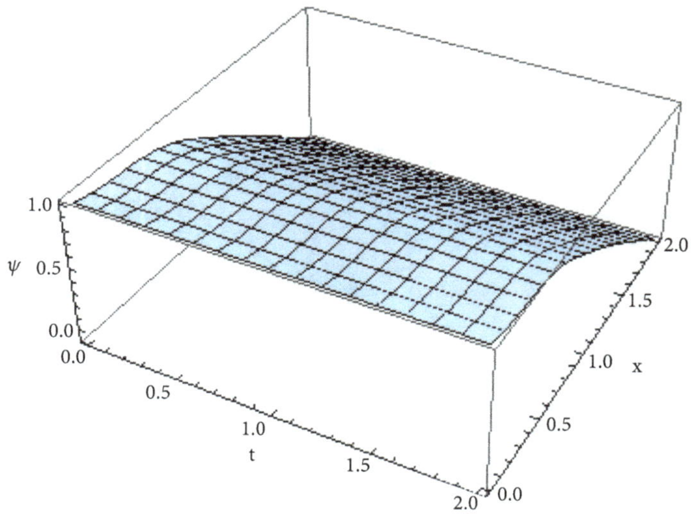

FIGURE 2: Real solution of the one-dimensional wave function obtained by NDsolve.

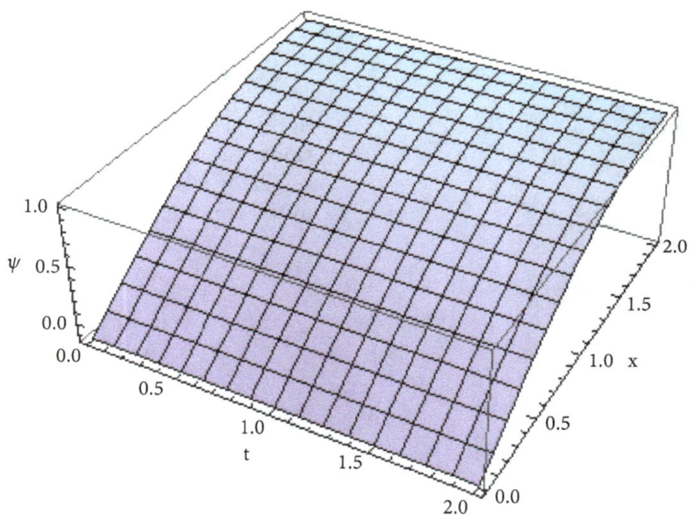

FIGURE 3: Imaginary solution of the one-dimensional wave function obtained by LADM and HPM.

and

$$|\Psi|^2 \Psi = \sum_{n=0}^{\infty} A_n (\Psi_0, \Psi_1, \dots, \Psi_n), \tag{40}$$

we get

$$\begin{aligned} \mathscr{L}\left\{\sum_{n=0}^{\infty} \Psi_n (x, y, t)\right\} &= \frac{e^{i(x+y)}}{s} + \frac{1}{s} \\ &\cdot \mathscr{L}\left\{\frac{i}{2m}\left(\frac{\partial^2}{\partial x^2}\left(\sum_{n=0}^{\infty} \Psi_n (x, y, t)\right)\right.\right. \\ &\left.\left. + \frac{\partial^2}{\partial y^2}\left(\sum_{n=0}^{\infty} \Psi_n (x, y, t)\right)\right)\right\} - \frac{1}{s}\mathscr{L}\left\{\frac{ik}{2}\left(x^2 + y^2\right)\right. \\ &\left.\cdot \sum_{n=0}^{\infty} \Psi_n (x, y, t)\right\} - \frac{1}{s}\mathscr{L}\left\{i\left(\sum_{n=0}^{\infty} A_n\right)\right\}. \end{aligned} \tag{41}$$

Applying inverse Laplace transform, we have

$$\Psi_0 = \mathscr{L}^{-1}\left\{\frac{e^{i(x+y)}}{s}\right\}, \tag{42}$$

$$\begin{aligned} \Psi_1 &= \mathscr{L}^{-1}\left\{\frac{i}{2ms}\mathscr{L}\left\{\frac{\partial^2 \Psi_0 (x, y, t)}{\partial x^2} + \frac{\partial^2 \Psi_0 (x, y, t)}{\partial y^2}\right\}\right\} \\ &- \mathscr{L}^{-1}\left\{\frac{ik}{2s}\mathscr{L}\left\{\left(x^2 + y^2\right) \Psi_0 (x, y, t)\right\}\right\} \\ &- \mathscr{L}^{-1}\left\{\frac{i}{s}\mathscr{L}\left\{A_0\right\}\right\}, \end{aligned} \tag{43}$$

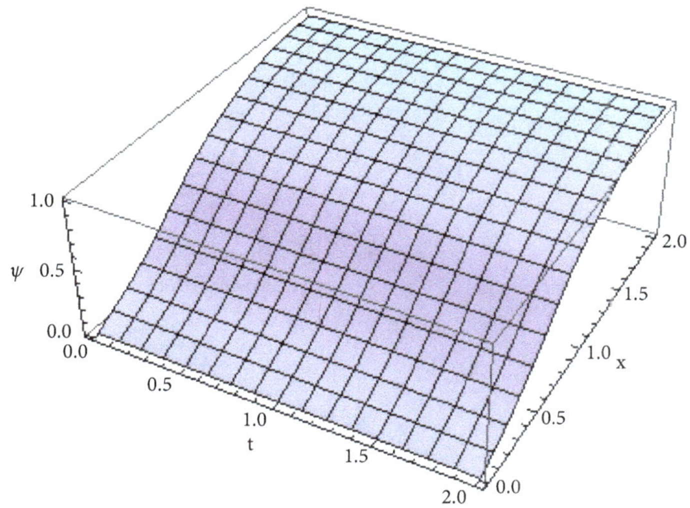

FIGURE 4: Imaginary solution of the one-dimensional wave function obtained by NDsolve.

$$\begin{aligned}\Psi_2 &= \mathscr{L}^{-1}\left\{\frac{i}{2ms}\mathscr{L}\left\{\frac{\partial^2\Psi_1(x,y,t)}{\partial x^2}+\frac{\partial^2\Psi_1(x,y,t)}{\partial y^2}\right\}\right\} \\ &\quad - \mathscr{L}^{-1}\left\{\frac{ik}{2s}\mathscr{L}\left\{\left(x^2+y^2\right)\Psi_1(x,y,t)\right\}\right\} \\ &\quad - \mathscr{L}^{-1}\left\{\frac{i}{s}\mathscr{L}\left\{A_1\right\}\right\},\end{aligned} \tag{44}$$

and thus

$$\begin{aligned}\Psi_n &= \mathscr{L}^{-1}\left\{\frac{i}{2ms}\right. \\ &\quad \left.\cdot\mathscr{L}\left\{\frac{\partial^2\Psi_{n-1}(x,y,t)}{\partial x^2}+\frac{\partial^2\Psi_{n-1}(x,y,t)}{\partial y^2}\right\}\right\} \\ &\quad - \mathscr{L}^{-1}\left\{\frac{ik}{2s}\mathscr{L}\left\{\left(x^2+y^2\right)\Psi_{n-1}(x,y,t)\right\}\right\} \\ &\quad - \mathscr{L}^{-1}\left\{\frac{i}{s}\mathscr{L}\left\{A_{n-1}\right\}\right\}.\end{aligned} \tag{45}$$

Solving the above system of equations, we get

$$\begin{aligned}\Psi_0(x,y,t) &= e^{i(x+y)}, \\ \Psi_1(x,y,t) &= \frac{-it}{m}\left(e^{i(x+y)}\right) - \frac{ikte^{i(x+y)}}{2}\left(x^2+y^2\right) \\ &\quad - ite^{i(x+y)}, \\ \Psi_2(x,y,t) &= \frac{-t^2e^{i(x+y)}}{(2!)\,m^2} - \frac{kt^2e^{i(x+y)}}{(2!)\,m}\left(x^2+y^2\right) \\ &\quad - \frac{2t^2e^{i(x+y)}}{(2!)\,m} + \frac{kt^2e^{i(x+y)}}{(2!)\,m} + \frac{ikt^2e^{i(x+y)}(x+y)}{(2!)\,m} \\ &\quad - \frac{k^2e^{i(x+y)}\left(x^2+y^2\right)^2t^2}{4\,(2!)} - \frac{t^2e^{i(x+y)}}{(2!)} \\ &\quad - \frac{ke^{i(x+y)}t^2\left(x^2+y^2\right)}{(2!)}, \\ \Psi_3(x,y,t) &= \frac{t^3}{3!}\left(\frac{ie^{i(x+y)}}{m^3} + \frac{3k(x+y)e^{i(x+y)}}{m^2}\right. \\ &\quad - \frac{5ike^{i(x+y)}}{m^2} + \frac{3k(x+y)e^{i(x+y)}}{m} \\ &\quad + \frac{3k^2\left(x^2+y^2\right)(x+y)e^{i(x+y)}}{2m} \\ &\quad + \frac{3ik\left(x^2+y^2\right)e^{i(x+y)}}{2m^2} - \frac{5ik^2\left(x^2+y^2\right)e^{i(x+y)}}{2m} \\ &\quad + \frac{3ik^2\left(x^2+y^2\right)^2e^{i(x+y)}}{4m} + \frac{2ie^{i(x+y)}}{m^2} - \frac{3ike^{i(x+y)}}{m} \\ &\quad + \frac{3ik\left(x^2+y^2\right)e^{i(x+y)}}{m} + \frac{3ie^{i(x+y)}}{m} \\ &\quad + \frac{ik^3\left(x^2+y^2\right)^3e^{i(x+y)}}{8} + \frac{3ik\left(x^2+y^2\right)e^{i(x+y)}}{2} \\ &\quad \left.+ \frac{3ik^2\left(x^2+y^2\right)^2e^{i(x+y)}}{4} + ie^{i(x+y)}\right).\end{aligned} \tag{46}$$

Therefore, the approximate solution can be written as

$$\Psi = \Psi_0 + \Psi_1 + \Psi_2 + \Psi_3 + \cdots$$

$$\begin{aligned}\Psi = e^{i(x+y)} \Bigg[& 1 + \left(\frac{-it}{m} - \frac{ikt}{2}\left(x^2+y^2\right) - it\right) \\ & + \frac{1}{2!}\left(\frac{-it}{m} - \frac{ikt}{2}\left(x^2+y^2\right) - it\right)^2 \\ & + \frac{t^2}{2!}\left(\frac{ik(x+y)}{m} + \frac{k}{m}\right) + \frac{1}{3!}\left(\frac{-it}{m}\right. \\ & \left. - \frac{ikt}{2}\left(x^2+y^2\right) - \mathrm{it}\right)^3 + \frac{t^3}{3!}\left(\frac{3k(x+y)}{m^2} - \frac{5ik}{m^2}\right. \\ & + \frac{3k(x+y)}{m} + \frac{3k^2\left(x^2+y^2\right)(x+y)}{2m} \\ & \left. - \frac{5ik^2\left(x^2+y^2\right)}{2m} - \frac{3ik}{m}\right) + \cdots \Bigg].\end{aligned} \tag{47}$$

5.2. HPM. In this section we apply the HPM to solve the two-dimensional equation; first we construct the following suitable homotopy:

$$\begin{aligned}H(v,p) = {} & (1-p)\left[\frac{\partial v}{\partial t} - \frac{\partial v_0}{\partial t}\right] + p\left[\frac{\partial v}{\partial t}\right. \\ & \left. - \frac{i}{2m}\left(\frac{\partial^2 v}{\partial x^2} + \frac{\partial^2 v}{\partial y^2}\right) + \frac{ik}{2}\left(x^2+y^2\right)v + i|v|^2 v\right] \\ = {} & 0.\end{aligned} \tag{48}$$

Using the same steps as in the one-dimensional case, we get the following system of equations:

$$\begin{aligned} & p^0 : \frac{\partial v}{\partial t} - \frac{\partial v_0}{\partial t} = 0, \\ & v_0(x,y,0) = e^{i(x+y)}, \\ & p^1 : \frac{\partial v_1}{\partial t} = \frac{i}{2m}\left(\frac{\partial^2 v_0}{\partial x^2} + \frac{\partial^2 v_0}{\partial y^2}\right) - \frac{ik}{2}\left(x^2+y^2\right)v_0 \\ & \quad - i\left|v_0\right|^2 v_0, \\ & v_1(x,y,0) = 0, \\ & \qquad \vdots \\ & p^j : \frac{\partial v_j}{\partial t} = \frac{i}{2m}\left[\frac{\partial^2 v_{j-1}}{\partial x^2} + \frac{\partial^2 v_{j-1}}{\partial y^2}\right] \\ & \quad - \frac{ik}{2}\left(x^2+y^2\right)v_{j-1} - i\sum_{i=0}^{i-1}\sum_{k=0}^{j-i-1}\left|v_i\right| v_k \left|v_{j-k-i-1}\right|, \\ & v_j(x,y,0) = 0, \qquad j > 0.\end{aligned} \tag{49}$$

The general j^{th} term can be obtained as follows:

$$\begin{aligned}v_j = \int_0^t \Bigg(& \frac{i}{2m}\left[\frac{\partial^2 v_{j-1}}{\partial x^2} + \frac{\partial^2 v_{j-1}}{\partial y^2}\right] - \frac{ik}{2}\left(x^2+y^2\right)v_{j-1} \\ & - i\sum_{i=0}^{i-1}\sum_{k=0}^{j-i-1}\left|v_i\right| v_k \left|v_{j-k-i-1}\right|\Bigg) d\tau.\end{aligned} \tag{50}$$

Therefore, we can now evaluate the solution to the above system of differential equations:

$$\begin{aligned} v_0(x,y,t) = {} & e^{i(x+y)}, \\ v_1(x,y,t) = {} & \frac{-it}{m}\left(e^{i(x+y)}\right) - \frac{ikte^{i(x+y)}}{2}\left(x^2+y^2\right) \\ & - ite^{i(x+y)}, \\ v_2(x,y,t) = {} & \frac{-t^2e^{i(x+y)}}{(2!)m^2} - \frac{kt^2e^{i(x+y)}}{(2!)m}\left(x^2+y^2\right) \\ & - \frac{2t^2e^{i(x+y)}}{(2!)m} + \frac{kt^2e^{i(x+y)}}{(2!)m} + \frac{ikt^2e^{i(x+y)}(x+y)}{(2!)m} \\ & - \frac{k^2e^{i(x+y)}\left(x^2+y^2\right)^2t^2}{4(2!)} - \frac{t^2e^{i(x+y)}}{(2!)} \\ & - \frac{ke^{i(x+y)}t^2\left(x^2+y^2\right)}{(2!)}, \\ v_3(x,y,t) = {} & \frac{t^3}{3!}\Bigg(\frac{ie^{i(x+y)}}{m^3} + \frac{3k(x+y)e^{i(x+y)}}{m^2} \\ & - \frac{5ike^{i(x+y)}}{m^2} + \frac{3k(x+y)e^{i(x+y)}}{m} \\ & + \frac{3k^2\left(x^2+y^2\right)(x+y)e^{i(x+y)}}{2m} \\ & + \frac{3ik\left(x^2+y^2\right)e^{i(x+y)}}{2m^2} - \frac{5ik^2\left(x^2+y^2\right)e^{i(x+y)}}{2m} \\ & + \frac{3ik^2\left(x^2+y^2\right)^2e^{i(x+y)}}{4m} + \frac{2ie^{i(x+y)}}{m^2} - \frac{3ike^{i(x+y)}}{m} \\ & + \frac{3ik\left(x^2+y^2\right)e^{i(x+y)}}{m} + \frac{3ie^{i(x+y)}}{m} \\ & + \frac{ik^3\left(x^2+y^2\right)^3e^{i(x+y)}}{8} + \frac{3ik\left(x^2+y^2\right)e^{i(x+y)}}{2} \\ & + \frac{3ik^2\left(x^2+y^2\right)^2e^{i(x+y)}}{4} + ie^{i(x+y)}\Bigg).\end{aligned} \tag{51}$$

Hence, the approximate solution to (36), (37) is

$$\begin{aligned}\Psi = e^{i(x+y)} \Bigg[& 1 + \left(\frac{-it}{m} - \frac{ikt}{2}\left(x^2+y^2\right) - it\right) \\ & + \frac{1}{2!}\left(\frac{-it}{m} - \frac{ikt}{2}\left(x^2+y^2\right) - \mathrm{it}\right)^2\end{aligned}$$

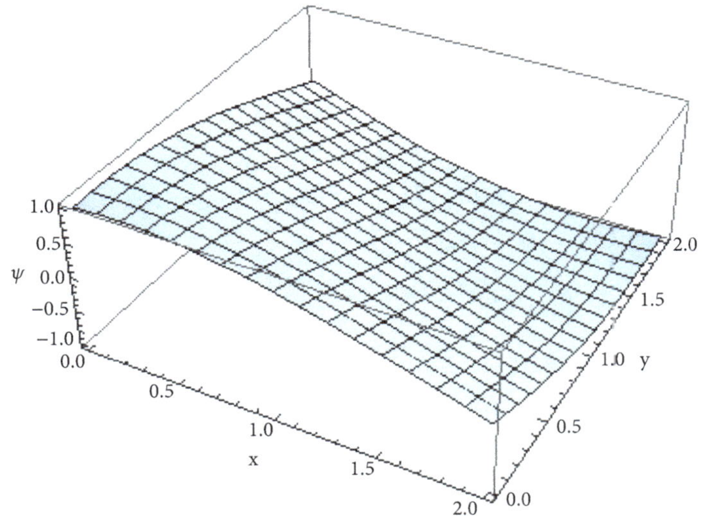

FIGURE 5: Real solution of the two-dimensional wave function obtained by LADM and HPM.

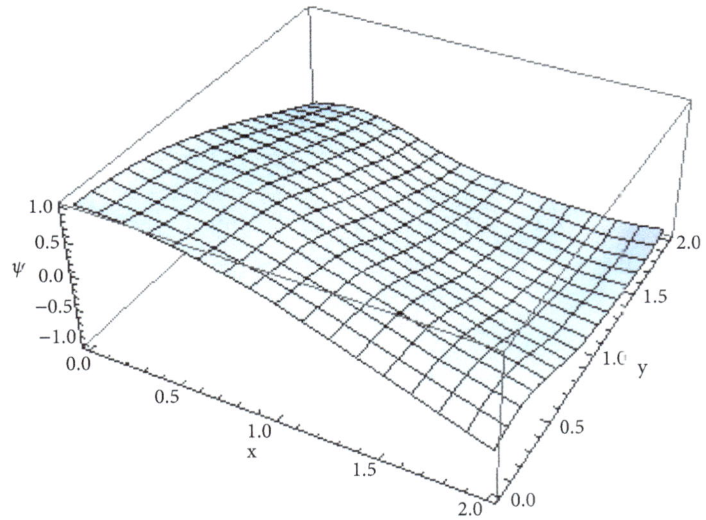

FIGURE 6: Real solution of the two-dimensional wave function obtained by NDsolve.

$$+\frac{t^2}{2!}\left(\frac{ik(x+y)}{m}+\frac{k}{m}\right)+\frac{1}{3!}\left(\frac{-it}{m}-\frac{ikt}{2}\left(x^2+y^2\right)-\mathrm{it}\right)^3+\frac{t^3}{3!}\left(\frac{3k(x+y)}{m^2}-\frac{5ik}{m^2}+\frac{3k(x+y)}{m}+\frac{3k^2\left(x^2+y^2\right)(x+y)}{2m}-\frac{5ik^2\left(x^2+y^2\right)}{2m}-\frac{3ik}{m}\right)+\cdots\Bigg]. \quad (52)$$

The graphs of the solution of the two-dimensional wave function are shown in Figures 5, 6, 7, and 8. Figure 5 is the graph of the third-order approximation for the solution of the real part of the wave function obtained by LADM and HPM. Figure 6 is the graph of the wave function obtained using Mathematica function NDsolve. Figures 7 and 8 show the imaginary part of the solution. We use $k = m = 1$ and $t = 0.1$ in the calculations.

6. Conclusion

In this paper, homotopy perturbation and Laplace-Adomian decomposition methods have proven successful when used to

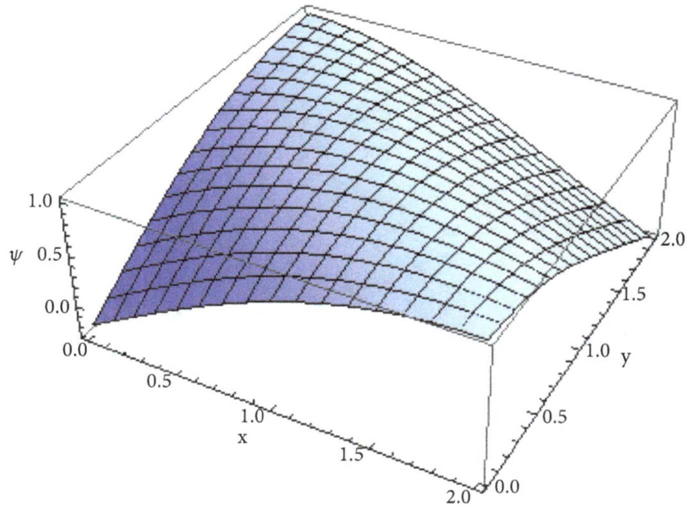

FIGURE 7: Imaginary solution of the two-dimensional wave function obtained by LADM and HPM.

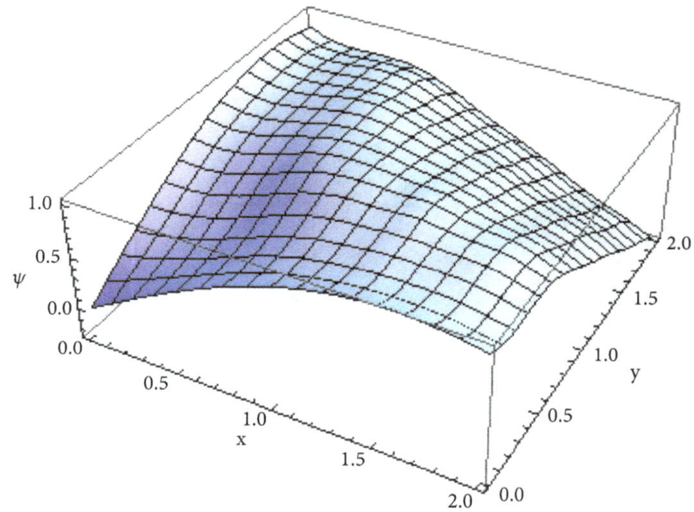

FIGURE 8: Imaginary solution of the two-dimensional wave function obtained by NDsolve.

find the approximate solution to the nonlinear Schrödinger equation with harmonic oscillator in one and in two dimensions. Our theoretical analyses have shown that both methods have given equivalent analytical approximate solutions successfully and efficiently. Comparison between HPM and LADM shows that although the results of these two methods when applied to solve the Schrödinger equation are in good agreement, HPM can overcome the difficulties arising in calculation of Adomian's polynomials. The solutions have been obtained and plotted for the real and imaginary wave function with the effect of adding the harmonic oscillator to the nonlinear Schrödinger equation in one and in two dimensions. HPM and LADM methods numerical results are in agreement with the solution obtained using Mathematica function NDsolve.

Conflicts of Interest

The authors declare no conflicts of interest.

References

[1] J. Cresser, *Quantum Physics Notes*, Department of Physics, Macquarie University, Australia, 2011, Quantum Physics Notes. Department of Physics, Macquarie University.

[2] D. J. Griffiths and D. F. Schroeter, *Introduction to Quantum Mechanics*, Cambridge University Press, 2018.

[3] J. Biazar, R. Ansari, K. Hosseini, and P. Gholamin, "Solution of the linear and non-linear schrodinger equations using

homotopy perturbation and Adomian decomposition methods," *International Mathematical Forum. Journal for Theory and Applications*, vol. 3, no. 37-40, pp. 1891–1897, 2008.

[4] A. Sadighi and D. D. Ganji, "Analytic treatment of linear and nonlinear schrodinger equations: a study with homotopy-perturbation and Adomian decomposition methods," *Physics Letters A*, vol. 372, no. 4, pp. 465–469, 2008.

[5] P. Moradweysi, R. Ansari, K. Hosseini, and F. Sadeghi, "Application of modified Adomian decomposition method to pull-in instability of nano-switches using nonlocal Timoshenko beam theory," *Applied Mathematical Modelling*, vol. 54, pp. 594–604, 2018.

[6] J. Biazar and K. Hosseini, "An effective modification of Adomian decomposition method for solving Emden-Fowler type systems," *National Academy of Science Letters*, vol. 40, no. 4, pp. 285–290, 2017.

[7] J. Biazar and K. Hosseini, "A modified Adomian decomposition method for singular initial value Emden-Fowler type equations," *International Journal of Applied Mathematics Research*, vol. 5, no. 1, p. 69, 2016.

[8] J. Biazar and K. Hosseini, "Analytic approximation of Volterra's population model," *Journal of Applied Mathematics & Informatics*, vol. 13, no. 1, pp. 5–17, 2017.

[9] R. I. Nuruddeen, "Elzaki decomposition method and its applications in solving linear and nonlinear Schrodinger equations," *Sohag Journal of Mathematics*, vol. 4, no. 2, pp. 1–5, 2017.

[10] A. Wazwaz, "A study on linear and nonlinear Schrodinger equations by the variational iteration method," *Chaos, Solitons & Fractals*, vol. 37, no. 4, pp. 1136–1142, 2008.

[11] A. Niknam, A. A. Rajabi, and M. Solaimani, "Solutions of D-dimensional Schrodinger equation for Woods–Saxon potential with spin–orbit, coulomb and centrifugal terms through a new hybrid numerical fitting Nikiforov–Uvarov method," *Journal of Theoretical and Applied Physics*, vol. 10, no. 1, pp. 53–59, 2016.

[12] A. K. Alomari, M. S. Noorani, and R. Nazar, "Explicit series solutions of some linear and nonlinear Schrodinger equations via the homotopy analysis method," *Communications in Nonlinear Science and Numerical Simulation*, vol. 14, no. 4, pp. 1196–1207, 2009.

[13] S. Liao, *Beyond perturbation: introduction to the homotopy analysis method*, CRC press, 2003.

[14] J. Jezierski and W. a. Marzantowicz, *Homotopy methods in topological fixed and periodic points theory*, vol. 3, Springer Science & Business Media, Dordrecht, Netherlands, 2006.

[15] J. H. He, "An elementary introduction to the homotopy perturbation method," *Computers & Mathematics with Applications*, vol. 57, no. 3, pp. 410–412, 2009.

[16] J. Biazar and H. Ghazvini, "Exact solutions for non-linear Schrödinger equations by He's homotopy perturbation method," *Physics Letters A*, vol. 366, no. 1-2, pp. 79–84, 2007.

[17] A. Borhanifar and R. Abazari, "Numerical study of nonlinear Schrödinger and coupled Schrödinger equations by differential transformation method," *Optics Communications*, vol. 283, no. 10, pp. 2026–2031, 2010.

[18] A. Shidfar, A. Molabahrami, A. Babaei, and A. Yazdanian, "A study on the d-dimensional Schrödinger equation with a power-law nonlinearity," *Chaos, Solitons & Fractals*, vol. 42, no. 4, pp. 2154–2158, 2009.

[19] A. Shidfar, A. Molabahrami, A. Babaei, and A. Yazdanian, "A series solution of the Cauchy problem for the generalized d-dimensional Schrödinger equation with a power-law nonlinearity," *Computers & Mathematics with Applications*, vol. 59, no. 4, pp. 1500–1508, 2010.

[20] L. Zheng, T. Wang, X. Zhang, and L. Ma, "The nonlinear Schrödinger harmonic oscillator problem with small odd or even disturbances," *Applied Mathematics Letters*, vol. 26, no. 4, pp. 463–468, 2013.

[21] G. Amador, K. Colon, N. Luna, G. Mercado, E. Pereira, and E. Suazo, "On solutions for linear and nonlinear Schrödinger equations with variable coefficients: A computational approach," *Symmetry*, vol. 8, no. 6, 2016.

[22] Y. Khan and Q. Wu, "Homotopy perturbation transform method for nonlinear equations using He's polynomials," *Computers & Mathematics with Applications*, vol. 61, no. 8, pp. 1963–1967, 2011.

[23] M. Kaplan, K. Hosseini, F. Samadani, and N. Raza, "Optical soliton solutions of the cubic-quintic non-linear Schrödinger's equation including an anti-cubic term," *Journal of Modern Optics*, vol. 65, no. 12, pp. 1431–1436, 2018.

[24] K. Hosseini, A. Zabihi, F. Samadani, and R. Ansari, "New explicit exact solutions of the unstable nonlinear Schrödinger's equation using the expa and hyperbolic function methods," *Optical and Quantum Electronics*, vol. 50, no. 2, 2018.

[25] K. Hosseini, J. Manafian, F. Samadani, M. Foroutan, M. Mirzazadeh, and Q. Zhou, "Resonant optical solitons with perturbation terms and fractional temporal evolution using improved tan $(\phi(\eta)/2)$-expansion method and exp function approach," *Optik - International Journal for Light and Electron Optics*, vol. 158, pp. 933–939, 2018.

[26] K. Hosseini, F. Samadani, D. Kumar, and M. Faridi, "New optical solitons of cubic-quartic nonlinear Schrödinger equation," *Optik - International Journal for Light and Electron Optics*, vol. 157, pp. 1101–1105, 2018.

[27] K. Hosseini, D. Kumar, M. Kaplan, and E. Y. Bejarbaneh, "New Exact Traveling Wave Solutions of the Unstable Nonlinear Schrödinger Equations," *Communications in Theoretical Physics*, vol. 68, no. 6, pp. 761–767, 2017.

[28] G. Adomian, "Solution of physical problems by decomposition," *Computers & Mathematics with Applications*, vol. 27, no. 9-10, pp. 145–154, 1994.

[29] G. Adomian, "A review of the decomposition method in applied mathematics," *Journal of Mathematical Analysis and Applications*, vol. 135, no. 2, pp. 501–544, 1988.

[30] E. K. Jaradat, W. A. Amer, and D. Aloqali, "Using Laplace Decomposition Method to Solve Nonlinear Klein- Gordon Equation," *UPB Scientific Bulletin*, vol. 80, no. 2, Series D, 2018.

[31] S. A. Khuri, "A Laplace decomposition algorithm applied to a class of nonlinear differential equations," *Journal of Applied Mathematics*, vol. 1, no. 4, pp. 141–155, 2001.

[32] S. Liao, "Homotopy analysis method: A new analytic method for nonlinear problems," *Applied Mathematics and Mechanics-English Edition*, vol. 19, no. 10, pp. 957–962, 1998.

[33] J. He, "Homotopy perturbation technique," *Computer Methods Applied Mechanics and Engineering*, vol. 178, no. 3-4, pp. 257–262, 1999.

16

The Convergence of Riemann Solutions to the Modified Chaplygin Gas Equations with a Coulomb-Like Friction Term as the Pressure Vanishes

Yongqiang Fan,[1] **Lihui Guo**,[1] **and Gan Yin**[2]

[1] *College of Mathematics and System Sciences, Xinjiang University, Urumqi 830046, China*
[2] *Department of Mathematics, Zhejiang University of Science & Technology, Hangzhou, Zhejiang 310023, China*

Correspondence should be addressed to Lihui Guo; lihguo@126.com

Academic Editor: Claudio Dappiaggi

This paper studies the convergence of Riemann solutions to the inhomogeneous modified Chaplygin gas equations as the pressure vanishes. The delta shock waves and vacuum states occur as the pressure vanishes. The Riemann solutions of inhomogeneous modified Chaplygin gas equations are no longer self-similar. It is obviously different from the Riemann solutions of homogeneous modified Chaplygin gas equations. When the pressure vanishes, the Riemann solutions of the modified Chaplygin gas equations with a coulomb-like friction term converge to the Riemann solutions of the pressureless Euler system with a source term.

1. Introduction

The inhomogeneous modified Chaplygin gas equations have the following form:

$$\begin{aligned} \rho_t + (\rho u)_x &= 0, \\ (\rho u)_t + (\rho u^2 + P)_x &= \beta \rho, \end{aligned} \tag{1}$$

where β is a constant and ρ, u denote the density and the velocity, respectively. The scalar pressure $P = P(\rho, \alpha)$ is the modified Chaplygin gas pressure $P = \alpha p$ satisfying $\lim_{\alpha \longrightarrow 0} P(\rho, \alpha) = 0$, where α is a sufficiently small positive parameter. Meanwhile, the pressure p satisfies the following equation of state:

$$p = A\rho - \frac{B}{\rho}, \tag{2}$$

where $A > 0$, $B > 0$ are two positive constants. Modified Chaplygin gas (MCG) model was proposed in [1] by Benaoum in 2002. MCG [2, 3] represents the evaluation of the cosmology starting from the radiation era to the Λ cold dark matter (ΛCDM) model mentioned in [2–5]. As an exotic fluid, the MCG plays an important role in describing the accelerated expansion of the universe. In recent years, some researchers made some studies on the thermal equation of state to MCG and found that it could cool down in some constraints of parameters [6]. To know more interesting results related to MCG, the readers are referred to [7–12].

If $\beta = 0$, the system (1) becomes the homogeneous modified Chaplygin gas equations. In [13], Yang and Wang considered the formation of delta shock waves and the vacuum states in the solutions of the homogeneous isentropic Euler equations for modified Chaplygin gas when the pressure vanishes.

Letting $A = 0$, $B = 1$ in (2), the equation of state is Chaplygin gas which was introduced by Chaplygin [14] in 1904, Tsien [15], and von Karman [16] as a suitable mathematical approximation for calculating the lifting force on a wing of an airplane in aerodynamics. The Chaplygin gas can be used to describe the dark energy. For the Riemann problem of homogeneous Chaplygin gas equations, there are lots of results. We refer the readers to [17–23]. For inhomogeneous Chaplygin gas equations, Shen [24] studied Riemann problem by introducing a new velocity:

$$v(x,t) = u(x,t) - \beta t, \tag{3}$$

which was introduced by Faccanoni and Mangeney in [25]. In 2016, Sun [26] studied the non-self-similar Riemann solution of inhomogeneous generalized Chaplygin gas equations. Guo, Li, Pan, and Han [27] considered the Riemann problem with delta initial data for the one-dimensional Chaplygin gas equations with a source term.

As $\alpha \longrightarrow 0$, system (1)-(2) becomes the inhomogeneous pressureless Euler system:

$$\begin{aligned} \rho_t + (\rho u)_x &= 0, \\ (\rho u)_t + (\rho u^2)_x &= \beta\rho. \end{aligned} \tag{4}$$

Shen [28] considered the Riemann problem of pressureless Euler system (4). Daw and Nedeljkov [29] studied the shadow waves for pressureless gas balance laws.

In this paper, we are concerned with the convergence of the Riemann solutions to the inhomogeneous modified Chaplygin gas equations as the pressure vanishes. Firstly, we give the Riemann solutions of the inhomogeneous modified Chaplygin gas equations. Then, we study the convergence of the Riemann solutions to the modified Chaplygin gas equations with a source term as the pressure vanishes. We find that the Riemann solutions of inhomogeneous modified Chaplygin gas equations converge to the corresponding Riemann solutions of the pressureless Euler system. We mainly use the method of vanishing pressure limits which was introduced by Li [30] and Chen and Liu [31, 32] in which they studied the formation of delta shocks and vacuum states in the vanishing pressure limit of solutions to the isentropic Euler equations and the concentration and cavitation in the vanishing pressure limit of solutions to the Euler equations for nonisentropic fluids. Shen [33] considered the limits of Riemann solutions to the isentropic magnetogasdynamics. Shen and Sun [34] studied the formation of delta shocks and vacuum states in the vanishing pressure limit of Riemann solutions to the perturbed Aw-Rascle model. Sheng, Wang, and Yin [35] studied the vanishing pressure limit of the generalized Chaplygin gas dynamics system. Yin and Sheng [36] considered the delta shocks and vacuum states in vanishing pressure limit of solutions to the relativistic Euler equations for polytropic gases. For inhomogeneous equations, Guo, Li, and Yin [37, 38] considered the vanishing pressure limits of Riemann solutions to the Chaplygin gas equations with a source term and the limit behavior of the Riemann solutions to the generalized Chaplygin gas equations with a source term.

We organize this article as follows: in Section 2, we give some preliminaries which include the consideration of Riemann solutions to system (1)-(2) and the review of Riemann solutions to system (4). In Section 3, we study the convergence of Riemann solutions to system (1)-(2).

2. Some Preliminaries

In this section, we consider the Riemann solutions of the inhomogeneous modified Chaplygin gas equations and briefly review the Riemann solutions of pressureless Euler system with a coulomb-like friction term.

2.1. Riemann Problem for (1)-(2). In this subsection, we are concerned with the Riemann problem of (1)-(2).

From (3), system (1)-(2) is turned into the following conservation laws:

$$\begin{aligned} \rho_t + (\rho(v+\beta t))_x &= 0, \\ (\rho v)_t + \left(\rho v(v+\beta t) + \alpha\left(A\rho - \frac{B}{\rho}\right)\right)_x &= 0. \end{aligned} \tag{5}$$

We consider the Riemann initial value as follows:

$$(v,\rho)(x,0) = (u_\pm, \rho_\pm), \quad \pm x \geq 0, \tag{6}$$

where $\rho_\pm \geq 0$ and $u_\pm$ are given constants. Letting $f(\rho,\alpha) = \sqrt{\alpha(A+B/\rho^2)}$, the two eigenvalues of system (5) are $\lambda_1 = v + \beta t - f(\rho,\alpha)$, $\lambda_2 = v + \beta t + f(\rho,\alpha)$. For arbitrary positive constants A, B, system (5) is strictly hyperbolic. The corresponding right eigenvectors are $\vec{r}_1 = (-\rho, f(\rho,\alpha))^T$, $\vec{r}_2 = (\rho, f(\rho,\alpha))^T$. By simple calculation, we obtain that $\nabla\lambda_i \cdot \vec{r}_i = A\alpha/f(\rho,\alpha) \neq 0$, $(i = 1,2)$, which implies that both the characteristic fields are genuinely nonlinear.

Given a state (u_-,ρ_-) in the phase plane, the curve of backward rarefaction wave is

$$R_1^\alpha(u_-,\rho_-): \begin{cases} \dfrac{dx}{dt} = \lambda_1 = v + \beta t - f(\rho,\alpha), \\ v = u_- - g(\rho_-,\rho,\alpha), \\ \lambda_1(u_-,\rho_-) < \lambda_1(v,\rho), \end{cases} \tag{7}$$

$$\rho < \rho_-, \quad v > u_-,$$

and the corresponding curve of forward rarefaction wave is

$$R_2^\alpha(u_-,\rho_-): \begin{cases} \dfrac{dx}{dt} = \lambda_2 = v + \beta t + f(\rho,\alpha), \\ v = u_- + g(\rho_-,\rho,\alpha), \\ \lambda_2(u_-,\rho_-) < \lambda_2(v,\rho), \end{cases} \tag{8}$$

$$\rho > \rho_-, \quad v > u_-,$$

where

$$\begin{aligned} g(\rho_-,\rho,\alpha) &= \int_{\rho_-}^{\rho} \frac{f(\rho,\alpha)}{\rho} d\rho = -f(\rho,\alpha) + f(\rho_-,\alpha) \\ &\quad + \sqrt{A\alpha}\left(\ln\left(f(\rho,1) + \sqrt{A}\right) + \ln\rho \right. \\ &\quad \left. - \ln\left(f(\rho_-,1) + \sqrt{A}\right) - \ln\rho_-\right). \end{aligned} \tag{9}$$

Denote the propagating speed of the bounded discontinuity $x = x(t)$ as $\sigma(t) = x'(t)$. The Rankine-Hugoniot conditions read

$$\begin{aligned} -\sigma(t)[\rho] + [\rho(v+\beta t)] &= 0, \\ -\sigma(t)[\rho v] + \left[\rho v(v+\beta t) + \alpha\left(A\rho - \frac{B}{\rho}\right)\right] &= 0. \end{aligned} \tag{10}$$

For convenience, we write $h(\rho,\rho_-,\alpha) = \sqrt{(\alpha/\rho\rho_-)(A + B/\rho\rho_-)}$. Then for a given state (u_-,ρ_-), from (10), we obtain two kinds of shock wave curves, i.e., the backward shock wave curve,

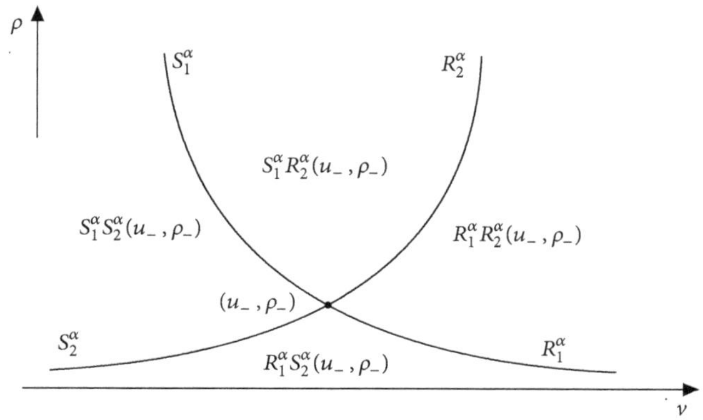

FIGURE 1: The phase plane (v, ρ).

$$S_1^\alpha(u_-,\rho_-):\begin{cases}\sigma_1^\alpha(t)=u_-+\beta t-\rho h(\rho,\rho_-,\alpha),\\ v=u_--h(\rho,\rho_-,\alpha)(\rho-\rho_-),\end{cases}\quad \rho>\rho_-,\ v<u_-, \tag{11}$$

and the forward shock wave curve,

$$S_2^\alpha(u_-,\rho_-):\begin{cases}\sigma_2^\alpha(t)=u_-+\beta t+\rho h(\rho,\rho_-,\alpha),\\ v=u_-+h(\rho,\rho_-,\alpha)(\rho-\rho_-),\end{cases}\quad \rho<\rho_-,\ v<u_-. \tag{12}$$

In the phase plane, given a state (u_-,ρ_-), the curves of $R_j^\alpha(u_-,\rho_-)$ and $S_j^\alpha(u_-,\rho_-),(j=1,2)$ divide the phase plane into four regions.

When $(u_+,\rho_+)\in(R_1^\alpha R_2^\alpha\cup R_1^\alpha S_2^\alpha\cup S_1^\alpha R_2^\alpha\cup S_1^\alpha S_2^\alpha)(u_-,\rho_-)$ (see Figure 1), the Riemann solutions of system (5) are

(1) $(u_+,\rho_+)\in R_1^\alpha R_2^\alpha(u_-,\rho_-):\ (u_-,\rho_-)+R_1^\alpha+(v_{*\alpha},\rho_{*\alpha})+R_2^\alpha+(u_+,\rho_+)$;

(2) $(u_+,\rho_+)\in R_1^\alpha S_2^\alpha(u_-,\rho_-):\ (u_-,\rho_-)+R_1^\alpha+(v_{*\alpha},\rho_{*\alpha})+S_2^\alpha+(u_+,\rho_+)$;

(3) $(u_+,\rho_+)\in S_1^\alpha R_2^\alpha(u_-,\rho_-):\ (u_-,\rho_-)+S_1^\alpha+(v_{*\alpha},\rho_{*\alpha})+R_2^\alpha+(u_+,\rho_+)$;

(4) $(u_+,\rho_+)\in S_1^\alpha S_2^\alpha(u_-,\rho_-):\ (u_-,\rho_-)+S_1^\alpha+(v_{*\alpha},\rho_{*\alpha})+S_2^\alpha+(u_+,\rho_+)$,

where $(v_{*\alpha},\rho_{*\alpha})$ denotes the intermediate state. By using (3), we obtain the Riemann solutions of (1)-(2) as follows:

(1) $(u_+,\rho_+)\in R_1^\alpha R_2^\alpha(u_-,\rho_-):\ (u_-+\beta t,\rho_-)+R_1^\alpha+(v_{*\alpha}+\beta t,\rho_{*\alpha})+R_2^\alpha+(u_++\beta t,\rho_+)$;

(2) $(u_+,\rho_+)\in R_1^\alpha S_2^\alpha(u_-,\rho_-):\ (u_-+\beta t,\rho_-)+R_1^\alpha+(v_{*\alpha}+\beta t,\rho_{*\alpha})+S_2^\alpha+(u_++\beta t,\rho_+)$;

(3) $(u_+,\rho_+)\in S_1^\alpha R_2^\alpha(u_-,\rho_-)\ (u_-+\beta t,\rho_-)+S_1^\alpha+(v_{*\alpha}+\beta t,\rho_{*\alpha})+R_2^\alpha+(u_++\beta t,\rho_+)$;

(4) $(u_+,\rho_+)\in S_1^\alpha S_2^\alpha(u_-,\rho_-)\ (u_-+\beta t,\rho_-)+S_1^\alpha+(v_{*\alpha}+\beta t,\rho_{*\alpha})+S_2^\alpha+(u_++\beta t,\rho_+)$.

2.2. Riemann Problem for (4). In this subsection, we restate the results on the Riemann solutions to system (4). The detailed results can be referred to in [28].

(1) For $u_->u_+$, the Riemann solution of system (4) has the following form:

$$\rho(x,t)=\begin{cases}\rho_-, & x<x(t),\\ \omega(t)\delta(x-x(t)), & x=x(t),\\ \rho_+, & x>x(t),\end{cases} \tag{13}$$

$$u(x,t)=\begin{cases}u_-, & x<x(t),\\ u_\delta(t), & x=x(t),\\ u_+, & x>x(t),\end{cases} \tag{14}$$

in which the delta shock wave is introduced into the Riemann solution (13)-(14). About the definition of delta shock wave, please refer to [31, 34, 39, 40], for example. The delta shock wave satisfies the generalized Rankine-Hugoniot conditions:

$$\begin{aligned}\frac{dx(t)}{dt}&=u_\delta(t),\\ \frac{d\omega(t)}{dt}&=u_\delta(t)[\rho]-[\rho u],\\ \frac{d(\omega(t)u_\delta(t))}{dt}&=u_\delta(t)[\rho u]-[\rho u^2]+\beta\omega(t),\end{aligned} \tag{15}$$

where $x(t)$, $\omega(t)$, and $u_\delta(t)$, respectively, denote the location, weight, and the velocity of delta shock wave, and $(x(0),w(0))=(0,0)$.

Letting $v_\delta=(\sqrt{\rho_-}u_-+\sqrt{\rho_+}u_+)/(\sqrt{\rho_-}+\sqrt{\rho_+})$, by simple calculation, we obtain

$$\begin{aligned}\omega(t)&=\sqrt{\rho_-\rho_+}(u_--u_+)t,\\ u_\delta(t)&=v_\delta+\beta t,\\ x(t)&=v_\delta t+\frac{1}{2}\beta t^2,\end{aligned} \tag{16}$$

for $[\rho]\neq 0$ and

$$\begin{aligned}\omega(t)&=-[\rho u]t,\\ u_\delta(t)&=\frac{1}{2}(u_++u_-)+\frac{1}{3}\beta t,\\ x(t)&=\frac{1}{2}(u_++u_-)t+\frac{1}{6}\beta t^2,\end{aligned} \tag{17}$$

for $[\rho]=0$.

(2) For $u_-<u_+$, the structure of Riemann solution is

$$(u_-+\beta t,\rho_-)+J_1+Vac+J_2+(u_++\beta t,\rho_+). \tag{18}$$

(3) For $u_-=u_+$, the structure of Riemann solution is

$$(u_-+\beta t,\rho_-)+J^*+(u_++\beta t,\rho_+). \tag{19}$$

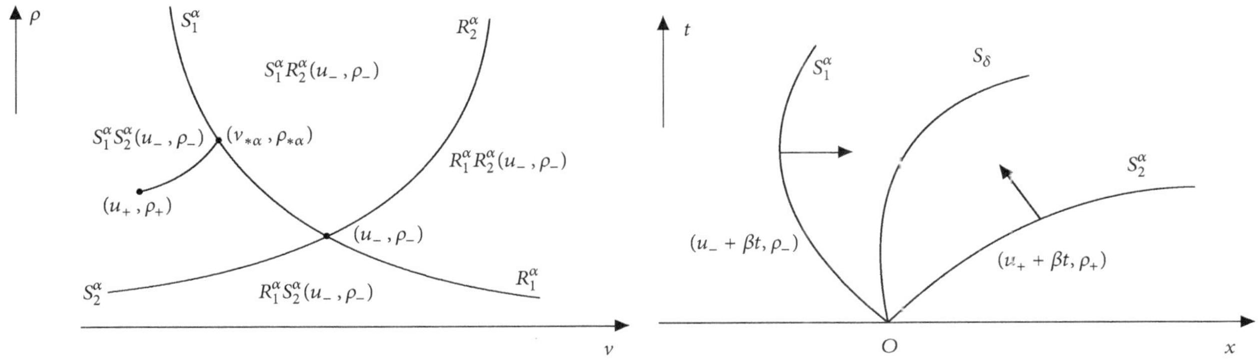

(a) The phase plane (b) The (x,t) plane

FIGURE 2: Riemann solution when $(u_+, \rho_+) \in S_1^\alpha S_2^\alpha(u_-, \rho_-)$.

3. Convergence of Riemann Solutions to (1)-(2)

In this section, we consider the convergence of the Riemann solutions to system (1)-(2) in the vanishing pressure. We divide our discussions into three parts: the concentration and delta shock wave for $u_- > u_+$, the occurrence of vacuum for $u_- < u_+$, and the formation of discontinuity for $u_- \le u_+$.

3.1. Concentration and Delta Shock Wave as the Pressure Vanishes. In this subcase, we show the phenomenon of the concentration and delta shock wave in the vanishing pressure limit of Riemann solutions to (1)-(2) when $u_- > u_+$ (see Figure 2(a)).

Lemma 1. *Assume $u_- > u_+$, then there exists certain value $\alpha_1 > 0$, such that $(u_+, \rho_+) \in S_1^\alpha S_2^\alpha(u_-, \rho_-)$ when $0 < \alpha < \alpha_1$.*

Proof. If $u_- > u_+$ and $(u_+, \rho_+) \in S_1^\alpha S_2^\alpha(u_-, \rho_-)$, we obtain (see Figure 2(a)):

$$u_+ < u_- - h(\rho_+, \rho_-, \alpha)(\rho_+ - \rho_-), \quad \rho_+ > \rho_-, \tag{20}$$

$$u_+ < u_- + h(\rho_+, \rho_-, \alpha)(\rho_+ - \rho_-), \quad \rho_+ < \rho_-. \tag{21}$$

From (20)-(21) and (2), we have

$$\sqrt{\alpha}\left|h(\rho_+, \rho_-, 1)(\rho_+ - \rho_-)\right| < u_- - u_+. \tag{22}$$

From (22), we derive that

$$\alpha < \frac{(u_- - u_+)^2}{h^2(\rho_+, \rho_-, 1)(\rho_+ - \rho_-)^2}. \tag{23}$$

Let $\alpha_1 = (u_- - u_+)^2/h^2(\rho_+, \rho_-, 1)(\rho_+ - \rho_-)^2$, then $(u_+, \rho_+) \in S_1^\alpha S_2^\alpha(u_-, \rho_-)$ when $u_- > u_+$ and $0 < \alpha < \alpha_1$. It is obvious that this conclusion is also true for $\rho_+ = \rho_-$, The proof is completed.

For $(u_+, \rho_+) \in S_1^\alpha S_2^\alpha(u_-, \rho_-)$, the Riemann solution of (5)-(6) contains two shock waves satisfying

$$S_1^\alpha: \begin{cases} v_{*\alpha} = u_- - h(\rho_{*\alpha}, \rho_-, \alpha)(\rho_{*\alpha} - \rho_-), \\ \sigma_1^\alpha(t) = u_- + \beta t - \rho_{*\alpha} h(\rho_{*\alpha}, \rho_-, \alpha), \end{cases} \tag{24}$$
$$\rho_{*\alpha} > \rho_-,$$

$$S_2^\alpha: \begin{cases} v_{*\alpha} = u_+ + h(\rho_{*\alpha}, \rho_+, \alpha)(\rho_{*\alpha} - \rho_+), \\ \sigma_2^\alpha(t) = v_{*\alpha} + \beta t + \rho_+ h(\rho_{*\alpha}, \rho_+, \alpha), \end{cases} \tag{25}$$
$$\rho_{*\alpha} > \rho_+,$$

where $(v_{*\alpha}, \rho_{*\alpha})$ is the intermediate state (see Figure 2(a)). From the first equation of (24)-(25), we obtain

$$\begin{aligned} u_- - u_+ &= h(\rho_{*\alpha}, \rho_-, \alpha)(\rho_{*\alpha} - \rho_-) \\ &\quad + h(\rho_{*\alpha}, \rho_+, \alpha)(\rho_{*\alpha} - \rho_+), \quad \rho_{*\alpha} > \rho_\pm. \end{aligned} \tag{26}$$

From (26), we can derive that $\lim_{\alpha \to 0} \rho_{*\alpha} = +\infty$ and $\lim_{\alpha \to 0} \alpha\rho_{*\alpha} = \rho_+\rho_-(u_- - u_+)^2/A(\sqrt{\rho_+} + \sqrt{\rho_-})^2$. From the above analysis, we can obtain Lemma 2 as follows.

Lemma 2. *As $u_- > u_+$ and $(u_+, \rho_+) \in S_1^\alpha S_2^\alpha(u_-, \rho_-)$, $\lim_{\alpha \to 0} \rho_{*\alpha} = +\infty$ and $\lim_{\alpha \to 0} \alpha\rho_{*\alpha} = \rho_+\rho_-(u_- - u_+)^2/A(\sqrt{\rho_+} + \sqrt{\rho_-})^2$, i.e., the phenomenon of concentration occurs.*

Next, we consider the formation of delta shock waves. Firstly, we analyze the velocity of intermediate state u_*. It follows from (3), (24) and Lemma 2 that

$$\begin{aligned} \lim_{\alpha \to 0} u_{*\alpha} &= \lim_{\alpha \to 0} (v_{*\alpha} + \beta t) \\ &= \lim_{\alpha \to 0} (u_- - h(\rho_{*\alpha}, \rho_-, \alpha)(\rho_{*\alpha} - \rho_-)) + \beta t \\ &= u_\delta(t). \end{aligned} \tag{27}$$

As the pressure vanishes, from the second equation of (24)-(25), the limits of the speeds of S_1^α and S_2^α are as follows:

$$\lim_{\alpha \to 0} \sigma_1^\alpha(t) = \lim_{\alpha \to 0} \sigma_2^\alpha(t) = u_\delta(t). \tag{28}$$

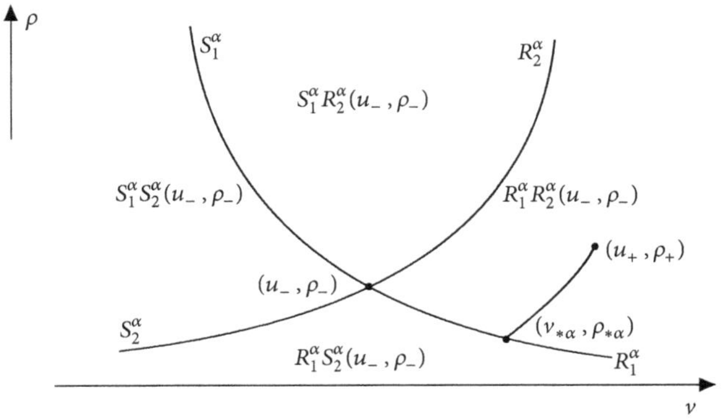

(a) The phase plane

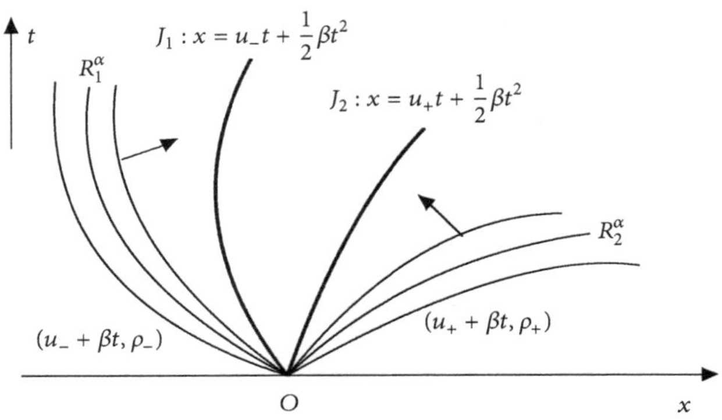

(b) The (x, t) plane

FIGURE 3: Riemann solution when $(u_+, \rho_+) \in R_1^\alpha R_2^\alpha(u_-, \rho_-)$.

Equation (27)-(28) shows the two shock waves S_1^α and S_2^α are merged into one shock wave with speed $u_\delta(t)$ (see Figure 2(b)).

By using the first equation of Rankine-Hugoniot conditions (10) for both S_1^α and S_2^α, we find

$$\begin{aligned} \sigma_1^\alpha(t)(\rho_{*\alpha} - \rho_-) &= \rho_{*\alpha}(v_{*\alpha} + \beta t) - \rho_-(u_- + \beta t), \\ \sigma_2^\alpha(t)(\rho_+ - \rho_{*\alpha}) &= \rho_+(u_+ + \beta t) - \rho_{*\alpha}(v_{*\alpha} + \beta t). \end{aligned} \tag{29}$$

Considering (28)-(29), we obtain

$$\lim_{\alpha \to 0} (\sigma_2^\alpha(t) - \sigma_1^\alpha(t)) \rho_{*\alpha} = u_\delta(t)[\rho] - [\rho(v + \beta t)]. \tag{30}$$

From (30), we obtain that

$$\begin{aligned} \lim_{\alpha \to 0} \int_{\sigma_1(t)t}^{\sigma_2(t)t} \rho_{*\alpha} dx &= \lim_{\alpha \to 0} (\sigma_2^\alpha(t) - \sigma_1^\alpha(t)) \rho_{*\alpha} t \\ &= (u_\delta(t)[\rho] - [\rho(v + \beta t)]) t \\ &= \omega(t). \end{aligned} \tag{31}$$

From above analysis and Lemma 2, we have Theorem 3.

Theorem 3. *For $u_- > u_+$, as $\alpha \to 0$, the two shock waves of system (1)-(2) converge to a delta shock solution which is the corresponding Riemann solution of system (4).*

We have considered the convergence of the Riemann solutions to system (1)-(2) for $u_- > u_+$ as the pressure vanishes. In the following, we study the convergence of the Riemann solutions to system (1)-(2) for $u_- \le u_+$.

3.2. The Occurrence of Vacuum as the Pressure Vanishes. In this subcase, as the pressure vanishes, we study the occurrence of vacuum when $(u_+, \rho_+) \in R_1^\alpha R_2^\alpha(u_-, \rho_-)$ and $u_- < u_+$ (see Figure 3(a)).

Suppose that $(u_+, \rho_+) \in R_1^\alpha R_2^\alpha(u_-, \rho_-)$, from (7)-(8) and (9), one can derive that

$$u_+ > u_- - g(\rho_-, \rho_+, \alpha), \quad \rho_+ < \rho_-, \tag{32}$$

$$u_+ > u_- + g(\rho_-, \rho_+, \alpha), \quad \rho_+ > \rho_-. \tag{33}$$

According to (32)-(33), we obtain that

$$\alpha < \left(\frac{u_+ - u_-}{f(\rho_+, 1) + f(\rho_-, 1) + k(\rho_+, \rho_-)} \right)^2, \tag{34}$$

where $k(\rho_+, \rho_-) = \ln((\sqrt{A\rho_+^2 + B} + \sqrt{A}\rho_+)/(\sqrt{A\rho_-^2 + B} + \sqrt{A}\rho_-))$. Letting $\alpha_0 = ((u_+ - u_-)/(f(\rho_+, 1) + f(\rho_-, 1) + k(\rho_+, \rho_-)))^2$, the right state $(u_+, \rho_+) \in R_1^\alpha R_2^\alpha(u_-, \rho_-)$ as $0 < \alpha < \alpha_0$. Next, we analyze the formation of vacuum when $(u_+, \rho_+) \in R_1^\alpha R_2^\alpha(u_-, \rho_-)$.

Lemma 4. *As $u_+ > u_-$ and $(u_+, \rho_+) \in R_1^\alpha R_2^\alpha(u_-, \rho_-)$, $\lim_{\alpha \to 0} \rho_{*\alpha} = 0$, i.e., the phenomenon of vacuum occurs.*

Proof. From the above analysis, there exists a constant α_0, such that $(u_+, \rho_+) \in R_1^\alpha R_2^\alpha(u_-, \rho_-)$ when $\alpha \in (0, \alpha_0)$. From (7)-(8), we obtain that $v = u_- - g(\rho_-, \rho, \alpha)$ on $R_1^\alpha(\rho_-, u_-)$ and $v = u_+ + g(\rho_+, \rho, \alpha)$ on $R_2^\alpha(\rho_+, u_+)$, which derive that

$$u_- - u_+ = g(\rho_+, \rho, \alpha) + g(\rho_-, \rho, \alpha). \tag{35}$$

If $\lim_{\alpha \to 0} \rho_{*\alpha} = c_0 \in (0, \min(\rho_-, \rho_+))$, as α tends to zero, from (35), we have $u_- = u_+$, which leads to a contradiction. Therefore, $\lim_{\alpha \to 0} \rho_{*\alpha} = 0$, which means the vacuum occurs as the pressure vanishes. □

In Sections 3.1 and 3.2, we discussed the concentration and delta shock wave and the occurrence of vacuum. In the following, we consider the formation of discontinuity.

3.3. The Formation of Discontinuity as the Pressure Vanishes. In this subsection, we are concerned with the formation of discontinuity for $u_- \le u_+$, as $\alpha \to 0$. Our argument is divided into three parts.

Case 1. $J_1 + Vac + J_2$ for $u_- < u_+$ and $\alpha \to 0$.

Lemma 4 shows that the phenomenon of vacuum occurs as $u_- < u_+$ and $\alpha \to 0$. Next, we introduce the formation of discontinuity for $u_- < u_+$.

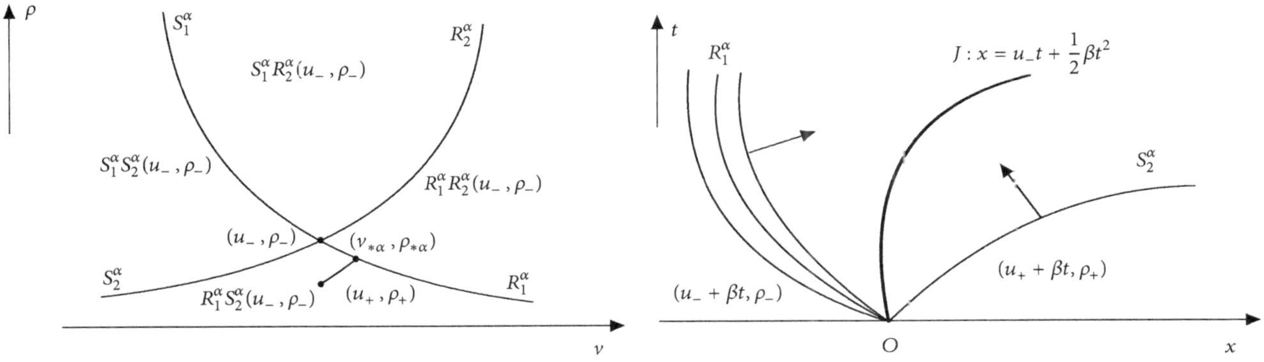

(a) The phase plane (b) The (x, t) plane

FIGURE 4: Riemann solution when $(u_+, \rho_+) \in R_1^\alpha S_2^\alpha(u_-, \rho_-)$.

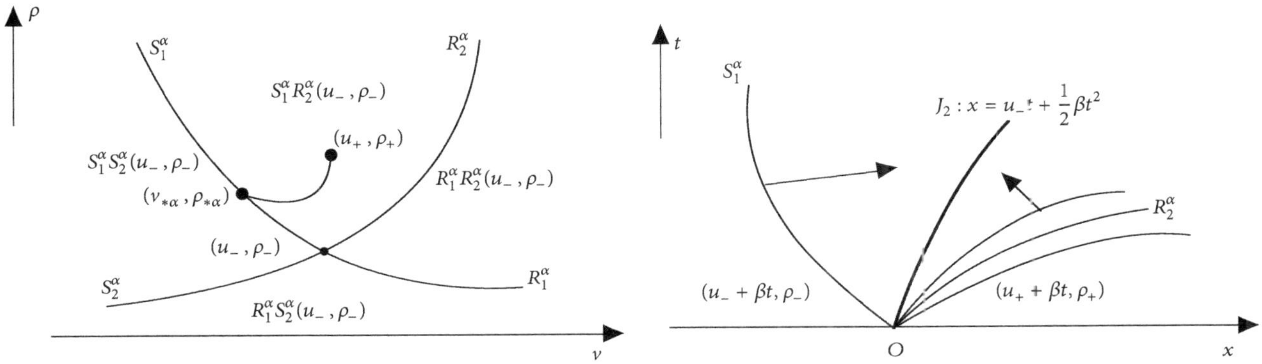

(a) The phase plane (b) The (x, t) plane

FIGURE 5: Riemann solution when $(u_+, \rho_+) \in S_1^\alpha R_2^\alpha(u_-, \rho_-)$.

When $(u_+, \rho_+) \in R_1^\alpha R_2^\alpha(u_-, \rho_-)$, from (7)-(8), we have

$$\begin{aligned} \lim_{\alpha \longrightarrow 0} v &= u_- \quad \text{on } R_1^\alpha(\rho_-, u_-), \\ \lim_{\alpha \longrightarrow 0} v &= u_+ \quad \text{on } R_2^\alpha(\rho_+, u_+) \end{aligned} \tag{36}$$

and

$$\begin{aligned} \lim_{\alpha \longrightarrow 0} \lambda_1 &= u_- + \beta t, \\ \lim_{\alpha \longrightarrow 0} \lambda_2 &= u_+ + \beta t. \end{aligned} \tag{37}$$

It follows from (36)-(37) that the two rarefaction waves R_1^α and R_2^α turn into two contact discontinuities as the pressure vanishes for $u_- < u_+$ (see Figure 3(b)). We summarize our conclusion as follows.

Theorem 5. *For $u_- < u_+$, as $\alpha \longrightarrow 0$, the two rarefaction waves of system (1)-(2) converge to two contact discontinuities which are the corresponding Riemann solution of system (4).*

Case 2. $J^\star$ for $u_- = u_+$ and $\alpha \longrightarrow 0$.

We have studied the convergence of the Riemann solution to system (1)-(2) for $u_- > u_+$ and $u_- < u_+$ as the pressure vanishes. Next, we discuss the situation for $u_- = u_+$. We divide our discussion into two parts for $(u_+, \rho_+) \in R_1^\alpha S_2^\alpha(u_-, \rho_-)$ (see Figure 4(a)) and $(u_+, \rho_+) \in S_1^\alpha R_2^\alpha(u_-, \rho_-)$ (see Figure 5(a)).

When $(u_+, \rho_+) \in R_1^\alpha S_2^\alpha(u_-, \rho_-)$, from (7) and (25), we deduce that

$$\lim_{\alpha \longrightarrow 0} v_{*\alpha} + \beta t = \lim_{\alpha \longrightarrow 0} \lambda_1 (v_{*\alpha}, \rho_{*\alpha}) = \sigma_2^\alpha(t) = u_- + \beta t, \tag{38}$$

which implies that the rarefaction wave $R_1(\rho_-, u_-)$ and the shock wave $S_2(\rho_+, u_+)$ become a contact discontinuity as the pressure vanishes (see Figure 4(b)).

Similarly, from (8) and (24), we know that

$$\lim_{\alpha \longrightarrow 0} v_{*\alpha} + \beta t = \lim_{\alpha \longrightarrow 0} \lambda_2 (v_{*\alpha}, \rho_{*\alpha}) = \sigma_1^\alpha(t) = u_- + \beta t, \tag{39}$$

which means that the rarefaction wave $R_2^\alpha(\rho_+, u_+)$ and the shock wave $S_1^\alpha(\rho_-, u_-)$ turn into a contact discontinuity as $\alpha \longrightarrow 0$ (see Figure 5(b)).

We have considered the convergence of the Riemann solution of system (1)-(2) for $u_- = u_+$, as the pressure vanishes. We summarize our main results as follows.

Theorem 6. *For $u_- = u_+$, as $\alpha \longrightarrow 0$, the Riemann solutions of system (1)-(2) converge to a contact discontinuity $J^\star$ which is the corresponding Riemann solution of system (4).*

4. Conclusion

We have considered the convergence of the Riemann solutions to the modified Chaplygin gas equations with a coulomb-like friction term. As the pressure vanishes, the concentration and delta shock wave are concerned. Meanwhile the occurrence of vacuum and formation of discontinuity are studied. We find that the Riemann solutions of (1)-(2) converge to the corresponding Riemann solutions of (4) as the pressure vanishes.

Conflicts of Interest

The authors declare that they have no conflicts of interest.

Acknowledgments

This work is partially supported by the Natural Science Foundation of Xinjiang (Grant no. 2017D01C053).

References

[1] H. B. Benaoum, "Accelerated universe from modified Chaplygin gas and tachyonic fluid," Physics, 2002.

[2] U. Debnath, "Constraining the parameters of modified chaplygin gas in einstein-aether gravity," *Advances in High Energy Physics*, vol. 2014, Article ID 653630, 8 pages, 2014.

[3] B. C. Paul, P. Thakur, and A. Saha, "Modified Chaplygin gas as scalar field and holographic dark energy model," High Energy Physics-Theory, 2008.

[4] S. Chakraborty and U. Debnath, "Role of a chameleon field in the presence of variable modified Chaplygin gas in Brans-Dicke theory," *Canadian Journal of Physics*, vol. 90, no. 2, pp. 131–135, 2012.

[5] J. Zschoche, "The Chaplygin gas equation of state for the quantized free scalar field on cosmological spacetimes," *Annales Henri Poincare. A Journal of Theoretical and Mathematical Physics*, vol. 15, no. 7, pp. 1285–1325, 2014.

[6] M. Szydłowski and W. Czaja, "Toward reconstruction of the dynamics of the Universe from distant type Ia supernovae," *Physical Review D: Particles, Fields, Gravitation and Cosmology*, vol. 69, no. 8, pp. 281–317, 2004.

[7] H. B. Benaoum, "Modified Chaplygin gas cosmology," *Advances in High Energy Physics*, vol. 2012, Article ID 357802, 12 pages, 2012.

[8] S. Bhattacharya and U. Debnath, "Thermodynamics of Modified Chaplygin Gas and Tachyonic Field," *International Journal of Theoretical Physics*, vol. 51, no. 2, pp. 565–576, 2012.

[9] S. Chattopadhyay and U. Debnath, "Interaction between phantom field and modified Chaplygin gas," *Astrophysics and Space Science*, vol. 326, no. 2, pp. 155–158, 2010.

[10] S. S. E. Costa, "Relations between the modified Chaplygin gas and a scalar field," Physics, 2008.

[11] E. O. Kahya and B. Pourhassan, "Observational constraints on the extended Chaplygin gas inflation," *Astrophysics and Space Science*, vol. 353, no. 2, pp. 677–682, 2014.

[12] P. Rudra, C. Ranjit, and S. Kundu, "How effective is new variable modified Chaplygin gas to play the role of dark energy-a dynamical system analysis in RS II brane model," *Astrophysics and Space Science*, vol. 347, no. 2, pp. 433–444, 2013.

[13] H. Yang and J. Wang, "Delta-shocks and vacuum states in the vanishing pressure limit of solutions to the isentropic Euler equations for modified Chaplygin gas," *Journal of Mathematical Analysis and Applications*, vol. 413, no. 2, pp. 800–820, 2014.

[14] S. A. Chaplygin, "On gas jets," *Scientific Memoirs, Moscow University Mathematic Physics*, vol. 21, no. 1063, pp. 1–121, 1904.

[15] H.-S. Tsien, "Two-dimensional subsonic flow of compressible fluids," *Journal of the Aeronautical Sciences*, vol. 6, pp. 399–407, 1939.

[16] T. von Karman, "Compressibility effects in aerodynamics," *Journal of the Aeronautical Sciences*, vol. 8, pp. 337–356, 1941.

[17] Y. Brenier, "Solutions with concentration to the Riemann problem for the one-dimensional Chaplygin gas equations," *Journal of Mathematical Fluid Mechanics*, vol. 7, no. suppl. 3, pp. S326–S331, 2005.

[18] L. Guo, W. Sheng, and T. Zhang, "The two-dimensional Riemann problem for isentropic Chaplygin gas dynamic system," *Communications on Pure and Applied Analysis*, vol. 9, no. 2, pp. 431–458, 2010.

[19] D. Kong and C. Wei, "Formation and propagation of singularities in one-dimensional Chaplygin gas," *Journal of North China Institute of Science & Technology*, vol. 80, pp. 58–70, 2014.

[20] G. Lai, W. Sheng, and Y. Zheng, "Simple waves and pressure delta waves for a Chaplygin gas in two-dimensions," *Discrete and Continuous Dynamical Systems - Series A*, vol. 31, no. 2, pp. 489–523, 2011.

[21] S. Chen and A. Qu, "Two-dimensional Riemann problems for Chaplygin gas," *SIAM Journal on Mathematical Analysis*, vol. 44, no. 3, pp. 2146–2178, 2012.

[22] Z. Wang and Q. Zhang, "The Riemann problem with delta initial data for the one-dimensional Chaplygin gas equations," *Acta Mathematica Scientia B*, vol. 32, no. 3, pp. 825–841, 2012.

[23] G. Wang, B. Chen, and Y. Hu, "The two-dimensional Riemann problem for Chaplygin gas dynamics with three constant states," *Journal of Mathematical Analysis & Applications*, vol. 393, pp. 544–562, 2012.

[24] C. Shen, "The Riemann problem for the Chaplygin gas equations with a source term," *ZAMM - Journal of Applied Mathematics and Mechanics/Zeitschrift für Angewandte Mathematik und Mechanik*, vol. 96, no. 6, pp. 681–695, 2016.

[25] G. Faccanoni and A. Mangeney, "Exact solution for granular flows," *International Journal for Numerical and Analytical Methods in Geomechanics*, vol. 37, no. 10, pp. 1408–1433, 2013.

[26] M. Sun, "The exact Riemann solutions to the generalized Chaplygin gas equations with friction," *Communications in Nonlinear Science and Numerical Simulation*, vol. 36, pp. 342–353, 2016.

[27] L. Guo, T. Li, L. Pan, and X. Han, "The Riemann problem with delta initial data for the one-dimensional Chaplygin gas equations with a source term," *Nonlinear Analysis: Real World Applications*, vol. 41, pp. 588–606, 2018.

[28] C. Shen, "The Riemann problem for the pressureless Euler system with the Coulomb-like friction term," *IMA Journal of Applied Mathematics*, vol. 81, no. 1, pp. 76–99, 2016.

[29] D. A. Daw and M. Nedeljkov, "Shadow waves for pressureless gas balance laws," *Applied Mathematics Letters*, vol. 57, pp. 54–59, 2016.

[30] J. Li, "Note on the compressible Euler equations with zero temperature," *Applied Mathematics Letters*, vol. 14, no. 4, pp. 519–523, 2001.

[31] G. Chen and H. Liu, "Formation of delta-shocks and vacuum states in the vanishing pressure limit of solutions to the Euler equations for isentropic fluids," *Journal of Speech & Hearing Research*, vol. 28, pp. 445–448, 2006.

[32] G.-Q. Chen and H. Liu, "Concentration and cavitation in the vanishing pressure limit of solutions to the Euler equations for nonisentropic fluids," *Physica D: Nonlinear Phenomena*, vol. 189, no. 1-2, pp. 141–165, 2004.

[33] C. Shen, "The limits of Riemann solutions to the isentropic magnetogasdynamics," *Applied Mathematics Letters*, vol. 24, no. 7, pp. 1124–1129, 2011.

[34] C. Shen and M. Sun, "Formation of delta shocks and vacuum states in the vanishing pressure limit of Riemann solutions to the perturbed Aw-RAScle model," *Journal of Differential Equations*, vol. 249, no. 12, pp. 3024–3051, 2010.

[35] W. Sheng, G. Wang, and G. Yin, "Delta wave and vacuum state for generalized Chaplygin gas dynamics system as pressure vanishes," *Nonlinear Analysis: Real World Applications*, vol. 22, pp. 115–128, 2015.

[36] G. Yin and W. Sheng, "Delta shocks and vacuum states in vanishing pressure limits of solutions to the relativistic Euler equations for polytropic gases," *Journal of Mathematical Analysis and Applications*, vol. 355, no. 2, pp. 594–605, 2009.

[37] L. Guo, T. Li, and G. Yin, "The vanishing pressure limits of Riemann solutions to the Chaplygin gas equations with a source term," *Communications on Pure and Applied Analysis*, vol. 16, no. 1, pp. 295–309, 2017.

[38] L. Guo, T. Li, and G. Yin, "The limit behavior of the Riemann solutions to the generalized Chaplygin gas equations with a source term," *Journal of Mathematical Analysis and Applications*, vol. 455, no. 1, pp. 127–140, 2017.

[39] C. Shen and M. Sun, "A distributional product approach to the delta shock wave solution for the one-dimensional zero-pressure gas dynamics system," *International Journal of Non-Linear Mechanics*, 8 pages, 2018.

[40] M. Sun, "Singular solutions to the Riemann problem for a macroscopic production model," *ZAMM - Journal of Applied Mathematics and Mechanics/Zeitschrift für Angewandte Mathematik und Mechanik*, vol. 97, no. 8, pp. 916–931, 2017.

17

Quantum Techniques for Reaction Networks

John C. Baez [1,2]

[1] *Department of Mathematics, University of California, Riverside, CA 92521, USA*
[2] *Centre for Quantum Technologies, National University of Singapore, Singapore 117543*

Correspondence should be addressed to John C. Baez; baez@math.ucr.edu

Academic Editor: Giampaolo Cristadoro

Reaction networks are a general formalism for describing collections of classical entities interacting in a random way. While reaction networks are mainly studied by chemists, they are equivalent to Petri nets, which are used for similar purposes in computer science and biology. As noted by Doi and others, techniques from quantum physics, such as second quantization, can be adapted to apply to such systems. Here we use these techniques to study how the "master equation" describing stochastic time evolution for a reaction network is related to the "rate equation" describing the deterministic evolution of the expected number of particles of each species in the large-number limit. We show that the relation is especially strong when a solution of master equation is a "coherent state", meaning that the numbers of entities of each kind are described by independent Poisson distributions. Remarkably, in this case the rate equation and master equation give the exact same formula for the time derivative of the expected number of particles of each species.

1. Introduction

A "reaction network" describes how various kinds of classical particles can interact and turn into other kinds. They are commonly used in chemistry. For example, this reaction network gives a very simplified picture of what is happening in a glass of water:

$$\begin{gathered} H_2O \rightleftarrows H^+ + OH^- \\ H_2O + H^+ \rightleftarrows H_3O^+ \end{gathered} \tag{1}$$

The reactions here are all reversible, but this is not required by the reaction network formalism. The relevant particles here are not atoms, but rather "species" including the water molecule H_2O, the proton H^+, the hydroxyl ion OH^-, and the hydronium ion H_3O^+.

While the dynamics of the reactions are fundamentally quantum-mechanical, reaction networks give a simplified description of what is going on. First, given a "rate constant" for each reaction, one can write down a differential equation called the *rate equation*, which describes the time evolution of the concentration of each species in a deterministic way. This is especially useful in the limit where there are many particles of each species. At a more fine-grained level, one can describe reactions stochastically, using the *master equation*. In the rate equation the concentration of each species is treated as a continuous variable, but, in the master equation, the number of particles of each species is an integer.

In this paper, we explain how techniques from quantum physics and in particular second quantization can be used to study reaction networks. This work is part of a broader program emphasizing the links between quantum mechanics and a subject we call "stochastic mechanics", where probabilities replace amplitudes [1–3].

Using second quantization methods to model classical stochastic systems goes back at least to 1976, with the work of Doi [4]. It has been further developed by Grassberger, Scheunert [5], and many others. The big surprise one meets at the very beginning of this subject is that the canonical commutation relations between creation and annihilation operators, long viewed as a hallmark of quantum theory, are also perfectly suited to systems of identical *classical* particles interacting in a probabilistic way.

In quantum theory, the simplest example of a Fock space is used to describe the states of a quantum harmonic oscillator. When such an oscillator is in its ℓth energy level, its state is a vector that we can formally write as z^ℓ for some formal variable z, or better, $z^\ell/\sqrt{\ell!}$, where the normalization factor comes in handy later. A general quantum state is a linear combination of such vectors, so we can write it as a power series

$$\Psi = \sum_{\ell=0}^{\infty} \psi_\ell \frac{z^\ell}{\sqrt{\ell!}} \tag{2}$$

where the coefficient $\psi_\ell \in \mathbb{C}$ is the amplitude for the oscillator to be in its ℓth energy level. We can increase the energy level using the "raising operator" $a^\dagger$, which multiplies any power series by z:

$$a^\dagger \Psi = z\Psi. \tag{3}$$

We can decrease the energy level using the "lowering operator" a, which differentiates any power series:

$$a\Psi = \frac{d}{dz}\Psi. \tag{4}$$

If we choose an inner product for which the vectors $z^\ell/\sqrt{\ell!}$ are orthonormal, these operators are indeed adjoint to each other:

$$\langle a\Psi, \Phi\rangle = \left\langle \Psi, a^\dagger \Phi\right\rangle. \tag{5}$$

In checking this one sees why the normalization factor is required.

We can think of z^ℓ as a state containing ℓ quanta of energy. In applications to quantum physics, we think of these quanta as actual particles, for example, photons. If we have k different kinds of particles or a single kind of particle with k different states, we describe the quantum state of a collection of particles using a power series in k variables. For example, if $k = 2$ there is a Fock space consisting of power series in two variables z_1 and z_2:

$$\Psi = \sum_{\ell_1=0}^{\infty}\sum_{\ell_2=0}^{\infty} \psi_{\ell_1,\ell_2} \frac{z_1^{\ell_1}}{\sqrt{\ell_1!}} \frac{z_2^{\ell_2}}{\sqrt{\ell_2!}} \tag{6}$$

where $\psi_{\ell_1,\ell_2} \in \mathbb{C}$ is the amplitude for having ℓ_1 particles of the first kind and ℓ_2 particles of the second kind. We have operators that create and annihilate particles of each kind:

$$\begin{aligned} a_i^\dagger \Psi &= z_i \Psi, \\ a_i \Psi &= \frac{\partial}{\partial z_i}\Psi. \end{aligned} \tag{7}$$

Each creation operator $a_i^\dagger$ is adjoint to the corresponding annihilation operator a_i, and these operators obey the "canonical commutation relations":

$$\begin{aligned} \left[a_i, a_j\right] &= 0 \\ \left[a_i^\dagger, a_j^\dagger\right] &= 0 \\ \left[a_i, a_j^\dagger\right] &= \delta_{ij}. \end{aligned} \tag{8}$$

These rules are the basis of many practical calculations in particle physics, where products of creation and annihilations operators are used to model processes in which particles interact and turn into other particles.

Doi noticed that all this has a stochastic analogue where we work with probabilities rather than amplitudes. For example, there is a stochastic Fock space whose elements are power series

$$\Psi = \sum_{\ell_1=0}^{\infty}\sum_{\ell_2=0}^{\infty} \psi_{\ell_1,\ell_2} z^{\ell_1} z^{\ell_2}. \tag{9}$$

Here, $\psi_{\ell_1,\ell_2} \in \mathbb{R}$ is the *probability* of having ℓ_1 particles of the first kind and ℓ_2 particles of the second kind. The formalism is somewhat simpler than in the quantum case; for example, the normalization factors are no longer required. For Ψ to describe a probability distribution, we require that all the coefficients be nonnegative and sum to 1. We call Ψ with these properties a "mixed state". A special case is a monomial $z_1^{\ell_1} z_2^{\ell_2}$. This is called a "pure state": it describes a completely definite situation where there are ℓ_1 particles of the first kind and ℓ_2 of the second kind.

This method of describing a probability distribution using a power series is far from new. It was introduced by DeMoivre [6] and perfected by Laplace [7] in a paper published in 1782. He dubbed these power series "generating functions", and they have been used in probability theory ever since. Doi's new realization was that spaces of generating functions are analogous to the Fock spaces in quantum theory. In particular, products of creation and annihilation operators can be used to describe stochastic processes in which objects interact and turn into other objects.

A natural example is chemistry, where reactions are often described stochastically using the master equation or deterministically using the rate equation. There is a long line of work on both of these equations, initiated by Horn and Jackson [8] along with Feinberg [9] in the late 1970s and continuing to this day. This line of work does not use quantum techniques, but it might profit from them and also contribute to their development. To illustrate how, here we use these techniques to study how the master equation reduces to the rate equation in the large-number limit. This is formally similar to the "classical limit" in which a quantum field theory reduces to a classical field theory.

In Theorem 8 we start with the master equation and derive an equation for the rate of change of the expected number of particles of each species. This equation is similar to the rate equation and reduces to it in a certain approximation. However, in Theorem 9 we show that this equation gives the rate equation *exactly* when the numbers of particles of each species are described by independent Poisson distributions. This relies on a nice fact: the generating function of a product of independent Poisson distributions is an eigenvector of all the annihilation operators.

In quantum mechanics, eigenvectors of the annihilation operators are called "coherent states". Many facts about classical mechanics have simple generalizations to quantum mechanics if we restrict attention to coherent states [10]. Here we are seeing a similar phenomenon in stochastic mechanics:

the deterministic dynamics given by the rate equation match the stochastic dynamics given by the master equation in the special case of coherent states. We explore this further in a companion paper [3].

The application of quantum techniques to reaction networks is just one of many interesting connections between quantum physics and network theory. For example, the evolution of networks displays phase transitions analogous to Bose–Einstein condensation [11, 12]. For more, see the review by Biamonte, Faccin, and De Domenico [13].

2. Reaction Networks

A reaction network, first defined by Aris [14], consists of a set of different kinds of particles—each kind being called a "species"—together with a set of processes, called "reactions", that turn collections of particles of various species into collections of particles of other species. As the name suggests, it is convenient to draw a reaction network as a graph.

While reaction networks first arose in chemistry, their uses are not limited to this context. Here is an example that arose in work on HIV: the human immunodeficiency virus is [15]

$$\begin{gathered} 0 \underset{\beta}{\overset{\alpha}{\rightleftarrows}} H \\ H + V \xrightarrow{\gamma} I \xrightarrow{\delta} I + V \\ I \xrightarrow{\epsilon} 0 \\ V \xrightarrow{\zeta} 0 \end{gathered} \tag{10}$$

Here we have three species:

(i) H: healthy white blood cells
(ii) I: infected white blood cells
(iii) V: virions (that is, individual virus particles)

We also have six reactions:

(i) α: the birth of one healthy cell, which has no input and one H as output
(ii) β: the death of a healthy cell, which has one H as input and no output
(iii) γ: the infection of a healthy cell, which has one H and one V as input, and one I as output
(iv) δ: "production", or the reproduction of the virus in an infected cell, which has one I as input and one I and one V as output
(v) ϵ: the death of an infected cell, which has one I as input and no output
(vi) ζ: the death of a virion, which has one V as input and no output

So, we have a finite set of species:

$$S = \{H, I, V\} \tag{11}$$

but reactions go between "complexes", which are finite linear combinations of species with natural number coefficients, such as $H + V$ or $2H + 3V$ or 0. (In the literature on reaction networks the complex 0, corresponding to nothing at all, is often denoted $\emptyset$.) We can think of complexes as elements of $\mathbb{N}^S$, the set of functions from S to the natural numbers. The reaction network is a graph with certain complexes as vertices and reactions as edges.

We can formalize this as follows. There are various kinds of graphs; the kind we want is sometimes called "directed multigraphs" or "quivers", but, to keep our terminology simple, we make the following definition.

Definition 1. A **graph** consists of a set E of **edges**, a set V of **vertices**, and **source** and **target** maps $s, t : E \longrightarrow V$ saying where each edge starts and ends.

Then we may say the following.

Definition 2. A **reaction network** (S, K, T, s, t) consists of

(i) a finite set S of **species**
(ii) a finite set of **complexes** $K \subseteq \mathbb{N}^S$
(iii) a graph $s, t : T \longrightarrow K$ with complexes as vertices and some finite set of **reactions** T as edges

In the chemistry literature this sort of thing is called a "chemical reaction network", but we want to emphasize that they are useful more generally, so we drop the adjective. Petri nets are an alternative formalism that is entirely equivalent to reaction networks, often used in subjects other than chemistry [16–18]. In the Petri net literature the species are called "places" and the reactions are called "transitions". For a more complete translation guide between reaction networks and Petri nets, see [3].

For convenience we shall often write $k = |S|$ for the number of species present in a reaction network and identify the set S with the set $\{1, \ldots, k\}$. This lets us write any complex as a k-tuple of natural numbers. In particular, we write the source and target of any reaction τ as

$$\begin{aligned} s(\tau) &= (s_1(\tau), \ldots, s_k(\tau)) \in \mathbb{N}^k, \\ t(\tau) &= (t_1(\tau), \ldots, t_k(\tau)) \in \mathbb{N}^k. \end{aligned} \tag{12}$$

3. The Rate Equation

The amount of each species is represented by a "state" of the chemical reaction network. There are various options here. A **pure state** is a vector $x \in \mathbb{N}^k$ with ith entry x_i specifying the number of instances of the ith species. This generalizes in two ways. First, we define a **classical state** to be any vector $x \in [0, \infty)^k$ of nonnegative real numbers and think of such a state as specifying the expected number or concentration of the species present. Second, we define a **mixed state** to be a probability distribution over the pure states. This assigns a probability ψ_ℓ to each $\ell \in \mathbb{N}^k$. The rate equation describes the time evolution of a classical state, while the master equation describes the time evolution of a mixed state.

To write down either of these equations, we must first specify the system's dynamics. In this paper we consider systems that evolve by the law of mass action [14]. Very roughly, this law states that the rate at which reactions occur is proportional to the product of the concentrations of all its input species. We call the constant of proportionality for each reaction its **rate constant**.

Definition 3. A **stochastic reaction network** (S, K, T, s, t, r) consists of a reaction network together with a map $r : T \longrightarrow (0, \infty)$ assigning a **rate constant** to each reaction.

This concept is equivalent to another concept in the literature, namely, that of a stochastic Petri net [3].

Given a stochastic reaction network, the rate equation says that for each reaction τ the time derivative of a classical state x is the product of the following:

(i) the vector $t(\tau) - s(\tau)$ whose ith component is the change in the number of the ith species due to the reaction τ

(ii) the concentration of each input species i of τ raised to the power given by the number of times it appears as an input, namely, $s_i(\tau)$

(iii) the rate constant $r(\tau)$ of τ

Definition 4. The **rate equation** for a stochastic reaction network (S, K, T, s, t, r) is

$$\frac{d}{dt}x(t) = \sum_{\tau \in T} r(\tau)(t(\tau) - s(\tau))\, x(t)^{s(\tau)} \tag{13}$$

where $x : \mathbb{R} \longrightarrow [0, \infty)^k$ and we have used multi-index notation to define

$$x^{s(\tau)} = x_1^{s_1(\tau)} \cdots x_k^{s_k(\tau)}. \tag{14}$$

For the HIV reaction network in (10), if we also use the Greek letter names of the reactions as names for their rate constants, we get this rate equation:

$$\begin{aligned} \frac{dH}{dt} &= \alpha - \beta H - \gamma HV \\ \frac{dI}{dt} &= \gamma HV - \epsilon I \\ \frac{dV}{dt} &= -\gamma HV + \delta I - \zeta V. \end{aligned} \tag{15}$$

This example involves only complexes where the coefficient of each species is at most 1. It is worthwhile comparing an example with larger coefficients, for example, the dissociation of a diatomic gas into atoms and the recombination of these atoms into a molecule:

$$\mathrm{Cl}_2 \underset{\beta}{\overset{\alpha}{\rightleftarrows}} 2\mathrm{Cl}. \tag{16}$$

This reaction network gives the following rate equation:

$$\begin{aligned} \frac{d\mathrm{Cl}_2}{dt} &= -\alpha \mathrm{Cl}_2 + \beta \mathrm{Cl}^2 \\ \frac{d\mathrm{Cl}}{dt} &= 2\alpha \mathrm{Cl}_2 - 2\beta \mathrm{Cl}^2. \end{aligned} \tag{17}$$

The recombination of two atoms of chlorine into a diatomic molecule occurs at a rate proportional to the square of the concentration of atomic chlorine, so both terms arising from reaction β are proportional to Cl^2. The reaction α produces two atoms of Cl while the reaction β uses up two atoms of Cl, so both terms in the second equation have a coefficient of 2. All this follows from (13).

4. The Master Equation

The master equation describes the time evolution of mixed states. For the rest of this paper we fix a stochastic reaction network (S, K, T, s, t, r). Recall that, for each $\ell \in \mathbb{N}^k$, a "mixed state" gives the probability ψ_ℓ that, for each i, exactly $\ell_i \in \mathbb{N}$ instances of the ith species are present. The **master equation** says that

$$\frac{d}{dt}\psi_{\ell'} = \sum_{\ell \in \mathbb{N}^k} H_{\ell'\ell}\psi_\ell \tag{18}$$

for some matrix of numbers H determined by the stochastic reaction network. The matrix element $H_{\ell'\ell}$ is the probability per time for the pure state ℓ to evolve to the pure state ℓ'. This probability is a sum over reactions:

$$H_{\ell'\ell} = \sum_{\tau \in T} H(\tau)_{\ell'\ell}. \tag{19}$$

To describe the matrix $H(\tau)$, we need "falling powers". For any natural numbers n and p, we define the pth **falling power** of n by

$$n^{\underline{p}} = n(n-1)\cdots(n-p+1). \tag{20}$$

This is the number of ways of choosing an ordered p-tuple of distinct elements from a set with n elements. Note that $n^{\underline{p}} = 0$ if $p > n$. More generally, for any multi-indices ℓ and m, we define

$$\ell^{\underline{m}} = \ell_1^{\underline{m_1}} \cdots \ell_k^{\underline{m_k}}. \tag{21}$$

This is the number of ways to choose, for each species $i = 1, \ldots, k$, an ordered list of m_i distinct things from a collection of ℓ_i things of that species. Thus, we expect a factor of this sort to appear in the master equation.

Using this notation, we define the matrix $H(\tau)$ as follows:

$$H(\tau)_{\ell'\ell} = r(\tau)\, \ell^{\underline{s(\tau)}} \left(\delta_{\ell', \ell+t(\tau)-s(\tau)} - \delta_{\ell'\ell}\right). \tag{22}$$

Here δ is the Kronecker delta, which equals 1 if its subscripts are equal and 0 otherwise. The first term describes the rate for the complex ℓ to become the complex ℓ' via the reaction τ: it equals the rate constant for this reaction multiplied by the number of ways this reaction can occur. The second term describes the rate at which ℓ goes away due to this reaction. Thus, the size of the second term is equal to that of the first, but its sign is opposite.

Putting together (19) and (22), we obtain this formula for the Hamiltonian:

$$H_{\ell'\ell} = \sum_{\tau\in T} r(\tau)\, \ell^{\underline{s(\tau)}} \left(\delta_{\ell',\ell+t(\tau)-s(\tau)} - \delta_{\ell'\ell}\right). \tag{23}$$

5. The Stochastic Fock Space

We can understand the Hamiltonian for the master equation at a deeper level using techniques borrowed from quantum physics. The first step is to introduce a stochastic version of Fock space. To do this, we write any mixed state as a formal power series in some variables $z_1, \ldots, z_k$, with the coefficient of

$$z^\ell = z_1^{\ell_1} \cdots z_k^{\ell_k} \tag{24}$$

being the probability ψ_ℓ. That is, we write any mixed state as

$$\Psi = \sum_{\ell\in\mathbb{N}^k} \psi_\ell z^\ell. \tag{25}$$

Because a mixed state is a probability distribution, the coefficients ψ_ℓ must be nonnegative and sum to 1. Indeed, in this formalism **mixed states** are precisely the formal power series Ψ with

$$\sum_{\ell\in\mathbb{N}^k} \psi_\ell = 1, \quad \psi_\ell \geq 0. \tag{26}$$

The simplest mixed states are the monomials z^ℓ; these are the **pure states**, where there is a definite number of things of each species.

In quantum mechanics we use a similar sort of Fock space, but with power series having complex rather than real coefficients. To obtain a Hilbert space, we often restrict attention to power series obeying for which a certain norm is finite. However, power series not obeying this condition are also useful, for example, to ensure that operators of interest are everywhere defined, instead of merely densely defined.

Thus, for the purposes of studying the master equation, we define the **stochastic Fock space** to be the vector space $\mathbb{R}[[z_1, \ldots, z_k]]$ of all real formal power series in the variables $z_1, \ldots, z_k$. We treat the Hamiltonian H for the master equation as the operator on stochastic Fock space with

$$H\Psi = \sum_{\ell,\ell'\in\mathbb{N}^k} H_{\ell'\ell}\psi_\ell z^{\ell'} \tag{27}$$

for all

$$\Psi = \sum_{\ell\in\mathbb{N}^k} \psi_\ell z^\ell, \tag{28}$$

where the matrix entries $H_{\ell'\ell}$ are given by (23).

This allows us to express the Hamiltonian in terms of certain special operators on the stochastic Fock space: the creation and annihilation operators. Our notation here follows that used in quantum physics, where the creation and annihilation operators are adjoints of each other. Let $1 \leq i \leq k$. The **creation operator** $a_i^\dagger$ is given by

$$a_i^\dagger \Psi = z_i \Psi. \tag{29}$$

This takes any pure state to the pure state with one additional instance of the ith species:

$$a_i^\dagger \left(z_1^{\ell_1} \cdots z_i^{\ell_i} \cdots z_k^{\ell_k}\right) = z_1^{\ell_1} \cdots z_i^{\ell_i+1} \cdots z_k^{\ell_k}. \tag{30}$$

The corresponding **annihilation operator** is given by formal differentiation:

$$a_i \Psi = \frac{\partial}{\partial z_i}\Psi. \tag{31}$$

This takes any pure state to the pure state with one fewer instance of the ith species, but multiplied by a coefficient ℓ_i:

$$a_i \left(z_1^{\ell_1} \cdots z_i^{\ell_i} \cdots z_k^{\ell_k}\right) = \ell_i z_1^{\ell_1} \cdots z_i^{\ell_i-1} \cdots z_k^{\ell_k}. \tag{32}$$

This represents the fact that there are ℓ_i ways to annihilate one of the instances of the ith species present.

In what follows we use multi-index notation to define, for any $n \in \mathbb{N}^k$,

$$a^n = a_1^{n_1} \cdots a_k^{n_k} \tag{33}$$

and

$$a^{\dagger n} = a_1^{\dagger n_1} \cdots a_k^{\dagger n_k}. \tag{34}$$

These expressions are well-defined because all the annihilation operators commute with each other and similarly for the creation operators.

Theorem 5. *For any stochastic reaction network, the Hamiltonian is given by*

$$H = \sum_{\tau\in T} r(\tau) \left(a^{\dagger t(\tau)} - a^{\dagger s(\tau)}\right) a^{s(\tau)}. \tag{35}$$

Proof. We start with a basic result on annihilation and creation operators.

Lemma 6. *For any multi-indices ℓ, m we have*

$$\begin{aligned} a^{\dagger m} z^\ell &= z^{\ell+m}, \\ a^m z^\ell &= \ell^{\underline{m}} z^{\ell-m}. \end{aligned} \tag{36}$$

Proof. The first equation follows inductively from the definition of the creation operators, which implies

$$a_i^\dagger z^\ell = z_1^{\ell_1} \cdots z_i^{\ell_i+1} \cdots z_k^{\ell_k}. \tag{37}$$

The second follows inductively from the definition of annihilation operators, which implies

$$a_i z^\ell = \ell_i z_1^{\ell_1} \cdots z_i^{\ell_i-1} \cdots z_k^{\ell_k}. \tag{38}$$

Note that $\ell^{\underline{m}} = 0$ if $m_i > \ell_i$ for any i, and in this case we also have $a^m z^\ell = 0$, so the equation $a^m z^\ell = \ell^{\underline{m}} z^{\ell-m}$ holds in this case because both sides vanish.

We now prove the theorem. For any multi-index ℓ, (23) gives

$$Hz^{\ell} = \sum_{\ell' \in \mathbb{N}^k} H_{\ell'\ell} z^{\ell'} = \sum_{\tau \in T} r(\tau)\, \ell^{\underline{s(\tau)}} \left(z^{\ell + t(\tau) - s(\tau)} - z^{\ell} \right). \quad (39)$$

Using Lemma 6 we obtain

$$Hz^{\ell} = \sum_{\tau \in T} r(\tau) \left(a^{\dagger t(\tau)} - a^{\dagger s(\tau)} \right) a^{s(\tau)} z^{\ell} \quad (40)$$

and thus

$$H = \sum_{\tau \in T} r(\tau) \left(a^{\dagger t(\tau)} - a^{\dagger s(\tau)} \right) a^{s(\tau)}. \quad (41)$$

The last step requires a little justification. Technically speaking, the monomials z^{ℓ} are not a basis of the stochastic Fock space $\mathbb{R}[[z_1, \ldots, z_k]]$, because not every formal power series is a *finite* linear combination of monomials. However, these monomials form a "topological basis" of $\mathbb{R}[[z_1, \ldots, z_k]]$, in the sense that every element of this vector space can be expressed as an *convergent infinite* linear combination of these monomials, with respect to a certain topology. (In this topology, a sequence of formal power series converges if each coefficient converges.) The annihilation and creation operators and indeed all the operators discussed in this paper are continuous in this topology. So, to check equations between these operators, it suffices to check them on the monomials z^{ℓ}. □

Next we investigate what the master equation says about the time evolution of expected values. For this, we start by defining

$$\langle \Psi \rangle = \sum_{\ell \in \mathbb{N}^k} \psi_{\ell} \quad (42)$$

for any formal power series

$$\Psi = \sum_{\ell \in \mathbb{N}^k} \psi_{\ell} z^{\ell}. \quad (43)$$

In general the sum defining $\langle \Psi \rangle$ may not converge, but it converges and equals 1 when Ψ is a mixed state. Suppose O is an operator on the space of formal power series in the variables $z_1, \ldots, z_k$. We define the **expected value** of O in the mixed state Ψ to be $\langle O\Psi \rangle$, assuming this converges. This generalizes the usual definition of the expected value of a random variable, in a way that emphasizes the analogy between stochastic mechanics and quantum mechanics. While in quantum mechanics we compute expected values of observables using expressions of the form $\langle \Psi, O\Psi \rangle$, where the brackets denote an inner product, in stochastic mechanics we use $\langle O\Psi \rangle$. For more details see [1, 2].

We are especially interested in the expected value of the **number operators**

$$N_i = a_i^{\dagger} a_i. \quad (44)$$

Note that

$$\langle N_i \Psi \rangle = \sum_{\ell \in \mathbb{N}^k} \ell_i \psi_{\ell}. \quad (45)$$

Since ψ_{ℓ} is the probability of being in a pure state where there are ℓ_i things of the ith species, the above formula indeed gives the expected value of the number of things of this species.

Our goal is to compute

$$\frac{d}{dt} \langle N_i \Psi(t) \rangle \quad (46)$$

assuming that $\Psi(t)$ obeys the master equation. For this, we need to define the falling powers of an operator A as follows:

$$A^{\underline{p}} = A(A-1)\cdots(A-p+1) \quad (47)$$

for any $p \in \mathbb{N}$. If m is a multi-index we define

$$N^{\underline{m}} = N_1^{\underline{m_1}} \cdots N_k^{\underline{m_k}}. \quad (48)$$

This is well-defined because the number operators N_i commute.

Lemma 7. *For any multi-indices ℓ and m we have*

$$N^{\underline{m}} z^{\ell} = \ell^{\underline{m}} z^{\ell}. \quad (49)$$

Proof. By Lemma 6 and the definition of the number operators we have

$$N_i z^{\ell} = \ell_i z^{\ell} \quad (50)$$

for all $1 \leq i \leq k$. The lemma follows directly from this.

There is a way to extract a classical state from a mixed state, which lets us connect the rate equation to the master equation. To do this, we write N for the vector, or list, of number operators $(N_1, \ldots, N_k)$. Then, given a mixed state Ψ, we set

$$\langle N\Psi \rangle = \left(\langle N_1 \Psi \rangle, \ldots, \langle N_k \Psi \rangle \right). \quad (51)$$

If the expected values here are well-defined, $\langle N\Psi \rangle \in [0, \infty)^k$ is a classical state. It is a simplified description of the mixed state Ψ, which only records the expected number of instances of each species.

The next result says how this sort of classical state evolves in time, assuming that the mixed state it comes from obeys the master equation.

Theorem 8. *For any stochastic reaction network and any mixed state $\Psi(t)$ evolving in time according to the master equation, we have*

$$\frac{d}{dt} \langle N\Psi(t) \rangle = \sum_{\tau \in T} r(\tau)\, (s(\tau) - t(\tau)) \left\langle N^{\underline{s(\tau)}} \Psi(t) \right\rangle \quad (52)$$

assuming the expected values and their derivatives exist.

Proof. First note using Lemmas 6 and 7 that for any multi-indices ℓ, m, n and any natural number i we have

$$\begin{aligned}\left\langle N_i a^{\dagger^m} a^n z^\ell \right\rangle &= \ell^{\underline{n}} \left\langle N_i a^{\dagger^m} z^{\ell-n} \right\rangle = \ell^{\underline{n}} \left\langle N_i z^{\ell-n+m} \right\rangle \\ &= \ell^{\underline{n}} \left(\ell_i - n_i + m_i\right).\end{aligned} \quad (53)$$

Thus if at any given time

$$\Psi(t) = \sum_{\ell \in \mathbb{N}^k} \psi_\ell(t) z^\ell \quad (54)$$

and $\Psi(t)$ evolves in time according to the master equation, we have

$$\begin{aligned}\frac{d}{dt}\left\langle N_i \Psi(t)\right\rangle &= \left\langle N_i H \Psi(t)\right\rangle \\ &= \sum_{\tau\in T}\sum_{\ell\in\mathbb{N}^k} r(\tau)\left\langle N_i\left(a^{\dagger^{t(\tau)}} - a^{\dagger^{s(\tau)}}\right)a^{s(\tau)}\psi_\ell(t) z^\ell\right\rangle \\ &= \sum_{\tau\in T}\sum_{\ell\in\mathbb{N}^k} r(\tau)\,\ell^{\underline{s(\tau)}}\left(s_i(\tau) - t_i(\tau)\right)\psi_\ell(t) \\ &= \sum_{\tau\in T}\sum_{\ell\in\mathbb{N}^k} r(\tau)\left(s_i(\tau) - t_i(\tau)\right)\left\langle N^{\underline{s(\tau)}} z^\ell\right\rangle \psi_\ell(t) \\ &= \sum_{\tau\in T} r(\tau)\left(s_i(\tau) - t_i(\tau)\right)\left\langle N^{\underline{s(\tau)}}\Psi(t)\right\rangle.\end{aligned} \quad (55)$$

Recall that N is our notation for the vector of number operators $(N_1, \ldots, N_k)$. Thus, the above equation is equivalent to

$$\frac{d}{dt}\left\langle N\Psi(t)\right\rangle = \sum_{\tau\in T} r(\tau)\left(s(\tau) - t(\tau)\right)\left\langle N^{\underline{s(\tau)}}\Psi(t)\right\rangle. \quad (56)$$

The above theorem makes it clear that the rate equation is closely related to the master equation, since the former says

$$\frac{d}{dt}x(t) = \sum_{\tau\in T} r(\tau)\left(t(\tau) - s(\tau)\right) x(t)^{s(\tau)} \quad (57)$$

while the latter implies

$$\frac{d}{dt}\left\langle N\Psi(t)\right\rangle = \sum_{\tau\in T} r(\tau)\left(s(\tau) - t(\tau)\right)\left\langle N^{\underline{s(\tau)}}\Psi(t)\right\rangle. \quad (58)$$

Suppose we let $x(t)$ be the classical state coming from the mixed state $\Psi(t)$:

$$x(t) = \left\langle N\Psi(t)\right\rangle. \quad (59)$$

The rate equation for $x(t)$ would then follow from the master equation for $\Psi(t)$ if we had

$$\left\langle N\Psi(t)\right\rangle^m = \left\langle N^{\underline{m}}\Psi(t)\right\rangle \quad (60)$$

for every multi-index m. This equation is not true in general. However, it should hold *approximately* in a suitable limit of large numbers. And surprisingly, it holds *exactly* for coherent states.

6. Coherent States

For any $c \in [0,\infty)^k$, we define the **coherent state** with expected value c to be the mixed state

$$\Psi_c = \frac{e^{c\cdot z}}{e^c} \quad (61)$$

where $c \cdot z$ is the dot product of the vectors c and z, and we set $e^c = e^{c_1+\cdots+c_n}$. Equivalently,

$$\Psi_c = \frac{1}{e^c}\sum_{n\in\mathbb{N}^k}\frac{c^n}{n!}z^n, \quad (62)$$

where c^n and z^n are defined as products in our usual way, and $n! = n_1!\cdots n_k!$. The name "coherent state" comes from quantum mechanics [10], where we think of the coherent state Ψ_c as the quantum state that best approximates the classical state c. In the state Ψ_c, the probability of having n_i things of the ith species is equal to

$$e^{-c_i}\frac{c_i^{n_i}}{n_i!}. \quad (63)$$

This is precisely the definition of a Poisson distribution with mean equal to c_i. The state Ψ_c is thus a product of independent Poisson distributions.

Theorem 9. *Given any stochastic reaction network, let $\Psi(t)$ be a mixed state evolving in time according to the master equation. If $\Psi(t)$ is a coherent state when $t = t_0$, then*

$$x(t) = \left\langle N\Psi(t)\right\rangle \quad (64)$$

obeys the rate equation when $t = t_0$.

Proof. We prove this using a series of lemmas.

Lemma 10. *For any multi-index m and any coherent state Ψ_c we have*

$$a^m\Psi_c = c^m\Psi_c \quad (65)$$

and

$$a^{\dagger^m}\Psi_c = z^m\Psi_c. \quad (66)$$

Proof. The second equation is immediate from the definition, while the first follows from

$$a_i\Psi_c = \frac{\partial}{\partial z_i}\frac{e^{c\cdot z}}{e^c} = c_i\Psi_c. \quad (67)$$

Lemma 11. *For any multi-index m we have*

$$N^{\underline{m}} = a^{\dagger^m}a^m. \quad (68)$$

Proof. For any multi-index ℓ, Lemmas 6 and 7 imply

$$a^{\dagger m} a^m z^\ell = \ell^{\underline{m}} a^{\dagger m} z^{\ell - m} = \ell^{\underline{m}} z^\ell = N^{\underline{m}} z^\ell. \quad (69)$$

Since the states z^ℓ form a topological basis of $\mathbb{R}[[z_1, \ldots, z_k]]$ and all the operators in question are continuous, the lemma follows.

Lemma 12. *For any multi-index m and any $\Psi \in \mathbb{R}[[z_1, \ldots, z_k]]$ we have*

$$\left\langle a^{\dagger m} \Psi \right\rangle = \langle \Psi \rangle. \quad (70)$$

Proof. We have

$$\left\langle a^{\dagger m} \Psi \right\rangle = \left\langle \sum_{\ell \in \mathbb{N}^k} \psi_\ell z^{\ell + m} \right\rangle = \sum_{\ell \in \mathbb{N}^k} \psi_\ell = \langle \Psi \rangle. \quad (71)$$

Lemma 13. *For any multi-index m and any coherent state Ψ_c we have*

$$\langle N^{\underline{m}} \Psi_c \rangle = \langle N \Psi_c \rangle^m. \quad (72)$$

Proof. By Lemmas 11 and 10 we have

$$N^{\underline{m}} \Psi_c = a^{\dagger m} a^m \Psi_c = c^m a^{\dagger m} \Psi_c \quad (73)$$

so by Lemma 12

$$\langle N^{\underline{m}} \Psi_c \rangle = \left\langle c^m a^{\dagger m} \Psi_c \right\rangle = c^m \langle \Psi_c \rangle = c^m. \quad (74)$$

As a special case we have $\langle N_i \Psi_c \rangle = c_i$, so we also have

$$\langle N \Psi_c \rangle^m = c^m \quad (75)$$

and it follows that $\langle N^{\underline{m}} \Psi_c \rangle = \langle N \Psi \rangle^m$.

To prove Theorem 9, we start by setting

$$x(t) = \langle N \Psi(t) \rangle \quad (76)$$

where $\Psi(t)$ is a solution of the master equation. Theorem 8 implies that the time derivative of x equals

$$\dot{x}(t) = \sum_{\tau \in T} r(\tau)(s(\tau) - t(\tau)) \left\langle N^{\underline{s(\tau)}} \Psi(t) \right\rangle. \quad (77)$$

If $\Psi(t)$ is a coherent state at time $t = t_0$, Lemma 13 implies that

$$\left\langle N^{\underline{s(\tau)}} \Psi(t_0) \right\rangle = \langle N \Psi(t_0) \rangle^{s(\tau)} = x(t_0)^{s(\tau)} \quad (78)$$

so we have

$$\dot{x}(t_0) = \sum_{\tau \in T} r(\tau)(s(\tau) - t(\tau)) x(t_0)^{s(\tau)} \quad (79)$$

which is the rate equation at time t_0, as desired.

In general, the above result applies at only one moment in time. The reason is that if a solution of the master equation is a coherent state at some time, it need not be a coherent state at later (or earlier) times. Still, Theorem 8 implies that, as long as the expected values $\langle N^{\underline{m}} \Psi(t) \rangle$ are *close* to the powers $\langle N \Psi(t) \rangle^m$, at least when m is the source of some reaction, the expected values $x(t) = \langle N \Psi(t) \rangle$ will *approximately* obey the rate equation. This could be studied in detail using either the techniques introduced here or those discussed by Anderson and Kurtz [19].

Furthermore, in some cases when $\Psi(t)$ is initially a coherent state it continues to be so for all times. In this case, $x(t)$ continues to exactly obey the rate equation. For example, this is true when all the complexes in the reaction network consist of a single species. It is also true for a large class of equilibrium solutions of the master equation, as shown by Anderson, Craciun, and Kurtz [20]. For more on how quantum techniques apply to this situation, see the companion to this paper [3].

Conflicts of Interest

The author declares that there are no conflicts of interest regarding the publication of this paper.

Acknowledgments

The author thank Jacob Biamonte, Brendan Fong, and the students at U. C. Riverside who helped come up with a preliminary version of the proof of Theorem 8, notably Daniel Estrada, Reeve Garrett, Michael Knap, Tu Pham, Blake Pollard, and Franciscus Rebro and thank John Rowlands and Blake Stacey for catching typos and other mistakes. This work was done at the Centre of Quantum Technology, in Singapore, and the Mathematics Department of U. C. Riverside.

References

[1] J. C. Baez and J. Biamonte, *Quantum Techniques for Stochastic Mechanics*, World Scientific, Singapore, 2018.

[2] J. C. Baez and B. Fong, "A Noether theorem for Markov processes," *Journal of Mathematical Physics*, vol. 54, no. 1, p. 013301, 2013.

[3] J. C. Baez and B. Fong, "Quantum techniques for studying equilibrium in reaction networks," *Journal of Complex Networks*, vol. 3, no. 1, pp. 22–34, 2015.

[4] M. Doi, "Second quantization representation for classical many-particle system," *Journal of Physics A: Mathematical and General*, vol. 9, no. 9, pp. 1465–1477, 1976.

[5] P. Grassberger and M. Scheunert, "Fock-space methods for identical classical objects," *Fortschritte der Physik*, vol. 28, no. 10, pp. 547–578, 1980.

[6] A. DeMoivre, in *Miscellanea Analytica de Seriebus et Quadraturis, Book VII*, pp. 194–197, Touson & Watts, London, UK, 1730.

[7] P.-S. Laplace, "Mémoire sur les suites," *Mémoires de l'Académie Royale des Sciences de Paris*, pp. 207–309, 1782.

[8] F. Horn and R. Jackson, "General mass action kinetics," *Archive for Rational Mechanics and Analysis*, vol. 47, pp. 81–116, 1972.

[9] M. Feinberg, *Lectures On Reaction Networks*, 1979, Available at https://crnt.osu.edu/LecturesOnReactionNetworks.

[10] J. R. Klauder and B. Skagerstam, *Coherent States*, World Scientific, Singapore, 1985.

[11] G. Bianconi and A.-L. Barabási, "Bose-Einstein condensation in complex networks," *Physical Review Letters*, vol. 86, no. 24, pp. 5632–5635, 2001.

[12] M. Alberto Javarone and G. Armano, "Quantum-classical transitions in complex networks," *Journal of Statistical Mechanics: Theory and Experiment*, no. 4, p. P04019, 2013.

[13] J. Biamonte, M. Faccin, and M. De Domenico, *Complex networks: from classical to quantum*, Available as https://arxiv.org/abs/1702.08459.

[14] R. Aris, "Prolegomena to the rational analysis of systems of chemical reactions," *Archive for Rational Mechanics and Analysis*, vol. 19, pp. 81–99, 1965.

[15] A. Korobeinikov, "Global properties of basic virus dynamics models," *Bulletin of Mathematical Biology*, vol. 66, no. 4, pp. 879–883, 2004.

[16] I. Koch, "Petri nets - A mathematical formalism to analyze chemical reaction networks," *Molecular Informatics*, vol. 29, no. 12, pp. 838–843, 2010.

[17] I. Koch, W. Reisig, and F. Schreiber, *Modeling in Systems Biology*, Springer, 2011.

[18] D. J. Wilkinson, *Stochastic Modelling for Systems Biology*, Taylor and Francis, New York, NY, USA, 2006.

[19] D. F. Anderson and T. G. Kurtz, "Continuous time Markov chain models for chemical reaction networks," in *Design and Analysis of Biomolecular Circuits: Engineering Approaches to Systems and Synthetic Biology*, H. Koeppl et al., Ed., pp. 3–42, Springer, 2011.

[20] D. F. Anderson, G. Craciun, and T. G. Kurtz, "Product-form stationary distributions for deficiency zero chemical reaction networks," *Bulletin of Mathematical Biology*, vol. 72, no. 8, pp. 1947–1970, 2010.

18

Dynamic Characteristics of Deeply Buried Spherical Biogas Digesters in Viscoelastic Soils

Hongwei Hou,[1,2] Shihu Gao,[3] Qianqian Guo,[4] Long Chen,[4] Bing Wu,[4] and Lei Wang[5]

[1]*Zhejiang Academy of Building Research and Design Co. Ltd., Hangzhou, Zhejiang 310000, China*
[2]*College of Civil Engineering and Architecture, Zhejiang University, Hangzhou, Zhejiang 310000, China*
[3]*College of Civil Engineering and Architecture, Wenzhou University, Wenzhou, Zhejiang 325035, China*
[4]*Key Laboratory of Ministry of Education for Geomechanics and Embankment Engineering, Hohai University, Nanjing 210098, China*
[5]*Shanghai City Construction and Design Research Institute, No. 1170 Tibet South Road, Huangpu District, Shanghai, China*

Correspondence should be addressed to Shihu Gao; lzccgsh@163.com

Academic Editor: Zhi-Yuan Sun

The harmonic vibration characteristics of a deeply buried spherical methane tank in viscoelastic soil subjected to cyclic loading in the frequency domain are investigated. The dynamic behavior of the soil is described based on the theory of fractional derivatives. By introducing potential functions, the closed-form expressions for the displacement and the stress of the viscoelastic soil surrounding the deeply buried spherical methane tank are obtained. Two die structures are considered: a homogeneous elastic medium and a shell structure. Based on the theory of elastic motion and the Flügge theory, analytic solutions for the dynamic responses of the spherical methane tank in a fractional-derivative viscoelastic soil are derived explicitly. Analytic solution expressions of the undetermined coefficients are determined by using the continuum boundary conditions. The system dynamic responses to the homogeneous elastic medium and the shell structure and the influences of the parameters of the fractional derivative, soil, and die on the dynamic characteristic of the system are compared and analyzed. The results indicate a significant difference between the dynamic responses of the die structures for the two models.

1. Introduction

Not only has an underground, deeply buried spherical digester the advantages of a small spherical mouth, good airtightness, and high gas production but also the fermentation of biogas slurry does not produce a crust and the structure is not affected by the external ambient temperature [1]. Therefore, spherical underground digesters are widely used and are promoted in China's rural areas.

The vibration response of biogas digesters has wide applications in the fields of seismic resistance, explosion-resistant design, and structural stability of the die. Forrestal and Sagartz studied the dynamic responses of thin-walled spherical shells and cylindrical shells in homogeneous elastic soil under the action of heavy loads [2]. Duffey obtained the transient dynamic response of viscoelastic and viscous-plastic spherical die structures under impulsive loading by using a Laplace transform technique [3]. Glenn regarded the soil as a linear elastic medium and deduced the vibration response of a deeply buried spherical digester under heavy loads and pulse loads [4]. The transient response of a spherical cavity has been studied under explosive loading by using displacement as an unknown quantity [5].

Many experimental studies have been carried out to investigate the clay behavior under dynamic loading; Lin Guo studied the deformation behavior of saturated soft soil under long-term cyclic loading and established two equations for the prediction of long-term resilient modulus and permanent strain [6, 7]. Strain and pore pressure development on soft marine soil have been studied under dynamic loadings and an empirical formula is established to predict the long-term permanent strain [8]. Chuan Gu studied the influence of cyclic loading history on small strain shear modulus of saturated soils by triaxial test; test results show that both cyclic strain history and cyclic stress history will induce reduction of Gmax compared to the corresponding Gmax values with noncyclic loading effects at the same effective stress [9]. Dynamic modulus characteristics of saturated soils under variable confining pressure have been studied by cyclic triaxial tests; tests results shows that both the phase differences and ratios between the cyclic deviatoric stress and cyclic confining pressure exerted a constant and pronounced influence on dynamic modulus [10–12].

Studies of the viscoelastic properties of soils have shown that a fractional-derivative constitutive model is well suited for the characterization of the soil's creep properties [13]. However, based on the fractional-derivative model, the dynamic equation is a singular integral-partial differential equation and its theoretical analysis and numerical calculation present many difficulties [14] and there have been few reports on the use of this approach for the study of soil dynamic characteristics [15]. What is more, the above studies ignore the thickness of the mold structure. In this paper, the soil element is regarded as a viscoelastic element with fractional-derivative properties and the exact expression of the displacement and stress of the infinite fractional-derivative viscoelastic soil surrounding the biogas digester is deduced based on the viscoelastic theory. In our research, the thickness of the mold structure also can be considered.

The gas fermentation and user consumption cause repeated changes in pressure in the pool and this pressure is uniformly applied to the inner side of the pool wall and is equivalent to the periodic load [16]. In this study, the soil body is regarded as a viscoelastic body with fractional-derivative properties and the exact expression of the displacement and stress of the infinite fractional-derivative viscoelastic soil surrounding the biogas digester is deduced based on the viscoelastic theory. Secondly, the die structure is considered a homogeneous elastic medium and thin-walled shell structure and the analytics solution for the vibration response of the die structure is obtained based on the theory of elastic motion [17] and the Flügge theory [18]. Finally, the concrete expression of the relevant undetermined coefficients is obtained according to the boundary conditions of displacement and stress in the mold structure and the contact surface of the soil body. A comparison of the results of this study and existing data demonstrates the correctness of the assumptions of this study. Finally, the influences of the parameters of the fractional derivative on the vibration response of the system are investigated, providing a theoretical basis for the design and construction of underground spherical digesters.

2. Mathematical Model

Because the spherical digester is deeply embedded in the soil, it is assumed that there is a spherical digester with an inner radius R_1 and an outer radius R_2, as shown in Figure 1; the distance from the center to the middle surface is a. The spherical digester has a circular frequency ω of a uniform radial harmony excitation load ($q_0 e^{i\omega t}$), where i is the imaginary unit. The shear modulus parameters of the soil and die structure are G^S and G^L, respectively, the material density parameters are ρ^S and ρ^L, respectively, and the Poisson's ratios of the soil and the model are ν^S and ν^L, respectively. The mold and the soil interface are in complete and close contact to meet the stress and displacement continuity.

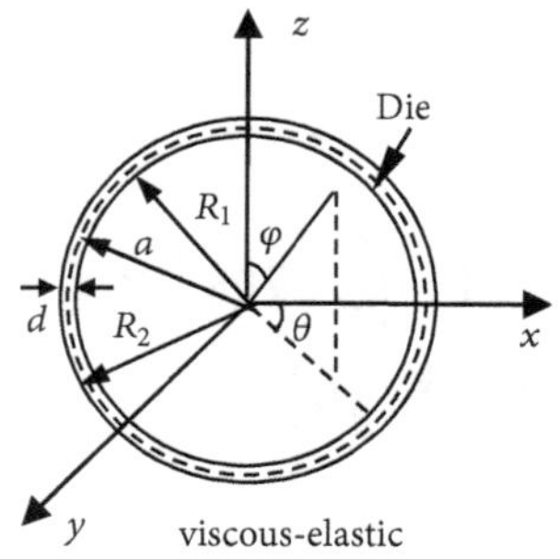

FIGURE 1: A spherical biogas digester in viscoelastic soil.

3. Establish and Solution of the Soil Control Equation

In the case of spherical symmetry, the stress-displacement constitutive relationship of the constitutive model with the fractional derivative is [11]

$$\begin{aligned}(1+\tau_\varepsilon^\alpha D^\alpha)\sigma_r^S &= (1+\tau_\sigma^\alpha D^\alpha)\left[\lambda^S\left(\frac{\partial u_r^S}{\partial r}+\frac{2u_r^S}{r}\right)+2\mu^S\frac{\partial u_r^S}{\partial r}\right]\\ (1+\tau_\varepsilon^\alpha D^\alpha)\sigma_\theta^S &= (1+\tau_\sigma^\alpha D^\alpha)\left[\lambda^S\left(\frac{\partial u_r^S}{\partial r}+\frac{2u_r^S}{r}\right)+2\mu^S\frac{u_r^S}{r}\right]\\ (1+\tau_\varepsilon^\alpha D^\alpha)\sigma_\varphi^S &= (1+\tau_\sigma^\alpha D^\alpha)\left[\lambda^S\left(\frac{\partial u_r^S}{\partial r}+\frac{2u_r^S}{r}\right)+2\mu^S\frac{u_r^S}{r}\right]\end{aligned} \tag{1}$$

Where σ_r is radial stress of soil layer, σ_θ is circumferential stress of soil layer, σ_φ is shear stress of soil layer, σ is stress tensor of soil, ε is strain tensor of soil, τ_ε and τ_σ are constitutive model parameters, λ^s and μ^s are Lame constant of soil skeleton and $\lambda^S = 2\nu^S/(1-2\nu^S)$, and ν is Poisson's ratio. $D^\alpha = d^\alpha/dt^\alpha$ is α-step Riemann-Liouville fractional derivative, which is defined as

$$D^\alpha[x(t)] = \frac{1}{\Gamma(1-\alpha)}\frac{d}{dt}\int_0^t \frac{x(\tau)}{(t-\tau)^\alpha}d\tau \tag{2}$$

in which $\Gamma(u) = \int_0^\infty t^{u-1}e^{-u}dt$ is Gamma Function.

Under the action of dynamic load, according to the theory of elasticity, the dynamic equation [16] of soil is as follows.

$$\frac{\partial\sigma_r^S}{\partial r}+\frac{2\sigma_r^S-\sigma_\theta^S-\sigma_\varphi^S}{r} = \rho^S\frac{\partial^2 u_r^S}{\partial t^2} \tag{3}$$

Substituting (1) into (3), we obtain

$$\left(\lambda^S+2\mu^S\right)\frac{1+\tau_\sigma^\alpha D^\alpha}{1+\tau_\varepsilon^\alpha D^\alpha}\frac{\partial}{\partial r}\left(\frac{\partial u_r^S}{\partial r}+\frac{2u_r^S}{r}\right) = \rho^S\frac{\partial^2 u_r^S}{\partial t^2} \tag{4}$$

Introduce potential function

$$u_r^{\mathrm{S}} = \frac{\partial}{\partial r}\left(\frac{\varphi^{\mathrm{S}}(r,t)}{r}\right) \tag{5}$$

For steady-state vibration, let

$$u_r^{\mathrm{S}} = R_2 U_\eta^{\mathrm{S}} e^{\mathrm{i}\omega t}, \quad \varphi^{\mathrm{S}} = R_2^3 \overline{\varphi}^{\mathrm{S}} e^{\mathrm{i}\omega t} \tag{6}$$

Introduce dimensionless quantity and constant as follows:

$$\eta = \frac{r}{R_2}, \quad \lambda = \frac{\omega R_2}{V^{\mathrm{S}}}, \quad T_\sigma = \frac{\tau_\sigma V^{\mathrm{S}}}{R_2}, \quad T_\varepsilon = \frac{\tau_\varepsilon V^{\mathrm{S}}}{R_2}, \quad \eta_0 = 1-\delta, \quad \delta = \frac{d}{R_2}. \tag{7}$$

Substituting (5), (6), and (7) into (4), we obtain (8) as follows:

$$\frac{d^2\overline{\varphi}^{\mathrm{S}}}{d\eta^2} - \beta^2\overline{\varphi}^{\mathrm{S}} = 0 \tag{8}$$

where

$$\beta^2 = -\frac{1+T_\varepsilon^\alpha(\mathrm{i}\lambda)^\alpha}{1+T_\sigma^\alpha(\mathrm{i}\lambda)^\alpha}\frac{\left(1-2\nu^{\mathrm{S}}\right)\lambda^2}{2\left(1-\nu^{\mathrm{S}}\right)} \tag{9}$$

According to (8), we obtain

$$\overline{\varphi}^{\mathrm{S}} = Ae^{-\beta\eta} + Be^{\beta\eta} \tag{10}$$

where A and B are undetermined coefficient; they are determined by boundary conditions. According to boundary conditions $r \longrightarrow \infty$, $u_r^{\mathrm{S}} = 0$, we obtain B=0,

$$\overline{\varphi}^{\mathrm{S}} = Ae^{-\beta\eta} \tag{11}$$

Substituting (5) into (11), we obtain (12) as follows:

$$U_\eta^{\mathrm{S}} = -\left(\frac{1}{\eta^2}+\frac{\beta}{\eta}\right)e^{-\beta\eta}A \tag{12}$$

Substituting (12) into (1), we obtain (13) as follows:

$$\begin{aligned} \sigma_r^{\mathrm{S}} &= \mu^{\mathrm{S}}\frac{1+T_\sigma^\alpha(\mathrm{i}\lambda)}{1+T_\varepsilon^\alpha(\mathrm{i}\lambda)} \cdot e^{\mathrm{i}\omega t}\left[\frac{2\nu^{\mathrm{S}}}{1-2\nu^{\mathrm{S}}}\frac{\beta^2}{\eta} + 2\left(\frac{2}{\eta^3}+\frac{2\beta}{\eta^2}+\frac{\beta^2}{\eta}\right)\right]e^{-\beta\eta}A \\ \sigma_\theta^{\mathrm{S}} &= \mu^{\mathrm{S}}\frac{1+T_\sigma^\alpha(\mathrm{i}\lambda)}{1+T_\varepsilon^\alpha(\mathrm{i}\lambda)}e^{\mathrm{i}\omega t}\left[\frac{2\nu^{\mathrm{S}}}{1-2\nu^{\mathrm{S}}}\frac{\beta^2}{\eta} - 2\left(\frac{1}{\eta^3}+\frac{\beta}{\eta^2}\right)\right] \cdot e^{-\beta\eta}A \end{aligned} \tag{13}$$

4. Establishment and Solution of Die Structure Motion Equation

4.1. Die Structure Equation of Type 1. We considered the mold structure of the spherical digester as a homogeneous elastic medium and assumed that the mold structure and soil interface are in complete and close contact with no relative slip to satisfy the stress and displacement continuity. Under the action of a dynamic load, the motion equation of the die junction is as follows [16].

$$\frac{\partial\sigma_r^{\mathrm{L}}}{\partial r} + \frac{2\sigma_r^{\mathrm{L}} - \sigma_\theta^{\mathrm{L}} - \sigma_\varphi^{\mathrm{L}}}{r} = \rho^{\mathrm{L}}\frac{\partial^2 u_r^{\mathrm{L}}}{\partial t^2} \tag{14}$$

Constitutive model is

$$\begin{aligned} \sigma_r^{\mathrm{L}} &= \lambda^{\mathrm{L}}\left(\frac{\partial u_r^{\mathrm{L}}}{\partial r}+\frac{2u_r^{\mathrm{L}}}{r}\right) + 2\mu^{\mathrm{L}}\frac{\partial u_r^{\mathrm{L}}}{\partial r} \\ \sigma_\theta^{\mathrm{L}} &= \lambda^{\mathrm{L}}\left(\frac{\partial u_r^{\mathrm{L}}}{\partial r}+\frac{2u_r^{\mathrm{L}}}{r}\right) + 2\mu^{\mathrm{L}}\frac{u_r^{\mathrm{L}}}{r} \\ \sigma_\varphi^{\mathrm{L}} &= \lambda^{\mathrm{L}}\left(\frac{\partial u_r^{\mathrm{L}}}{\partial r}+\frac{2u_r^{\mathrm{L}}}{r}\right) + 2\mu^{\mathrm{L}}\frac{u_r^{\mathrm{L}}}{r} \end{aligned} \tag{15}$$

Substituting (15) into (14), we obtain (16) as follows:

$$\left(\lambda^{\mathrm{L}}+2\mu^{\mathrm{L}}\right)\frac{\partial}{\partial r}\left(\frac{\partial u_r^{\mathrm{L}}}{\partial r}+\frac{2u_r^{\mathrm{L}}}{r}\right) = \rho^{\mathrm{L}}\frac{\partial^2 u_r^{\mathrm{L}}}{\partial t^2} \tag{16}$$

For steady-state vibration, let $u_r^{\mathrm{L}} = R_2 U_\eta^{\mathrm{L}} e^{\mathrm{i}\omega t}$, $\varphi^{\mathrm{L}} = R_2^3\overline{\varphi}^{\mathrm{L}} e^{\mathrm{i}\omega t}$. Introduce displacement potential function

$$u_r^{\mathrm{L}} = \frac{\partial}{\partial r}\left(\frac{\varphi^{\mathrm{L}}(r,t)}{r}\right) \tag{17}$$

Substituting (17) into (16), we obtain

$$\frac{d^2\overline{\varphi}^{\mathrm{L}}}{d\eta^2} - q^2\overline{\varphi}^{\mathrm{L}} = 0 \tag{18}$$

where

$$G^{SL} = \frac{G^S}{G^L}, \quad \rho^{LS} = \frac{\rho^L}{\rho^S}, \quad q^2 = -\frac{(1-2\nu^L)\lambda^2\rho^{LS}G^{SL}}{2(1-\nu^L)} \tag{19}$$

According to (18), we obtain

$$\overline{\varphi}^L = C_5e^{-q\eta} + C_6e^{q\eta} \tag{20}$$

where C_5 and C_6 are undetermined coefficient; they are determined by boundary conditions. Substituting (20) into (17), we obtain

$$U_\eta^L = -\left(\frac{1}{\eta^2} + \frac{q}{\eta}\right)e^{-q\eta}C_5 - \left(\frac{1}{\eta^2} - \frac{q}{\eta}\right)e^{q\eta}C_6 \tag{21}$$

According to (15), we obtain radial stress amplitude

$$\sigma_r^L = \left(\frac{2\nu^LG^L}{1-2\nu^L}\frac{q^2}{\eta} + 2G^L\vartheta_1\right)e^{-q\eta}e^{i\omega t}C_5 + \left(\frac{2\nu^LG^L}{1-2\nu^L}\frac{q^2}{\eta} + 2G^L\vartheta_2\right)e^{q\eta}e^{i\omega t}C_6 \tag{22}$$

where

$$\vartheta_1 = \frac{2}{\eta^3} + \frac{2q}{\eta^2} + \frac{q^2}{\eta}, \quad \vartheta_2 = \frac{2}{\eta^3} - \frac{2q}{\eta^2} + \frac{q^2}{\eta} \tag{23}$$

Satisfy the boundary conditions as follows:

(1) Displacement continuous in the mold structure and soil interface ($r = R_2, \eta = 1$); that is,

$$u_r^S = u_r^L \quad r = R_2 \tag{24}$$

(2) Stress continuity in the mold structure and soil interface ($r = R_2, \eta = 1$); that is,

$$\sigma_r^S = \sigma_r^L \quad r = R_2 \tag{25}$$

(3) Internal hemorrhage loading equal to the radial stress in the mold structure and soil interface ($r = R_1, \eta = 1$) of the die structure; that is,

$$\sigma_r^L = q_0e^{i\omega t} \quad r = R_1 \tag{26}$$

According to (24)~(26), we obtain the actual formulas of the undetermined coefficients A, C_5, C_6,

$$C_6 = \frac{a_2b_1 - a_1b_2}{e_1(a_1b_3 - a_3b_1) + e_2(a_2b_1 - a_1b_2)}Q, \quad C_5 = \frac{Q}{e_1} - \frac{e_2}{e_1}C_6, \quad A = -\frac{a_2}{a_1}C_5 - \frac{a_3}{a_1}C_6, \tag{27}$$

where

$$\begin{aligned} a_1 &= -(1+\beta)e^{-\beta}, \\ a_2 &= (1+q)e^{-q}, \\ a_3 &= (1-q)e^{q}, \\ b_1 &= G^{SL} \cdot \frac{1+T_\sigma^\alpha(i\lambda)^\alpha}{1+T_\varepsilon^\alpha(i\lambda)^\alpha}\left[\frac{2\nu^S\beta^2e^{-\beta}}{1-2\nu^S} + 2e^{-\beta}(2+2\beta+\beta^2)\right], \\ b_2 &= -\left[\frac{2\nu^Lq^2}{1-2\nu^L} + 2(2+2q+q^2)\right]e^{-q}, \\ b_3 &= -\left[\frac{2\nu^Lq^2}{1-2\nu^L} + 2(2-2q+q^2)\right]e^{q}, \\ e_1 &= \left[\frac{2\nu^Lq^2}{(1-2\nu^L)\eta_0} + 2\left(\frac{2}{\eta_0^3} + \frac{2q}{\eta_0^2} + \frac{q^2}{\eta_0}\right)\right]e^{-q\eta_0}, \\ e_2 &= \left[\frac{2\nu^Lq^2}{(1-2\nu^L)\eta_0} + 2\left(\frac{2}{\eta_0^3} - \frac{2q}{\eta_0^2} + \frac{q^2}{\eta_0}\right)\right]e^{q\eta_0}. \end{aligned} \tag{28}$$

4.2. Die Structure Equation of Type 2. Since the thickness of the die structure is much smaller than the radius of the spherical digester, it can be assumed that the mold structure has little effect on the interface $r = R_2$ of the soil and the middle surface radius [6]$r = a$. According to the Hamilton principle, the motion equation of the die structure [7] is as follows:

$$u_r^L\frac{2G^Ld}{(1-\nu^L)R_2^2} + \rho^Ld\frac{\partial^2u_r^L}{\partial t^2} = Q_a \tag{29}$$

where Q_a is net pressure.

For steady-state vibration, we assumed

$$u_r^L = R_2U_\eta^Le^{i\omega t}, \quad Q_a = G^S\overline{Q}_ae^{i\omega t}, \quad G^{SL} = \frac{G^S}{G^L}, \quad \rho^{LS} = \frac{\rho^L}{\rho^S} \tag{30}$$

Meanwhile, satisfy the boundary conditions as follows:

(1) Stress compatibility in the mold structure and soil interface ($r = R_2, \eta = 1$); that is,

$$G^{SL}\overline{Q}_ae^{i\omega t} = Qe^{i\omega t} - \frac{\sigma_r^S}{G^L} \tag{31}$$

(2) Displacement continuous in the mold structure and soil interface ($r = R_2, \eta = 1$); that is,

$$u_r^S = u_r^L \quad r = R_2 \tag{32}$$

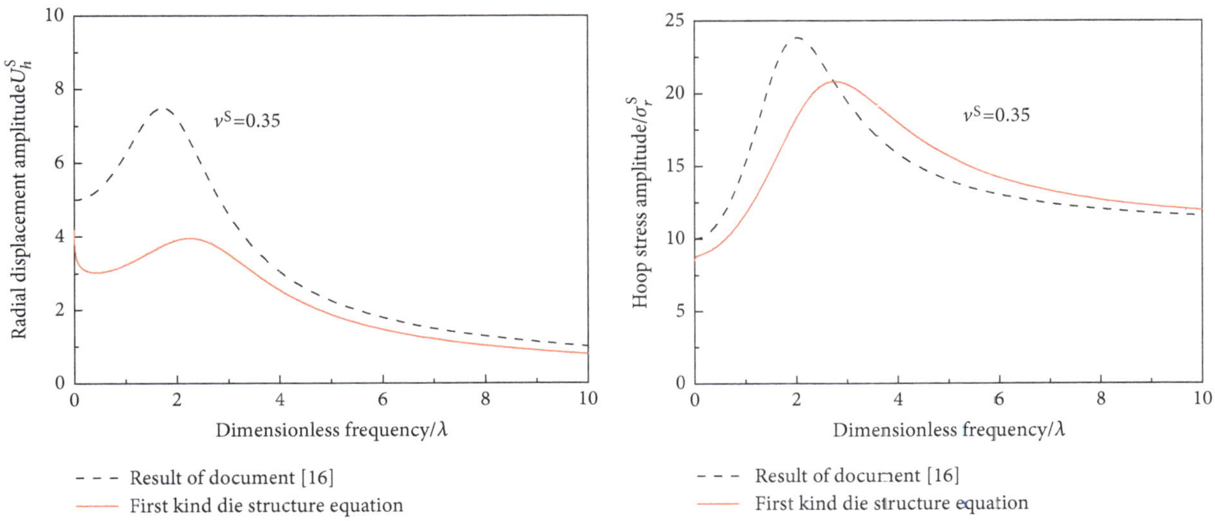

(a) Radial displacement amplitude

(b) Hoop stress amplitude

FIGURE 2: Comparison between present results and [16] results.

Substituting (29) and (13) into (31) and (32), we obtain

$$A = \frac{Q}{G^{SL}\left(\left(1+T_\sigma^\alpha (i\lambda)^\alpha\right)/\left(1+T_\varepsilon^\alpha (i\lambda)^\alpha\right)\right)\vartheta_3 - \vartheta_4 (1+\beta) e^{-\beta}} \quad (33)$$

where

$$\vartheta_3 = \frac{2\nu^S\beta^2}{1-2\nu^S}e^{-\beta} + 2e^{-\beta}\left(2+2\beta+\beta^2\right)$$
$$\vartheta_4 = \frac{2\delta}{1-2\nu^L} - \rho^{LS}G^{SL}\delta\lambda^2 \quad (34)$$

5. Comparison with Existing Results

In order to verify the correctness of the calculation, we compared the results with existing results, which include the ideal elastic model of [16], the classical elastic constitutive model [17], and classical viscoelastic constitutive model [18].

The ideal elastic model of [16] ignores the influence of the die structure firstly. Besides, it establishes the dynamic characteristics of the spherical digester in a general and ideal elastic soil. Substituting the equation for the classical elastic constitutive method into (3) and assuming that the material parameters of the soil are the same as in the die structure, we determined that the results of this study were in agreement with other published solutions [16]. The parameter values were as follows: η=1, vS=0.35, ρLS=1.5, GSL=0.05, vL=0.25, δ=0.005, Tσ/Tε=3, Tε=10, and α=0.5. Figure 2 shows a curve for the radial displacement and hoop stress amplitude with a dimensionless frequency for the two models of a uniform elastic medium. It can be seen that, in [16], the resonance effect is significantly greater for the model than for the homogeneous elastic dielectric model and that the base frequency is also higher than the base frequency of the resonance effect of the system. There are two main reasons for this: firstly, the author uses the fractional-derivative constitutive model to describe the soil viscosity and [16] uses a general and ideal elastic soil body; secondly, the author considers the influence of the die structure which has a greater stiffness than the soil body.

Thirdly, the fractional-derivative constitutive model used in this study is compared with the classical elastic model [17] and the classical viscoelastic constitutive model [18]. Substituting the classical elastic constitutive model and classical viscoelastic constitutive model used in the literature [19] into the soil equation of motion (3), respectively, changes the results of [19, 20]. Figure 3 shows the changes in the curves of the radial displacement amplitude and circumference displacement amplitude as a result of the changing frequency of the three models. It can be seen that, for the elastic model in [20], the peak value of the displacement amplitude of the soil is maximum; the resonance effect is clear and the resonance phenomenon decreases with increasing. The peak value of the classical viscoelastic model used in [19] is the minimum, while, for the fractional-derivative viscoelastic model, it is in the middle.

6. Analysis of Example and Graphic Analysis

The effects of the shear modulus ratio G^{SL} of the soil and die, the Poisson ratio ν^s of the soil material, the fractional-derivative order α, the material parameter ratio T_σ/T_ε on the radial displacement, and the hoop stress amplitude were investigated.

Figure 4 shows that the values of the radial displacement and circumferential stress amplitude change with the frequency f as the shear modulus ratio G^{SL} of the soil body and die structure variation change. It is evident that when the modulus ratio G^{SL} is comparatively small, the displacement amplitudes are very different for the model and the second type of mold model and the resonance effect produced by the

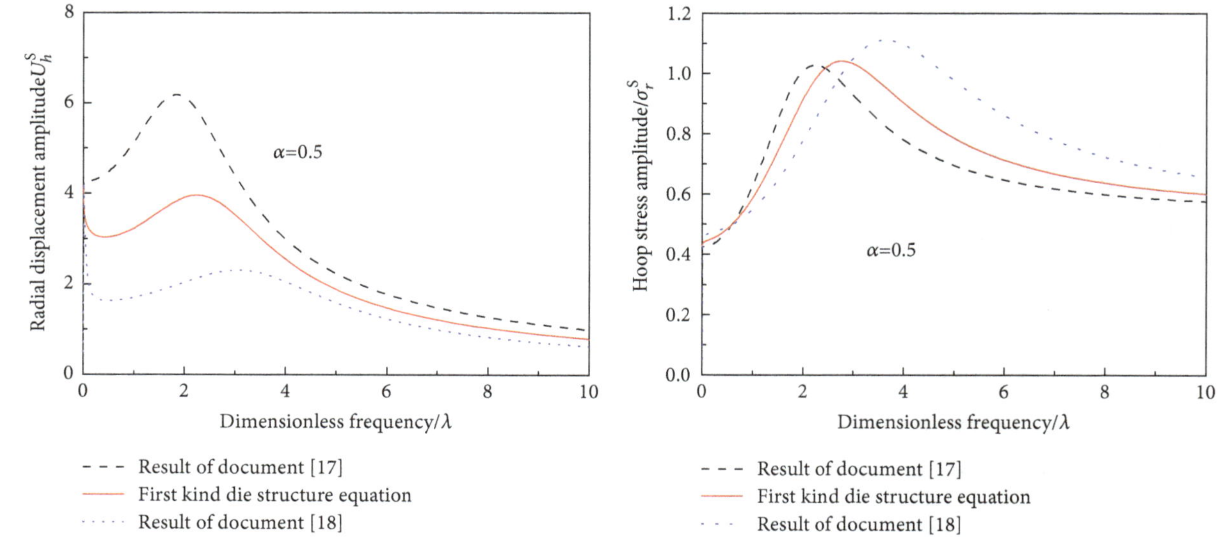

(a) Radial displacement amplitude (b) Hoop stress amplitude

FIGURE 3: Comparison analysis of three models.

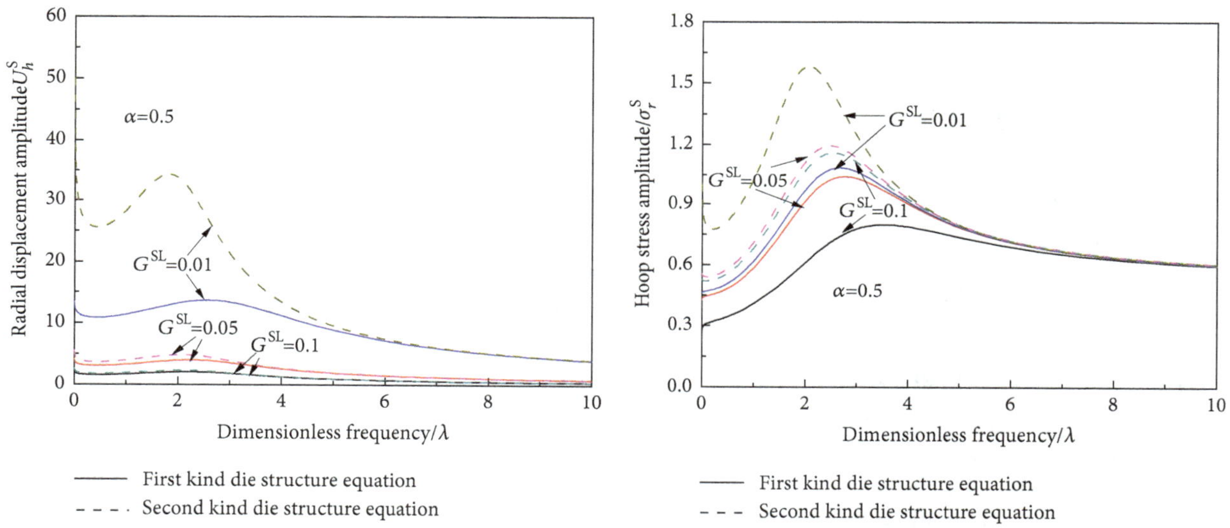

(a) Radial displacement amplitude (b) Hoop stress amplitude

FIGURE 4: Influence of modulus ratio on radial displacement and hoop stress amplitude.

latter system is significant. With the increase in the modulus ratio λ, the difference in the displacement amplitude for the two models is small. The stress amplitude is clearly greater for the spherical shell die structure than for the uniform elastic medium. As the frequency increases, the difference between the two models decreases and the displacement and the stress amplitude decrease with an increase in the modulus ratio.

Figure 5 shows the effect of the changes in Poisson's ratio ν^s on the radial displacement and the amplitude of the circumferential stress. As Poisson's ratio ν^s increases, the radial displacement and amplitude of the hoop stress increase gradually. The difference in the displacement and stress amplitude between the two types of die structures at high frequency is very small.

Figure 6 shows the influence of the fractional-derivative order α on the radial displacement and circumferential stress amplitude. It is evident that the amplitude of radial displacement decreases with an increase in the order number α in genera; with the increase in the frequency λ, the difference in the displacement amplitude for the two models is small. The hoop stress changed with the fractional-derivative order number α being interesting. Its change has a certain

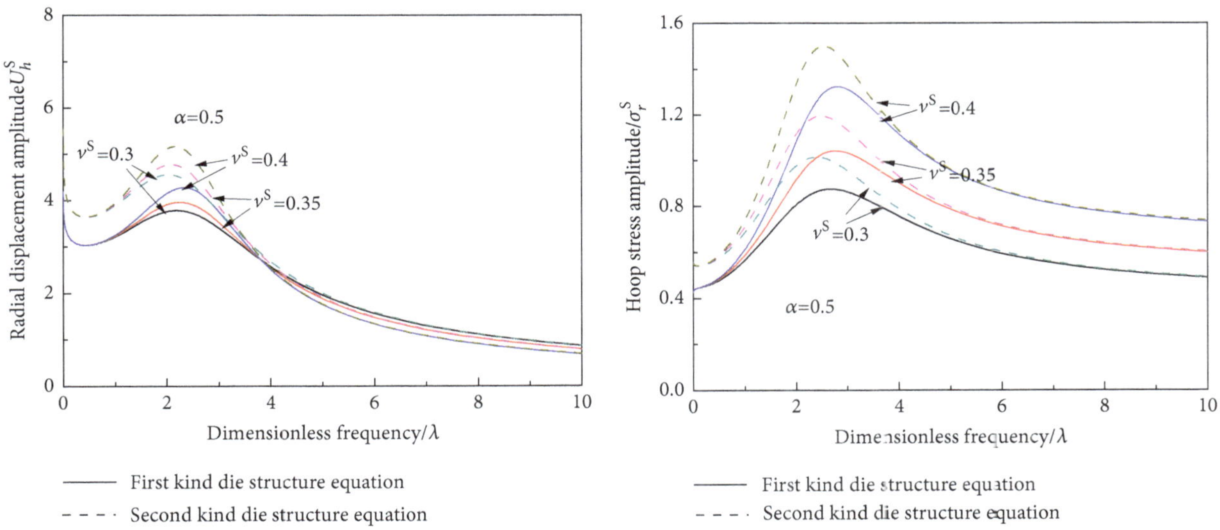

(a) Radial displacement amplitude (b) Hoop stress amplitude

FIGURE 5: Influence of Poisson's ratio on radial displacement and hoop stress amplitude.

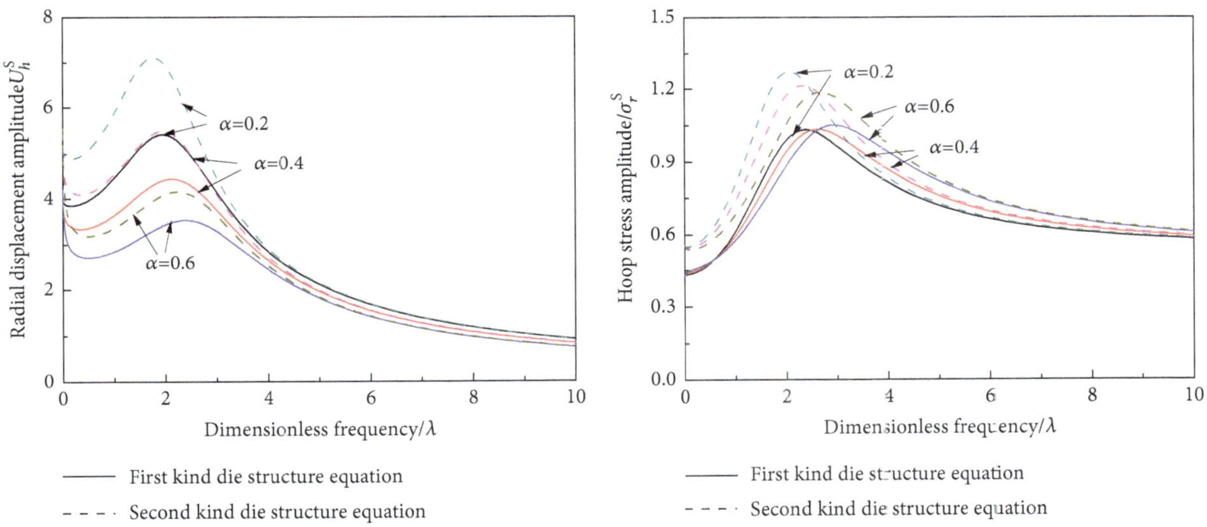

(a) Radial displacement amplitude (b) Hoop stress amplitude

FIGURE 6: Influence of order on radial displacement and hoop stress amplitude.

relationship with the frequency λ: when the frequency λ is between 0 and 3, the hoop stress amplitude decreases with an increase in the order number α; when the frequency λ is greater than 3, the circumferential stress amplitude increases with an increase in the order number α.

Figure 7 shows the changes in the radial displacement and circumferential stress amplitude as a result of changes in the frequency λ when the material parameter ratio T_σ/T_ε of the constitutive model is changed. The data indicate that the radial displacement and circumferential stress amplitude decrease as the material parameter ratio T_σ/T_ε increases; this is caused by the increase in the impedance of the soil body.

7. Conclusions

In this paper, the simplified vibration characteristics of a deeply buried spherical digester in viscoelastic soil under uniform internal pressure in the frequency domain are studied using an analytics method; the effects of relative physical properties and geometrical parameters on the dynamic characteristics of the system are considered. The main conclusions are summarized as follows:

(1) Under the condition of T_σ/T_ε =3, the steady-state response of the viscoelastic soil is greater for the fractional-derivative viscoelastic soil than for the classical viscoelastic soil but less than for the classical elastic soil

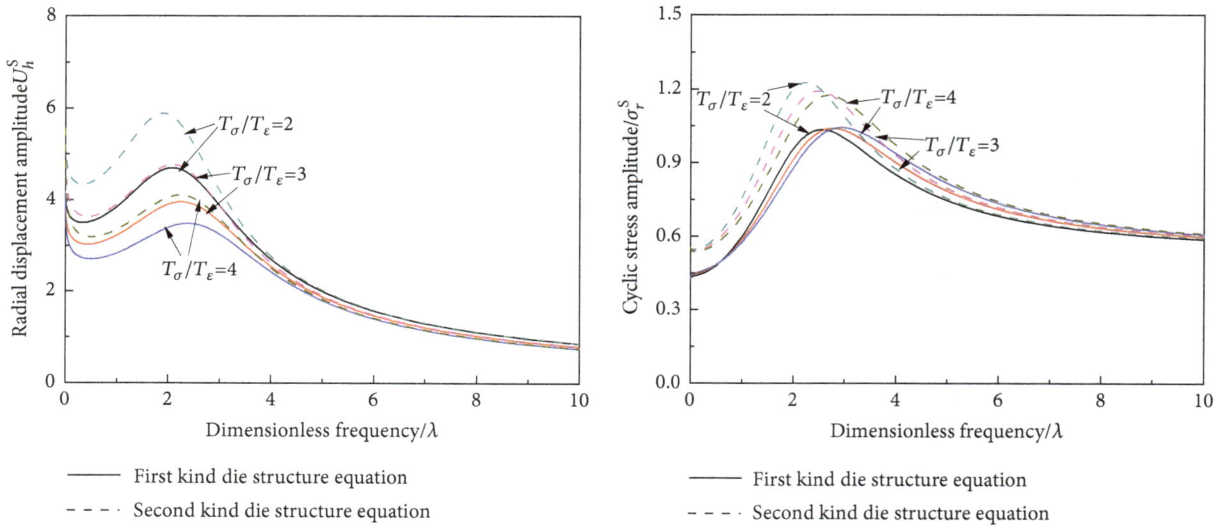

(a) Radial displacement amplitude

(b) Hoop stress amplitude

FIGURE 7: Influence of material parameter ratio on radial displacement and hoop stress amplitude.

(2) The steady-state response of the spherical digester is clearly greater for the spherical shell mold structure model than for the homogeneous elastic medium model

(3) When the modulus is comparatively small, the displacement amplitude is very different for the two types of mold type models and the difference between them decreases with increases in G^{SL}

(4) With the increase in Poisson's ratio ν^S of the soil material, the displacement amplitude and the hoop stress amplitude increases gradually

(5) With the increase in the order number α, the radial displacement and the hoop stress amplitude decrease and the system resonance effect weakens

(6) With the increase in the material parameter ratio T_σ/T_ε, the radial displacement and the hoop stress amplitude decrease

Conflicts of Interest

The authors declare that they have no conflicts of interest.

Acknowledgments

The work presented in this paper was supported by the Department of Construction of Zhejiang Province Science and Technology Project (no. 2015K31), National Natural Science Foundation of China under (Grant no. 51708175), and Business Funds of Central Colleges and Universities (Grant no. 2018B00814).

References

[1] W. Gong, M. Ding, Q. Xu, and X. Jiang, "Earth pressure on pool wall and stress analysis on pool cover of underground biogas digester," *Nongye Gongcheng Xuebao*, vol. 27, no. 4, pp. 270–276, 2011.

[2] M. J. Forrestal and M. J. Sagartz, "Radiated pressure in an acoustic medium produced by pulsed cylindrical and spherical shells," *Journal of Applied Mechanics*, vol. 38, no. 4, pp. 1057–1060, 1971.

[3] T. A. Duffey, "Transient response of viscous-elastic and viscous-elastic shells submerged in fluid media," *AMSE Journal of Applied Mechanics*, vol. 98, pp. 137–143, 1976.

[4] A. L. Glenn and R. E. Kidder, "Blast loading of a spherical container surrounded by an infinite elastic medium," *Journal of Impact Engineering*, vol. 22, pp. 984–992, 2000.

[5] Z. Qingyuan and Z. Renrui, "The dynamic response of spherical hollow chamber under explosive loading," *Explosion and Shock Waves*, vol. 14, no. 2, pp. 182–185, 1994.

[6] C. Gu, J. Wang, Y. Cai, L. Sun, P. Wang, and Q. Dong, "Deformation characteristics of overconsolidated clay sheared under constant and variable confining pressure," *Soils and Foundations*, vol. 56, no. 3, pp. 427–439, 2016.

[7] L. Guo, J. Wang, Y. Cai, H. Liu, Y. Gao, and H. Sun, "Undrained deformation behavior of saturated soft clay under long-term cyclic loading," *Soil Dynamics and Earthquake Engineering*, vol. 50, pp. 28–37, 2013.

[8] J. Wang, L. Guo, Y. Cai, C. Xu, and C. Gu, "Strain and pore pressure development on soft marine clay in triaxial tests with a large number of cycles," *Ocean Engineering*, vol. 74, pp. 125–132, 2013.

[9] C. Gu, J. Wang, Y. Cai, and L. Guo, "Influence of cyclic loading history on small strain shear modulus of saturated clays," *Soil Dynamics and Earthquake Engineering*, vol. 66, pp. 1–12, 2014.

[10] C. Gu, Z. Gu, Y. Cai, J. Wang, and D. Ling, "Dynamic modulus characteristics of saturated clays under variable confining pressure," *Canadian Geotechnical Journal*, vol. 54, no. 5, pp. 729–735, 2017.

[11] Y. Cai, T. Wu, L. Guo, and J. Wang, "Stiffness degradation and plastic strain accumulation of clay under cyclic load with principal stress rotation and deviatoric stress variation," *Journal of Geotechnical and Geoenvironmental Engineering*, vol. 144, no. 5, Article ID 04018021, 2018.

[12] T. Wu, Y. Cai, L. Guo, D. Ling, and J. Wang, "Influence of shear stress level on cyclic deformation behaviour of intact Wenzhou soft clay under traffic loading," *Engineering Geology*, vol. 228, pp. 61–70, 2017.

[13] H. X. Gao and M. J. Wen, "Dynamic characteristics of a spherical cavity in nearly saturated viscoelastic soil," *Chinese Journal of Theoretical and Applied Mechanics*, vol. 44, no. 4, pp. 753–761, 2012.

[14] L.-C. Liu and X. Yang, "Analysis of vertical vibrations of a pile in saturated soil described by fractional derivative model," *Rock and Soil Mechanics*, vol. 32, no. 2, pp. 526–532, 2011.

[15] H. Lijun, K. Lingwei, W. Wenjun et al., "A description of creep model for soft soil with fractional derivative," *Rock and Soil Mechanics*, vol. 32, supplement 2, pp. 239–244, 2011.

[16] P. Weiland, "Biogas production: current state and perspectives," *Applied Microbiology and Biotechnology*, vol. 85, no. 4, pp. 849–860, 2010.

[17] X. Changjie and W. Shiming, "Spherical wave propagation in saturated soils," *Applied Mathematics and Mechanics*, vol. 19, no. 3, pp. 243–252, 1998.

[18] X. Changjie and C. Yuanqiang, "Dynamic response of spherical cavity in viscous-elastic saturated soils," *China Civil Engineering Journal*, vol. 34, no. 4, pp. 88–92, 2001.

[19] U. Zakout and N. Akkas, "Transient response of a cylindrical cavity with and without a bonded shell in an infinite elastic medium," *International Journal of Engineering Science*, vol. 35, no. 13-12, pp. 1203–1220, 1997.

[20] S. P. Timoshenko and J. N. Goodier, *Theory of Elasticity*, McGraw-Hill, New York, NY, USA, 1977.

19

Noise-Induced Truth Seeking of Heterogeneous Hegselmann-Krause Opinion Dynamics

Yan Liu[1,2] **and Lipo Mo**[3]

[1]*School of Automation and Electrical Engineering, University of Science and Technology, Beijing 100083, China*
[2]*Key Laboratory of Knowledge Automation for Industrial Processes, Ministry of Education, Beijing 100083, China*
[3]*School of Science, Beijing Technology and Business University, Beijing 100048, China*

Correspondence should be addressed to Lipo Mo; beihangmlp@126.com

Academic Editor: Carlo Bianca

In this paper, we study the noise-induced truth seeking for heterogeneous Hegselmann-Krause (HK) model in opinion dynamics. It has been proved that small noise could induce the group to achieve truth in homogeneous HK model; however, for the more practical heterogeneous HK model, the theoretical conclusion is absent. Here, we prove that small noise could also induce the group to achieve truth in heterogenous HK model, and, moreover, we first theoretically prove that large noise could drive some agents to deviate from the truth. These theoretical findings evidently reveal how the free information flow spreading in the media determines the social truth seeking.

1. Introduction

Recently opinion dynamics has attracted increasing attention from researchers of various fields [1–6]. In opinion dynamics, a basic model is the widely known Hegselmann-Krause (HK) model, which is built based on the so-called *bounded confidence* mechanism [7–9]. In HK model, each agent possesses a confidence threshold and updates its opinion by averaging those of its neighbors (agents whose opinions locate within its confidence threshold). If all the agents have an identical confidence threshold, the model is said to be homogeneous; otherwise, the model is said to be heterogeneous. Though it looks simple, the HK dynamics captures some basic opinion phenomena such as consensus and cleavage and also provides a fundamental mechanism that could develop more opinion models [10–12].

In social dynamics, how the social groups seek and achieve truth interests much studies. In [13], a truth seeking model was built based on the HK dynamics. Using the model, some factors are found to be related to whether the group achieves the truth. One is the number of truth seekers in the group. If all the agents in the group are truth seekers, it is proved that the group will finally achieve the truth. If there are agents which are not truth seekers in the group, whether the truth can be acquired is highly dependent on the initial conditions of the group and most likely the truth cannot be achieved by the whole group. Very lately, using the model in [13] a study found that noise plays a dominant role for the group to achieve the truth [14]. It is rigorously proved that, in the presence of small noise, the group will almost surely achieve the truth for any initial conditions and even there is only one truth seeker in the group. The noise is appropriately explained as the intensive free information we randomly received from media each day in this information era, and the finding actually reveals that free information flow spreading in the media benefits the truth seeking in a society.

The truth seeking model in [14] is homogeneous; i.e., all the agents possess an identical confidence threshold and all the truth seekers have the same preference for the truth. However, as we know, the real agents in a society are usually heterogeneous and have different confidence threshold and different preference for truth. Moreover, [14] only reveals the influence of properly small noise which could induce the truth seeking. But, in real social network, there always exist some "strong" noises, such as some fabricated rumors which may violently perturb people's opinion. Thus, there is a great need to investigate theoretically how small and large noise determine the truth seeking of heterogeneous HK model.

In this paper, we completely establish the theoretical analysis for the truth seeking of heterogeneous HK model with noise. Similar to [14], we also prove that small noise could drive the group to the truth from any initial state; what is more important, we for the first time rigorously prove that when noise is large, the truth cannot be achieved in the group no matter what the initial state is. These results reveal completely how free information flow determines the truth seeking in social network.

The rest of this paper is organized as follows: Section 2 gives the heterogeneous truth seeking model with noise; Section 3 proves the main theoretical results about how small and large noise determine the truth seeking; Section 4 provides some simulation studies to verify the main theoretical results; and at last Section 5 concludes the paper.

2. Preliminary and Formulation

2.1. Heterogeneous Truth Seeking Model. Let $\mathcal{S} = \{1, \ldots, n\}$ be a group of agents which consist of truth seekers and neutral followers and $\mathcal{K} \subset \mathcal{S}$ be the set of truth seekers. Then the heterogeneous truth seeking model is modelled as

$$\begin{aligned} y_i(t+1) = {} & \lambda_i I_{\{i\in\mathcal{K}\}} Q \\ & + \left(1 - \lambda_i I_{\{i\in\mathcal{K}\}}\right) \sum_{j\in\mathcal{N}_i(y(t))} \frac{y_j(t)}{|\mathcal{N}_i(y(t))|} \\ & + \eta_i(t+1), \quad i \in \mathcal{S}, \end{aligned} \tag{1}$$

where $y_i(t) \in [0,1]$ is the state of agent i at time t, $Q \in [0,1]$ is the truth value, $\lambda_i \in [0,1]$ is the attraction strength of truth Q for agent i, and

$$\mathcal{N}_i(y(t)) = \left\{ j \in \mathcal{S} \mid \left| y_j(t) - y_i(t) \right| \le r_i \right\} \tag{2}$$

is the neighbor set of i at time t with $r_i \in (0,1]$ being the confidence threshold. In addition, $I_{\{i\in\mathcal{K}\}}$ is the indicator function which takes value 1 or 0 according to $i \in \mathcal{K}$ or not and $|\cdot|$ represents the absolute value of a real number or the cardinal number of a set accordingly. $\{\xi_i(t)\}_{i\in\mathcal{S},t>0}$ are noises which are independent and uniformly distributed on $[-\delta, \delta]$ with $\delta > 0$. If $r_i = r \in (0,1], \lambda_i = \lambda \in [0,1]$ for all $i \in \mathcal{S}$, the heterogeneous model (1) degenerates to a homogeneous one studied in [14]. For a truth seeker i, it has $\lambda_i > 0$; i.e., $\mathcal{K} = \{i \in \mathcal{S} : 0 < \lambda_i \le 1\}$, while a neutral follower has $\lambda_i = 0$.

In system (1), noise usually models the free information flow spread in social network. Each day we are exposed to intensive free information which will affect our opinions. However, we do not know what information we will meet, whether and how much they will change our opinions. This results in a stochastic effect of the update of our opinions. Mathematically, the introduction of noise will make opinion values exceed the bounded opinion space $[0,1]$. Similar to previous studies, we also make an assumption that the noisy opinion values are limited in $[0,1]$. Subsequently, model (1) is modified as

$$y_i(t+1) = \begin{cases} 1, & y_i^*(t) > 1 \\ y_i^*(t), & y_i^*(t) \in [0,1], \\ 0, & y_i^*(t) < 0 \end{cases} \qquad \forall i \in \mathcal{S},\ t \ge 0, \tag{3}$$

where

$$\begin{aligned} y_i^*(t) = {} & \lambda_i I_{\{i\in\mathcal{K}\}} Q \\ & + \left(1 - \lambda_i I_{\{i\in\mathcal{K}\}}\right) \sum_{j\in\mathcal{N}_i(y(t))} \frac{y_j(t)}{|\mathcal{N}_i(y(t))|} \\ & + \eta_i(t+1). \end{aligned} \tag{4}$$

and

$$\begin{aligned} & 1 \le |\mathcal{K}| \le n, \\ & \{\eta_i(t), i \in \mathcal{S}, t \\ & \quad > 0\} \text{ are independent and uniformly distributed on } [-\delta, \delta]. \end{aligned} \tag{5}$$

2.2. Achieving Truth. The definition of achieving truth for the noisy system (3)-(5) is as follows.

Definition 1 (see [14]). Let $d_{\mathcal{S}}^{Q}(t) = \max_{i\in\mathcal{S}} |y_i(t) - Q|$; if $\limsup_{t\to\infty} d_{\mathcal{S}}^{Q}(t) \le \phi$ a.s., we say system (3)-(5) achieves the truth Q with ϕ-precision.

3. Main Results

3.1. Small Noise Benefits Truth Seeking. In [14], it was proved that small noise could drive the homogeneous HK model to achieve the truth. In this part, we will show how small noise affects the truth seeking of heterogeneous HK model. Actually, we obtain the same conclusion for the heterogeneous HK model.

Let $m = |\mathcal{K}|$, $\underline{\lambda} = \min_{i\in\mathcal{S}} \lambda_i$, $\underline{r} = \min_{i\in\mathcal{S}} r_i$ and denote

$$\begin{aligned} \delta_1 &= \frac{n(1-\underline{\lambda})}{m\underline{\lambda}}\delta + \delta, \\ \delta_2 &= \frac{n}{m\underline{\lambda}}\delta + \delta, \\ \overline{\delta} &= \max\{\delta_1, \delta_2\}, \\ \underline{\delta} &= \min\left\{ \frac{m\underline{\lambda}}{2n + (2m-n)\underline{\lambda}}\underline{r}, \frac{m}{n+2m}\underline{r} \right\}, \end{aligned} \tag{6}$$

and then, for system (3)-(5), we have the following main result.

Theorem 2. *Let $0 < \underline{\lambda} < 1$ in model (3)-(5), and then, for any initial state $y_i(0) \in [0,1], r_i \in (0,1], i \in \mathcal{S}$ and $\delta \in (0, \underline{\delta}]$, the model will a.s. achieve the truth Q with $\overline{\delta}$-precision.*

Also, to demonstrate Theorem 2 better, we present the following corollary like that in [14].

Corollary 3. *Suppose* $m = \lceil n/2 \rceil, \underline{\lambda} = 0.5$, *and then a.s.* $\limsup_{t\to\infty} d^{Q}_{\mathcal{S}}(t) \le 5\delta$ *for all* $\delta \in (0, \underline{r}/8]$.

The key of the proof of Theorem 2 is the following lemma.

Lemma 4. *Let* $0 < \underline{\lambda} \le 1$, *and if there is a moment* $t_0 < \infty$ *a.s. such that* $d^{Q}_{\mathcal{K}}(t_0) \le \delta_1, d^{Q}_{\overline{\mathcal{K}}}(t_0) \le \delta_2 (\overline{\mathcal{K}} = \mathcal{S} - \mathcal{K})$, *then, for* $0 < \delta \le \underline{\delta}$, *it holds* $\limsup_{t\to\infty} d^{Q}_{\mathcal{K}}(t) \le \delta_1, \limsup_{t\to\infty} d^{Q}_{\overline{\mathcal{K}}}(t) \le \delta_2$ *a.s.*

Proof. It can be checked that, at moment t_0, it holds that

$$\begin{aligned} \max_{i,j\in\mathcal{S}} \left| y_i(t_0) - y_j(t_0) \right| &\le \max_{i\in\mathcal{S}} \left| y_i(t_0) - Q \right| \\ &\quad + \max_{j\in\mathcal{S}} \left| y_j(t_0) - Q \right| \le \delta_1 + \delta_2 \\ &= \frac{2n + (2m - n)\underline{\lambda}}{m\underline{\lambda}} \le \underline{r}. \end{aligned} \tag{7}$$

Thus all agents are neighbors to each other at moment t_0 and hence $|\mathcal{N}_i(y(t_0))| = n, i \in \mathcal{S}$. By (3) and (4), it has

$$\begin{aligned} y_i(t_0+1) &= \lambda_i Q + (1-\lambda_i) \frac{\sum_{k\in\mathcal{S}} y_k(t_0)}{n} \\ &\quad + \eta_i(t_0+1), \quad i \in \mathcal{K}, \\ y_j(t_0+1) &= \frac{\sum_{k\in\mathcal{S}} y_k(t_0)}{n} + \eta_j(t_0+1), \quad j \in \overline{\mathcal{K}}. \end{aligned} \tag{8}$$

Thus

$$\begin{aligned} &\left| y_i(t_0+1) - Q \right| \\ &= \left| \lambda_i Q - Q + (1-\lambda) \frac{\sum_{k\in\mathcal{S}} y_k(t_0)}{n} + \eta_i(t_0+1) \right| \\ &\le \frac{1-\lambda_i}{n} \sum_{k\in\mathcal{S}} \left| Q - y_k(t_0) \right| + \delta \\ &\le \frac{1-\underline{\lambda}}{n} (m\delta_1 + (n-m)\delta_2) + \delta = \frac{1-\underline{\lambda}}{n} \frac{n^2\delta}{m\underline{\lambda}} + \delta \\ &= \delta_1, \quad i \in \mathcal{K}. \end{aligned} \tag{9}$$

And also

$$\begin{aligned} \left| y_j(t_0+1) - Q \right| &= \left| \frac{\sum_{k\in\mathcal{S}} y_k(t_0)}{n} - Q + \eta_i(t_0+1) \right| \\ &\le \frac{1}{n} \sum_{k\in\mathcal{S}} \left| Q - y_k(t_0) \right| + \delta \\ &\le \frac{m\delta_1 + (n-m)\delta_2}{n} + \delta = \frac{n\delta}{m\underline{\lambda}} + \delta \\ &= \delta_2, \quad j \in \overline{\mathcal{K}}. \end{aligned} \tag{10}$$

Consequently

$$\begin{aligned} d^{Q}_{\mathcal{K}}(t_0+1) &\le \delta_1, \\ d^{Q}_{\overline{\mathcal{K}}}(t_0+1) &\le \delta_2. \end{aligned} \tag{11}$$

Repeating the above procedure obtains the conclusion. □

The proof of Lemma 4 quite follows the line of Lemma 3.2 in [14]. It tells that there is an "attraction region" for the system to achieve the truth, and once the opinion of the group enters the region, small noise cannot drive them out. The main point of the proof of Lemma 4 is to find the "attraction region" in the heterogeneous model. With Lemma 4, the rest of the proof of Theorem 2 is similar to that of Theorem 3.1 in [14] and we omit it here.

3.2. Large Noise Harms Truth Seeking. Theorem 2 reveals that small noise could promote the truth seeking of the heterogeneous HK model. The noise in social networks mainly models the free information flow spreading in social media that influences people's opinion every day. Usually most information softly changes our opinions and brings small noise to the opinion dynamics; however, there are always some breaking news or rumors in social networks that could change our opinion shockingly and bring severe noise to the opinion dynamics. It is absent in [14] how large noise or breaking information affects the truth seeking in opinion dynamics. Here, we for the first time investigate this problem with the heterogeneous HK model and reveal theoretically that large noise is detrimental to the truth seeking of opinion dynamics.

Theorem 5. *Suppose* $|\mathcal{K}| < n$ *in system (3)-(5) and* $\underline{r}' < \overline{D}$, *where* $\underline{r}' = \min_{i\in\overline{\mathcal{K}}} r_i, \overline{D} = \max\{Q, 1-Q\}$ *with* $Q \in [0,1]$. *If* $\delta > \underline{r}'$, *then, for all* $y(0) \in [0,1]^n$, *it has* $d^{Q}_{\mathcal{S}} = \overline{D}$.

Remark 6. In Theorem 5, we suppose $|\mathcal{K}| < n$; i.e., there exist agents which are not truth seekers in the system. If all agents are truth seekers, it can be proved that the opinions will stay near the truth value Q within an area dependent on the noise strength δ. However, if there are neutral followers, Theorem 5 actually implies that some opinions will deviate from the truth value with any far distance within the opinion space. Theorem 2 shows an order phenomenon when noise is small while Theorem 5 shows a disorder phenomenon when noise is large. The measure of order is by the ϕ-*precision* in Definition 1 and Theorem 2 reveals that when noise strength diminishes the precision increases. Theorem 5 shows that when large noise makes a disorder, the precision is independent of the noise strength and can be worst.

Proof. Let $Q \in [0, 1/2]$ without loss of generality, and then we only need to prove that there exists $i \in \mathcal{S}$ such that $\limsup_{t\to\infty} y_i(t) = 1$ a.s.; i.e.

$$P\left\{ \bigcup_{t_0=0}^{\infty} \{ y_i(t) < 1, t > t_0 \} \right\} = 0. \tag{12}$$

Given any $t_0 \geq 0$ consider the following noise protocol: for $t \geq t_0$,

$$\eta_j(t) \in \left(\frac{\delta + \underline{r}'}{2}, \delta\right], \quad j \in \mathcal{S}. \tag{13}$$

Let $\mathcal{M} = \{i \in \overline{\mathcal{K}} \mid r_i = \min_{j \in \overline{\mathcal{K}}} r_j\}$, and then, for $i \in \mathcal{M}$, by (4), it has under protocol (13) that

$$\begin{aligned} y_i^*(t_0+1) &= \frac{\sum_{k \in \mathcal{N}_i(y(t_0))} y_k(t_0)}{|\mathcal{N}_i(y(t_0))|} + \eta_i(t_0+1) \\ &> \frac{\sum_{k \in \mathcal{N}_i(y(t_0))} y_k(t_0)}{|\mathcal{N}_i(y(t_0))|} + \frac{\delta + \underline{r}'}{2}, \quad t \geq 0. \end{aligned} \tag{14}$$

If $y_i^*(t_0) \geq 1$, it has $y_i(t_0+1) = 1$; otherwise, $y_i(t_0+1) = y_i^*(t_0+1) < 1$. Since $|y_k(t_0) - y_i(t_0)| \leq \underline{r}'$, for $k \in \mathcal{N}_i(y(t_0))$, by (14)

$$\begin{aligned} y_i(t_0+1) - y_i(t_0) &> \frac{\sum_{k \in \mathcal{N}_i(y(t_0))} (y_k(t_0) - y_i(t_0))}{|\mathcal{N}_i(y(t_0))|} \\ &\quad + \frac{\delta + \underline{r}'}{2} \geq -\underline{r}' + \frac{\delta + \underline{r}'}{2} \\ &= \frac{\delta - \underline{r}'}{2} \end{aligned} \tag{15}$$

This implies, at moment t_0+1, the opinion value of agent i will increase by no less than $(\delta - \underline{r}')/2$ under protocol (13). Note that $P\{\eta_j(t) \in ((\delta+\underline{r}')/2, \delta]\} = (\delta - \underline{r}')/4\delta$ for all $j \in \mathcal{S}, t \geq t_0$, and, by independence of $\eta_i(t)$, it has

$$P\left\{y_i(t_0+1) - y_i(t_0) > \frac{\delta - \underline{r}'}{2}\right\} \geq \left(\frac{\delta - \underline{r}'}{4\delta}\right)^n > 0. \tag{16}$$

Denote $L = \lceil 2/(\delta - \underline{r}') \rceil$ and repeat the above procedure for L times, and we know that, under protocol (13), $y_i(t_0 + L) = 1$. By (16) and with independence of $\eta_i(t), i \in \mathcal{S}, t \geq 1$, it has

$$P\{y_i(t_0+L) = 1\} \geq \left(\frac{\delta - \underline{r}'}{4\delta}\right)^{nL} > 0, \tag{17}$$

and thus

$$P\{y_i(t_0+L) < 1\} < 1 - \left(\frac{\delta - \underline{r}'}{4\delta}\right)^{nL} < 1. \tag{18}$$

Similarly, for all $t > t_0$

$$P\left\{y_i(t) < 1 \mid \bigcap_{t_0 \leq l < t} \{y_i(l) < 1\}\right\} \leq 1 - \left(\frac{\delta - \underline{r}'}{4\delta}\right)^{nL}. \tag{19}$$

Note that $P\{y_i(t_0) < 1\} \leq 1$, and it has

$$\begin{aligned} P\{y_i(t) < 1, t > t_0\} &= P\left\{\bigcap_{t=t_0+1}^{\infty} \{y_i(t) < 1\}\right\} \\ &= \lim_{m \to \infty} P\left\{\bigcap_{t=t_0+1}^{m} \{y_i(t) < 1\}\right\} \\ &\leq \lim_{m \to \infty} \prod_{t=t_0+1}^{m} P\left\{y_i(t) < 1 \mid \bigcap_{t_0 \leq l < t} \{y_i(l) < 1\}\right\} \\ &\quad \cdot P\{y_i(t_0) < 1\} \leq \lim_{m \to \infty} \left(1 - \left(\frac{\delta - \underline{r}}{4\delta}\right)^{nL}\right)^m \\ &= 0. \end{aligned} \tag{20}$$

This completes the proof. □

4. Simulations

In this section, we provide some simulation results to demonstrate the effect of small and large noise on truth seeking. Theorem 2 reveals that the phenomena of noise-induced truth seeking are independent of the group size n, the number of truth seekers $0 < m \leq n$, and the truth value Q and can be observed as long as the noise strength is properly small. Take $n = 120$ agents, and 30 agents are truth seekers. The truth value $Q = 0.5$, and the initial opinion values $y(0) = [y_1(0), \ldots, y_{120}(0)]$ are randomly generated on $[0, 1]$, the heterogeneous confidence thresholds $r = [r_1, \ldots, r_{120}]$ are randomly generated on $(0, 0.2]$, and the truth attraction strength $\lambda = [\lambda_1, \ldots, \lambda_{30}]$ is randomly generated on $[0, 1]$. Figure 1 shows when there is no noise, only part of agents achieve the truth while some agents stay away from the truth. Then we add small noise with strength $\delta = 0.0008$ to the group, and Figure 2 shows that all the agents finally achieve the truth. Next we add large noise $\delta = 0.02$ to the group, and Figure 3 clearly shows that some agents frequently deviate from the truth value.

5. Conclusions and Discussions

In this paper, we investigated the noise-induced truth seeking of the heterogeneous HK model. Compared to the existing results about the homogeneous case, we completely revealed how the free information flow with different strength affects the truth seeking in the heterogeneous model. On the one hand we showed that noise with proper strength will almost surely bring the group to achieve the truth, while, on the other hand, we further proved that large noise, which implies drastic information perturbation, will make some agents deviate from the truth. These results not only reveal the role free information flow plays in the social truth seeking but also provide prospective intervention design for promoting the truth seeking of social group.

At last we want to have a further discussion for the order/disorder transition which appears in this paper. It can

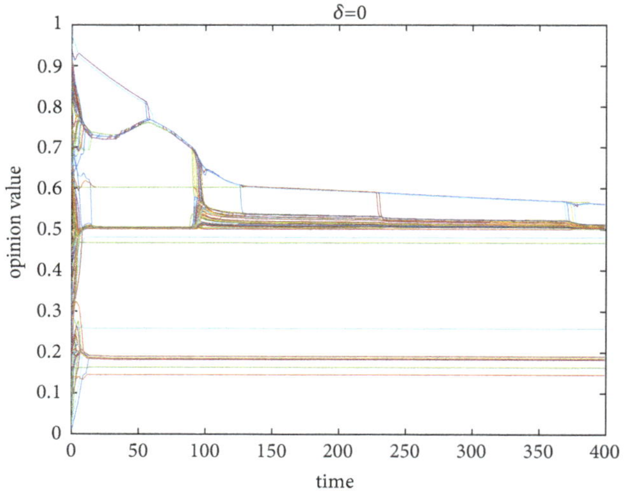

FIGURE 1: The truth seeking of model (3)-(5) without noise. It shows that some agents acquire the truth while the others stay away from the truth.

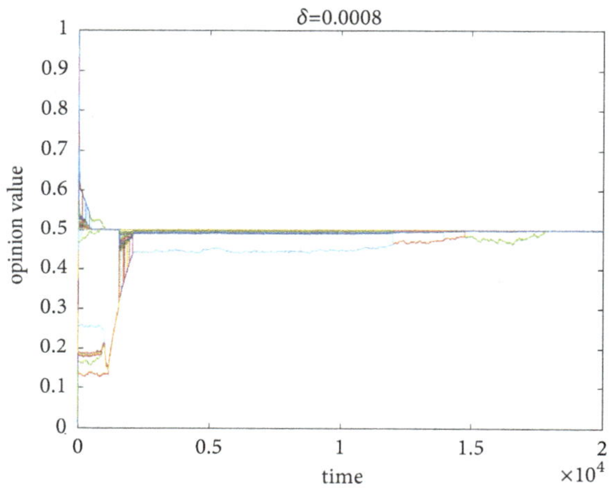

FIGURE 2: The truth seeking of model (3)-(5) with small noise. Here the noise strength is taken $\delta = 0.0076$ and it shows that all the agents finally achieve the truth.

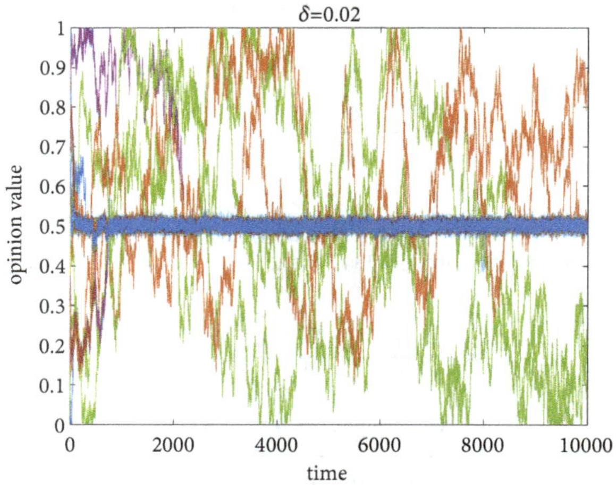

FIGURE 3: The truth seeking of model (3)-(5) with large noise. The noise strength is taken $\delta = 0.045$ and it shows that the truth cannot be achieved in the group while some agents frequently deviate from the truth.

be clearly found that an order/disorder transition happens with the variation of noise strength. However, whether the transition is a critical or just a crossover phenomenon is yet unclear in this study. In [11], a critical noise strength is obtained by analysis for the homogeneous HK model, but the analysis for the heterogeneous case is far more complicated. Simulation studies can show that the noise strength bound for order (δ in Theorem 2) and for disorder (δ in Theorem 5) is conservative and a fuzzy region of noise strength exists between the noise strength bounds of order and disorder.

Conflicts of Interest

The authors declare that there are no conflicts of interest regarding the publication of this paper.

Acknowledgments

This work is supported by the General Plan of Scientific Research Project of Beijing Educational Committee and the National Natural Science Foundation (NNSF) of China (Grant no. 61304155).

References

[1] C. Castellano, S. Fortunato, and V. Loreto, "Statistical physics of social dynamics," *Reviews of Modern Physics*, vol. 81, no. 2, pp. 591–646, 2009.

[2] S. Galam, *Sociophysics-A Physicist's Modeling of Psycho-political Phenomena*, Springer, New York, NY, USA, 2012.

[3] P. Sen and B. K. Chakrabarti, *Sociophysics: An Introduction*, Oxford University Press, 2013.

[4] A. V. Proskurnikov and R. Tempo, "A tutorial on modeling and analysis of dynamic social networks. Part I," *Annual Reviews in Control*, vol. 43, pp. 65–79, 2017.

[5] A. V. Proskurnikov and R. Tempo, "A tutorial on modeling and analysis of dynamic social networks. Part II," *Annual Reviews in Control*, vol. 45, pp. 166–190, 2018.

[6] Yong Ding and Lipo Mo, "Improved finite time in eliminating disagreement of opinion dynamics via noise," *Advances in Mathematical Physics*, vol. 2017, Article ID 1098169, 6 pages, 2017.

[7] G. Deffuant, D. Neau, F. Amblard, and G. Weisbuch, "Mixing beliefs among interacting agents," *Advances in Complex Systems (ACS)*, vol. 3, no. 1, pp. 87–98, 2000.

[8] U. Krause, "A discrete nonlinear and non-autonomous model of consensus formation," in *Communications in Difference Equations*, S. Elaydi, G. Ldas, J. Popenda, and J. Rakowski, Eds., pp. 227–238, Gordon and Breach, Amsterdam, 2000.

[9] R. Hegselmann and U. Krause, "Opinion dynamics and bounded confidence models, analysis, and simulation," *Artificial Societies and Social Simulation*, vol. 5, no. 3, pp. 1–33, 2002.

[10] V. D. Blondel, J. M. Hendrickx, and J. N. Tsitsiklis, "On Krause's multi-agent consensus model with state-dependent connectivity," *IEEE Transactions on Automatic Control*, vol. 54, no. 11, pp. 2586–2597, 2009.

[11] W. Su, G. Chen, and Y. Hong, "Noise leads to quasi-consensus of Hegselmann–Krause opinion dynamics," *Automatica*, vol. 85, pp. 448–454, 2017.

[12] W. Su, J. Guo, X. Chen, and G. Chen, *Robust Fragmentation Modeling of Hegselmann-Krause-Type Dynamics*, 2018.

[13] R. Hegselmann and U. Krause, "Truth and cognitive division of labour: first steps towards a computer aided social epistemology," *Journal of Artificial Societies and Social Simulation*, vol. 9, no. 3, pp. 1–28, 2006.

[14] W. Su and Y. Yu, "Free information flow benefits truth seeking," *Journal of Systems Science & Complexity*, vol. 31, no. 4, pp. 964–974, 2018.

20

Nanobeams and AFM Subject to Piezoelectric and Surface Scale Effects

Julio Claeyssen,[1,2] Rosemaira Copetti,[1] Leticia Tonetto,[3] and Joao Carvalho[2]

[1] *PGMAT, Departamento de Matemática, CCNE, Universidade Federal de Santa Maria, Av. Roraima 1000, 97105-900, Santa Maria, RS, Brazil*
[2] *PPGMAp-IME, Universidade Federal do Rio Grande do Sul, Av. Bento Gonçalves, 9500, 91509-900, Porto Alegre, RS, Brazil*
[3] *Centro de Engenharias, CENG, Universidade Federal de Pelotas, Almirante Barroso, 1734, Pelotas, RS, 96010-280, Brazil*

Correspondence should be addressed to Julio Claeyssen; jcrclaeyssen@yahoo.com

Academic Editor: Phuc Phung-Van

Vibration dynamics of elastic beams that are used in nanotechnology, such as atomic force microscope modeling and carbon nanotubes, are considered in terms of a fundamental response within a matrix framework. The modeling equations with piezoelectric and surface scale effects are written as a matrix differential equation subject to tip-sample general boundary conditions and to compatibility conditions for the case of multispan beams. We considered a quadratic and a cubic eigenvalue problem related to the inclusion of smart materials and surface effects. Simulations were performed for a two stepped beam with a piezoelectric patch subject to pulse forcing terms. Results with Timoshenko models that include surface effects are presented for micro- and nanoscale. It was observed that the effects are significant just in nanoscale. We also simulate the frequency effects of a double-span beam in which one segment includes rotatory inertia and shear deformation and the other one neglects both phenomena. The proposed analytical methodology can be useful in the design of micro- and nanoresonator structures that involve deformable flexural models for detecting and imaging of physical and biochemical quantities.

1. Introduction

Atomic force microscopy (AFM) is a scanning probe microscopy (SPM) technique to obtain images of surface topography at the atomic scale, in a noninvasive manner, from a wide variety of samples on a scale from angstroms to 100 microns [1]. A typical AFM consists of a microcantilever with a sharp tip, a sample positioning system, a detection system, and a control system. The associated length scales are sufficiently small to call the applicability of classical continuum models into question [2]. In this work, we seek to develop a vibration dynamics framework for beams that include smart materials and subject to surface effects. This framework is also considered in the case of a two-span beam in which the first segment is governed by the Timoshenko model and the second segment is an Euler-Bernoulli beam model [3].

Recently, new generations of active microcantilevers have included piezoelectric materials locally attached at the microbeam with the role of sensors and/or actuators linear and nonlinear [4–9], among others. This has led to the study of multispan beams for AFM. The inclusion of smart materials layers will modify material properties between neighboring segments, producing discontinuities and fulfilling compatibility conditions for the continuity of the displacement and rotation of the beam and for the equilibrium of bending moment and shear at the discontinuity points.

Surface effects often play a significant role in the physical properties of micro- and nanosized materials and structures. Since the atoms within a very thin layer near surfaces experience a different local environment form that is experienced by atoms in bulk, the physical properties and mechanical response of surfaces will be distinct from those of bulk materials. Contrary to macroscopic structures, surface effects can strongly influence the stress and deformation properties of nanodevices. This latter is due to the increasing ratio between surface/interface area and volume. For instance, for a bulk deformation energy $U_b \approx Ed^3$ (E elastic stiffness) and $U_s \approx \gamma d^2$ (γ surface energy per area), we shall have $U_s > U_b$ with distance $d < \gamma/E$. In [10], it has been shown that with a modified Timoshenko beam model the dynamics of nanoscaled tubes with surface effects deviate considerably from results obtained with classic theories.

In some models, considering effective properties (D) of nanosized beams when compared with standard continuum mechanics (D_c), the relative error $(D - D_c)/D_c$ scale is found as $\alpha S/Ed$, where α is a geometrical constant depending upon the nanobeam considered, S is a surface elastic constant, E is a bulk elastic modulus, and d is a length defining the size of the structural element [11]. Here we shall consider free vibration of a nanoscale continuum beam models which incorporates surface energy [12–14]. The search for time exponential solutions leads to a quadratic and cubic eigenvalue problems associated with a second-order modal differential system and a singular third-order differential system, respectively.

Simulations have been performed for triple span beams with and without a patch in the first segment and with loads in the first and last segment with tip-sample interaction. These loads included a uniform pulse load and its second derivative associated with the moment due to a piezoelectric patch. It is observed how the inclusion of piezoelectric materials absorbs vibrations when compared with classical multispan beams. Also, the robustness of the impulse response method with varying parameters and its influence in the behavior of the responses are observed.

The size dependence effects in Timoshenko microbeams with surface effects (TMB) have been simulated for the nondimensional natural frequency and compared with those of the classical Timoshenko beam model (TB). It has been observed that for beam length on the order of nanometer to microns, the difference between natural frequencies is apparent, and by increasing the length of the microbeam, the results tend to Timoshenko classical theory; that is, the surface effects are significant only in nanoscale. This same behavior was observed in [14] for a microbeam simply supported.

This work involves a methodology that may be adequately adapted to study recent advances in micro-/nanosized structures that are intensively used to design advanced micro-/nanosensors and molecular transportation systems devices for various engineering and medical applications. These sensors due to their ultrahigh resonant frequencies are important in sensitive sensing, molecular transportation, molecular separation, high-frequency signal processing, and biological imaging. Some deposited processes in the formation of thin-film composite membranes can be similar to types of multispan vibrating beams. However, the complex configuration of materials used in real world devices requires the study of material properties stemming from their atomically thin layered structures. Chemical deposited processes in the manufacturing of thin films on semiconductors can involve low temperatures with silicon rich nitride. Conventional structural analysis methods assume ideal structures (free of irregularities) but material and/or geometrical variations in a structure may result in drastic changes in its dynamical behaviors. Diffusion- and reaction-controlled interfacial polymerization is an important and practical topic that is beyond our scope [15–19].

This paper is organized as follows. In Section 2, we consider the Euler-Bernoulli multispan beam model, their compatibility conditions, and the boundary conditions in a matrix form. In Section 3, the dynamic response of the matrix Euler-Bernoulli multispan beam model subject to tip-sample interactions and external forcing is given in terms of the distributed matrix impulse response. The case of a microcantilever with a piezoelectric layer is discussed in Section 4. In Section 5 are presented the results of a double-span beam, obtained by the expansion of the first segment of the triple piezoelectric beam. In Sections 6 and 7 is considered the eigenanalysis of Timoshenko beam models with surface effects and their comparison with the Timoshenko classical model, for micro- and nanoscale. In Section 8, we consider the multispan Timoshenko beam model. AFM-based nanoscale processing with continuum surrounding media such as that found in biology and nanomachining applications [20–22] suggests observing frequency effects that arise with an academic two span beam model in which one segment includes rotatory inertia and shear deformation and the other one neglects both effects.

2. Flexural Vibrations Using the Euler-Bernoulli Multispan Beam Model

We consider a multispan microcantilever of length L composed of N segments in which the displacement $v_j(t,x)$ in the segment $x_{j-1} \le x \le x_j$ is governed by the Euler-Bernoulli model

$$m_j \frac{\partial^2 v_j(t,x)}{\partial t^2} + \mathrm{k}_j \frac{\partial^4 v_j(t,x)}{\partial x^4} = f_j(t,x), \qquad (1)$$

where $m_j = \rho_j A_j$ is the mass per unit length, A_j is the cross section area, ρ_j is the mass density, $\mathrm{k}_j = E_j I_j$ is the flexural stiffness, and $f_j(t,x)$ is the transverse dynamic load at the $j-th$ station, $j = 1:N$. Here $x_0 = 0$ and $X_N = L$.

2.1. Matrix Formulation. The stepped Euler-Bernoulli model (1) can be written as a second-order differential equation

$$\mathsf{M}\ddot{\mathsf{v}} + \mathsf{K}\mathsf{v} = \mathsf{F}, \qquad (2)$$

with diagonal matrix coefficients

$$\mathsf{M} = \begin{pmatrix} \mathsf{M}_1 & & 0 \\ & \ddots & \\ 0 & & \mathsf{M}_N \end{pmatrix},$$
$$\mathsf{K} = \mathsf{EI}\frac{\partial^4}{\partial x^4} = \begin{pmatrix} \mathsf{EI}_1 & & 0 \\ & \ddots & \\ 0 & & \mathsf{EI}_N \end{pmatrix} \frac{\partial^4}{\partial x^4}, \qquad (3)$$

where

$$\mathsf{M}_j = \rho_j A_j,$$
$$\mathsf{EI}_j = E_j I_j. \qquad (4)$$

Here the displacement and the dynamic load $\mathsf{v}(t,x)$ and $\mathsf{F}(t,x)$, at the multispan beam, are given in vector form as

$$\mathsf{v}(t,x)=\begin{pmatrix} v_1(t,x)\\ v_2(t,x)\\ \vdots\\ v_N(t,x)\end{pmatrix},\qquad \mathsf{F}=\begin{pmatrix} f_1(t,x)\\ f_2(t,x)\\ \vdots\\ f_N(t,x)\end{pmatrix}. \tag{5}$$

2.2. Boundary and Internal Conditions for a Multispan Cantilever Beam. For a cantilever beam, the boundary conditions can be written in a compact matrix form

$$\mathbf{B}_0\mathsf{v}_0=\mathbf{0},\qquad \mathbf{B}_L\mathsf{v}_L=\mathbf{0}, \tag{6}$$

where

$$\mathbf{B}_0=\begin{pmatrix}1&0&0&0\\0&1&0&0\end{pmatrix},\qquad \mathbf{B}_L=\begin{pmatrix}0&0&1&0\\0&0&0&1\end{pmatrix}, \tag{7}$$

$$\mathsf{v}_0=\begin{pmatrix} v_1(t,0)\\ v_1'(t,0)\\ v_1''(t,0)\\ v_1'''(t,0)\end{pmatrix},\qquad \mathsf{v}_L=\begin{pmatrix} v_N(t,L)\\ v_N'(t,L)\\ v_N''(t,L)\\ v_N'''(t,L)\end{pmatrix}. \tag{8}$$

At the points x_i, $i=1,2,\ldots,N-1$, located between two consecutive segments, the compatibility conditions for the displacement, slope, bending moment, and shear lead to the compatibility matrix relationship

$$\mathscr{C}_{1,i}\mathsf{v}_i(t,x_i)=\mathscr{C}_{2,i}\mathsf{v}_{i+1}(t,x_i), \tag{9}$$

with

$$\mathscr{C}_{1,i}=\begin{pmatrix}1&0&0&0\\0&1&0&0\\0&0&1&0\\0&0&0&1\end{pmatrix},\qquad \mathscr{C}_{2,i}=\begin{pmatrix}1&0&0&0\\0&1&0&0\\0&0&\alpha_i&0\\0&0&0&\alpha_i\end{pmatrix},\qquad \mathsf{v}_i=\begin{pmatrix} v_i(t,x_i)\\ v_i'(t,x_i)\\ v_i''(t,x_i)\\ v_i'''(t,x_i)\end{pmatrix}, \tag{10}$$

being $\alpha_i=EI_{i+1}/EI_i$ for $i=1,\ldots,N-1$.

3. Forced Vibrations

The dynamic response of a forced multispan Euler-Bernoulli model $\mathsf{M}\ddot{\mathbf{v}}+\mathsf{K}\mathbf{v}=\mathsf{F}$ given in (2), subject to homogeneous boundary conditions, could be described in terms of the $N\times N$ diagonal matrix impulse response or initial-value Green function $\mathbf{h}(t,x,\xi)$ satisfying the initial-value problem

$$\begin{gathered}\mathsf{M}\ddot{\mathbf{h}}(t,x,\xi)+\mathsf{K}\mathbf{h}(t,x,\xi)=\mathbf{0},\\ 0<x<L,\ 0<\xi<L,\ t>0,\\ \mathbf{h}(0,x,\xi)=\mathbf{0},\\ \mathsf{M}\mathbf{h}_t(0,x,\xi)=\delta(x-\xi)\mathbf{I},\\ \mathbf{B}_0\mathsf{h}_0=\mathbf{0},\\ \mathbf{B}_L\mathsf{h}_L=\mathbf{0},\\ \mathscr{C}_{1,i}\mathsf{h}_i(t,x_i^-,\xi)=\mathscr{C}_{2,i}\mathsf{h}_i(t,x_i^+,\xi),\end{gathered} \tag{11}$$

where $\mathbf{I}$ is the $N\times N$ identity matrix and $\mathscr{C}_{1,i},\mathscr{C}_{2,i},\mathsf{h}_i$ are as given in (10), by using the i-th diagonal component of $\mathbf{h}$.

To work on the frequency domain, we need to introduce transfer function $\mathbf{H}(s,x,\xi)$ as being the Laplace transform of $\mathbf{h}(t,x,\xi)$ with respect to time (For inputs $e^{\lambda t}F(x)$ we seek outputs $e^{\lambda t}V(x)$ where $V(x)=\mathsf{H}F(x)$ with H defined in (17)). Thus

$$\begin{gathered}\left(s^2\mathsf{M}+\mathsf{K}\right)\mathbf{H}(s,x,\xi)=\delta(x-\xi)\mathbf{I},\quad 0<x,\xi<L,\\ \mathbf{B}_0\mathsf{H}_0=\mathbf{0},\\ \mathbf{B}_L\mathsf{H}_L=\mathbf{0},\\ \mathscr{C}_{1,i}\mathsf{H}_i(t,x_i^-,\xi)=\mathscr{C}_{2,i}\mathsf{H}_i(t,x_i^+,\xi).\end{gathered} \tag{12}$$

It turns out that $\mathbf{h}(t-\tau,x,\xi)$ acts as an integrating factor in Lagrange's adjoint method for the forced equation (2).

Multiplying (2) by $\mathbf{h}(t-\tau,x,\xi)$ and integrating by parts, we obtain the dynamic response

$$\mathsf{v}(t,x)=\int_0^t\int_0^L\left(\mathbf{h}_t(\tau,x,\xi)\,\mathsf{M}\mathsf{v}_o(\xi)+\mathbf{h}(\tau,x,\xi)\,\mathsf{M}\mathsf{v}_1(\xi)\right)d\xi d\tau+\int_0^t\int_0^L\mathbf{h}(t-\tau,x,\xi)\,\mathsf{F}(\tau,\xi)\,d\xi d\tau, \tag{13}$$

where $\mathsf{v}_o(x)=\mathsf{v}(0,x)$ and $\mathsf{v}_1(x)=\mathsf{v}_t(0,x)$ are given initial conditions.

The procedure mentioned above is also related to the Riemann function method for integrating partial differential equations. Dynamic responses have been considered in the field of control of distributed systems and in elastodynamics in connection with vibrations and cracking problems [23–26].

In practice, when computing the convolution integral, which corresponds to the forced response $\mathsf{v}(t,x)$, with null initial values $\mathsf{v}(0,x)=0,\mathsf{v}_t(0,x)=0$, we actually have $\mathsf{v}(t,x)=\mathsf{v}_h(t,x)+\mathsf{v}_p(t,x)$, where $\mathsf{v}_h(t,x)$ is a free vibration introduced by the system, whose initial values are a priori unknown, and $\mathsf{v}_p(t,x)$ is a particular response. When $\mathsf{v}_p(t,x)$ can be determined by other means, then those initial values can be supplied as $\mathsf{v}_h(0,x)=-\mathsf{v}_p(0,x)$ and $(\mathsf{v}_h)_t(0,x)=-(\mathsf{v}_p)_t(0,x)$ [27]. Thus, the induced system free response, due to a permanent response $\mathsf{v}_p(t,x)$, is given by

$$\mathsf{v}_h(t,x)=-\int_0^L\mathbf{h}(t,x,\xi)\,\mathsf{M}\dot{\mathsf{v}}_p(0,\xi)\,d\xi-\int_0^L\frac{\partial\mathbf{h}(t,x,\xi)}{\partial t}\mathsf{M}\mathsf{v}_p(0,\xi)\,d\xi. \tag{14}$$

3.1. Frequency Response. Harmonic and piecewise linear forcing are of interest in frequency analysis. When seeking a response of the same type, the transfer function is introduced. Given the harmonic input

$$f(t,x)=e^{i\omega t}v(x), \tag{15}$$

we have the harmonic output response

$$\mathsf{v}_p(t,x)=e^{i\omega t}\mathsf{H}(i\omega)\,v(x), \tag{16}$$

where

$$\mathsf{H}(i\omega)\,v(x)=\int_0^L\mathbf{H}(i\omega,x,\xi)\,v(\xi)\,d\xi. \tag{17}$$

In particular, for a concentrated force at a point $x=a$, of spatial amplitude $\mathsf{v}(x)=v(x)\delta(x-a)$, we have the permanent response

$$\mathsf{v}_p(t,x)=e^{i\omega t}\mathbf{H}(i\omega,x,a)\,v(a). \tag{18}$$

With the initial values $\mathsf{v}_p(0,\xi)=\mathbf{H}(i\omega,\xi,a)v(a)$, $\dot{\mathsf{v}}_p(0,\xi)=i\omega\mathsf{v}_p(0,\xi)$, the induced free response is given by

$$\mathsf{v}_h(t,x)=-\int_0^L\mathsf{r}(t,x,\xi,\omega)\,\mathsf{M}\mathbf{H}(i\omega,a,\xi)\,v(a)\,d\xi, \tag{19}$$

where

$$\mathsf{r}=\mathbf{h}_t(t,x,\xi)+i\omega\mathbf{h}(t,x,\xi). \tag{20}$$

For a pulse amplitude

$$\mathsf{v}(x)=\mathsf{v}_o(u(x-a)-u(x-b)), \tag{21}$$

where $u(x)$ is the unit step function, the permanent response turns out

$$\mathsf{v}_p(t,x)=e^{i\omega t}\int_a^b\mathbf{H}(i\omega,x,\xi)\,\mathsf{v}_o d\xi. \tag{22}$$

As before, by substituting the initial values in (14), the induced free response will now be

$$\mathsf{v}_h(t,x)=-\int_0^L\mathsf{r}(t,x,\xi,\omega)\,\mathsf{M}\mathbf{H}(i\omega,x,\xi)\,\mathsf{v}(\xi)\,d\xi, \tag{23}$$

with r given as in (20).

3.2. Closed Form of the Transfer Function for a Multispan Cantilever Beam. The Green function of a multispan Euler-Bernoulli can be determined by using an appropriate solution basis in each segment. From (12), $\mathbf{H}(s,x,\xi)$ is a diagonal matrix, with entries H_j that satisfy the differential equation

$$X^{(iv)}(x)-\varepsilon_j^4X(x)=\delta(x-\xi), \tag{24}$$

where

$$\varepsilon_j^4=-\frac{\rho_jA_js^2}{E_jI_j},\quad j=1,2,\cdots,N. \tag{25}$$

The functions $\{h(x-x_{j-1}),h'(x-x_{j-1}),h''(x-x_{j-1}),h'''(x-x_{j-1})\}$ constitute a solution basis of the equation $X^{(iv)}(x)-\varepsilon_j^4X(x)=0$. Here

$$h(x)=\frac{\sinh(\varepsilon_jx)-\sin(\varepsilon_jx)}{2\varepsilon_j^3} \tag{26}$$

is the solution of the initial-value problem

$$\begin{aligned} h^{(iv)}(x) - \varepsilon_j^4 h(x) &= 0, \\ h(0) &= 0, \\ h'(0) &= 0, \\ h''(0) &= 0, \\ h'''(0) &= 1. \end{aligned} \tag{27}$$

By denoting

$$\begin{aligned} h_0(x) = h'''(x) &= \frac{1}{2}\left(\cosh \varepsilon_j x + \cos \varepsilon_j x\right), \\ h_1(x) = h''(x) &= \frac{1}{2\varepsilon}\left(\sinh \varepsilon_j x + \sin \varepsilon_j x\right), \\ h_2(x) = h'(x) &= \frac{1}{2\varepsilon_j^2}\left(\cosh \varepsilon_j x - \cos \varepsilon_j x\right), \\ h_3(x) = h(x) &= \frac{1}{2\varepsilon_j^3}\left(\sinh \varepsilon_j x - \sin \varepsilon_j x\right), \end{aligned} \tag{28}$$

in each segment we have

$$\mathbf{H}(s,x,\xi) = \begin{pmatrix} H_0 & 0 & \cdots & 0 \\ 0 & H_1 & \cdots & 0 \\ \vdots & \vdots & \vdots & \vdots \\ 0 & 0 & \cdots & H_{N-1} \end{pmatrix}, \tag{29}$$

for $j = 0, .., N-1$

$$\begin{aligned} H_j(s,x,\xi) = {} & h_0\left(x - x_j\right) H\left(s, x_j, \xi\right) \\ & + h_1\left(x - x_j\right) H_x\left(s, x_j, \xi\right) \\ & + h_2\left(x - x_j\right) H_{xx}\left(s, x_j, \xi\right) \\ & + h_3\left(x - x_j\right) H_{xxx}\left(s, x_j, \xi\right) \\ & + h_3\left(x - x_j\right) u(x - \xi), \\ & \qquad x_j \le x \le x_{j+1}. \end{aligned} \tag{30}$$

In order to determine $\mathbf{H}(s,x,\xi)$, we need to find the initial-value vectors

$$\mathsf{H} = \begin{pmatrix} H\left(s, x_j, \xi\right) \\ H_x\left(s, x_j, \xi\right) \\ H_{xx}\left(s, x_j, \xi\right) \\ H_{xxx}\left(s, x_j, \xi\right) \end{pmatrix}. \tag{31}$$

By using the compatibility conditions

$$\mathscr{C}_{1,i}\mathsf{H}\left(s, x_j^-, \xi\right) = \mathscr{C}_{2,i}\mathsf{H}\left(s, x_j^+, \xi\right), \qquad j = 1, 2, \cdots, N-1, \tag{32}$$

given similarly to (9), and by deriving (30) with respect to x.

The above process will be carried out for the case of a two stepped cantilever beam with the intermediate discontinuity point $x_1 = l_1$.

The boundary conditions and the initial values of $h(x)$ lead to a simplification of H in the first and last segments. We have two null initial values at $x = 0$, and we can use the boundary conditions at $x = L$ as initial values, instead of those at x_{N-1}. This is possible by changing the solution basis at the last segment in (30) to the one generated by $h(L - x)$.

Thus, for $0 \le \xi \le L$

$$\begin{aligned} H(s,x,\xi) = {} & h_2(x) H_{xx}(s,0,\xi) + h_3(x) H_{xxx}(s,0,\xi) \\ & + h_3(x-\xi) u(x-\xi), \quad 0 \le x \le x_1, \\ H(s,x,\xi) = {} & h_0(L-x) H(s,L,\xi) \\ & + h_1(L-x) H_x(s,L,\xi) \\ & + h_3(\xi - x) u(\xi - x), \quad x_1 \le x \le L. \end{aligned} \tag{33}$$

By applying the compatibility conditions

$$\begin{aligned} H\left(s, l_1^-, \xi\right) &= H\left(s, l_1^+, \xi\right), \\ H_x\left(s, l_1^-, \xi\right) &= H_x\left(s, l_1^+, \xi\right), \\ E_1 I_1 H_{xx}\left(s, l_1^-, \xi\right) &= E_2 I_2 H_{xx}\left(s, l_1^+, \xi\right), \\ E_1 I_1 H_{xxx}\left(s, l_1^-, \xi\right) &= E_2 I_2 H_{xxx}\left(s, l_1^+, \xi\right), \end{aligned} \tag{34}$$

we get the algebraic system

$$\begin{pmatrix} h_2(l_1) & h_3(l_1) & -\left(h_0(L-x)\right)\big|_{x=l_1} & -\left(h_1(L-x)\right)\big|_{x=l_1} \\ h_2'(l_1) & h_3'(l_1) & -\left(h_0'(L-x)\right)\big|_{x=l_1} & -\left(h_1'(L-x)\right)\big|_{x=l_1} \\ h_2''(l_1) & h_3''(l_1) & -\left(\dfrac{E_2 I_2}{E_1 I_1} h_0''(L-x)\right)\bigg|_{x=l_1} & -\left(\dfrac{E_2 I_2}{E_1 I_1} h_1''(L-x)\right)\bigg|_{x=l_1} \\ h_2'''(l_1) & h_3'''(l_1) & -\left(\dfrac{E_2 I_2}{E_1 I_1} h_0'''(L-x)\right)\bigg|_{x=l_1} & -\left(\dfrac{E_2 I_2}{E_1 I_1} h_1'''(L-x)\right)\bigg|_{x=l_1} \end{pmatrix} \begin{pmatrix} H''(s,0,\xi) \\ H'''(s,0,\xi) \\ H(s,L,\xi) \\ H'(s,L,\xi) \end{pmatrix}$$

$$= \begin{pmatrix} h(l_1 - \xi)\, u(l_1 - \xi) - h(l_1 - \xi)\, u(l_1 - \xi) \\ h'(l_1 - \xi)\, u(l_1 - \xi) - h'(l_1 - \xi)\, u(l_1 - \xi) \\ \frac{E_2 I_2}{E_1 I_1} \left(h''(\xi - x)\right)\Big|_{x=l_1} u(\xi - l_1) - h''(l_1 - \xi)\, u(l_1 - \xi) \\ \frac{E_2 I_2}{E_1 I_1} \left(h'''(\xi - x)\right)\Big|_{x=l_1} u(\xi - l_1) - h'''(l_1 - \xi)\, u(l_1 - \xi) \end{pmatrix}, \quad (35)$$

for determining $H_j(s, x, \xi)$ from (30).

Remark. The case of axially loaded stepped beams has been investigated in [28] with nonclassical conditions. There, the Green function was obtained by working with a solution basis that considers hyperbolic and trigonometric functions in the general solution. These functions depend upon simple roots of a complete quadratic polynomial due to the inclusion of a second derivative spatial term in (24). When the axial load is removed, the solution basis reduces to the one considered for particular cases of vibrating beams. We observe that through a limit process the use of $h(x)$, which involves a division by root parameters, can also handle the case of repeated roots, which is of interest in the static case or critical frequencies.

4. A Cantilever Beam with a Piezoelectric Layer in AFM

Active microcantilever beams due to their structural flexibility and sensitivity to atomic and molecular forces have received increased attention in a variety of nanoscale sensing and measuring applications, including atomic force microscopy, thermal scanning microscopy, and biomass sensing.

In [9], a microcantilever model was proposed for studying atomic force microscope. It was formulated as a three stepped beam with a piezoelectric layer patch in the first segment and the other two segments were simple beams with different cross-sectional areas according to Figure 1(a).

The governing equations can be written as a system of three second-order partial differential equations with constant coefficients

$$\begin{aligned} m_1 \frac{\partial^2 u_1(t,x)}{\partial t^2} + (EI)_1 \frac{\partial^4 u_1(t,x)}{\partial x^4} &= \frac{\partial^2 M_p(t,x)}{\partial x^2}, \\ & \qquad 0 \le x \le l_1, \\ m_2 \frac{\partial^2 u_2(t,x)}{\partial t^2} + (EI)_2 \frac{\partial^4 u_2(t,x)}{\partial x^4} &= 0, \quad l_1 < x \le l_2, \\ m_3 \frac{\partial^2 u_3(t,x)}{\partial t^2} + (EI)_3 \frac{\partial^4 u_3(t,x)}{\partial x^4} &= 0 \quad l_2 < x \le L, \end{aligned} \quad (36)$$

subject to cantilever boundary conditions (7) and compatibility conditions (9) at the discontinuity points x_1, x_2. The parameters m_j and $(EI)_j$ for each segment of the beam are given by

$$\begin{aligned} m_1 &= (\rho A)_1 = \rho_p w_p t_p + \rho_b w_{b1} t_b, \\ m_2 &= (\rho A)_2 = \rho_b w_{b1} t_b, \\ m_3 &= (\rho A)_3 = \rho_b w_{b2} t_b, \end{aligned} \quad (37)$$

$$\begin{aligned} (EI)_1 &= E_p I_{pp} + E_b I_{pb} + E_b I_{b1}, \\ (EI)_2 &= E_b I_{b1}, \\ (EI)_3 &= E_b I_{b2}. \end{aligned} \quad (38)$$

The parameters ρ_p, ρ_b are the mass density of piezoelectric material and mass density of beam material; w_p, w_{b1}, and w_{b2} are the piezoelectric layer width, the width of the second, and third beam segments, respectively; t_p and t_b are the piezoelectric layer thickness and beam thickness. E_p and E_b are the Young modulus of piezoelectric layer material and Young modulus of second and third segment material. I_{b1}, I_{b2}, I_{pp}, I_{pb} are the beam moment of inertia, layer moment of inertia, and layer-beam moment of inertia and are calculated as $I_{b1} = w_{b1} t_b^3/12$, $I_{b2} = w_{b2} t_b^3/12$, $I_{pp} = z_n^2 t_p w_p + ((1/3)t_p^3 + (1/2)t_b t_p^2 + (1/4)t_b^2 t_p) w_p - z_n (t_p^2 + t_b t_p) w_p$; $I_{pb} = z_n^2 t_b w_{b1}$. And z_n is the neutral axis of the beam on the composite portion expressed by

$$z_n = \frac{1}{2} \frac{E_p t_p w_p (t_p + t_b)}{E_p t_p w_p + E_b t_b w_{b1}}. \quad (39)$$

The values of these parameters [9] are presented in Table 1.

The distributed cross-sectional moment $M_p(x,t) = M_{po}(t)(1 - u(x - l_1))$ in the first segment induces a forcing term. Its time amplitude is given by

$$M_{po}(t) = \frac{1}{2} w_p E_p d_{31} (t_b + t_p)\, v(t), \quad (40)$$

where d_{31} and $v(t)$ are the coefficients of the converse piezoelectric effect and the applied voltage, respectively.

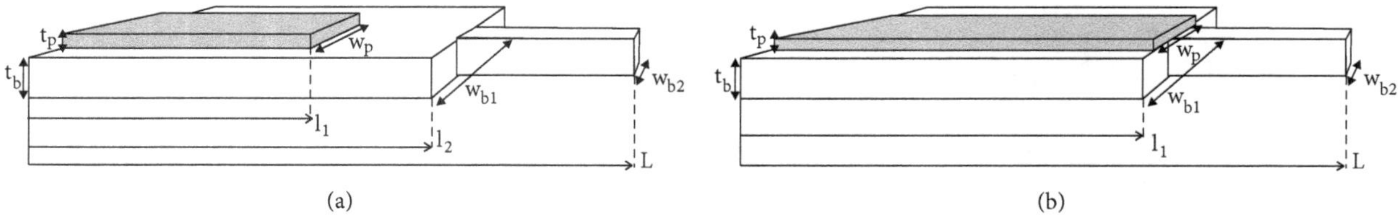

FIGURE 1: Schematic representation of the (a) triple span and (b) double-span piezoelectric beams.

4.1. Forced Responses. For homogeneous boundary conditions and null initial data, we have from (13) the forced response

$$\mathsf{v}(\mathsf{t},\mathsf{x}) = \int_0^{\mathsf{t}}\int_0^{\mathsf{L}} \mathbf{h}(\mathsf{t}-\tau,\mathsf{x},\xi)\,\mathsf{F}(\tau,\xi)\,d\xi d\tau, \qquad \mathsf{F}(\mathsf{t},\mathsf{x}) = \begin{pmatrix} \dfrac{\partial^2 M_p(t,x)}{\partial x^2} \\ 0 \\ 0 \end{pmatrix}, \tag{41}$$

where $\partial^2 M_p(\tau,\xi)/\partial\xi^2 = -M_{po}(\tau)\delta'(\xi - l_1)$. Thus

$$\begin{aligned} \mathsf{v}(\mathsf{t},\mathsf{x}) &= \int_0^t\int_0^L \mathbf{h}(t-\tau,x,\xi)\,\frac{\partial^2 M_p(\tau,\xi)}{\partial\xi^2}\,d\xi d\tau \\ &= \int_0^t M_{po}(\tau)\,\frac{\partial \mathbf{h}}{\partial\xi}(t-\tau,x,l_1)\,d\tau. \end{aligned} \tag{42}$$

By assuming a discrete complete spectrum of natural frequencies ω_j and the normal mode property, the Galerkin method can be used for determining approximate forced vibrations. We have the modal series

$$\mathbf{h}(t,x,\xi) = \sum_{n=1}^{\infty} X^{(n)}(x,\xi)\,\frac{\sin\omega_n t}{\omega_n}, \quad 0 \le x,\xi \le L, \tag{43}$$

and the spatial Green function

$$\mathbf{H}(t,x,\xi) = \sum_{n=1}^{\infty} X^{(n)}(x,\xi)\,\frac{1}{s^2+\omega_n^2}, \tag{44}$$

where

$$X^{(n)}(x,\xi) = \begin{pmatrix} X_1^{(n)}(x)X_1^{(n)}(\xi) & 0 & \cdots & 0 \\ 0 & X_2^{(n)}(x)X_2^{(n)}(\xi) & \cdots & 0 \\ \vdots & \vdots & \vdots & \vdots \\ 0 & 0 & \cdots & X_{N-1}^{(n)}(x)X_{N-1}^{(n)}(\xi) \end{pmatrix}. \tag{45}$$

Here $X(x)$

$$X(x) = \begin{pmatrix} X_1(x) \\ X_2(x) \\ \vdots \\ X_N(x) \end{pmatrix} \tag{46}$$

is the spatial amplitude of exponential solutions $\mathsf{v}(t,x) = e^{\lambda t}X(x)$ of the unforced Euler-Bernoulli model $\mathbf{M}\ddot{\mathbf{v}} + \mathbf{K}\mathbf{v} = 0$ that satisfy the boundary conditions (6), and the compatibility conditions (9) at the discontinuity points x_1 and x_2.

The modal approximation

$$\frac{\partial \mathbf{h}}{\partial\xi}(t-\tau,x,\xi) = \sum_{n=1}^{m}\frac{\partial X^{(n)}}{\partial\xi}(x,\xi)\,\frac{\sin\omega_n(t-\tau)}{\omega_n} \tag{47}$$

has been used in simulations with and without the inclusion of a piezoelectric layer. Expansions were truncated with a small number m of terms, usually between 5 and 10 [30].

In Table 2 are showed the first six natural frequencies obtained in this work for a triple span beam with piezoelectric patch and without it. The first three frequencies are compared with those obtained experimentally in [9]. There is an agreement with the methodology of this work and it can be observed that the elimination of piezoelectric layers diminishes the frequency values. In Figure 2 are shown the corresponding vibration modes; in the first segment the modes of piezoelectric beam have smaller amplitude.

The effects of the inclusion of a piezoelectric patch are appreciated by applying a harmonic rectangular pulse load in the first segment and in the third segment for a three stepped beam with and without piezoelectric layer. It is observed

TABLE 1: Parameters of piezoelectric multispan beam.

Parameter	Numeric Value	Unity
Length L	486×10^{-6}	m
Length l_1	325×10^{-6}	m
Length l_2	360×10^{-6}	m
Width w_{b1}	250×10^{-6}	m
Width w_{b2}	55×10^{-6}	m
Width w_p	130×10^{-6}	m
Thickness t_p	4×10^{-6}	m
Thickness t_b	4×10^{-6}	m
Density ρ_b	2330	Kg/m^3
Density ρ_p	6390	Kg/m^3
Young's modulus E_b	105	GPa
Young's modulus E_p	104	GPa

TABLE 2: Comparative natural frequencies between cases of double and triple span beams with and without piezoelectric patch on the first segment.

Case	Piezoelectric path	Natural frequencies ω_i (Hz)					
		ω_1	ω_2	ω_3	ω_4	ω_5	ω_6
Triple span [9]∗	with piezoelectric path	52.3	203.0	382.5	-	-	-
Triple span (Figure 1(a))	with piezoelectric path	50.6	209.2	393.7	910.7	1412.5	2016.7
Double span (Figure 1(b))	with piezoelectric path	46.2	222.9	376.2	875.3	1537.5	1948.6
Double/Triple span	without piezoelectric path	27.3	142.9	320.1	618.9	1073.0	1579.4

∗Experimentally obtained

that the spatial amplitude diminishes with the inclusion of a piezoelectric patch when the pulse is positioned in the piezoelectric layer segment; this effect does not appear when the pulse is positioned in the third segment (Figure 3).

In Figure 4 are compared the forced responses due to forcing terms only in the first segment. The first output tracks the forcing input $f_1(t,x) = M_{p0}(t)(1 - u(x - l_1))$. For the original forcing term proposed in [9] $f_1(t,x) = \partial^2 M_p(t,x)/\partial x^2 = -M_{p0}(t)\delta'(x - l_1)$, we can observe that the effect of doublet due to derivation of Heaviside function is to concentrate the output at the point $x = l_1$. Also, the amplitudes of forced response are smaller for three stepped beam with piezoelectric layer.

5. Double-Span Microbeam with Material and Geometric Discontinuities

We consider a double-span beam of length $L = 486 \times 10^{-6} m$, formed for two segments with different rectangular transversal section area, according to Figure 1(b). For simulation purposes, we consider the same parameters utilized in [9] for the three stepped beam with a piezoelectric patch in the first segment, and now for the double-span beam, obtained by removing the second segment of the first three stepped beam, which means extending the piezoelectric layer from l_1 to l_2. In our methodology, this amounts to consider the limit case when l_1 tends to l_2, so that the flexural stiffness $(EI)_1$ and mass density $(\rho A)_1$ will be valid on the first segment $[0, l_2]$, since l_1 was extended to l_2, and on the new second segment $[l_2, L]$ we will have $(EI)_3$ and $(\rho A)_3$, described in (37) and (38).

The first six natural frequencies of this beam, with and without piezoelectric patch, were given in the third and fourth rows of Table 2 and were obtained by the modal method. We observed that the reduction of three segments for two segments, maintaining the others material and geometric properties, influences the magnitude of the natural frequencies. When the piezoelectric patch is removed, then the frequencies also decrease. The results were the same as those of the case of three stepped beam when removing the piezoelectric patch.

In Figure 5, by using the methodology described from (24) to (35), is presented a 3D graphic of the Green function $H(s, x, \xi)$ of the two stepped piezoelectric beam, for the sixth natural frequency. The amplitude of this Green function has peaks at the same natural frequencies obtained by the modal method (Table 2, third row).

6. Surface Elasticity and Residual Surface Tension in the Timoshenko Beam Model

The Timoshenko beam model has been modified with the inclusion of the residual tension and surface elastic modulus [14]. By using the energy method with the elastic strain

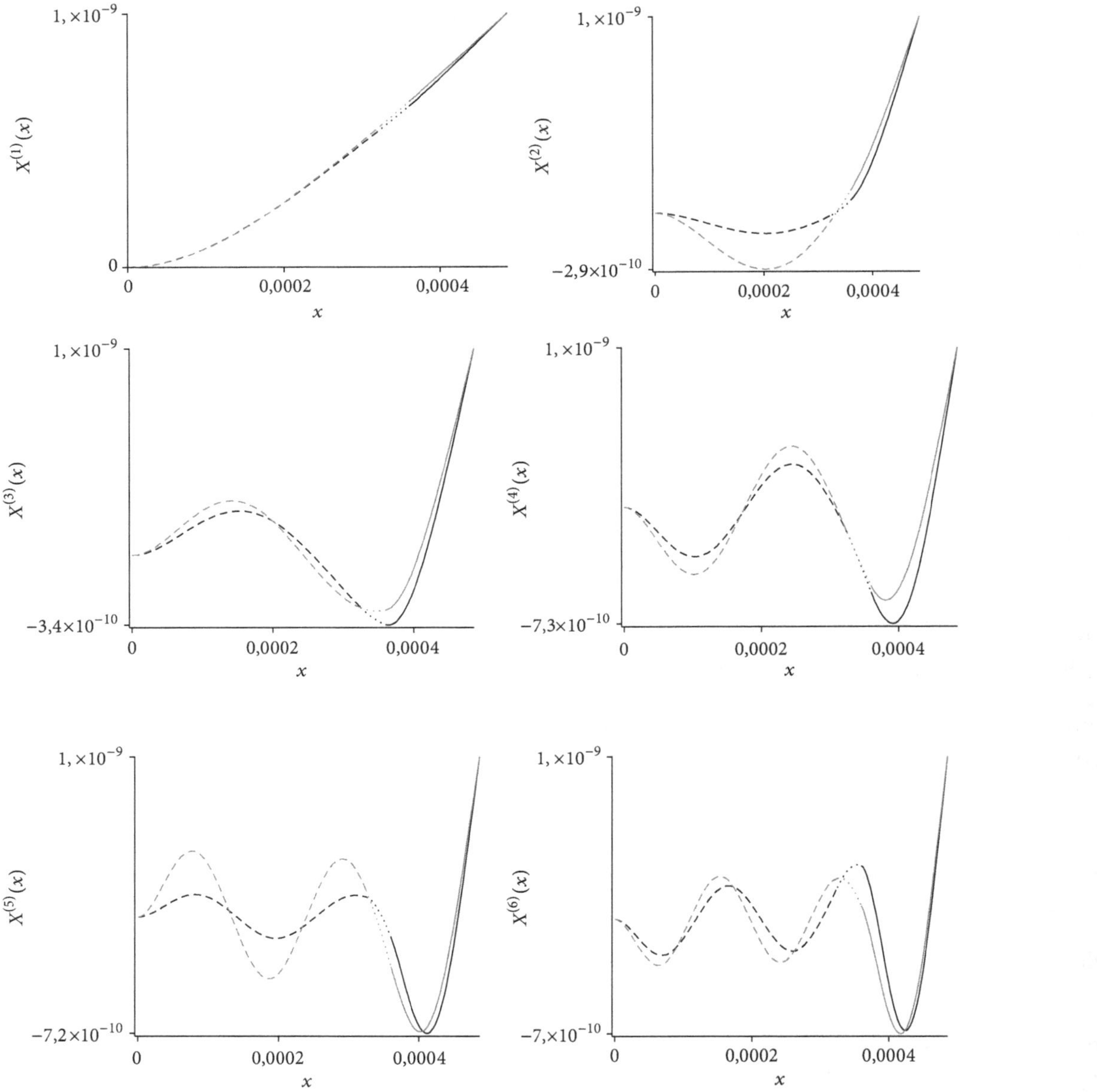

FIGURE 2: First six beam vibration modes: beam with piezoelectric layer (black line); beam (gray line) 1st segment (dash) $[0, l_1]$, $l_1 = 325 \times 10^{-6} m$; 2nd segment (dot) $[l_1, l_2]$, $l_2 = 360 \times 10^{-6} m$; 3rd segment (solid) $[l_2, L]$, $L = 486 \times 10^{-6} m$.

energy of surface induced in the potential energy and residual surface tensions, expressed by Laplace-Young equations, it was obtained the modified Timoskenko model

$$\begin{aligned} \rho A w_{tt} - \kappa G A w_{xx} + \overline{\kappa G A} \psi_x &= f(t, x), \\ \rho I \psi_{tt} - \overline{EI} \psi_{xx} - \kappa G A (w_x - \psi) &= q(t, x), \end{aligned} \tag{48}$$

for the flexural deflection $w(t, x)$ of the beam and the rotation angle $\psi(t, x)$ of the cross section of the beam. Here $f(t, x)$ denotes a transverse dynamic load and $q(t, x)$ a moment load. The values

$$\begin{aligned} \overline{\kappa G A} &= \kappa G A - (\tau_u + \tau_b) b, \\ \overline{EI} &= \left(EI + 2bh^2 E_s\right) \end{aligned} \tag{49}$$

are the effective curvature effect and flexural rigidity, respectively. Here τ_u and τ_b denote the upper and lower surfaces residual tensions and E_s is a surface elastic modulus. Without these later parameters, the model (48) that includes surface effects becomes the classical Timoshenko model. As usual, Young's modulus, moment of inertia of the cross section area, shear modulus, mass density area, and shear deformation

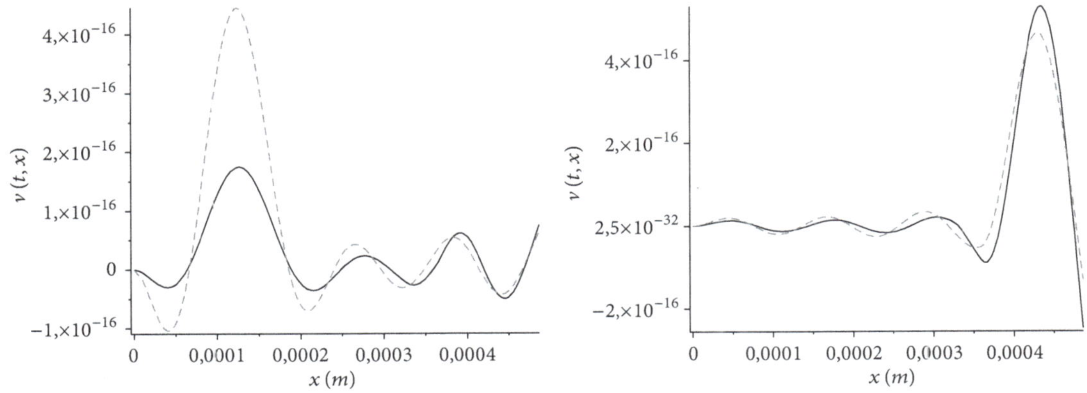

FIGURE 3: Left: forced response v(t, x) with fixed t for $f_1(x) = \sin(\omega t)(u(x-a) - u(x-b))$, $a = 100 \times 10^{-6} m$, and $b = 150 \times 10^{-6} m$. Right: forced response $v(t,x)$ with fixed t for $f_3(t,x) = \sin(\omega t)(u(x-c) - u(x-d))$, $c = 430 \times 10^{-6} m$, and $d = 440 \times 10^{-6} m$. Solid black line: beam with piezoelectric layer. Dashed gray line: beam without piezoelectric layer.

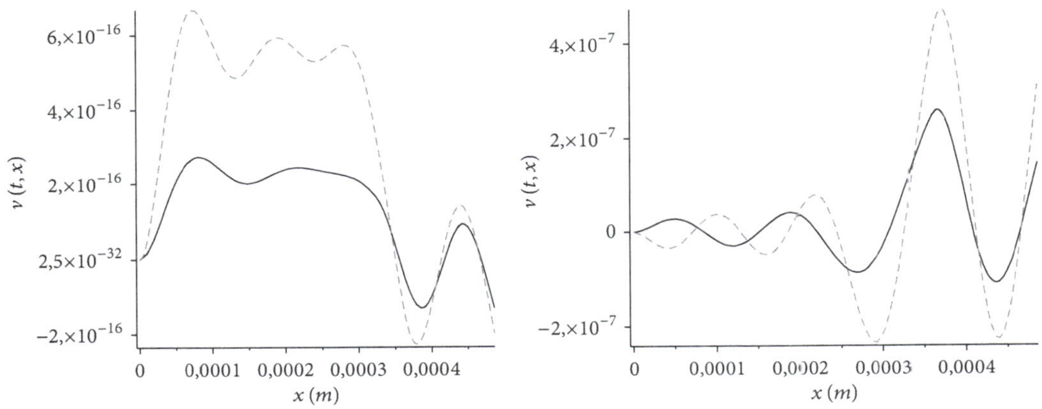

FIGURE 4: Left: forced response $v(t,x)$ for $f_1(t,x) = v(t)(1 - u(x - l_1))$. Right: forced response $v(t,x)$ for $f_1(t,x) = v(t)(\partial^2 M_p(x)/\partial x^2)$. Beam without piezoelectric layer (dashed line) and beam with piezoelectric layer (solid line), $v(t) = \sin(\omega t)$.

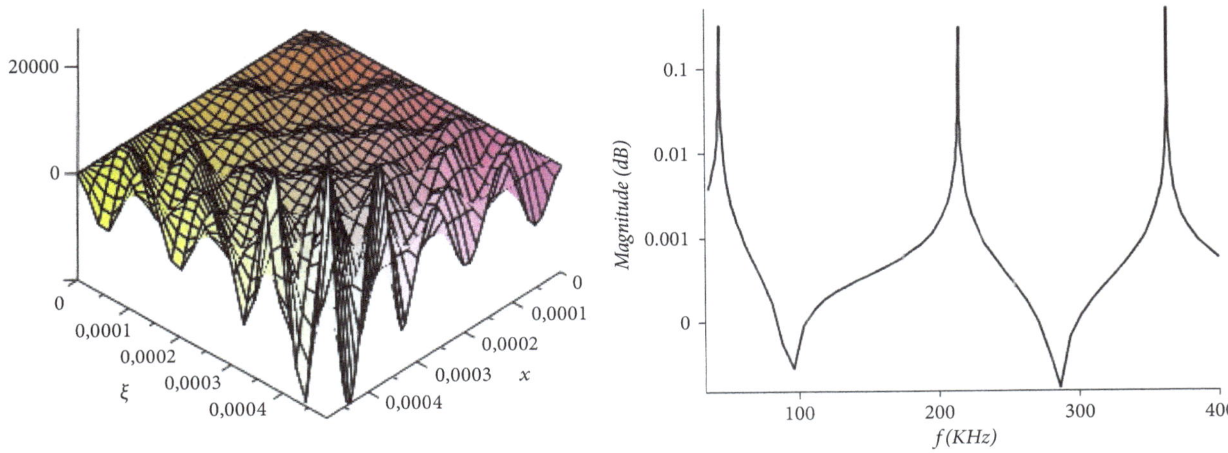

FIGURE 5: Left: frequency function $H(i\omega_6, x, \xi)$. Right: modal frequency response $H(i\omega, x_o, \xi_o)$ with the first three natural frequencies: 46.2KHz, 222.9KHz, and 376.2Khz at a given point.

where

$$d(x) = \frac{\delta \operatorname{senh}(\varepsilon x) - \varepsilon \operatorname{sen}(\delta x)}{ab_m \varepsilon \delta (\delta^2 + \varepsilon^2)} \quad (65)$$

is the solution of the initial-value problem

$$\begin{aligned} & ab_m d^{(iv)}(x) + \left(-ae\lambda^2 - c\lambda^2 b_m - a^2 + a_m a\right) d''(x) \\ & \quad + \left(c\lambda^2 a + c\lambda^4 e\right) d(x) = 0, \\ & d(0) = d'(0) = d''(0) = 0, \\ & ab_m d'''(0) = 1, \end{aligned} \quad (66)$$

with

$$\begin{aligned} \varepsilon &= \frac{1}{2}\sqrt{-2g^2 + 2\sqrt{g^4 + 4\overline{r^4}}}, \\ \delta &= \frac{1}{2}\sqrt{2g^2 + 2\sqrt{g^4 + 4\overline{r^4}}}, \end{aligned} \quad (67)$$

$$\begin{aligned} g^2 &= \overline{g_m^2} + \overline{s}^2, \\ \overline{g_m^2} &= -\left(\frac{e}{b_m} + \frac{c}{a}\right)\lambda^2, \\ \overline{s}^2 &= \frac{1}{b_m}(a_m - a), \\ \overline{r^4} &= -c\lambda^2\left(\frac{a + e\lambda^2}{ab_m}\right). \end{aligned} \quad (68)$$

Here

$$\begin{aligned} a &= \kappa GA, \\ c &= \rho A, \\ e &= \rho I, \\ a_m &= \overline{\kappa GA} = \kappa GA - (\tau_u + \tau_b) b, \\ b_m &= \overline{EI} = \left(EI + 2bh^2 E_s\right). \end{aligned} \quad (69)$$

From (62), the natural frequencies $\lambda = i\omega$ will satisfy

$$\begin{aligned} \Delta(i\omega) &= -\omega^4 ce\left(d'(L)\right)^2 \\ &\quad + \omega^2\left[acd(L)d''(L) - (ae + cb_m)d'(L)d'''(L)\right] \\ &\quad + \left(a^2 - aa_m\right)\left(d''(L)\right)^2 - ab_m\left(d'''(L)\right)^2 = 0. \end{aligned} \quad (70)$$

The fundamental matrix response $\mathbf{h}(x)$ is obtained in a similar way to the classical case [32]

$$\mathbf{h}(x) = \begin{pmatrix} \left(a + e\lambda^2\right)d(x) - b_m d''(x) & -a_m d'(x) \\ ad'(x) & -ad''(x) + c\lambda^2 d(x) \end{pmatrix}. \quad (71)$$

We observe from Figure 6 that for frequencies not above the critical frequency, the fundamental response $d(x)$ has lower amplitude when considering surface effects, while for higher frequencies the oscillatory behavior of $d(x)$ follows the classical one.

For a microcantilever beam described by the Timoshenko model with surface effects, the size dependence in the natural frequency of Timoshenko classical model and Timoshenko model including surface effects is illustrated in Figure 7. The solutions based on classical Timoshenko beam theory and Timoshenko beam theory including surface effects are denoted by TB and TMB, respectively. The fundamental natural frequencies are normalized to fundamental frequency of cantilever Euler-Bernoulli beam. In this figure are considered the parameters utilized in [14] for the same purposes. The parameters of surface elasticity and residual surface tension can be determined by molecular dynamics simulations or experiments. Residual surface stresses can be either positive or negative, depending on the crystallographic structure [33]. For an anodic alumina Al (Young's modulus $E = 70GPa$, Poisson's ratio $\nu = 0.3$ and $\rho = 2700kg/m^3$) are considered two types of crystallographic direction

$$\begin{aligned} Al[100]: E_s &= -7.9253N/m \text{ and} \\ \tau &= 0.5689N/m, \\ Al[111]: E_s &= 5.1882N/m \text{ and} \\ \tau &= 0.9108N/m. \end{aligned} \quad (72)$$

In Figure 7, the size dependence effects in the nondimensional natural frequency of TMB microbeams in comparison to the classical TB are illustrated. We can observe that for beam length on the order of nanometer to microns, the difference between natural frequencies is apparent and, by increasing the length of the microbeam, the results tend to Timoshenko classical theory; that is, the surface effects are significant only in nanoscale. This same behavior was observed in [14] for a microbeam simply supported.

7. Timoshenko Model with Surface Energy in Thick Nanobeams

In [13], the surface effects are included in the Timoshenko beam model following Gurtin-Murdoch continuum theory [34], in which is considered an elastic surface with zero thickness fully bonded to its bulk material. This elastic surface adds a set of specific constitutive equations relative to distinct material properties of the surface and surface energy effects. Let us denote an isotropic material beam, with rectangular transversal section, length L, width b, thickness

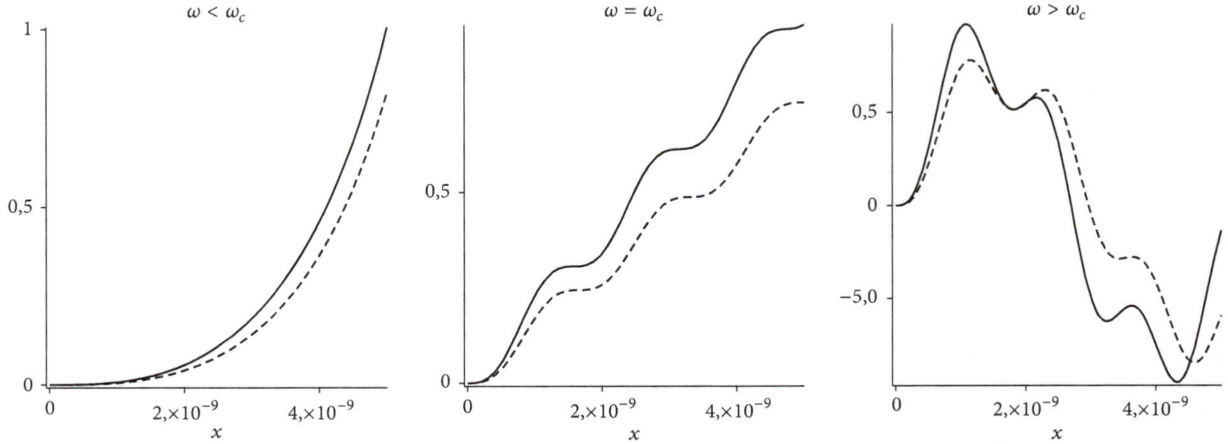

FIGURE 6: Behavior of the fundamental solution $d(x)$, according to critical natural frequency $\omega_c = \sqrt{a/e}$, using $L = 5 \times 10^{-9}$ and $Al[111]$ beam configuration, including surface effects (dashed line) and classical case (solid line).

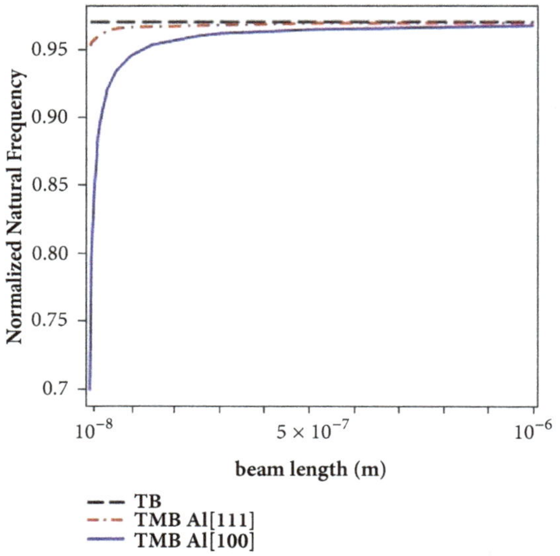

FIGURE 7: Influence of surface effects and size dependence on the normalized fundamental natural frequency of the microcantilever for 2h=0.2L, b=0.4L, and $\kappa = 5/6$.

$H = 2h$, moment of inertia $I = 2bh^3/3$, and $A = 2bh$ which is the transversal sectional area. The equations for the bulk are the same constitutive equations of classical theory of elasticity. The government equations of transverse and angular displacement, $w(t,x)$ and $\phi(t,x)$, respectively, are then given by

$$\begin{gathered}(\rho A + \rho_0 s^*) w_{tt} - (\kappa GA + \tau_0 s^*) w_{xx} + \kappa GA\phi_x = 0, \\ (\rho I + \rho_0 I^*) \phi_{tt} - (EI + (2\mu_0 + \lambda_0) I^*) \phi_{xx} \\ - \frac{2\nu I\rho_0}{H} w_{xtt} + \frac{2\nu I\tau_0}{H} w_{xxx} - \kappa GA (w_x - \phi) = 0.\end{gathered} \tag{73}$$

In a compact notation we have

$$\mathbf{M}\ddot{\mathrm{v}} + \mathbf{K}\mathrm{v} = 0, \tag{74}$$

where

$$\begin{gathered}\mathrm{v} = \begin{pmatrix} w \\ \phi \end{pmatrix}, \\ \mathbf{M} = M_0 + M_1 \frac{\partial}{\partial x}, \\ M_0 = \begin{pmatrix} c_m & 0 \\ 0 & e_m \end{pmatrix}, \\ M_1 = \begin{pmatrix} 0 & 0 \\ -I^0 & 0 \end{pmatrix},\end{gathered} \tag{75}$$

and $\mathbf{K}$ is a sum of matrix differential operators

$$\mathbf{K} = K_3 \frac{\partial^3}{\partial x^3} + K_2 \frac{\partial^2}{\partial x^2} + K_1 \frac{\partial}{\partial x} + K_0, \tag{76}$$

$$\begin{gathered}K_3 = \begin{pmatrix} 0 & 0 \\ I^1 & 0 \end{pmatrix}, \\ K_2 = \begin{pmatrix} -a_m & 0 \\ 0 & -b_m \end{pmatrix}, \\ K_1 = \begin{pmatrix} 0 & a \\ -a & 0 \end{pmatrix}, \\ K_0 = \begin{pmatrix} 0 & 0 \\ 0 & a \end{pmatrix},\end{gathered} \tag{77}$$

with $c_m = \rho A + \rho_0 s^*, a_m = \kappa GA + \tau_0 s^*, a = \kappa GA, e_m = \rho I + \rho_0 I^*, I^0 = 2\nu I\rho_0/H, I^1 = 2\nu I\tau_0/H$.

The material parameters of bulk are E, ν, and ρ, which represent Young's modulus, Poisson's ratio, and mass density, respectively. Also, we have beam surface parameters $\lambda_0, \mu_0, \tau_0, \rho_0$, Lamé constants, surface residual stress, and mass density of surface. I^* is the moment of inertia relative to the surface

$I^* = 2bh^2 + 4h^3/3$, and for rectangular transversal section $s^* = 2b$. κ is the shear correction coefficient.

Rewriting (74), we have

$$\begin{pmatrix} c_m \frac{\partial^2}{\partial t^2} - a_m \frac{\partial^2}{\partial x^2} & a\frac{\partial}{\partial x} \\ -I^0 \frac{\partial^3}{\partial t^2 \partial x} + I^1 \frac{\partial^3}{\partial x^3} - a\frac{\partial}{\partial x} & e_m \frac{\partial^2}{\partial t^2} - b_m \frac{\partial^2}{\partial x^2} + a \end{pmatrix} \mathrm{v} = 0. \tag{78}$$

Considering exponential solutions in the form $\mathrm{v} = e^{\lambda t + \beta x}\mathbf{w}$, then

$$\mathbf{L}\mathrm{v} = 0 \tag{79}$$

where

$$\mathbf{L} = \begin{pmatrix} c_m\lambda^2 - a_m\beta^2 & a\beta \\ -I^0\lambda^2\beta + I^1\beta^3 - a\beta & e_m\lambda^2 - b_m\beta^2 + a \end{pmatrix}. \tag{80}$$

In order to guarantee $\mathbf{w}$ nonzero, we must have $\det(\mathbf{L}) = 0$. Thus we obtain the characteristic equation

$$\begin{aligned} \Delta(\beta,\lambda) &= c_m e_m \lambda^4 \\ &\quad + \left[\left(aI^0 - b_m c_m - a_m e_m\right)\beta^2 + ac_m\right]\lambda^2 \\ &\quad + \left(a^2 - aa_m\right)\beta^2 + \left(a_m b_m - aI^1\right)\beta^4 = 0. \end{aligned} \tag{81}$$

7.1. Singular Eigenvalue Problem. By assuming exponential solutions $\mathrm{v}(t,x) = e^{\lambda t}\mathbf{w}(x)$, $\mathbf{w}(x) = \binom{W(x)}{\Psi(x)}$ in (74), it follows the eigenvalue problem

$$\left(\lambda^2\mathbf{M} + \mathbf{K}\right)\mathbf{w}(x) = \mathbf{0}, \tag{82}$$

which can be written as a singular third-order differential matrix equation

$$\begin{aligned} &\mathbb{R}\mathbf{w}'''(x) + \mathbb{M}\mathbf{w}''(x) + \mathbb{C}(\lambda)\mathbf{w}'(x) + \mathbb{K}(\lambda)\mathbf{w}(x) \\ &\quad = \mathbf{0}, \end{aligned} \tag{83}$$

with

$$\begin{aligned} \mathbb{R} &= \begin{pmatrix} 0 & 0 \\ I^1 & 0 \end{pmatrix}, \\ \mathbb{M} &= \begin{pmatrix} -a_m & 0 \\ 0 & -b_m \end{pmatrix}, \\ \mathbb{C}(\lambda) &= \begin{pmatrix} 0 & a \\ -\left(\lambda^2 I^0 + a\right) & 0 \end{pmatrix}, \\ \mathbb{K}(\lambda) &= \begin{pmatrix} \lambda^2 c_m & 0 \\ 0 & \lambda^2 e_m + a \end{pmatrix}. \end{aligned} \tag{84}$$

We observe that the nature of the matrix $\mathbb{R}$ characterizes the problem (83) as being singular. Following the spectral methodology, we seek exponential solutions $\mathbf{w}(x) = e^{\beta x}\mathrm{v}$, $\mathrm{v} \neq 0$, of (83). It follows that

$$\Gamma(\beta,\lambda)\mathrm{v} = \left(\beta^3\mathbb{R} + \beta^2\mathbb{M} + \beta\mathbb{C}(\lambda) + \mathbb{K}(\lambda)\right)\mathrm{v} = \mathbf{0}, \tag{85}$$

where

$$\Gamma(\beta,\lambda) = \begin{pmatrix} -\beta^2 a_m c_m \lambda^2 & \beta a \\ \beta^3 I^1 - \beta\left(I^0\lambda^2 + a\right) & -\beta^2 b_m + e_m\lambda^2 + a \end{pmatrix}. \tag{86}$$

For each λ, non-zero solutions v can be obtained whenever β is a root of the characteristic polynomial

$$P(\beta,\lambda) = \det\left(\beta^3\mathbb{R} + \beta^2\mathbb{M} + \beta\mathbb{C}(\lambda) + \mathbb{K}(\lambda)\right). \tag{87}$$

The characteristic equation

$$\Delta(\beta,\lambda) = P(\beta,\lambda) = \beta^4 + g^2(\lambda)\beta^2 - r^4(\lambda) = 0, \tag{88}$$

with

$$\begin{aligned} g^2(\lambda) &= \frac{b_2}{b_0} = \frac{\left(I^0 a - b_m c_m - a_m e_m\right)\lambda^2 + a^2 - aa_m}{\left(a_m b_m - aI^1\right)}, \\ r^4(\lambda) &= -\frac{b_4}{b_0} = -\frac{\lambda^2\left(ac_m + c_m e_m \lambda^2\right)}{\left(a_m b_m - aI^1\right)} \end{aligned} \tag{89}$$

has the roots $\beta_{1,2} = \pm\varepsilon$ and $\beta_{3,4} = \pm i\delta$, where

$$\begin{aligned} \varepsilon &= \sqrt{\frac{-g^2(\lambda) + \sqrt{\left(g^2(\lambda)\right)^2 + 4r^4(\lambda)}}{2}}, \\ \delta &= \sqrt{\frac{g^2(\lambda) + \sqrt{\left(g^2(\lambda)\right)^2 + 4r^4(\lambda)}}{2}}. \end{aligned} \tag{90}$$

The case of simple roots can be characterized from the condition $P' = 2\beta^3 + g^2(\lambda)$, $\beta \neq 0$. This excludes the possibility of $\beta = 0$ to be a simple root. Moreover, we can determine the eigenvectors satisfying (85). We summarize these results in the following.

Lemma 1. *β is a simple root of $P(\beta,\lambda)$ if and only if $\beta \neq 0$ and $\beta^2 \neq -g^2(\lambda)/2$. For simple roots, the eigenvectors satisfying (85) are given by*

$$\mathrm{v}_j = \begin{pmatrix} v_{1j} \\ \left(\dfrac{\beta_j^2 a_m - \lambda^2 c_m}{\beta_j a}\right) v_{1j} \end{pmatrix}, \tag{91}$$

where v_{1j} is an arbitrary constant.

For the case of distinct roots β, the general solution

$$\mathbf{w}(x) = c_1 e^{\beta_1 t}\mathrm{v}_1 + c_2 e^{\beta_2 t}\mathrm{v}_2 + c_3 e^{\beta_2 t}\mathrm{v}_3 + c_4 e^{\beta_4 t}\mathrm{v}_4$$

$$= \begin{pmatrix} c_1 & c_2 & c_3 & c_4 \\ c_1\left(\dfrac{\varepsilon^2 a_m - \lambda^2 c_m}{a\varepsilon}\right) & c_2\left(\dfrac{\varepsilon^2 a_m - \lambda^2 c_m}{-a\varepsilon}\right) & \dfrac{c_3}{i}\left(\dfrac{-\delta^2 a_m - \lambda^2 c_m}{a\delta}\right) & \dfrac{c_4}{i}\left(\dfrac{-\delta^2 a_m - \lambda^2 c_m}{-a\delta}\right)\end{pmatrix}\begin{pmatrix} e^{\varepsilon x} \\ e^{-\varepsilon x} \\ e^{\delta i x} \\ e^{-\delta i x}\end{pmatrix}, \quad (92)$$

and can be also presented in the form

$$\mathbf{w}(x) = \begin{pmatrix} C_1 & C_2 & C_3 & C_4 \\ C_2 K_a & C_1 K_a & C_4 K_b & -C_3 K_b \end{pmatrix}\begin{pmatrix}\cosh(\varepsilon x) \\ \sinh(\varepsilon x) \\ \cos(\delta x) \\ \sin(\delta x)\end{pmatrix}, \quad (93)$$

where $C_1 = c_1 + c_2$, $C_2 = c_1 - c_2$, $C_3 = c_3 + c_4$, $C_4 = (c_3 - c_4)i$ *and*

$$K_a = \frac{a_m \varepsilon^2 - \lambda^2 c_m}{a\varepsilon}, \qquad K_b = \frac{a_m \delta^2 - \lambda^2 c_m}{a\delta}. \quad (94)$$

The above representation of $\mathbf{w}(\mathbf{x})$ in terms of hyperbolic and trigonometric functions follows by introducing in (92) the basic change

$$\begin{pmatrix} e^{\varepsilon x} \\ e^{-\varepsilon x} \\ e^{i\delta x} \\ e^{-i\delta x}\end{pmatrix} = \begin{pmatrix} 1 & 1 & 0 & 0 \\ 1 & -1 & 0 & 0 \\ 0 & 0 & 1 & i \\ 0 & 0 & 1 & -i \end{pmatrix}\begin{pmatrix}\cosh(\varepsilon x) \\ \sinh(\varepsilon x) \\ \cos(\delta x) \\ \sin(\delta x)\end{pmatrix}. \quad (95)$$

7.1.1. Repeated Roots Analysis. From the given characterization of simple roots, it is clear that the repeated roots can be only $\beta = 0$ and $\beta^2 = -g^2(\lambda)/2$. The conditions $P''' = 12\beta = 0$ and $P^{(iv)} = 12$ exclude the possibility of having nonzero third roots. In such a case, $\beta = 0$ will be a quadruple root. More precisely, we have the following.

Lemma 2. (1) *If $\beta = 0$ is a root of $P(\beta, \lambda)$, then it is a double root if $g^2(\lambda) \neq 0$, or quadruple if $g^2(\lambda) = 0$.*

(2) If $\beta = \sqrt{-g^2(\lambda)/2}$ and $\beta = -\sqrt{-g^2(\lambda)/2}$ are roots, with $g^2(\lambda) \neq 0$, then they will be double roots.

(3) The characteristic polynomial $P(\beta, \lambda)$ does not have triple roots.

From the above lemmas, critical frequencies values will arise for natural frequencies $\lambda = i\omega$.

(i) $\beta = 0$ is a root of $P(\beta, i\omega)$, if and only if $r^4(\omega) = 0$. That is, $r^4(i\omega) = 0$ for $\omega = 0$ or $\omega = \omega_c$, with $\omega_c = \sqrt{a/em}$.

(ii) $\beta = \pm\sqrt{-g^2(\omega)/2}$, $g^2(\omega) \neq 0$, is a root of $P(\beta, i\omega)$ if and only if $(g^2(i\omega))^2 + 4r^4(\omega) = 0$, with $r^4(\omega) \neq 0$. We have $g^2(\omega) = 0$ when $\omega = \omega_g = \sqrt{(a^2 - aa_m)/(b_m c_m + a_m e_m - I^0 a)}$.

7.2. Frequency Equations. Frequency equations will arise when applying boundary conditions as given below.

7.2.1. Simply Supported Beam. In this case we have the following boundary conditions, at $x = 0$

$$w(0) = 0, \qquad \omega^2 I^0 w(0) - b_m \phi'(0) + I^1 w''(0) = 0, \quad (96)$$

and at $x = L$

$$w(L) = 0, \qquad \omega^2 I^0 w(L) - b_m \phi'(L) + I^1 w''(L) = 0. \quad (97)$$

Using the first boundary condition at $x = 0$, we obtain $C_2 + C_4 = 0 \Longrightarrow C_4 = -C_2$. From the second boundary condition we have

$$I^1 w''(0) + b_m \phi'(0) = 0, \qquad -C_2 I^1\left(\delta^2 + \varepsilon^2\right) + C_2 b_m\left(K_a \delta + K_b \varepsilon\right) = 0, \quad (98)$$

replacing K_a and K_b and using that $a_m b_m - I^1 a > 0$ [13], $\varepsilon^2 + \delta^2 \neq 0$, results in $C_2 = C_4 = 0$.

From the boundary conditions at $x = L$, written in matrix form

$$\begin{pmatrix} 1 & 0 & 0 \\ \omega^2 I^0 & I^1 & -b_m \end{pmatrix}\begin{pmatrix} W(L) \\ W''(L) \\ \psi'(L)\end{pmatrix} = \mathbf{0} \quad (99)$$

the system $\mathbf{Uc} = 0$ is obtained, described by

$$\begin{pmatrix} 1 & 0 & 0 \\ \omega^2 I^0 & I^1 & -b_m \end{pmatrix}\begin{pmatrix} \sin(\delta L) & \sinh(\varepsilon L) \\ -\sin(\delta L)\,\delta^2 & \varepsilon^2 \sinh(\varepsilon L) \\ -K_a \sin(\delta L)\,\delta & \varepsilon K_b \sinh(\varepsilon L)\end{pmatrix}\begin{pmatrix} C_1 \\ C_3 \end{pmatrix} = 0. \quad (100)$$

Since $b_m a_m - I^1 a > 0$ and $\varepsilon^2 + \delta^2 \neq 0$, the frequency equation

$$\begin{aligned}\Delta(\omega) &= \det(\mathbf{U}) \\ &= \frac{\left(I^1 a - b_m a_m\right)\left(\varepsilon^2 + \delta^2\right)}{a} \sinh(\varepsilon L)\sin(\delta L) \\ &= 0\end{aligned} \quad (101)$$

reduces to

$$\sinh(\varepsilon L)\sin(\delta L) = 0. \quad (102)$$

7.2.2. Cantilever Beam. In this case the boundary conditions are given by

$$\begin{aligned} w(0) &= 0 \\ \psi(0) &= 0, \\ a_m w'(L) - a\psi(L) &= 0 \\ -b_m\psi'(L) + I^1 w''(L) + I^0\omega^2 w(L) &= 0. \end{aligned} \quad (103)$$

Similarly to the previous case, by using the boundary conditions at $x = 0$, we obtain

$$\begin{aligned} W(x) &= C_1\left(\sin(\delta x) - \frac{K_a}{K_b}\sinh(\varepsilon x)\right) \\ &\quad + C_2(\cos(\delta x) - \cosh(\varepsilon x)), \\ \Psi(x) &= C_1 K_a(\cos(\delta x) - \cos(\varepsilon x)) \\ &\quad - C_2(K_a\sin(\delta x) + K_b\sinh(\varepsilon x)), \end{aligned} \quad (104)$$

The boundary conditions in matrix form at $x = L$

$$\begin{pmatrix} 0 & a_m & 0 & -a & 0 \\ \omega^2 I^0 & 0 & I^1 & 0 & -b_m \end{pmatrix}\begin{pmatrix} W(L) \\ W'(L) \\ W''(L) \\ \psi(L) \\ \psi'(L)\end{pmatrix}\begin{pmatrix} C_1 \\ C_2\end{pmatrix} = 0 \quad (105)$$

lead to a system $\mathbf{Uc} = 0$ from which we obtain the characteristic equation.

In Table 3 and Figure 8, we present numerical results about natural frequencies and mode shapes. Both cases of supported-supported (S-S) and cantilever (C) boundary conditions are considered. The simulation was done with a silicon beam having the following material and surface parameters [13]:

$$\begin{aligned} E &= 107GPa, \\ \nu &= 0.33, \\ \rho &= 2.33\times 10^3 kg/m^3, \\ \rho_0 &= 3.17\times 10^{-7} kg/m^2, \\ \mu_0 &= -2.7779N/m, \\ \lambda_0 &= -4.4939N/m, \\ \tau_0 &= 0.6056N/m. \end{aligned} \quad (106)$$

Comments

(i) From Table 3 and Figure 8, it is observed that the results agree with those found in the literature [13, 14].

(ii) When the beam length increases, from nano- to microscale, the results tend to the classical results.

(iii) Natural frequencies can increase or decrease compared to classical results, depending on elastic surface constants signs and boundary conditions (cantilever beam is more influenceable than simply supported beam).

(iv) According to Figure 8, the mode shape component $W(x)$ of a simply supported beam with the inclusion of surface effects is the same of a classical beam. Perceptible differences occur for the cantilever beam case, principally for the first one mode shape, and just in the nanoscale.

8. Multispan Timoshenko Beams

The same formulation, which was presented in Section 2 for Euler-Bernoulli multispan beams, can be used with the Timoshenko multispan beam model. We replace the scalar components that appear in the block matrices defined after (1), by the 2×2 matrix components M_j and K_j

$$\begin{aligned} \mathsf{M}_j &= \begin{pmatrix} \rho_j A_j & 0 \\ 0 & \rho_j I_j \end{pmatrix}, \\ \mathsf{K}_j &= \begin{pmatrix} -\kappa_j G_j A_j \dfrac{\partial^2}{\partial x^2} & \kappa_j G_j A_j \dfrac{\partial}{\partial x} \\ -\kappa_j G_j A_j \dfrac{\partial}{\partial x} & -E_j I_j \dfrac{\partial^2}{\partial x^2} + \kappa_j G_j A_j \end{pmatrix}. \end{aligned} \quad (107)$$

The components v_j become 2×1 vectors, $v_j = \begin{pmatrix} w_j(t,x) \\ \phi_j(t,x)\end{pmatrix}$, where $w_j(t,x)$ is the flexural displacement and $\phi_j(t,x)$ is the angular displacement. Similarly, the components of F_j are external loads and moments, respectively, f_j and q_j. Then

TABLE 3: Comparison between natural frequencies of Timoshenko beam models including and not including surface effects, considering nano- and microscale, of a Silicon beam. Nanobeam geometric parameters: $L = 50 \times 10^{-9} m$, $H = 2h = 6 \times 10^{-9} m$, and $b = 3 \times 10^{-9} m$. Microbeam geometric parameters: $L = 50 \times 10^{-6} m$, $H = 2h = 6 \times 10^{-6} m$, and $b = 3 \times 10^{-6} m$.

BC	Model	Nano Natural frequencies ω_i(GHz)				Micro Natural frequencies ω_i(MHz)			
		ω_1	ω_2	ω_3	ω_4	ω_1	ω_2	ω_3	ω_4
S-S	TMB	7.0808	25.063	51.327	82.922	7.1996	27.010	55.669	89.756
	TB	7.1998	27.012	55.673	89.763	7.1998	27.012	55.673	89.763
	TMB [13]	7.08	25.07	51.33	82.92	-	-	-	-
	TB [13]	7.20	27.02	55.70	89.76	-	-	-	-
C	TMB	3.0168	14.674	36.484	64.514	2.5976	15.273	39.297	69.725
	TB	2.597	15.273	39.300	69.730	2.5971	15.273	39.300	69.730
	TMB [13]	3.02	14.67	36.38	64.46	-	-	-	-
	TB [13]	2.60	15.28	39.31	69.74	-	-	-	-

$$\mathsf{v} = \begin{pmatrix} w_1 \\ \phi_1 \\ w_2 \\ \phi_2 \\ \vdots \\ w_N \\ \phi_N \end{pmatrix}, \quad \mathsf{F} = \begin{pmatrix} f_1 \\ q_1 \\ f_2 \\ q_2 \\ \vdots \\ f_N \\ q_N \end{pmatrix}. \tag{108}$$

General boundary conditions can be written as

$$\begin{aligned} \mathbf{B}_0 \mathbf{w}_1 (t, 0) &= \mathbf{n}_0, \\ \mathbf{B}_L \mathbf{w}_N (t, L) &= \mathbf{n}_L, \end{aligned} \tag{109}$$

where

$$\begin{aligned} \mathbf{B}_0 &= \begin{pmatrix} a_{11} & b_{11} & c_{11} & d_{11} \\ a_{12} & b_{12} & c_{12} & d_{12} \end{pmatrix}, \\ \mathbf{n}_0 &= \begin{pmatrix} n_1 \\ n_2 \end{pmatrix}, \\ \mathbf{B}_L &= \begin{pmatrix} a_{21} & b_{21} & c_{21} & d_{21} \\ a_{22} & b_{22} & c_{22} & d_{22} \end{pmatrix}, \\ \mathbf{n}_L &= \begin{pmatrix} n_3 \\ n_4 \end{pmatrix}, \end{aligned} \tag{110}$$

where n_j corresponds to devices or interaction of extremities with the external medium, when they occur, and

$$\mathbf{w}_j (t, x) = \begin{pmatrix} w_j (t, x) \\ \phi_j (t, x) \\ w'_j (t, x) \\ \phi'_j (t, x) \end{pmatrix}, \quad j = 1 : N. \tag{111}$$

The continuity conditions for displacement, gyro, flexural moment, and shear force in the transversal section discontinuity points $x = x_j$, $j = 1 : N - 1$, including intermediate devices can be written as

$$\mathscr{C}_{1,i} \mathbf{w}_i (t, x_i) = \mathscr{C}_{2,i} \mathbf{w}_{i+1} (t, x_i), \quad i = 1 : N - 1, \tag{112}$$

where

$$\begin{aligned} \mathscr{C}_{1,i} &= \begin{pmatrix} e_{11}^{(i)} & f_{11}^{(i)} & g_{11}^{(i)} & h_{11}^{(i)} \\ e_{21}^{(i)} & f_{21}^{(i)} & g_{21}^{(i)} & h_{21}^{(i)} \\ e_{31}^{(i)} & f_{31}^{(i)} & g_{31}^{(i)} & h_{31}^{(i)} \\ e_{41}^{(i)} & f_{41}^{(i)} & g_{41}^{(i)} & h_{41}^{(i)} \end{pmatrix}, \\ \mathscr{C}_{2,i} &= \begin{pmatrix} e_{12}^{(i)} & f_{12}^{(i)} & g_{12}^{(i)} & h_{12}^{(i)} \\ e_{22}^{(i)} & f_{22}^{(i)} & g_{22}^{(i)} & h_{22}^{(i)} \\ e_{32}^{(i)} & f_{32}^{(i)} & g_{32}^{(i)} & h_{32}^{(i)} \\ e_{42}^{(i)} & f_{42}^{(i)} & g_{42}^{(i)} & h_{42}^{(i)} \end{pmatrix}, \end{aligned} \tag{113}$$

and the vector $\mathbf{w}_i$ defined in (111).

In what follows, we shall consider double-span beams in which the first segment is modeled by the Timoshenko beam model and the second segment can be modeled by a Timoshenko (TB) or an Euler-Bernoulli (EB) model.

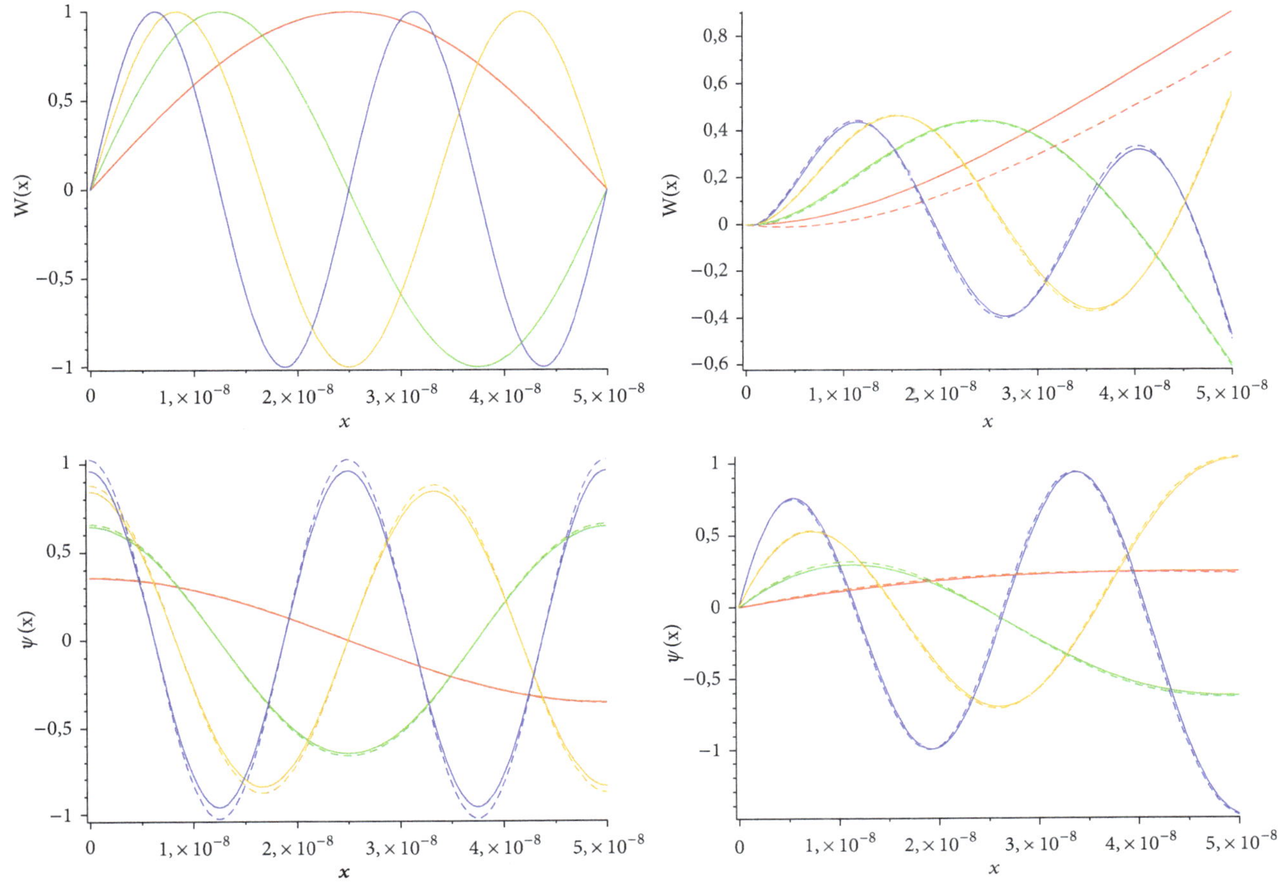

FIGURE 8: Mode shapes of simply supported beam (first column) and cantilever beam (second column), classical case (solid line) and including surface effects (dashed line).

TB-TB. In this case, $w_j(t,x)$ and $\phi_j(t,x)$ satisfy for $j = 1, 2$ the equations

$$\begin{aligned}
&\rho_j A_j \frac{\partial^2 w_j(t,x)}{\partial t^2} \\
&\quad - \kappa_j G_j A_j \left(\frac{\partial^2 w_j(t,x)}{\partial x^2} - \frac{\partial \phi_j(t,x)}{\partial x} \right) = 0, \\
&\rho_j I_j \frac{\partial^2 \phi_j(t,x)}{\partial t^2} - E_j I_j \frac{\partial^2 \phi_j(t,x)}{\partial x^2} \\
&\quad - \kappa_j G_j A_j \left(\frac{\partial w_j(t,x)}{\partial x} - \phi_j(t,x) \right) = 0.
\end{aligned} \tag{114}$$

For a simply supported beam we have

$$\begin{aligned}
w_1(t,0) &= 0, \\
\phi_1{}'(t,0) &= 0, \\
w_2(t,L) &= 0, \\
\phi_2{}'(t,L) &= 0,
\end{aligned} \tag{115}$$

and for a cantilever beam

$$\begin{aligned}
w_1(t,0) &= 0, \\
\phi_1(t,0) &= 0, \\
\phi_2'(t,L) &= 0, \\
w_2'(t,L) - \phi_2(t,L) &= 0.
\end{aligned} \tag{116}$$

The compatibility conditions, at the point $x = x_1$, without including intermediate devices, are given by

$$\begin{aligned}
&w_1(t,x_1) = w_2(t,x_1), \\
&\phi_1(t,x_1) = \phi_2(t,x_1), \\
&E_1 I_1 \phi_1'(t,x_1) = E_2 I_2 \phi_2'(t,x_1), \\
&\kappa_1 G_1 A_1 \left(w_1'(t,x_1) - \phi_1(t,x_1) \right) \\
&\quad = \kappa_2 G_2 A_2 \left(w_2'(t,x_1) - \phi_2(t,x_1) \right).
\end{aligned} \tag{117}$$

The spatial amplitude of a free transversal vibration, at each segment, $\mathsf{v}_j(t,x) = e^{\lambda t}X_j(x)$, where $X_j(x) = \begin{pmatrix} W_j(x) \\ \Psi_j(x) \end{pmatrix}$, for $j = 1,2$, satisfies

$$\left(\lambda^2 \mathsf{M}_j + \lambda \mathsf{C}_j\right) X_j(x) + \mathsf{K}_j X_j(x) = 0, \tag{118}$$

or in a differential form

$$\mathbb{M}_j X_j'' + \mathbb{C}_j X_j' + \mathbb{K}_j(\lambda) X_j = 0, \tag{119}$$

with

$$\mathbb{M}_j = \begin{pmatrix} -\kappa_j G_j A_j & 0 \\ 0 & -E_j I_j \end{pmatrix},$$
$$\mathbb{C}_j = \begin{pmatrix} 0 & \kappa_j G_j A_j \\ -\kappa_j G_j A_j & 0 \end{pmatrix}, \tag{120}$$
$$\mathbb{K}_j = \begin{pmatrix} \lambda^2 \rho_j A_j & 0 \\ 0 & \rho_j I_j \lambda^2 + \kappa G A \end{pmatrix}.$$

The boundary conditions, obtained from (109), can be conveniently written as

$$\begin{aligned} \mathbf{B}_0 \mathbf{X}_1(0) &= 0, \\ \mathbf{B}_L \mathbf{X}_2(L) &= 0, \end{aligned} \tag{121}$$

and the compatibility conditions, obtained from (112), at $x = x_1$

$$\begin{aligned} \mathscr{C}_{1,1}\mathbf{X}_1(x_1) &= \mathscr{C}_{2,1}\mathbf{X}_2(x_1), \\ \mathbf{X}_j(x) &= \begin{pmatrix} X_j(x) \\ X_j'(x) \end{pmatrix}. \end{aligned} \tag{122}$$

where

$$\mathbf{X}(x) = \begin{cases} X_1(x), & 0 \le x \le x_1, \\ X_2(x), & x_1 < x \le L. \end{cases} \tag{123}$$

We choose the fundamental basis at each segment, considering convenient translations,

$$\begin{aligned} \Psi_1(x) &= \left[\mathbf{h}_1(x), \mathbf{h}_1'(x)\right], \\ \Psi_2(x) &= \left[\mathbf{h}_2(L-x), \mathbf{h}_2'(L-x)\ )\right], \end{aligned} \tag{124}$$

with $\mathbf{h}_j(x)$ being the 2×2 matrix solution of the initial-value problem

$$\begin{aligned} \mathbb{M}_j \mathbf{h}_j''(x) + \mathbb{C}_j(\lambda)\mathbf{h}_j'(x) + \mathbb{K}_j(\lambda)\mathbf{h}_j(x) &= 0, \\ j &= 1,\dots,N. \\ \mathbb{M}_j \mathbf{h}_j'(0) &= I, \\ \mathbf{h}_j(0) &= 0. \end{aligned} \tag{125}$$

The matrix fundamental solution $\mathbf{h}_j$ is given by [25]

$$\mathbf{h}_j(x) = \begin{pmatrix} a_j d_j(x) + b_j d_j''(x) & -s_j d_j(x) \\ s_j d_j(x) & r_j d_j(x) - s_j d_j''(x) \end{pmatrix}, \tag{126}$$

where

$$\begin{aligned} a_j &= \kappa_j G_j A_j + \lambda^2 \rho_j I_j, \\ b_j &= -E_j I_j, \\ r_j &= \lambda^2 \rho_j A_j, \\ s_j &= \kappa_j G_j A_j. \end{aligned} \tag{127}$$

Then

$$\begin{aligned} X_1(x) &= \mathbf{h}_1(x)\mathbf{c}_{11} + \mathbf{h}_1'(x)\mathbf{c}_{12}, \\ X_2(x) &= \mathbf{h}_2(x-x_1)\mathbf{c}_{21} + \mathbf{h}_2'(x-x_1)\mathbf{c}_{22}. \end{aligned} \tag{128}$$

Replacing (128) in the boundary and compatibility conditions we obtained a system $\mathrm{Uc} = 0$, with $\mathrm{U} = \mathbf{B}\Phi$,

$$\mathbf{B} = \begin{pmatrix} \mathscr{B}_0 & 0 & 0 & 0 \\ 0 & [\mathscr{C}_{1,1}] & -[\mathscr{C}_{2,1}] & 0 \\ 0 & 0 & \mathbf{0} & \mathscr{B}_L \end{pmatrix}_{8\times 16}. \tag{129}$$

For a cantilever beam we have

$$\begin{aligned} \mathscr{B}_0 &= \begin{pmatrix} 1 & 0 & 0 & 0 \\ 0 & 1 & 0 & 0 \end{pmatrix}, \\ \mathscr{B}_L &= \begin{pmatrix} 0 & 0 & 0 & 1 \\ 0 & -1 & 1 & 0 \end{pmatrix}. \end{aligned} \tag{130}$$

The matrix Φ is composed of the elements of the solution basis

$$\Phi = \begin{pmatrix} \Phi_1(0) & \mathbf{0} \\ \Phi_1(l_1) & \mathbf{0} \\ \mathbf{0} & \Phi_2(l_1) \\ \mathbf{0} & \Phi_2(L) \end{pmatrix}_{16\times 8}, \tag{131}$$

where

$$\begin{aligned} \Phi_j(x) &= \begin{pmatrix} \mathbf{h}_j(x-a_j) & \mathbf{h}_j'(x-a_j) \\ \mathbf{h}_j'(x-a_j) & \mathbf{h}_j''(x-a_j) \end{pmatrix}, \quad j = 1,2, \\ a_1 &= 0, \\ a_2 &= L. \end{aligned} \tag{132}$$

TB-EB. We now consider that the movement in the second segment is governed by the Euler-Bernoulli beam model

(TB-EB). For simplicity, we shall consider Euler-Bernoulli compatibility conditions at the discontinuity point.

In this case, for $0 \le x \le x_1$ the transversal displacement w_T and angular displacement ϕ_T are governed by the Timoshenko equations

$$\begin{aligned}
&\rho_1 A_1 \frac{\partial^2 w_T(t,x)}{\partial t^2} \\
&\quad - \kappa_1 G_1 A_1 \left(\frac{\partial^2 w_T(t,x)}{\partial x^2} - \frac{\partial \phi_T(t,x)}{\partial x} \right) = 0, \\
&\rho_1 I_1 \frac{\partial^2 \phi_T(t,x)}{\partial t^2} - E_1 I_1 \frac{\partial^2 \phi_T(t,x)}{\partial x^2} \\
&\quad - \kappa_1 G_1 A_1 \left(\frac{\partial w_T(t,x)}{\partial x} - \phi_T(t,x) \right) = 0,
\end{aligned} \tag{133}$$

while for $x_1 \le x \le L$, the displacement w_E follows the Euler-Bernoulli model

$$\rho A_2 \frac{\partial^2 w_E(t,x)}{\partial t^2} + E I_2 \frac{\partial^4 w_E(x)}{\partial x^4} = 0. \tag{134}$$

The problem can be also described in matrix form as $\mathsf{M}\ddot{\mathsf{v}} + \mathsf{K}\mathsf{v} = 0$, where the block matrix coefficients are

$$\begin{aligned}
\mathsf{M} &= \begin{pmatrix} M_1^T & 0 \\ 0 & M_2^E \end{pmatrix}, \\
\mathsf{K} &= \begin{pmatrix} K_1^T & 0 \\ 0 & K_2^E \end{pmatrix}, \\
\mathsf{v} &= \begin{pmatrix} \mathsf{v}_1^{\mathsf{T}} \\ \mathsf{v}_2^{\mathsf{E}} \end{pmatrix} = \begin{pmatrix} w_T(t,x) \\ \phi_T(t,x) \\ w_E(t,x) \end{pmatrix},
\end{aligned} \tag{135}$$

with M_1^T, C_1^T, and K_1^T being the 2×2 matrix coefficients corresponding to Timoshenko equations and M_2^E, K_2^E are scalar components corresponding to Euler-Bernoulli equations.

The general boundary conditions are given in $x = 0$ according to Timoshenko model, in terms of the state vector $\mathbf{w}_T$. And in $x = L$ the general boundary conditions are given as Euler-Bernoulli model; that is,

$$\begin{aligned}
&\mathbf{B}_0 \mathbf{w}^T(t,0) = \mathbf{0}, \\
&\mathbf{w}_T(t,x) = \begin{pmatrix} w_T(t,x) \\ \phi_T(t,x) \\ w_T'(t,x) \\ \phi_T'(t,x) \end{pmatrix}, \\
&\mathbf{B}_L \mathbf{w}_E(t,L) = \mathbf{0}, \\
&\mathbf{w}_E(t,x) = \begin{pmatrix} w_E(t,x) \\ w_E'(t,x) \\ w_E''(t,x) \\ w_E'''(t,x) \end{pmatrix}.
\end{aligned} \tag{136}$$

In the discontinuity point $x = x_1$, point of section transversal area change, we have chosen compatibility conditions referring to Euler-Bernoulli beam model, having

$$\begin{aligned}
w_T(t,x_1) &= w_E(t,x_1), \\
w_T'(t,x_1) &= w_E'(t,x_1), \\
E_1 I_1 w_T''(t,x_1) &= E_2 I_2 w_E''(t,x_1), \\
E_1 I_1 w_T'''(t,x_1) &= E_2 I_2 w_E'''(t,x_1).
\end{aligned} \tag{137}$$

The eigensolutions are of the type $\mathsf{v} = \mathsf{e}^{\lambda t}\mathbf{X}(\mathsf{x})$ with

$$\begin{aligned}
\mathbf{X}(x) &= \begin{pmatrix} X_1^T(x) \\ X_2^E(x) \end{pmatrix}, \\
X_1^T &= \begin{pmatrix} W_T(x) \\ \Psi_T(x) \end{pmatrix}, \\
X_2^E &= W_E(x).
\end{aligned} \tag{138}$$

The eigenvalue problem $(\lambda^2 \mathsf{M} + \mathsf{K})\mathbf{X}(\mathsf{x}) = 0$ is equivalent to the differential system

$$\begin{aligned}
&\mathbb{M} X_T''(x) + \mathbb{C} X_T'(x) + \mathbb{K}(\lambda) X_T(x) = 0, \\
&\qquad\qquad 0 \le x \le x_1, \\
&X_E^{(iv)}(x) - \varepsilon_2^4 X_E(x) = 0, \\
&\qquad\qquad x_1 < x \le L,
\end{aligned} \tag{139}$$

where $\mathbb{M}$, $\mathbb{C}$, and $\mathbb{K}$ were defined in (120) and ε_2 was described in (25).

The boundary conditions can be written as

$$\begin{aligned}
\widetilde{B}_C \widetilde{\mathbf{X}}_T(0) &= \mathbf{0}, \\
B_L \widetilde{\mathbf{X}}_E(0) &= \mathbf{0},
\end{aligned} \tag{140}$$

where

$$\begin{aligned}
\widetilde{\mathbf{X}}_T(x) &= \begin{pmatrix} W_T(x) \\ \Psi_T(x) \\ W_T'(x) \\ \Psi_T'(x) \\ W_T''(x) \\ W_T'''(x) \end{pmatrix}, \\
\widetilde{\mathbf{X}}_E &= \begin{pmatrix} W_E(x) \\ W_E'(x) \\ W_E''(x) \\ W_E'''(x) \end{pmatrix}.
\end{aligned} \tag{141}$$

The compatibility conditions (137) can be written as

$$\begin{pmatrix} 1 & 0 & 0 & 0 & 0 & 0 & -1 & 0 & 0 & 0 \\ 0 & 0 & 1 & 0 & 0 & 0 & 0 & -1 & 0 & 0 \\ 0 & 0 & 0 & 0 & 1 & 0 & 0 & 0 & -\alpha & 0 \\ 0 & 0 & 0 & 0 & 0 & 1 & 0 & 0 & 0 & -\alpha \end{pmatrix} \widetilde{\mathbf{X}}(x_1) = \mathbf{0}, \quad (142)$$

with respect to the state span vector $\widetilde{\mathbf{X}}(x) = \begin{pmatrix} \widetilde{\mathbf{X}}_T(x) \\ \widetilde{\mathbf{X}}_E(x) \end{pmatrix}$.

The solutions $X_T(x)$ and $X_E(x)$ can be written using the solutions basis, which were introduced in the previous sections:

$$\begin{aligned} X_T(x) &= \begin{pmatrix} W_T(x) \\ \Psi_T(x) \end{pmatrix} = \mathbf{h}(x)\begin{pmatrix} c_1 \\ c_2 \end{pmatrix} + \mathbf{h}'(x)\begin{pmatrix} c_3 \\ c_4 \end{pmatrix}, \\ X_E(x) &= d(L-x)c_5 + d'(L-x)c_6 + d''(L-x)c_7 \\ &\quad + d'''(L-x)c_8. \end{aligned} \quad (143)$$

By replacing (143) in the boundary conditions and compatibility equations, we obtain a system $\mathbf{Uc} = 0$, $\mathbf{c} = [c_1\ c_2\ c_3\ c_4\ c_5\ c_6\ c_7\ c_8]^T$ and the characteristic equation $\Delta(\lambda) = \det(\mathbf{U}) = 0$. As before, we decomp$\mathbf{U}$ as $\mathbf{U} = \mathbf{B}\phi$, where

$$\phi = \begin{pmatrix} \Phi_1(0) & 0 \\ \Phi_1(x_1) & 0 \\ 0 & \Phi_2(x_1) \\ 0 & \Phi_2(L) \end{pmatrix}_{20\times 8}, \quad (144)$$

with

$$\begin{aligned} \phi_1 &= \begin{pmatrix} h(x)_{2\times2} & h'(x)_{2\times2} \\ h'(x)_{2\times2} & h''(x)_{2\times2} \\ \widetilde{h}(x)_{2\times2} & \widetilde{h}'(x)_{2\times2} \end{pmatrix}_{6\times4}, \\ \widetilde{h}(x) &= \begin{pmatrix} h''(x)[1,1] & h''(x)[1,2] \\ h'''(x)[1,1] & h'''(x)[1,2] \end{pmatrix}, \\ \phi_2 \\ &= \begin{pmatrix} d(L-x) & d'(L-x) & d''(L-x) & d_2'''(L-x) \\ d'(L-x) & d''(L-x) & d'''(L-x) & d''''(L-x) \\ d''(L-x) & d'''(L-x) & d^{(iv)}(L-x) & d^{(v)}(L-x) \\ d'''(L-x) & d^{iv}(L-x) & d^{(v)}(L-x) & d_2^{(vi)}(L-x) \end{pmatrix}_{4\times4}, \end{aligned} \quad (145)$$

$$\mathbf{B} = \begin{pmatrix} [\widetilde{B}_0]_{2\times6} & 0 & 0 \\ 0 & [B_{x_1}]_{4\times10} & 0 \\ 0 & 0 & [B_L]_{2\times4} \end{pmatrix}. \quad (146)$$

8.1. Simulations. The two span beam models considered have been compared with results of Euler-Bernoulli and Timoshenko double-span beam models in dimensional and unidimensional formulations.

Table 4: Parameters of the double-span beam [29].

Parameters	Numeric value	Unity
Length l_1	0,254	m
Length l_2	0,140	m
Width w	0,02545	m
Thickness t_1	0,01905	m
Thickness t_2	0,00549	m
Mass density ρ	2830	Kg/m^3
Young's modulus E	71,7	GPa
Moment of inertia I_1	$1{,}4633 \times 10^{-8}$	m^4
Moment of inertia I_2	$3{,}502 \times 10^{-10}$	m^4

Dimensional Model. We assume a double-span beam, with transversal section discontinuity due to variable thickness. This beam has constant width, but different thickness. The geometric and materials properties of this beam are presented in Table 4.

We present results about natural frequencies, of three cases: two Euler-Bernoulli segments (EB-EB), two Timoshenko segments (TB-TB), and coupled case (TB-EB), in Table 5.

We observe from Table 5 that the values of natural frequencies of the coupled beam (TB-EB) are between the respective values obtained by TB-TB and EB-EB cases, wherein the EB-EB natural frequencies, as expected, are greater than the others, in all boundary conditions cases presented: S-S (simply supported), C-F (clamped free), and F-F (free-free). Moreover, in the case of free-free boundary condition, as expected, TB-TB produces the most similar results to those obtained experimentally in [29].

Dimensionless Model. We now consider the two span models with classical dimensionless variables. For the Euler-Bernoulli beam model, the spatial problem in a dimensionless form is given by

$$\frac{d^4\overline{X}_j(X_j)}{dX_j} - \Omega_j^4\overline{X}_j(X_j) = 0, \quad (147)$$

where the dimensionless parameters are $X_j = x_j/L, \overline{X}_j(X_j) = X_j/L, \Omega_j^4 = \rho_j A_j \omega^2/E_j I_j, \Omega_j^4 = \mu_j \Omega_1^4$.

The Timoshenko model in a dimensionless matrix form is given by

$$\begin{aligned} &\begin{bmatrix} \xi_j & 0 \\ 0 & 1 \end{bmatrix}\begin{bmatrix} Y_j''(X) \\ \Psi_j''(X) \end{bmatrix} + \begin{bmatrix} 0 & -\xi_j \\ \xi_j & 1 \end{bmatrix}\begin{bmatrix} Y_j'(X) \\ \Psi_j'(X) \end{bmatrix} \\ &\quad + \begin{bmatrix} \Omega_j^2 & 0 \\ 0 & -\xi_j + \eta_j\Omega_j^2 \end{bmatrix}\begin{bmatrix} Y_j(X) \\ \Psi_j(X) \end{bmatrix} = \begin{bmatrix} 0 \\ 0 \end{bmatrix}, \end{aligned} \quad (148)$$

TABLE 5: Comparison of natural frequencies from the double-span beam. EB-EB: two segments of Euler-Bernoulli beam, TB-TB: two segments of Timoshenko beam, and TB-EB: first segment using Euler-Bernoulli beam model and second segment using Timoshenko beam model.

Case	Model	Natural frequencies ω_i (Hz)			
		ω_1	ω_2	ω_3	ω_4
S-S	EB-EB	101.77	782.09	1300.56	2874.69
	TB-EB	101.74	777.85	1291.65	2816.98
	TB-TB	101.64	775.85	1285.65	2796.89
C-F	EB-EB	139.56	326.69	1170.27	1797.41
	TB-EB	139.39	325.84	1154.90	1771.78
	TB-TB	139.31	325.45	1150.82	1761.06
F-F	EB-EB	292.42	1181.28	1804.01	3603.04
	TB-EB	292.14	1775.94	1775.8	3512.95
	TB-TB	291.77	1167.89	1786.63	3483.40
	[Ref]	286-291	1159-1165	1759-1771	-

TABLE 6: The effects of the change of discontinuity point $R_1 = x_1/L$ in the dimensionless natural frequencies for a cantilever double span beam with variations of thickness.

Case	Model	Configuration A				Configuration B			
		Dimensionless natural frequencies $\Omega^{[i]}$							
		$\Omega^{[1]}$	$\Omega^{[2]}$	$\Omega^{[3]}$	$\Omega^{[4]}$	$\Omega^{[1]}$	$\Omega^{[2]}$	$\Omega^{[3]}$	$\Omega^{[4]}$
$R_1 = 0.2$	TB-TB	1.4093	3.4619	5.6243	7.4545	1.7184	4.0974	6.5148	8.7633
	TB-EB	1.4098	3.4721	5.6637	7.5438	1.7193	4.1198	6.5990	8.9637
	EB-EB	1.4101	3.4769	5.6854	7.6159	1.7209	4.1381	6.6554	9.077
$R_1 = 0.5$	TB-TB	1.9972	3.5642	5.9015	8.2377	2.0150	4.0913	6.8545	9.1581
	TB-EB	1.9985	3.5725	5.9372	8.3063	2.0162	4.1061	6.9121	9.3086
	EB-EB	2.0004	3.5901	5.9831	8.5546	2.0198	4.1331	7.050	9.5862
$R_2 = 0.8$	TB-TB	2.0982	4.7587	6.6143	9.1855	1.9918	4.7325	7.3808	9.7778
	TB-EB	2.1000	4.7673	6.6436	9.2371	1.9933	4.7395	7.4083	9.8619
	EB-EB	2.1046	4.8267	6.7614	9.6670	1.9973	4.8113	7.6241	10.329

with the dimensionless parameters defined by

$$X = \frac{x}{L};$$

$$Y_j(X) = \frac{Y_j(x)}{L};$$

$$\Psi_j(X) = \psi_j(x);$$

$$\Omega_j^2 = \frac{(\rho A)_j \omega^2 l^4}{(EI)_j};$$

$$\eta_j = \frac{I_j}{A_j L^2};$$

$$\xi_j = \frac{\kappa G_j A_j L^2}{(EI)_j} = \frac{\kappa}{2\eta_j (1 + \nu_j)}. \tag{149}$$

Then we have the dimensionless frequencies by each segment

$$\Omega_j^2 = \frac{(\rho A)_j \omega^2 L^4}{(EI)_j} \quad (Timoshenko),$$
$$\Omega_j^4 = \frac{(\rho A)_j \omega^2 L^4}{(EI)_j} \quad (Euler-Bernoulli). \tag{150}$$

These frequencies can be written as a relation of Ω_1. $\Omega_j = (\mu_j)^{1/2}\Omega_1$, for Timoshenko beam, and $\Omega_j = (\mu_j)^{1/4}\Omega_1$ for Euler-Bernoulli beam, being $\mu_j = ((\rho A)_j/(\rho A)_1)((EI)_1/(EI)_j)$.

In Table 6 we present results about dimensionless natural frequencies; we have considered variations in the position of the discontinuity point $R_j = x_j/L$ and variations of thickness of the second segment; constant width is considered, according to the following configurations.

Configuration A: double-span beam with variable thickness and constant width, $t_1 = 0.08, t_2 = 0.03, w_1 = w_2 = 0.03$.

Configuration B: double-span beam with variable thickness and constant width, $t_1 = 0.08, t_2 = 0.05, w_1 = w_2 = 0.03$.

Comparative results of dimensionless natural frequencies are presented in Table 6.

factor of the beam are denoted as E, I, G, ρ, and κ, respectively. It is assumed that the beam has length L and area $A = 2bh$, where b, $2h$ are the width and thickness of the beam and $I = 2bh^3/3$. The boundary conditions are those of a cantilever beam or subject to balance of the moment and shear at the free end $x = L$ [31].

6.1. Matrix Formulation. The coupled Timoshenko model (48) can be written as a second-order differential equation with matrix coefficients

$$\mathsf{M}\ddot{\mathsf{v}} + \mathsf{K}\mathsf{v} = \mathsf{F}, \tag{50}$$

where

$$\begin{aligned}\mathsf{v} &= \begin{pmatrix} w(t,x) \\ \psi(t,x) \end{pmatrix}, \\ \mathsf{F} &= \begin{pmatrix} f(t,x) \\ q(t,x) \end{pmatrix}, \\ \mathsf{M} &= \begin{pmatrix} \rho A & 0 \\ 0 & \rho I \end{pmatrix}, \\ \mathsf{K} &= \mathsf{K}_2 \frac{\partial^2}{\partial x^2} + \mathsf{K}_1 \frac{\partial}{\partial x} + \mathsf{K}_0,\end{aligned} \tag{51}$$

with

$$\begin{aligned}\mathsf{K}_2 &= \begin{pmatrix} -\kappa GA & 0 \\ 0 & -\overline{EI} \end{pmatrix}, \\ \mathsf{K}_1 &= \begin{pmatrix} 0 & \overline{\kappa GA} \\ -\kappa GA & 0 \end{pmatrix}, \\ \mathsf{K}_0 &= \begin{pmatrix} 0 & 0 \\ 0 & \kappa GA \end{pmatrix}.\end{aligned} \tag{52}$$

For a microcantilever beam of length L, we have the boundary conditions in a matrix formulation

$$\begin{aligned}\begin{pmatrix} 1 & 0 \\ 0 & 1 \end{pmatrix} \mathsf{v}(t,0) + \begin{pmatrix} 0 & 0 \\ 0 & 0 \end{pmatrix} \mathsf{v}_x(t,0) &= 0, \\ \begin{pmatrix} 0 & 0 \\ 0 & -1 \end{pmatrix} \mathsf{v}(t,L) + \begin{pmatrix} 0 & 1 \\ 1 & 0 \end{pmatrix} \mathsf{v}_x(t,L) &= 0,\end{aligned} \tag{53}$$

or in a more compact form

$$\begin{aligned}\mathsf{A}\mathsf{v}(t,0) + \mathsf{B}\mathsf{v}_x(t,0) &= 0, \\ \mathsf{P}\mathsf{v}(t,L) + \mathsf{Q}\mathsf{v}_x(t,L) &= 0.\end{aligned} \tag{54}$$

6.2. Eigenanalysis. The search of exponential solutions

$$\begin{aligned}\mathsf{v}(t,x) &= e^{\lambda t}\mathbf{v}(x), \\ \mathbf{v}(x) &= \begin{pmatrix} w(x) \\ \psi(x) \end{pmatrix},\end{aligned} \tag{55}$$

of the unforced Timoshenko model

$$\mathsf{M}\frac{\partial^2 \mathsf{v}}{\partial t^2} + \mathsf{K}\mathsf{v} = 0, \tag{56}$$

subject to general separated homogeneous boundary conditions (54), results in that $\mathbf{v}$ is the general solution of a second-order matrix differential equation and is given by

$$\mathbf{v}(x) = \mathbf{h}(x)\mathbf{c}_1 + \mathbf{h}'(x)\mathbf{c}_2, \tag{57}$$

for constant 2×1 vectors $\mathbf{c}_1$ and $\mathbf{c}_2$. Here $\mathbf{h}(x)$ is the 2×2 matrix solution of the initial-value problem

$$\begin{aligned}M\mathbf{h}''(x) + C\mathbf{h}'(x) + K(\lambda)\mathbf{h}(x) &= \mathbf{0}, \\ \mathbf{h}(0) &= \mathbf{0}, \\ M\mathbf{h}'(0) &= \mathbf{I},\end{aligned} \tag{58}$$

where $\mathbf{0}$ denotes the 2×2 null matrix and $\mathbf{I}$ the 2×2 identity matrix. The matrix coefficients are

$$\begin{aligned}M &= \begin{pmatrix} -\kappa GA & 0 \\ 0 & -\overline{EI} \end{pmatrix}, \\ C &= \begin{pmatrix} 0 & \overline{\kappa GA} \\ -\kappa GA & 0 \end{pmatrix}, \\ K(\lambda) &= \begin{pmatrix} \rho A\lambda^2 & 0 \\ 0 & \lambda^2 \rho I + \kappa GA \end{pmatrix}.\end{aligned} \tag{59}$$

By using the initial values of $\mathbf{h}(\mathrm{x})$ in (57) and the clamped boundary condition $\mathbf{v}(0) = \mathbf{0}$, b, it turns out that $\mathbf{c}_2 = \mathbf{0}$. Thus we have to determine λ so that

$$\mathbf{v}(x) = \mathbf{h}(x,\lambda)\mathbf{c}_1 \tag{60}$$

satisfies the boundary condition at the free end $x = L$. By assuming homogeneous boundary conditions, we have the nonlinear eigenmatrix problem

$$\mathbf{U}(\lambda)\mathbf{c} = \left(\mathsf{P}\mathbf{h}(L,\lambda) + \mathsf{Q}\mathbf{h}'(L,\lambda)\right)\mathbf{c}_1 = \mathbf{0}. \tag{61}$$

From this, it turns out the characteristic equation

$$\Delta(\lambda) = \det(\mathbf{U}) = \mathbf{0}. \tag{62}$$

We thus have the reduced system

$$\begin{aligned}\mathcal{U}_D\mathbf{c}_1 &= \mathbf{0}, \\ \mathbf{c}_1 &= \begin{pmatrix} c_{11} \\ c_{12} \end{pmatrix}^T,\end{aligned} \tag{63}$$

with

$$\mathcal{U}_D = \begin{pmatrix} ad''(L) & -ad'''(L) + \lambda^2 cd'(L) \\ -b_m d'''(L) + \lambda^2 ed'(L) & ad''(L) - \lambda^2 cd(L) - a_m d''(L) \end{pmatrix}. \tag{64}$$

We observe from Table 6 that the values of dimensionless natural frequencies of the coupled beam (TB-EB) are between the respective values obtained by TB-TB and EB-EB cases, wherein the EB-EB natural frequencies, as expected, are greater than the others. Also, we can observe that when the thickness of the second segment increases, the natural frequencies become greater, mainly the higher frequencies (third and fourth frequencies), except for the case $R_1 = 0.8$ in which the first and second frequencies become greater. When R_1 goes from 0.2 to 0.5, all frequencies increase, but when R_1 goes from 0.5 to 0.8, this occurs only for Configuration A, and for Configuration B the first one frequency does not increase in this case.

9. Conclusions

This paper formulated in matrix form the time domain determination of forced responses and in the frequency domain the obtention of modes of elastic models that can be used in AFM when subjected to tip-sample interactions.

Forced responses are determined by convolution of the input load with the time domain impulse matrix function. The corresponding matrix transfer function and modes of a multispan cantilever beam were determined in terms of solution basis which have the same shape and are generated by a fundamental solution.

It was observed that the spatial amplitude diminishes with the inclusion of a piezoelectric patch when the pulse is positioned in the piezoelectric layer segment; this effect does not appear when the pulse is positioned in the third segment. The reduction of three segments for two segments, maintaining the others material and geometric properties, influences the magnitude of the natural frequencies. When the piezoelectric patch is removed, then the frequencies decrease.

We worked with two Timoshenko beam models including surface effects through different approaches and changing the classical Timoshenko equations in different manners. For both models were presented the results which evidence that, in agreement with the literature [13, 14], for beam length on the order of nanometer to microns, the difference between natural frequencies is apparent and, by increasing the length of the microbeam, the results tend to Timoshenko classical theory; that is, the surface effects are significant only in nanoscale.

Simulations were performed with double-span beams in which the first segment is modeled by the Timoshenko beam model and the second segment can be modeled by a Timoshenko (TB) or an Euler-Bernoulli (EB) model. Consistent results of TB-EB beams were obtained, in dimensional and dimensionless forms, when compared with EB-EB and TB-TB cases. The value of frequencies for this model (TB-EB) is located between the natural frequencies of uniform double-span models TB-TB and EB-EB.

Conflicts of Interest

We have no conflicts of interest.

Acknowledgments

This work was performed as part of the employment of the authors at their institutions and the third author held a fellowship no. 140823/2014-0 from CNPq of Brazil.

References

[1] B. Voigtlander, *Scanning Probe Microscopy:Atomic Force Microscopy and Scanning Tunneling Microscopy*, Springer, 2016.

[2] V. M. Harik, "Mechanics of carbon nanotubes: applicability of the continuum-beam models," *Computational Materials Science*, vol. 24, no. 3, pp. 328–342, 2002.

[3] E. B. Magrab, *Vibrations of elastic systems: with applications to MEMS and NEMS*, vol. 184, Springer Science and Business Media, 2012.

[4] F. Ebrahimi and M. R. Barati, "A modified nonlocal couple stress-based beam model for vibration analysis of higher-order FG nanobeams," *Mechanics of Advanced Materials and Structures*, pp. 1–12, 2017.

[5] M. K. Moutlana and S. Adali, "Fundamental frequencies of a nano beam used for atomic force microscopy (AFM) in tapping mode," *MRS Advances*, vol. 3, no. 42-43, pp. 2617–2626, 2018.

[6] R. Gholami, R. Ansari, and Y. Gholami, "Size-dependent bending, buckling and vibration of higher-order shear deformable magneto-electro-thermo-elastic rectangular nanoplates," *Materials Research Express*, vol. 4, no. 6, 2017.

[7] A. M. Dehrouyeh-Semnani and A. Bahrami, "On size-dependent Timoshenko beam element based on modified couple stress theory," *International Journal of Engineering Science*, vol. 107, pp. 134–148, 2016.

[8] N. Togun and S. M. Bağdatli, "Size dependent nonlinear vibration of the tensioned nanobeam based on the modified couple stress theory," *Composites Part B: Engineering*, vol. 97, pp. 255–262, 2016.

[9] A. Salehi-Khojin, S. Bashash, and N. Jalili, "Modeling and experimental vibration analysis of nanomechanical cantilever active probes," *Journal of Micromechanics and Microengineering*, vol. 18, no. 8, 2008.

[10] B. Farshi, A. Assadi, and A. Alinia-Ziazi, "Frequency analysis of nanotubes with consideration of surface effects," *Applied Physics Letters*, vol. 96, no. 9, 2010.

[11] R. E. Miller and V. B. Shenoy, "Size-dependent elastic properties of nanosized structural elements," *Nanotechnology*, vol. 11, no. 3, pp. 139–147, 2000.

[12] H. S. Zhao, Y. Zhang, and S. T. Lie, "Frequency equations of nonlocal elastic micro/nanobeams with the consideration of the surface effects," *Applied Mathematics and Mechanics-English Edition*, vol. 39, no. 8, pp. 1089–1102, 2018.

[13] C. Liu and R. K. N. D. Rajapakse, "Continuum models incorporating surface energy for static and dynamic response of nanoscale beams," *IEEE Transactions on Nanotechnology*, vol. 9, no. 4, pp. 422–431, 2010.

[14] S. Abbasion, A. Rafsanjani, R. Avazmohammadi, and A. Farshidianfar, "Free vibration of microscaled Timoshenko beams," *Applied Physics Letters*, vol. 95, no. 14, 2009.

[15] R. Ghaffari and R. A. Sauer, "Modal analysis of graphene-

based structures for large deformations, contact and material nonlinearities," *Journal of Sound and Vibration*, vol. 423, pp. 161–179, 2018.

[16] Z. Guo, M. Sheng, and T. Wang, "Flexural wave attenuation in a periodic laminated beam," *American Journal of Engineering Research (AJER)*, vol. 5, no. 6, pp. 258–265, 2016.

[17] C. Abels, V. Mastronardi, F. Guido et al., "Nitride-Based Materials for Flexible MEMS Tactile and Flow Sensors in Robotics," *Sensors*, vol. 17, no. 5, p. 1080, 2017.

[18] F. L. Forgerini and R. Marchiori, "A brief review of mathematical models of thin film growth and surfaces," *Biomatter*, vol. 4, no. 1, 2014.

[19] M. Shaat, "Mode localization phenomenon in microbeams due to surface roughness," https://arxiv.org/abs/1803.02638, pp. 2-18, 2018.

[20] M. Marrese, V. Guarino, and L. Ambrosio, "Atomic Force Microscopy: A Powerful Tool to Address Scaffold Design in Tissue Engineering," *Journal of Functional Biomaterials*, vol. 8, no. 1, p. 7, 2017.

[21] Y. F. Dufrêne, T. Ando, R. Garcia et al., "Imaging modes of atomic force microscopy for application in molecular and cell biology," *Nature Nanotechnology*, vol. 12, no. 4, pp. 295–307, 2017.

[22] W. Chang, H. Lee, and Y. Yang, "Mathematical modeling of nanomachining with atomic force microscope cantilevers," *Journal of Physics: Conference Series*, vol. 490, article no. 012161, 2014.

[23] A. V. Aksenov, "Method of construction of the Riemann function for a second-order hyperbolic equation," *Journal of Physics: Conference Series*, vol. 937, article no. 012001, 2017.

[24] M. H. Korayem, S. Zafari, A. Amanati, M. Damircheli, and N. Ebrahimi, "Analysis and control of micro-cantilever in dynamic mode AFM," *The International Journal of Advanced Manufacturing Technology*, vol. 50, no. 9-12, pp. 979–990, 2010.

[25] J. R. Claeyssen, "The timoshenko beam model in vibrating afm cantilevers," in *Proceedings of the 9th Brazilian Conference on Dynamics Control and their Applications*, J. M. Balthazar, Ed., pp. 60–70, 2010.

[26] H. K. Hong and J. T. Chen, "On the dual representation and its application to vibration problems," in *Boundary Elements XV Stress Analysis*, vol. 2, 1988.

[27] J. R. Claeyssen and I. F. Ferreira, *A resposta impulso em modelos evolutivos e estacionários*, Universidade Federal de Santa Maria, Santa Maria,Brazil, 2009.

[28] S. Kukla and I. Zamojska, "Frequency analysis of axially loaded stepped beams by Green's function method," *Journal of Sound and Vibration*, vol. 300, no. 3-5, pp. 1034–1041, 2007.

[29] S. C. Stanton and B. P. Mann, "On the dynamic response of beams with multiple geometric or material discontinuities," *Mechanical Systems and Signal Processing*, vol. 24, no. 5, pp. 1409–1419, 2010.

[30] N. Jalili, *Piezoelectric-Based Vibration Control From Macro to Micro/Nano Scale Systems*, Springer, 2010.

[31] M. Zhu, W. Zhe, and Z. M. Juan, "Dynamic analysis of timoshenko beam with arbitrary constraints and a further optimization based on least energy principle," *Mathematical Problems in Engineering*, vol. 2018, Article ID 1269738, 19 pages, 2018.

[32] J. R. Claeyssen and S. N. J. Costa, "Modes for the coupled Timoshenko model with a restrained end," *Journal of Sound and Vibration*, vol. 296, no. 4-5, pp. 1053–1058, 2006.

[33] G.-F. Wang and X.-Q. Feng, "Effects of surface elasticity and residual surface tension on the natural frequency of microbeams," *Applied Physics Letters*, vol. 90, no. 23, Article ID 231904, 2007.

[34] M. E. Gurtin and A. Ian Murdoch, "Surface stress in solids," *International Journal of Solids and Structures*, vol. 14, no. 6, pp. 431–440, 1978.

Permissions

All chapters in this book were first published in AMP, by Hindawi Publishing Corporation; hereby published with permission under the Creative Commons Attribution License or equivalent. Every chapter published in this book has been scrutinized by our experts. Their significance has been extensively debated. The topics covered herein carry significant findings which will fuel the growth of the discipline. They may even be implemented as practical applications or may be referred to as a beginning point for another development.

The contributors of this book come from diverse backgrounds, making this book a truly international effort. This book will bring forth new frontiers with its revolutionizing research information and detailed analysis of the nascent developments around the world.

We would like to thank all the contributing authors for lending their expertise to make the book truly unique. They have played a crucial role in the development of this book. Without their invaluable contributions this book wouldn't have been possible. They have made vital efforts to compile up to date information on the varied aspects of this subject to make this book a valuable addition to the collection of many professionals and students.

This book was conceptualized with the vision of imparting up-to-date information and advanced data in this field. To ensure the same, a matchless editorial board was set up. Every individual on the board went through rigorous rounds of assessment to prove their worth. After which they invested a large part of their time researching and compiling the most relevant data for our readers.

The editorial board has been involved in producing this book since its inception. They have spent rigorous hours researching and exploring the diverse topics which have resulted in the successful publishing of this book. They have passed on their knowledge of decades through this book. To expedite this challenging task, the publisher supported the team at every step. A small team of assistant editors was also appointed to further simplify the editing procedure and attain best results for the readers.

Apart from the editorial board, the designing team has also invested a significant amount of their time in understanding the subject and creating the most relevant covers. They scrutinized every image to scout for the most suitable representation of the subject and create an appropriate cover for the book.

The publishing team has been an ardent support to the editorial, designing and production team. Their endless efforts to recruit the best for this project, has resulted in the accomplishment of this book. They are a veteran in the field of academics and their pool of knowledge is as vast as their experience in printing. Their expertise and guidance has proved useful at every step. Their uncompromising quality standards have made this book an exceptional effort. Their encouragement from time to time has been an inspiration for everyone.

The publisher and the editorial board hope that this book will prove to be a valuable piece of knowledge for researchers, students, practitioners and scholars across the globe.

List of Contributors

Pierre Gaillard
Université de Bourgogne, Institut de Mathématiques de Bourgogne, 9 avenue Alain Savary, BP 47870, 21078 Dijon Cedex, France

Xiaoyun Tang, Zhanying Yang and Jie Zhang
School of Mathematics and Statistics, South-Central University for Nationalities,Wuhan, Hubei 430074, China

J. Socorro and Omar E. Núñez
Departamento de Física, DCeI, Universidad de Guanajuato-Campus León, 37150 León, GTO, Mexico

Rafael Hernández-Jiménez
School of Physics and Astronomy, University of Edinburgh, Edinburgh EH9 3FD, UK

Xinli Han
School of Science, Nanjing University of Posts and Telecommunications, Nanjing 210023, China

Lijun Pan
Department of Mathematics, Nanjing University of Aeronautics and Astronautics, Nanjing 211106, China

Francesco Toppan
CBPF, Rua Dr. Xavier Sigaud 150, Urca, cep 22290-180, Rio de Janeiro (RJ), Brazil

Mauricio Valenzuela
Facultad de Ingeniería y Tecnología, Universidad San Sebastián, General Lagos 1163, Valdivia 5 110693, Chile

Sumayah Batwa
Department of Mathematics and Statistics, University of South Florida, Tampa, FL 33620-5700, USA
Mathematics Department, King Abdulaziz University, Jeddah 21589, Saudi Arabia

Wen-Xiu Ma
Department of Mathematics and Statistics, University of South Florida, Tampa, FL 33620-5700, USA
Department of Mathematics, Zhejiang Normal University, Jinhua 321004, China
College of Mathematics and Systems Science, Shandong University of Science and Technology, Qingdao, Shandong 266590, China
College of Mathematics and Physics, Shanghai University of Electric Power, Shanghai 200090, China
International Institute for Symmetry Analysis and Mathematical Modelling, Department of Mathematical Sciences, North-West University, Mafikeng Campus, Private Bag X2046, Mmabatho 2735, South Africa

Imre F. Barna
Wigner Research Center for Physics of the Hungarian Academy of Sciences, Konkoly-Thege Miklóśut 29-33, 1121 Budapest, Hungary
ELI-HU Nonprofit Kft., Dugonics Tér 13, H-6720 Szeged, Hungary

Mihály A. Pocsai
Wigner Research Center for Physics of the Hungarian Academy of Sciences, Konkoly-Thege Miklóśut 29-33, 1121 Budapest, Hungary
University of Pécs, Institute of Physics, Ifíuśag Útja 6, H-7624 Pécs, Hungary

L. Mátyás
Sapientia University, Department of Bioengineering, Libertătii Sq. 1, 530104 MiercureaCiuc, Romania

Penghong Zhong
Department of Mathematics, Guangdong University of Education, Guangzhou 510640, China

Yujin Liu and Wenhua Sun
School of Mathematics and Statistics, Shandong University of Technology, Zibo, Shandong 255000, China

Oscar Bohórquez and A. M. S. Macêdo
Departamento de Física, Laboratório de Física Teórica e Computacional, Universidade Federal de Pernambuco, 50670-901 Recife, Pernambuco, Brazil

Massimo Tessarotto
Department of Mathematics and Geosciences, University of Trieste, Via Valerio 12, 34127 Trieste, Italy
Institute of Physics, Faculty of Philosophy and Science, Silesian University in Opava, Bezručovo nám 13, 74601Opava, Czech Republic

Claudio Cremaschini
Institute of Physics and Research Center for Theoretical Physics and Astrophysics, Faculty of Philosophy and Science, Silesian University in Opava, Bezručovo nám 13, 74601 Opava, Czech Republic

Karthikeyan Rajagopal
Center for Nonlinear Dynamics, Defense University, Ethiopia

Sifeu Takougang Kingni
Department of Mechanical, Petroleum and Gas Engineering, Faculty of Mines and Petroleum Industries, University of Maroua, Cameroon

Gaetan Fautso Kuiate
Department of Physics, Higher Teacher Training College, The University of Bamenda, Cameroon

Victor Kamdoum Tamba
Laboratory of Electronics and Signal Processing (LETS), Department of Physics, Faculty of Science, University of Dschang, Cameroon

Viet-Thanh Pham
School of Electronics and Telecommunications, Hanoi University of Science and Technology, 01 Dai Co Viet, Hanoi, Vietnam

M. Irshad Khodabocus
Department of Mathematics, Faculty of Science, University of Mauritius, Réduit 80837, Mauritius

Mathieu Sellier
Department of Mechanical Engineering, University of Canterbury, Private Bag 4800, Christchurch 8140, New Zealand

Volker Nock
Department of Electrical and Computer Engineering, University of Canterbury, Private Bag 4800, Christchurch 8140, New Zealand

Yuanyuan Chen and Liangyun Zhang
College of Science, Nanjing Agricultural University, Nanjing, Jiangsu 210095, China

Emad K. Jaradat and Ala'a M. Al-Faqih
Department of physics, Mutah University, Jordan

Omar Alomari and Mohammad Abudayah
German Jordanian University, Amman, Jordan

Yongqiang Fan and Lihui Guo
College of Mathematics and System Sciences, Xinjiang University, Urumqi 830046, China

Gan Yin
Department of Mathematics, Zhejiang University of Science & Technology, Hangzhou, Zhejiang 310023, China

John C. Baez
Department of Mathematics, University of California, Riverside, CA 92521, USA
Centre for Quantum Technologies, National University of Singapore, Singapore 117543

Hongwei Hou
Zhejiang Academy of Building Research and Design Co. Ltd., Hangzhou, Zhejiang 310000, China
College of Civil Engineering and Architecture, Zhejiang University, Hangzhou, Zhejiang 310000, China

Shihu Gao
College of Civil Engineering and Architecture, Wenzhou University, Wenzhou, Zhejiang 325035, China

Qianqian Guo, Long Chen and Bing Wu
Key Laboratory of Ministry of Education for Geomechanics and Embankment Engineering, Hohai University, Nanjing 210098, China

Lei Wang
Shanghai City Construction and Design Research Institute, No. 1170 Tibet South Road, Huangpu District, Shanghai, China

Yan Liu
School of Automation and Electrical Engineering, University of Science and Technology, Beijing 100083, China
Key Laboratory of Knowledge Automation for Industrial Processes, Ministry of Education, Beijing 100083, China

Lipo Mo
School of Science, Beijing Technology and Business University, Beijing 100048, China

Julio Claeyssen
PGMAT, Departamento de Matemática, CCNE, Universidade Federal de Santa Maria, Av. Roraima 1000, 97105-900, Santa Maria, RS, Brazil

PPGMAp-IME, Universidade Federal do Rio Grande do Sul, Av. Bento Gonc¸alves, 9500, 91509-900, Porto Alegre, RS, Brazil

Joao Carvalho
PPGMAp-IME, Universidade Federal do Rio Grande do Sul, Av. Bento Gonçalves, 9500, 91509-900, Porto Alegre, RS, Brazil

Rosemaira Copetti
PGMAT, Departamento de Matemática, CCNE, Universidade Federal de Santa Maria, Av. Roraima 1000, 97105-900, Santa Maria, RS, Brazil

Leticia Tonetto
Centro de Engenharias, CENG, Universidade Federal de Pelotas, Almirante Barroso, 1734, Pelotas, RS, 96010-280, Brazil

Index

www.ingramcontent.com/pod-product-compliance
Lightning Source LLC
LaVergne TN
LVHW080708210926
848485LV00003B/433
* 9 7 8 1 6 4 7 2 8 3 5 4 4 *